HIGHWAY ENGINEERING

HIGHWAY ENGINEERING

Planning, Design, and Operations

DANIEL J. FINDLEY PhD, PE

BASTIAN J. SCHROEDER PhD, PE

CHRISTOPHER M. CUNNINGHAM MS, PE

THOMAS H. BROWN Jr. PhD, PE

ELSEVIER

AMSTERDAM • BOSTON • HEIDELBERG • LONDON
NEW YORK • OXFORD • PARIS • SAN DIEGO
SAN FRANCISCO • SINGAPORE • SYDNEY • TOKYO
Butterworth-Heinemann is an imprint of Elsevier

BH

Butterworth-Heinemann is an imprint of Elsevier
The Boulevard, Langford Lane, Kidlington, Oxford OX5 1GB, UK
225 Wyman Street, Waltham, MA 02451, USA

ISBN: 978-0-12-801248-2

British Library Cataloguing-in-Publication Data
A catalogue record for this book is available from the British Library.

Library of Congress Cataloging-in-Publication Data
A catalog record for this book is available from the Library of Congress.

For Information on all Butterworth-Heinemann publications
visit our website at http://store.elsevier.com/

CONTENTS

PART 1

Introduction

Daniel J. Findley, PhD, PE
Institute for Transportation Research and Education, North Carolina State University

Contents

1.1 INTRODUCTION

Highway engineering is a subset of transportation engineering, which itself is typically a component of civil engineering. The presence of more than 4 million miles of public roads in the United States (BTS, 2014) serving widely varying traffic volumes and trip purposes, emphasizes the need for qualified and capable professionals to address problems and improve the system. Two primary metrics of quality of highways are efficiency (measured by delay, travel time, speed, or other operational characteristics) and safety (measured by collisions or fatalities). An inefficient highway can have detrimental effects on local and regional economies and drivers, by burdening the movement of goods and people with additional costs and loss of productivity. The continual improvement of highways is also essential to reduce deaths resulting from collisions on roadways, which, in the United States, totaled 33,561 fatalities in 2012 (NHTSA, 2014).

All users and modes on the transportation system play important roles in the efficient and effective movement of goods and people. This book focuses on highway infrastructure and the operations of that system, and documents the methodologies and analysis practices for the design, operational analysis, and safety assessment of the system. A fundamental consideration in highway engineering is the human element. There is a need for meeting drivers' expectations or effectively communicating any disruptions to their expectations. This is illustrated throughout highway engineering, as violations to a driver's expectation without proper notification results in operational inefficiency or safety concerns.

Highway engineering is a multidisciplinary field with interconnected subdisciplines that include planning, safety, operations, design, and related fields such as structural, hydraulic, and geotechnical engineering. This book presents thematic topics within highway engineering and the holistic system required to develop a highway from initial planning through the full design process and operations. Ultimately, to meet overall efficiency and effectiveness goals, highway engineers must understand how their role fits into the larger process, and apply flexibility with the implementation of standard practice to maximize the overall final product. Decisions made throughout the design process must consider impacts on safety and operations. For instance, in the alignment of a highway, an engineer should attempt tangential and perpendicular crossings of water features or overpasses to minimize the complexity and cost of structural elements. Similarly, operational treatments may affect design decisions. Figure 1.1 shows a roadway setup with reversible lanes, allowing for peak direction traffic to have additional lanes during peak travel times. Reversible lanes are particularly useful in areas with very unbalanced traffic flows, such as during entertainment or sporting events or into and out of a central business district. This simple example illustrates how the design and operations of a roadway are closely interrelated and how clearly communicating those principles to the driver—the human element—is critical to assure safe operations.

Figure 1.1 Reversible lanes.

There are several questions that might be asked about the topic of highway engineering, and these are addressed in the following parts of this book.

- How do we know when we need to build a new highway or make improvements to an existing highway?
- How do we know that a highway is functioning as designed (from an efficiency and safety perspective)?
- What geometric components are necessary to produce an efficient and safe highway?
- How does an individual highway engineer's role fit into the overall completion of a highway project?
- What other engineering is needed in a highway project?
- What aspects related to those areas should highway engineers consider in their efforts?

REFERENCES

National Highway Traffic Safety Administration (NHTSA), 2014. Fatality Analysis Reporting System (FARS) Encyclopedia. National Highway Traffic Safety Administration.

Bureau of Transportation Statistics (BTS), 2014. Table 1-1: System mileage within the United States. Bureau of Transportation Statistics. Office of the Assistant Secretary for Research and Technology. United States Department of Transportation.

1.2 ORGANIZATION OF THE BOOK

This book is organized into nine parts, each addressing one aspect of the field of highway engineering. Each part presents a standalone overview of a component of highway engineering, details analysis methodologies, defines key concepts, and presents applications and examples. However, all nine parts interrelate; for example, design decisions can impact safety, or the forecast of traffic demands through transportation planning is closely tied to the expected operations of an intersection or facility. These correlations and interactions will be discussed throughout each part.

1.2.1 Part 2: Transportation Planning

Highway engineers need to be able to recognize when a highway has reached its service life and which improvements and modifications should be made to that facility, or if a new facility is needed. Part 2 describes the long-term planning and forecasting process and presents the methodologies used to predict when and where transportation improvements are needed. Many planning applications are closely tied to new developments on a local or regional scale that are expected to impact traffic patterns. The planning methods are used to predict how many trips a new development generates, where those drivers are expected to go, what facilities or routes they are expected to take, and even what mode of transportation they are likely to choose. This part of the book presents and discusses the use of planning tools in the four key steps of trip generation, trip distribution, traffic assignment, and modal split, and advises the engineer on making informed decisions.

1.2.2 Part 3: Horizontal and Vertical Alignment

Part 3 describes the decisions related to choosing an optimal highway alignment given substantial environmental and design considerations, including: corridor selection, traverses, sight distance, horizontal alignment, and vertical alignment. Corridor selection follows transportation planning, which identifies the broader transportation needs of a community. Corridor selection is comprised of the broad task of choosing a highway location through decisions relating to minimizing costs and impacts to the human and natural environment. The engineering computations of such corridors are derived from consideration of the highway segments as a traverse. The horizontal and vertical components each affect the highway location and require an iterative process to balance the various quantitative measures and tradeoffs of a particular alternative, as well as including feedback gathered from stakeholders in the public involvement process. At any point along a highway drivers should be able to perceive an obstruction or change in alignment and react by changing their speed, direction, or path. The distance required to perform this maneuver—the sight distance—is an integral part of highway alignment.

1.2.3 Part 4: Highway Geometric Design

Part 4 details the process of choosing appropriate geometric features for a highway. Design controls govern key aspects of highway design and are essential for safety and efficiency. The geometric features considered in this section include the basic components that guide horizontal and vertical alignment, including curvature and grades, and elements that form the cross section of the highway, including lanes, shoulders, and medians. Intersections and interchanges are important in highway design due to their significant impact on safety performance and operational efficiency.

Designers need to be able to work with transportation planners and operations staff to determine which locations and designs work best considering all tradeoffs. Designers must be able to translate a vision or concept into a horizontal and vertical design with appropriate geometric proportions given various field constraints, while considering all stakeholders (i.e., transportation system users, adjacent land uses, people affected by the roadway, and the effect on other aspects of the community).

1.2.4 Part 5: Traffic Operations

Highway engineers need to clearly understand the basic functions of all facility types and apply those concepts to real-world designs. Part 5 provides details of uninterrupted flow facilities on freeways or rural highways, and interrupted flow facilities, which include the analysis of traffic signals, roundabouts, stop signs, and yield signs. These features interrupt or control the flow of two or more intersecting traffic streams to assure a safe and efficient operation of the highway junction point or intersection. This part presents the tools for evaluating those facility types and methods for measuring the impact of these facilities for all users, including nonmotorized travelers. The methods are used to predict delay, travel time, and other operational performance measures, and are often summarized in a level-of-service (LOS) score for a movement, approach, or overall intersection.

1.2.5 Part 6: Traffic Safety

Safety is a primary focus area of transportation agencies, as traffic collisions are a primary contributor to injuries and deaths for citizens in most countries. The availability of analysis techniques allows for predictive analysis of crash problems for a location, as well as reactive analysis of newly emerging crash and safety patterns across a region. Part 6 of this book provides guidance on safety analysis tools that can be used during the preliminary stages of design (e.g., countermeasure selection, site selection, etc.), basic safety tools for analyzing designs or treatments after implementation, and supplemental tools that can be used for safety analysis. The safety performance of an intersection or roadway can be closely tied to its design and alignment, as well as its operations in terms of the volume or speed of traffic traversing it. As such, highway design, traffic operations, and safety interrelate, and are tied together by the human element, the driver traveling on a roadway.

1.2.6 Part 7: Geotechnical

Coordination with other engineers should be considered throughout the entire project. The last three parts of the book provides a basic understanding of other engineering fields as they apply to the field of transportation. These aspects will influence the design of the highway or vice versa (i.e., geotechnical concerns might influence the selection of the corridor, while highway alignment needs to account for structural design when a water crossing is necessary), which provides an overview of these related field to expose the reader to these important additional considerations in the design and operation of a transportation facility. Part 7 details the geotechnical field and its relationship with highway engineering. Settlement is a primary concern of geotechnical analysis which requires adequate soil sampling, classification, testing, and estimation of settlement rates.

1.2.7 Part 8: Structures

As part of the final three parts of the book, Part 8 focuses on bridge structures due to their importance and prevalence in highway networks. The topic is presented from a top down approach, starting with the bridge superstructure dominated by the roadway down to the design of the support footing.

1.2.8 Part 9: Hydraulics

Part 9 completes the book and the discussion of transportation-related engineering fields. This part focuses on hydraulics – primarily the prediction, collection, and direction of storm water runoff from highway facilities.

1.3 FUNCTIONAL CLASSIFICATIONS OF HIGHWAY

Highways can be classified by their function, which generally relates to the amount of mobility and access they provide. Mobility and access are competing objectives of highways. On a highway that prioritizes mobility, impediments to the flow of traffic should be minimized, while highways with a purpose of providing access to adjacent land uses allow for more frequent access points. The tradeoffs between mobility and access impact the operation and safety of the highway and should be planned carefully to fit the context of the overall highway network. Highways with a mobility focus generally sustain higher traffic volumes and comprise a small portion of the overall mileage of the system. Each type of highway is essential for a well-operating and efficient overall network that facilitates higher-speed, long-distance travel and lower-speed, short-distance trips. When classified by mobility, arterials offer the highest level of mobility,

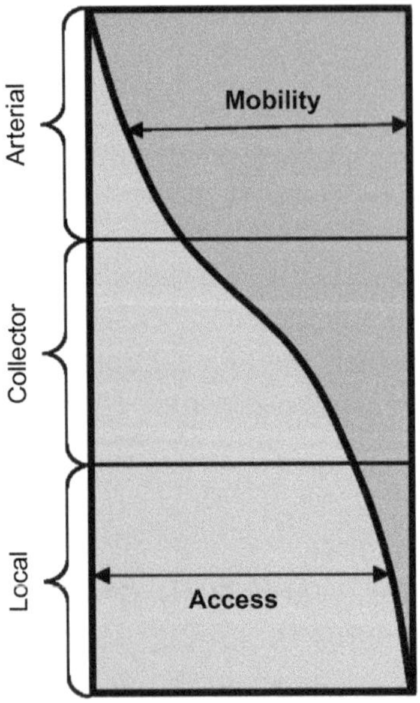

Figure 1.2 Relationship between mobility and access.

while collectors provide more balance between access and mobility, and local roads favor access over mobility. Figure 1.2 shows the relationship of access and mobility based on the type of roadway.

1.3.1 Arterials

Arterials focus primarily on mobility with an emphasis on providing high-speed, uninterrupted flow. Long-distance trips are most practical on arterials. As a subset of arterials, freeways are the highest functional classification of highways and carry a significant portion of traffic volumes, based on lane-miles of road. Freeways are an essential part of the highway network, particularly for travel that occurs between cities, regions, and states. Well-designed freeways have the ability to support economic development through the safe and efficient travel of goods and people. The characteristics of freeways can vary tremendously depending on their setting. Figure 1.3 shows an interstate in a suburban location with six lanes and wide shoulders on the outside and inside edges of pavement. Figure 1.4 shows a depressed interstate, which reduces noise effects and allows for crossroads to occur at street level, in an urban environment. Figure 1.5 shows an urban arterial that serves traffic from suburban zones

Figure 1.3 Suburban interstate.

Figure 1.4 Urban interstate.

Figure 1.5 Urban arterial.

into the central business district. Figure 1.6 shows a rural two-lane highway that is a primary route for commerce and recreation in a rural area.

1.3.2 Collectors

Collectors have a blended objective of maintaining mobility and access. Collectors facilitate travel between local roads and arterials by collecting traffic and distributing it to local roads or to higher mobility arterials. Collectors cover a wide spectrum of needs and vary depending on the type and quantity of access that is provided to adjacent land uses and potential future land uses (Figure 1.7).

1.3.3 Local Streets

Local streets provide direct connectivity to businesses, residences, and other land uses. Local streets can be designed to provide access while

Figure 1.6 Rural two-lane highway.

Figure 1.7 Collector roadway.

Figure 1.8 Local street.

minimizing speeds. A prevalence of turning movements and nonmotorized usage on local streets accentuates the need for controlling speeds. Figure 1.8 shows a developed area with a pedestrian crossing and on-street parking on a local street.

1.4 TYPES OF INTERSECTIONS

Similar to the classification and groupings of highways and highway segments, different intersection types are distinguished. Intersections are nodes in the transportation network, the point at which two roads meet to form an at-grade junction. The traffic control type of the intersection governs the rules for how the traffic streams from these two roads interact. General intersection forms include yield-controlled intersections, stop-controlled intersections, signalized intersections, and modern roundabouts.

The type of intersection control impacts most, if not all, aspects of highway engineering described in this book, including planning for the adequate size of the intersection, geometric design and appropriate alignment of the intersection and its approaches, operational characteristics and capacity of the intersection, safety performance of the intersection, and other engineering considerations such as geotechnical and hydraulics aspects. These considerations interact and impact one another in roadway design and should be considered jointly in intersection design as well.

1.4.1 Unsignalized Intersections

Unsignalized intersections are controlled by either yield or stop signs, and often represent relatively low-volume junctions. The example intersection in Figure 1.9 shows an all-way stop-controlled (AWSC) intersection on a university campus. Stop signs are also used to control minor, low-volume approaches at two-way stop-controlled (TWSC) intersections, although major roads or arterials in this case often carry considerably higher volumes.

Other unsignalized intersections are controlled by yield signs, which differ from stop-controlled intersections in that drivers do not have to come to a full stop (but still have to yield the right of way to one or more conflicting approaches). Yield-controlled intersections are more common in other countries than in the United States, where stop signs are used more prevalently. Although stop signs in the United States may be considered the standard control treatment at many minor and unsignalized

Figure 1.9 Stop controlled intersection.

Figure 1.10 Modern roundabout.

intersections, stop signs in many other countries are limited to approaches that have sight distance restrictions. The stop sign therefore primarily serves as a safety treatment.

Yield control is also commonly used at modern roundabout intersections, which are circular intersections that feature generally low design speeds and yield control at entry. An example of a two-lane roundabout is shown in Figure 1.10. Roundabouts are very common intersection treatments in many countries, including the United States.

Roundabouts are a very attractive intersection form due to their impressive safety record and low rate of serious injury and fatal crashes compared to signalized intersections or two-way stop-controlled intersections. The positive safety record of roundabouts is due in large part to the low design speed, and to the fact that vehicle conflict types are reduced to merge conflicts at entry and potential rear-end conflicts, both of which occur at low speeds. Modern roundabouts eliminate the potential for high-speed angle and T-bone crashes, which tend to be the most severe type of crash at other intersection forms.

Roundabouts are fundamentally different from traffic circles, which were a common intersection form built in the middle of the twentieth century in the United States, and are still prevalent in some parts of the country (and internationally in some locations). Traffic circles, such as the ones found in the northeastern United States, typically feature a much larger footprint than a modern roundabout, and this leads to higher speeds and, in some cases, more frequent and more severe collisions. Traffic circles also tend to be controlled by a merge or weaving maneuver on entry, as opposed to the low-speed, unambiguous yield control of a roundabout. The weaving maneuver at entry is another contributor to the poor safety (and operational performance) of many traffic circles, and further can lead to driver confusion and low public opinion of this particular intersection type.

Finally, some unsignalized intersections can be unsigned, or entirely uncontrolled. Unsigned intersections exist in some European countries in low-speed, low-volume neighborhood environments and are governed by a "right before left" rule (in countries where traffic travels on the right side of the road). Other countries feature entirely uncontrolled intersections, where driver courtesy and interpersonal communication between all road users (including pedestrians and cyclists) governs the operations of the intersections.

1.4.2 Signalized Intersections

For intersections with elevated traffic volumes, traffic signals are commonly used to control the interaction and order of movements from different approaches. The traffic signal, and its alternating red-yellow-green indication to conflicting approaches, controls which movement is allowed in what order. An example of a signalized intersection is shown in Figure 1.11.

Signalized intersections come in a wide range of sizes and configurations, and the study of different control and timing strategies consumes entire books and manuals. The study of signalized intersections includes a range of topics, including estimating the capacity of each approach, optimizing the signal timing for an intersection to balance the needs of different phases, optimizing the signal timing in the context of a corridor to coordinate movements from one intersection to the next, and a host of topics including location and configuration of signal displays themselves.

1.4.3 Alternative Intersections

A special category of intersections is referred to as alternative intersections, or sometimes unconventional or innovative intersections.

Figure 1.11 Signalized intersection.

Alternative intersections typically aim to enhance the safety and operations of an intersection in ways that don't require extensive construction or right-of-way costs (i.e., from building a grade-separated interchange). Alternative intersection types often gain operational efficiency over "traditional" intersections by removing the number of phases or reducing the number of conflicting movements. For example, several alternative intersection forms gain capacity by moving or modifying the way left turns are progressed through the intersection, or in some cases by eliminating through movements (e.g., from a minor approach).

The most common alternative intersection forms include the restricted crossing U-turn (RCUT) or "superstreet," the median U-turn intersection (MUT), the displaced left-turn intersection (DLT), and the quadrant roadway. There are also several forms of alternative interchanges (essentially grade-separated intersections at a freeway junction), with the diverging diamond interchange (DDI), also known as the double crossover diamond (DCD), being the most popular form in the United States today.

PART 2

Transportation Planning

Bastian J. Schroeder, PhD, PE
Director of Highway Systems, Institute for Transportation Research and Education (ITRE)

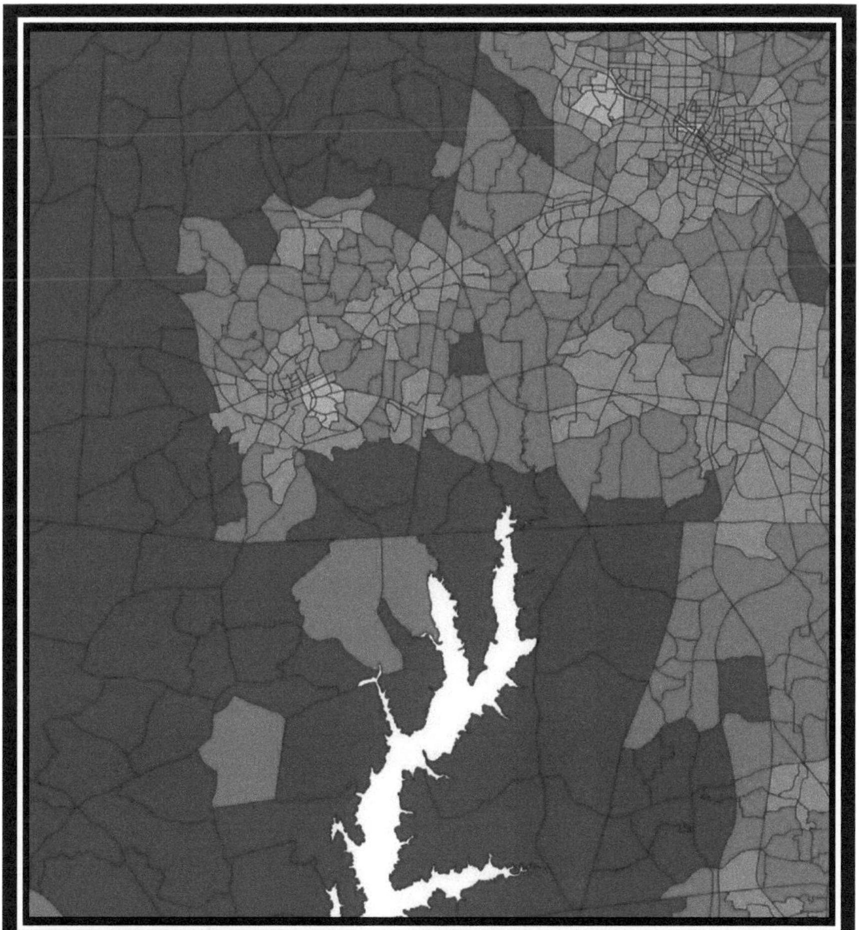

Figure by Joe Huegy, ITRE – used with permission

Contents

2.1 INTRODUCTION

Planning can be defined as the activity or process that examines the potential of future actions to guide a situation or system toward a desired direction. Transportation planning, which encompasses all modes of transportation, is a very important function of the state Departments of Transportation (DOTs), as well as planning organization at the metropolitan or regional levels.

This part describes the basic process of transportation planning. Transportation professionals use these planning methods to forecast how much traffic is expected in future-year analyses, where that traffic is expected to go, what routes that traffic is expected to take, and what modes of transportation travelers are most likely to choose. These methods aim to predict and estimate how travelers, the human element in any transportation problem, are likely to behave and act based on prior observations and historic trends. As such, planning methods borrow principles from other disciplines, ranging from psychology and sociology, to operations research and network optimization, to estimated behavioral patterns by travelers across a localized area or broader region.

The text presents critical planning concepts and terms and then introduces the reader to the four-step forecasting process used in transportation planning: trip generation, trip distribution, mode choice, and traffic assignment. The discussion includes data needs and sources for planning analyses, as well as mathematical models used to complete various steps in the forecasting process. The part concludes with a discussion of planning applications and software use, as well as practice problems to further explore and apply the concepts presented in the text.

2.1.1 Purpose of Transportation Planning

The most important aspect of planning is that it is oriented toward the future. Although planning may increase the likelihood that a recommended action will take place, it does not guarantee that it will. Transportation planning relies on the use of models and forecasts, which are often wrong, because even the best models can, at best, only approximate future conditions. As such, transportation planning is as much an art form as it is science, placing a constant burden on the analyst to properly interpret model results and apply good professional judgment to any analysis.

State transportation planning organizations are responsible for conducting long-range transportation planning, also known as systems planning, for all areas of the state, including metropolitan areas, small urban areas, counties, and multicounty regions. The metropolitan areas and regions in turn have their own planning organizations that offer a more detailed and in-depth look at transportation plans under their jurisdiction. Ideally, state and metropolitan planning organizations (MPOs) work jointly and in concert to develop robust future-year forecasts based on their combined expertise.

A typical long-range planning study involves several steps, including:

- Initial meetings with the local area citizens, staff, and elected officials
- Data collection and analysis
- Discussions of findings with the local area
- Development of several alternative future scenarios
- Public information workshops and public hearings
- Selection of a preferred plan

The Federal Highway Administration (FHWA) identifies eight critical factors in the transportation planning process, as outlined in Figure 2.1. The FHWA publication emphasizes early on that transportation planning is a collaborative process between all systems of the users. Stakeholders in transportation problems include the business community, the residential community, environmental groups, the traveling public, freight operators, and the general public. As such, the process in Figure 2.1 emphasizes feedback and public input throughout the planning process. Similarly, the reality of fiscal constraints, as well as recognition of sensitive environmental, historic, or socioeconomic areas, make transportation planning a continually evolving and dynamic process.

To accommodate diverse interests and stakeholder groups, planning organizations at the state, regional, and municipal levels often collaborate and exchange data and predictions to arrive at a more accurate and comprehensive final transportation plan to benefit all stakeholders. By first determining the transportation deficiencies, the process attempts to determine the 20–30-year long-range transportation needs for each area at the state level, with shorter planning horizons at the regional and metropolitan level. Table 2.1 illustrates typical responsibilities, time horizons, content, and update requirements in the planning process for different public entities.

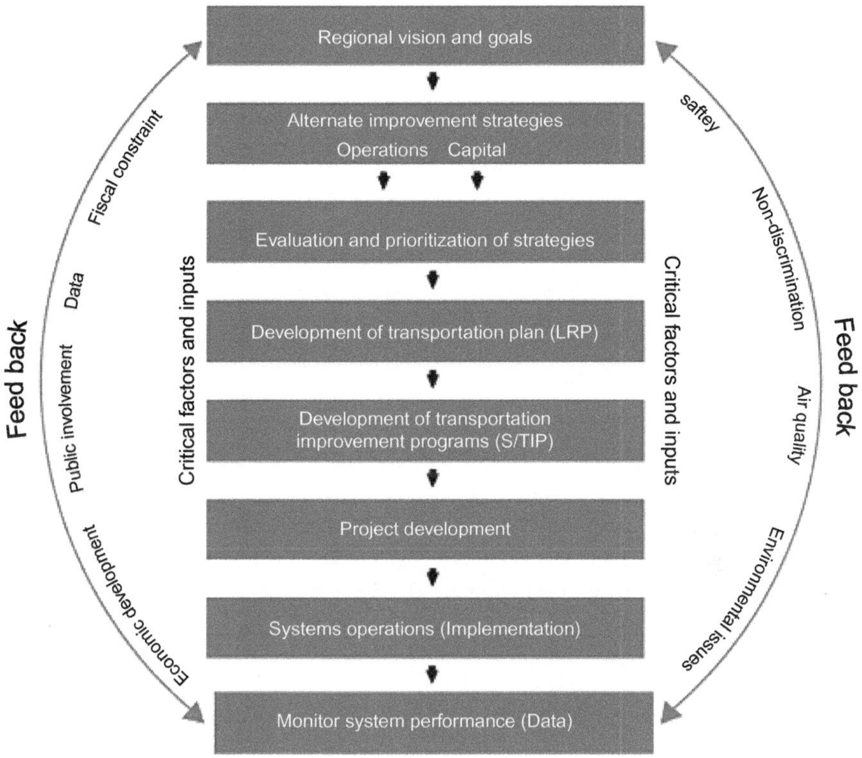

Figure 2.1 The transportation planning process. *Source: FHWA, http://www.planning.dot.gov/documents/briefingbook/bbook.htm.*

Metropolitan Planning Organization

A metropolitan planning organization (MPO) may include the urbanized area (UA), areas expected to become urbanized within the next 20 years, and additional areas determined by political boundaries (e.g., a county) or geographic boundaries (e.g., an air basin). A UA is an area that contains a city of 50,000 or more in population, plus the incorporated surrounding areas meeting size or density criteria as defined by the U.S. Census Bureau. The formation of an MPO is required for any UA, based on federal legislation passed in the 1970s. That legislation emphasizes the "3-C" planning process, assuring that the process used to program and allocate funds for transportation improvements was *continuous, cooperative,* and *comprehensive.*

Table 2.1 Key planning products

Planning document or product	Who develops?	Who approves?	Time horizon	Content	Update requirements
Unified Planning Work Program (UPWP)	MPO	MPO	1 or 2 years	Planning studies and tasks	Annually
Metropolitan Transportation Plan (MTP)	MPO	MPO	20 years	Future goals, strategies, and projects	Every 5 years (4 years for nonattainment and maintenance areas)
Transportation Improvement Program (TIP)	MPO	MPO/governor	4 years	Transportation investments	Every 4 years
Long-Range Statewide Transportation Plan (LRSTP)	State DOT	State DOT	20 years	Future goals, strategies, and projects	Not specified
Statewide Transportation Improvement Program (STIP)	State DOT	U.S. DOT	4 years	Transportation investments	Every 4 years

Adapted from FHWA, http://www.planning.dot.gov/documents/briefingbook/bbook.htm.

The MPO is responsible for five core functions including:

1. Establishing a setting for effective decision making in planning in a metropolitan area
2. Identifying and evaluating alternative transportation investment options in a Unified Planning Work Program (UPWP)
3. Preparing and maintaining a Metropolitan Transportation Plan (MTP)
4. Developing a Transportation Improvement Program (TIP)
5. Involving the public and other stakeholders in the previous four areas

State Department of Transportation Planning Organization

Each U.S. state is required to have an agency or department that is tasked with planning, programming, and project implementation for transportation improvements at the state level. This state Department of Transportation (DOT) interacts with the MPOs and other entities to provide a comprehensive, cooperative, and continuous planning process in the state. Specific to the planning process, each state DOT has three key responsibilities:

1. Prepare and maintain a long-range statewide transportation plan (LRSTP)
2. Develop a Statewide Transportation Improvement Program (STIP)
3. Involve the public and other stakeholders in both the LRSTP and STIP for a transparent planning process

Different state DOTs will vary on the level of detail of the LRSTP, ranging from broad policy documents to providing a specific list of projects. The STIP is generally more specific, providing a prioritized list of projects that are targeted to serve the state's key transportation goals and address areas of critical transportation needs. The STIP is fiscally anchored, taking into consideration available spending and financial resources in the project prioritization and programming. The STIP should incorporate the TIPs developed by the MPOs across the state.

Regional Transportation Planning Organization

In rural states and areas, a regional transportation planning organization (RTPO) may be formed to serve a similar function as an MPO. The RTPO is an association of local governments within a county or contiguous counties. RTPO members may include counties, (small) cities, transportation service providers, tribal government, and others.

Similar to MPOs, RTPOs are tasked with developing a regional transportation plan and to assure that local or county policies are consistent with that broader regional plan. The plan is typically developed for a specified period (e.g., six years) and is updated and evaluated at regular intervals. Overall, RTPOs and MPOs serve very similar transportation planning functions, including the development of a long-range plan, coordinating within a region, and preparing a Transportation Improvement Program that prioritizes funding and investment in the region.

RTPOs are different from MPOs in that they are often created through state legislation, while MPOs are federally mandated. RTPOs are also different in that they cover both urban and rural areas, and therefore serve somewhat expanded functions than MPOs. The stakeholders for an RTPO are therefore also likely to be more diverse, and include rural interests in addition to the urban focus.

Prioritizing Transportation Investments

Because many MPOs rely on state and federal funding for critical transportation infrastructure improvements, the prioritization and funding formula can become a sensitive political issue across a state. For example, state prioritization formulas may tend to favor urbanized areas or regions that require high levels of mobility and reliability of transportation to serve the broader economic interests of the state. As such, routes connecting ports, airports, and large urbanized areas may be looked on favorably by prioritization formulas.

But these urbanized areas also tend to be the regions in a state with the highest population densities and therefore the highest tax base. In other words, these urbanized areas also tend to have a larger internal funding level for transportation investments. Urban areas also often have the ability to pass bond measures for specific transportation projects or packages (e.g., a bond to improve multimodal transportation connectivity), or to levy additional taxes to support strategic investments (e.g., a citywide sales tax increase to fund public transportation infrastructure).

Because rural areas can be at a disadvantage when it comes to passing transportation bonds or tax measures, state prioritization formulas often include an *equity* component, to where even rural, nonmobility focus areas receive a share of the state transportation dollars. So while these rural areas may not experience the same level of traffic or congestion as facilities in urbanized areas, they nonetheless require funding for maintenance, bridge replacement, or for strategic investments that may spur future development.

Rural areas may also be key contributors to the tourism industry of a state, whether it be coastal regions, state parks, mountain resorts, or other features that attract visitors (and their money). Planning organizations may therefore consider additional factors in their prioritization formulas to support tourism and seasonal traffic, even if these roads score lower in terms of total traffic demand or volume-to-capacity ratio.

Emphasizing Agency Coordination

Each transportation plan is mutually adopted by the state and the local area or region to serve as a guide in the development of an area transportation system. A state-level plan can then be used by the local area or region to develop their requests for project funding in the Transportation Improvement Program (TIP). The TIP documents the schedule and funding sources of transportation projects that are expected to be funded over the near future. At the same time, the state takes the MPO TIPs into consideration in developing and streamlining statewide goals.

The cooperation and coordination of the different entities is critical as no single agency has responsibility for and jurisdiction over the entire transportation system. Roads that are part of the Interstate Highway System (IHS) are usually maintained by state DOTs and are subject to federal standards and requirements. Some DOTs further own and maintain a significant portion of the secondary road system, while in other states, county or local municipalities are responsible for design, operation, and maintenance of vast portions of the transportation system. Often, urban transportation projects involve intersections of a state road with a municipal road. Traffic signal system operation and maintenance are also often under municipal authority, even if the roads themselves are owned by the state. Transit systems are often under the control of an entirely separate entity. In short, cooperation and coordination between these agencies is a critical aspect of transportation planning.

Together, these various agencies continuously analyze the transportation system to identify deficiencies and determine the future transportation needs for either new construction projects or to enhance existing transportation facilities. These enhancements vary from nonconstruction alternatives, such as improvements to an area's traffic signal system, to construction alternatives, such as widening a road or highway or the creation of a new interchange.

Many transportation plans also include a distinct multimodal component, as agencies look to nonauto modes to support the long-term sustainability of the transportation system, especially in an urban environment. As such, transportation master plans often feature aspects of transit provisions, as well as nonmotorized transportation. In addition, specific plans for other modes of transportation may be developed to supplement an overall strategic plan. These include pedestrian master plans, bicycle master plans, or public transportation plans, which, depending on the state, may be a requirement for MPOs.

Planning Applications and Topics

FHWA provides guidance on a wide range of planning topics and considerations, which go well beyond the impact of traffic growth on road congestion[1]. A comprehensive planning process encompasses a range of considerations and integrates them in the overall planning process. The following is a list of additional considerations and topics in planning applications:

- Air quality
- Congestion management process (CMP)
- Financial planning and programming
- Freight movement
- Land use and transportation
- Performance measures
- Planning and environment linkages
- Public involvement
- Safety
- Security
- System management and operations (M&O)
- Technology applications for planning: models, geographic information system (GIS), and visualization
- Title VI/environmental justice (EJ)
- Transportation asset management

Each of these topics is described in detail in guidance put forth by FHWA, which is available online[1]. The same resource provides definitions of planning terms and includes fact sheets with key planning-related topics. Additional guidance on transportation planning topics and best practices is available through the American Planning Association (APA)[2].

[1] http://www.planning.dot.gov/documents/briefingbook/bbook.htm
[2] https://www.planning.org/

2.1.2 Accuracy and Error in Forecasting

Transportation planning looks toward the future, and as such is a process that by definition is unverifiable. Transportation planners rely on historical data, model predictions, comparable studies, and many assumptions to make estimates about future conditions and predictions about future transportation needs. Each of these components, from historical data to model predictions, has internal errors; and the combination of multiple models and data sources introduces more error and variability.

It is not uncommon to hear phrases such as "All models are wrong, but some are useful," and "Planning is as much an art as it is a science." It is the job of a transportation planner to carefully weigh and scrutinize these potential sources of errors and apply engineering judgment to any final planning recommendations.

Calibration and validation techniques exist to assist the transportation planner in evaluating the performance of forecasting models. Following a well-defined planning process helps bring consistency in planning predictions across MPOs and other agencies. It is equally important to document assumptions and data sources, and strive for a clear and transparent planning process. Ideally, that process should include feedback from stakeholder groups throughout, as well as an assessment of prior planning predictions against actual observations, to help inform and refine the planning process in future TIP iterations.

2.2 PLANNING CONCEPTS AND FOUR-STEP PROCESS OVERVIEW

This section introduces the planning process, consisting of four basic methodological steps. Each of these is then described in greater detail in the following sections.

2.2.1 Regional and Statewide Planning Process

Transportation planning at a citywide, regional, or statewide level involves a large transportation network, many diverse stakeholder groups, a full spectrum of land uses and their impacts, and a very elaborate planning process. Generally, we think of transportation planning at those levels in terms of the traditional four-step transportation planning process:

1. *Trip generation*: An estimate of future trips generated based on land use type, land use size, and travel patterns.

2. *Trip distribution*: Distributing the estimated trips to traffic analysis zones (TAZs) and distinguishing trip productions and attractions in each zone in various trip categories (e.g., home-to-work trips).
3. *Mode choice*: Determining the preference of people within zones to make a trip by personal vehicle, public transportation, carpooling, bicycling, walking, etc., and dividing the total trips among those modes.
4. *Trip assignment*: Assigning all the trips to the transportation system to identify actual routes taken, consider capacity constraints on the network, and to ultimately determine the impact to the transportation network.

Conceptually, trip generation forecasts how many trips an analysis zone generates; trip distribution predicts where those trips end up; mode choice estimates how those trips are completed (car, transit, bike, walking, etc.); and traffic assignments project what specific roads or sequence of roads those trips are assigned to.

These steps are completed within a transportation planning model, traditionally referred to as a *travel demand forecasting model*, or a *travel demand model* (TDM). Each of these steps will be described in detail in later sections.

Some of the planning steps are very complex and cover large metropolitan areas and surrounding communities. Planners use them to help determine and evaluate long range transportation plans (LRTP) or comprehensive transportation plans (CTP), which can test various transportation system alternatives to accommodate the future travel demand. The most sophisticated models cover all modes of transportation, including toll facilities, high occupancy vehicle (HOV) lanes, high occupancy toll (HOT) lanes, and so on. However, depending on funding and staff availability, and political will (or lack thereof), many models are much simpler and often autocentric.

The study area for a planning study is divided into traffic analysis zones (or TAZs). Within these planning models several inventories and population demographics are associated with each TAZ. Some common inventories include:

- Population (number of residents)
- Land use (type of development)
- Economic activity (retail, office, medical, education, manufacturing, etc.)
- Transportation systems (road network, transit routes, multiuse paths, etc.)
- Travel patterns
- Laws and ordinances (e.g., speed limits, zoning restrictions)
- Financial resources
- Community values

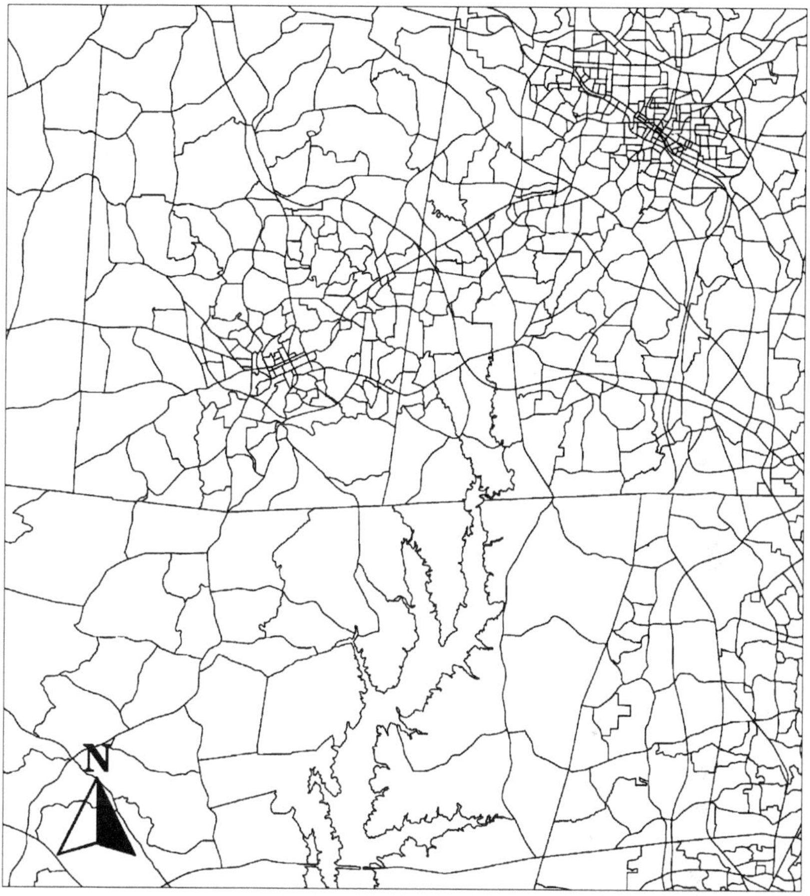

Figure 2.2 Example planning network with TAZs. *Figure by Joe Huegy, ITRE – used with permission.*

These inventories are periodically updated and new models calibrated and validated for new base conditions and then used for forecasting future travel demand. An example of a planning network with TAZs is shown in Figure 2.2.

The figure shows a number of TAZs representing urban and rural areas. Note that highly populated urban areas are more likely to have smaller TAZs, while suburban or rural areas are likely to have larger TAZs in terms of geographic area. But these larger TAZs may still be smaller in terms of the number of trips generated, due to lower population density and/or commercial and employment centers contained therein. In the figure it is evident (just from the small size of the TAZs) that there are two small towns toward the northern side of the figure, with more rural development to the south of the figure around a large lake.

2.2.2 Traffic Impact Analyses

In addition to citywide, regional, or statewide planning efforts, almost all jurisdictions that control land use require a traffic impact study (or traffic impact analysis, TIA) when land is being developed. These can be thought of as smaller-scale planning studies, where the impacts are focused on the portion of the road network surrounding a particular development site. TIAs can be limited to a few intersections, but can also cover a large portion of the city transportation network, depending on the size of the land use under study. For example, a new grocery store or big-box retail store is likely to require a larger-scale TIA than a new gas station or isolated restaurant. But in general, a TIA is much reduced in scale and effort from the more comprehensive planning studies described in the previous subsection.

TIAs can be for development that fits within the current zoning, as well as for development that will require a change in the zoning. Based on the outcome of the TIA, the developer of a site may be required to mitigate the impacts from the new development on the surrounding road network. The controlling agency has leverage in the development process because it must approve any rezoning request, special use permit request, site plan, driveway access permit request, request for any changes to the road network, and implement any impact fee ordinance.

Typically the developer hires a consultant (private engineer or engineering firm) to create the TIA, which is then reviewed by the controlling agency. A traffic impact study takes the planned development (at "full build out") and generates future traffic. The traffic is assigned to the highway network using what can amount to significant judgment for large developments. In some cases, all intersections and interchanges within the local influence area of the development may need to be analyzed. But it is rare to go beyond a one-mile radius from a development, even for large projects. The adjacent street system, or area network, is analyzed for capacity impacts using methodologies described in Part 5 of this book.

Most agencies do not want to see any degradation on the highway system and will require the developer to mitigate estimated impacts before issuing a construction or development permit. If the project involves staged development, then intermediate impact analyses are also performed for their individual impacts.

Traditionally, TIAs have been performed in an isolated fashion, meaning that a TIA for any one development is reviewed by itself, without explicitly considering other developments in the vicinity that are planned or proposed. But more recently, agencies have started to look at the *cumulative impact* of various TIAs and associated development proposals. Facilitated by the use of planning software, a new development may be evaluated not only relative to existing conditions and background traffic growth, but also under consideration of specific other developments in the vicinity of the proposed site. This form of cumulative TIA is likely to assure better coordination between projects, and help mitigate potentially confounding traffic impacts of parallel development applications.

Scope

TIAs are generally required for larger projects and development, such as a 160 + unit single-family subdivision, 220 + unit multifamily complexes, and most commercial or industrial sites. Some small projects, such as a fast-food restaurant, will need a TIA depending on agency practice. The planning horizon for a TIA is generally short term, with a lookout period of 2–5 years, or until build out.

For most TIAs, the analysis is usually confined to peak periods (am peak hour, pm peak hour, lunch peak hour, or weekend peak hour). The geographic focus includes the site plus surrounding major intersections or interchanges, with agency practices dictating the study area and geographical extent. In most cases the analysis is almost always limited to traffic operational measures and level of service using *Highway Capacity Manual* (HCM) methodologies (TRB, 2010), or their equivalent. Safety impacts are usually not quantified, although some agencies are starting to embrace user perception–based quality-of-service metrics in the TIA process. The specific operational analysis methodologies from the HCM are detailed in Part 5 of this book.

Process

Agency practices and requirements for traffic impact analyses (TIAs) differ, and analysts should always consult specific guidelines for the area or region they are working in. But a typical TIA often involves the following 10-step process, which is intended here as a sample illustration of potential agency practices.

1. *Establish scope*: Decide on the timeframe of analysis (typically 2–5 years), analysis area (road network surrounding study site, up to a one-mile radius), etc.
2. *Data collection*: Gather current traffic counts and roadway geometric data. If no current traffic data are available, driveway counts and intersection turning movement counts may need to be conducted before proceeding.
3. *Trip generation*: Using references such as the Institute of Transportation Engineers (ITE) *Trip Generation Manual*, estimate trips generated by proposed development. The ITE manual provides estimates by land use classification abd time of day, as well as guidance for mixed-use development as discussed in further detail in following sections.
4. *Trip distribution*: The analyst distributes trips generated from proposed development to adjacent zones or directions. This may involve the use of planning models, as described later. For small developments, the analyst may sometimes assume that new trips are distributed proportionally to existing trips in terms of their origins and destinations.
5. *Background traffic*: The new development is evaluated not relative to existing traffic, but considers growth in background traffic in the local area over the planning horizon (2–5 years). The analyst estimates the volumes of background traffic in forecast year by applying growth rates. Growth rates typically range from zero to 5% per year, and are dictated by local agency practice. Growth rates are applied using a standard compound interest formula, where future traffic, F, is estimated from present year traffic, P, the annual growth rate, i, and the number of years in the forecast, n.

$$F = P(1+i)^n$$

6. *Traffic assignment*: Trips are then assigned to routes, which places them on the links and segments in the surrounding network. Usually all trips are assumed to use the quickest possible route (see

shortest path assignment, Section 2.6.2), but may utilize more advanced assignment algorithms depending on the scale of the study. The result of the traffic assignment is a future-year set of driveway counts and turning movement volumes at all intersections in the study area.

7. *Background impacts*: Using methods in the *Highway Capacity Manual* (see Part 5 of the book), the analyst computes levels of service at key points in the forecast year for background traffic only. This will be the reference against which the development impacts are evaluated.
8. *Development impacts*: Compute levels of service at key points in the forecast year for background plus the new site traffic estimated in the TIA. This development impact is compared to the background impacts to evaluate the changes in levels of service due to the proposed development.
9. *Reexamine assignment*: Depending on the results in step 8, the assignment of site traffic may be reexamined to potentially avoid resulting bottlenecks and congestion points. If necessary, reassign traffic and recompute levels of service, although the assumptions for this new assignment have to be well documented and justified.
10. *Mitigation strategies*: Depending on the evaluation results for the future year, the analyst may propose a series of mitigation strategies to assure that an adequate level of service is maintained even with the addition of the proposed site to the network. The analyst recomputes level of service with these mitigation strategies, and shows satisfactory performance. Several options for mitigation may be considered, both in terms of traffic impacts as well as cost. The cost for mitigation is typically carried by the developer, and the agency may require mitigation before a permit is issued.

Agencies may require variations of the 10 steps mentioned, or may require additional or fewer steps. Further, the specific methodologies used to apply each step, as well as any default assumptions (e.g., the growth factor used for background traffic in step 5), are based on agency practice.

Typical Mitigation Strategies

At the completion of the estimation of traffic impacts, a TIA typically generates one or more mitigation strategies to help offset the predicted traffic impacts. Common mitigation strategies include adding exclusive left-turn/right-turn lanes (for unsignalized and/or signalized intersections), widening the roadway (e.g., two-lane to five-lane with two-way left turn lanes (TWLTL) or four-lane divided cross section), installing or upgrading signals at major intersections, and considering alternative intersection and interchange treatments. Alternative intersections and interchanges include modern roundabouts, restricted crossing U-turns (RCUT) or superstreets, diverging diamond interchanges (DDI), median U-turns (MUT), displaced left-turn intersections (DLT), quadrant left-turn intersections, and others. Specific guidance for the design and application of these applications can be found elsewhere (FHWA, 2014).

2.2.3 Highway Functional Classification System

Functional classification is a process by which streets and highways are grouped into classes, or systems, according to the character or service they are intended to provide. Individual roads and streets do not serve travelers independently in any major way; rather, travel involves a network of roads. Functional classification defines the roles that any road, street, or highway should serve in the network.

Rural and urban areas have different characteristics as to the density and types of land use, density of street and highway networks, nature of travel patterns, and the way in which all these elements are related in terms of highway function. Figure 2.3 exhibits the hierarchy of the Highway Functional Classification system for U.S. roads.

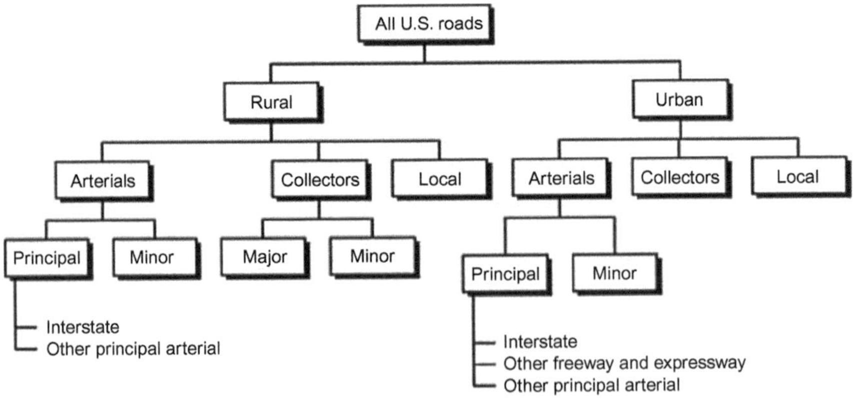

Figure 2.3 Highway functional classification system. *http://www.fhwa.dot.gov/policy/2006cpr/chap2.htm*

Arterials provide the highest level of mobility, at the highest speed, for long, uninterrupted travel. The Interstate Highway System is an arterial network. Arterials generally have higher design standards than other roads, often with multiple lanes and some degree of access control. The arterial network provides interstate and intercounty service so that all developed areas are within a reasonable distance of an arterial highway. This network is further broken down into principal and minor arterials:

- *Principal arterials*: For rural locations, this system is a connected network of continuous routes that serve corridor movements having substantial statewide or interstate travel characteristics, such as long trip lengths and travel times and relatively high densities. For urban systems, principal arterials are the key arteries that carry traffic to and from major residential and commercial centers.
- *Minor arterials*: This system forms a network that links cities, larger towns, and other major traffic generators such as large resorts. The rural minor arterial system generally serves intrastate and intercounty travel, and travel corridors with trip lengths and travel densities somewhat less than the principal arterial system. In an urban system, minor arterials serve key mobility needs between traffic attractors and generators, but at a slightly reduced scale than a principal arterial.

Collectors provide a lower degree of mobility than arterials. They are designed for travel at lower speeds and for shorter distances. Collectors are typically two-lane roads that collect and distribute traffic from the arterial system. The collector road system is further classified into major and minor collector roads:

- *Major collector roads*: In rural locations, these roads provide service to the larger towns not directly served by the higher systems. They also provide service to other traffic generators of equivalent importance, such as schools, shipping points, county parks, and important industrial, agricultural, or cultural areas. Major collector roads also link these places to routes of higher classification and serve the more important intracounty travel corridors. In urban settings, collector roads can connect neighborhoods or provide access to local businesses.
- *Minor collector roads*: Rural minor collectors are spaced at intervals, consistent with population density, to collect traffic from local roads and to insure that all urbanized areas are within a reasonable distance of a collector road. Urban minor connectors are found within neighborhoods and business districts to provide access to residential communities and employment centers.

Local roads are all roads that are not on a higher system. Local residential subdivision streets and residential collector streets are elements of the local road system. Local residential streets are either cul-de-sacs, loop streets less than 2500 ft (762.2 m) in length, or streets less than one mile (1.6 km) in length. They do not connect thoroughfares, serve major traffic generators, or collect traffic from areas with more than 100 dwelling units. Residential collectors serve as the connecting street system between local residential streets and the thoroughfare system.

An example of an urban street network is shown in Figure 2.4, with different functional classes represented with different line weights. In a network representation, higher functional classes such as primary arterials are represented as thicker lines, and local collector streets are shown as thin lines.

In a network model representation or in a travel demand model, the analyst generally starts with the higher functional classes and then continues to add detail. It is generally impractical to represent all roads in an area in the modeled network. Local roads and those without significant traffic volumes can thus be excluded to facilitate network coding and processing. As a rule of thumb, the sum of all omitted links should carry less than 10% of all trips in the network.

Figure 2.4 Illustration of a street network in an urban area. *http://www.fhwa.dot.gov/planning/processes/statewide/related/highway_functional_classifications/section03.cfm*

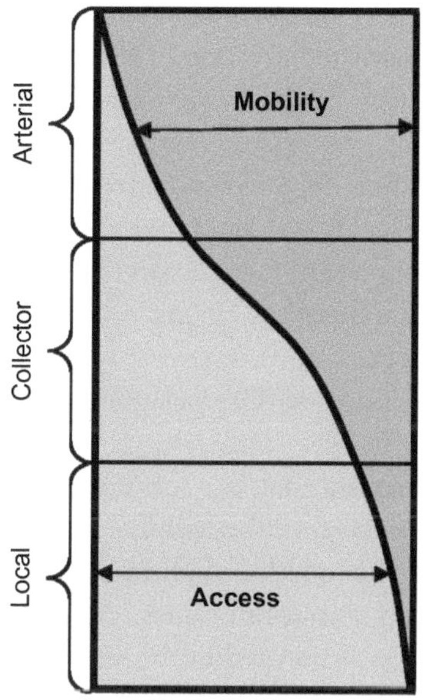

Figure 2.5 Relationship of functionally classified highway systems in serving traffic mobility and land access. *Source: Adapted from FHWA, http://www.fhwa.dot.gov/environment/publications/flexibility/ch03.cfm*

All roads serve a dual purpose: (1) mobility, to allow traffic to travel between zones and destinations, and (2) land access, to allow traffic to enter neighborhoods and access businesses. In the classic diagram on the tradeoffs between land access and mobility in Figure 2.5, it is evident the higher functional classes such as arterials generally serve a predominant mobility purpose, local roads predominately serve a land-access purpose, and collectors tend to balance the two.

2.2.4 Planning Data and Data Sources

The first step in transportation planning is the collection and analysis of transportation-related data. Transportation models are created that use data such as traffic counts, population, housing, employment, and vehicle ownership to simulate existing (base-year) and future (design-year) travel. Anticipating future demands for travel is accomplished by projecting current travel levels and demands into the future using traffic-growth factors. These factors are based on various sources of information.

First, the planner must be familiar with the local land use plan and the existing and future development in the area. Historic traffic trends, such as the average daily traffic volumes (ADT) must also be known. These traffic count volumes are typically recorded by state DOTs for many years, and they are a source for determining the historic (and the future) traffic trends.

Also, transportation models are used to simulate travel patterns in an area. The planner will create an existing (base-year) model of the area and then project the model into the future (design-year). These models may use traffic-growth factors from mathematical equations or may borrow factors from other similar transportation models.

Traffic Counts

Traffic counts are part of the factual information that can be collected prior to the design process. Existing and projected traffic volumes affect the geometric features of design such as the lane widths, alignments, and grades. Traffic volumes are the loads placed on a street network. A roadway design process that did not include traffic volumes would be as senseless as the design of a bridge without the proposed vehicle weights.

State agencies typically collect traffic volume data at numerous locations throughout the state. These data include variations due to different times of the year, different times of the day, and the distribution of vehicle types or weights. Either a temporary counter or a permanent counter can record these counts. A temporary counter (traditionally a rubber road tube, more recently microwave or video detectors) is located with the intention of studying a particular area or street. For the most accurate results, the counters are usually put down during a time period when the travel for the area is average. For instance, at the coast and in tourist destinations, a major holiday weekend would be avoided, because the travel is unusually high then (unless, of course, the focus of such study is on peak seasonal loads).

Permanent control stations are also located throughout all states at key locations, collecting data 365 days a year. These permanent count stations are used to generate adjustment factors to account for seasonality, month-by-month variation, and time-of-day patterns for other, more short-term counts.

Traffic engineers use these count data along with future land use proposals to estimate future traffic volumes. The design year used for the planning process is usually either 10 or 20 years beyond the beginning of construction. Whenever a project is delayed, the traffic forecasts need to be updated at the beginning of the preliminary design.

Average Daily Traffic

The most commonly used measure of traffic is the average daily traffic (ADT) volume. Theoretically, this is the volume of traffic moving in both directions on a highway for the most average traffic day in the year for 24 hours (h). It can be visualized as if there were an individual counting the number of vehicles passing a given point 24 h a day for 365 days a year, and then dividing the sum of all counts by 365. This would then be the ADT of that highway.

ADTs can be calculated from any sample of repeated daily counts of traffic volumes, with duration as short as one week. Because that short-duration count may be subject to seasonal fluctuation or other sources of bias, ADTs are often *annualized* by applying adjustment factors from nearby permanent count stations. The resulting average annual daily traffic (AADT) is often used to describe traffic volume characteristics of a roadway in a planning context.

AADTs can be calculated from continuous count stations using federal guidance (FHWA, 2013). The FHWA formula for estimating AADT from a long-term or whole-year traffic count is:

$$\mathrm{AADT} = \frac{1}{7}\sum_{i=1}^{7}\left[\frac{1}{12}\sum_{j=1}^{12}\left(\frac{1}{n}\sum_{k=1}^{n} VOL_{ijk}\right)\right]$$

where

VOL = daily traffic for day k, of day of the week i, and month j

i = day of the week

j = month of the year

k = the occurrence order number of day of the week

n = the number of days of that day of the week during that month

The aforementioned AADT estimation method can overcome isolated days of missing traffic data resulting from the use of an automatic traffic data collection environment.

AADT data are also published in an annual report by recording volumes for each studied roadway on a county or city map book. An example of an AADT map is shown in Figure 2.6.

Many agencies are also migrating AADT data to online mapping tools with a linked database, to provide ready access for analysts. The traffic volumes can be easily found and compared with previous years' volumes. Although AADT is important, it is typically not used for geometric design because it does not reflect the variations of traffic over the months of the year, days of the week, or hours of the day. Therefore, the design of most major highways is based on the design hourly volume (DHV).

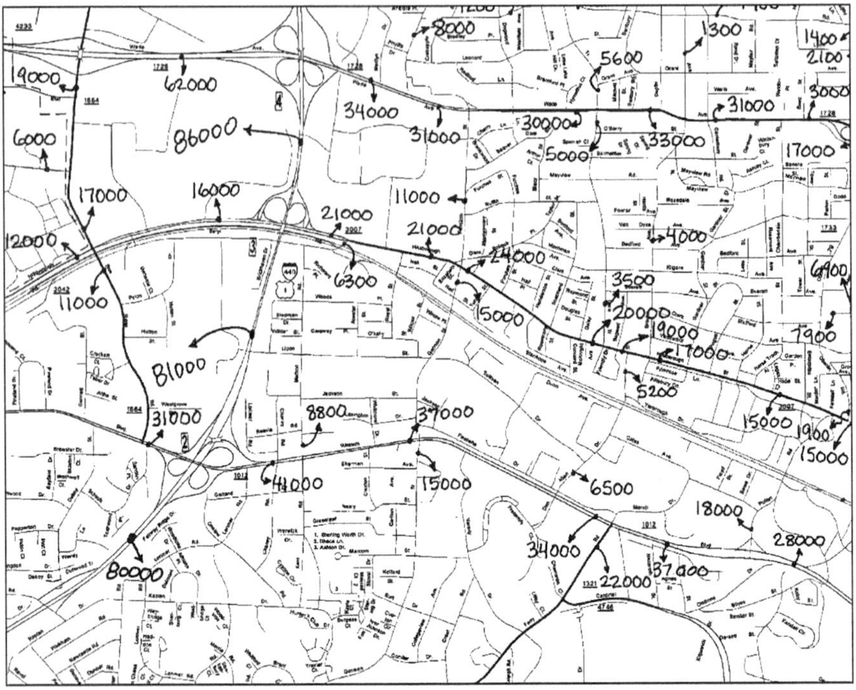

Figure 2.6 Sample AADT map for Raleigh, North Carolina. *Source: NCDOT http://www.ncdot.gov/travel/statemapping/trafficvolumemaps/urban/*

Design Hourly Volume and Directional Design Hourly Volume

Traffic volumes are a major factor in selecting design criteria. Most design criteria are based on a design hourly volume (DHV) factor and more specifically, the directional design hourly volume (DDHV). The DHV is based on the 30th highest hourly volume out of 365 counts (i.e., just over the 90% level). This essentially means that a road with a design based on this DHV will be able to handle traffic demand 90% of the time. DHV is determined by multiplying the ADT by the design hourly volume factor, also known as the *K factor.* If the DHV is not given, then an assumed 10% (use the decimal form, 0.10 for problems), K factor can be applied to the ADT to determine DHV.

Directional design hourly volume (DDHV) is calculated by multiplying the DHV by the directional movement factor (D). The D factor tells you the maximum percentage of the DHV that is moving in one direction during the day. If the directional movement is not given, the student should assume 60% (use the decimal form, 0.60 for problems).

The following equations are given for the conversion between the various traffic count metrics, followed by several examples illustrating their use.

$$DHV = ADT \times K$$
$$DDHV = DHV \times D$$

or

$$DDHV = ADT \times K \times D$$

where

DHV = design hourly volume
DDHV = directional design hourly volume (vehicles per hour, veh/h)
ADT = average daily traffic (vehicles per day, veh/d)
K = design hourly volume factor (0.10 typically)
D = directional movement factor (0.60 typically)

EXAMPLE 2.1

The following traffic counts were taken along an urban freeway:

Day 1: 1900 vehicles
Day 2: 2150 vehicles, D = 55%
Day 3: 2300 vehicles, K = 12%
Day 4: 1950 vehicles
Day 5: 2000 vehicles

Find the ADT, DHV, and DDHV.

Solution

$$ADT = \frac{1900 + 2150 + 2300 + 1950 + 2000}{5}$$

$$ADT = 2060 \text{ veh/d}$$

$$DHV = ADT \times K$$

$$DHV = 2060 \text{ veh/d} \times 0.12 = 247 \text{ veh/h}$$

$$DDHV = DHV \times D$$

$$DDHV = 247 \text{ veh/h} \times 0.55$$

$$DDHV = 136 \text{ veh/h in the peak direction}$$

EXAMPLE 2.2

Compute the DDHV for each movement shown based on the ADT for the intersection shown here; K = 10%.

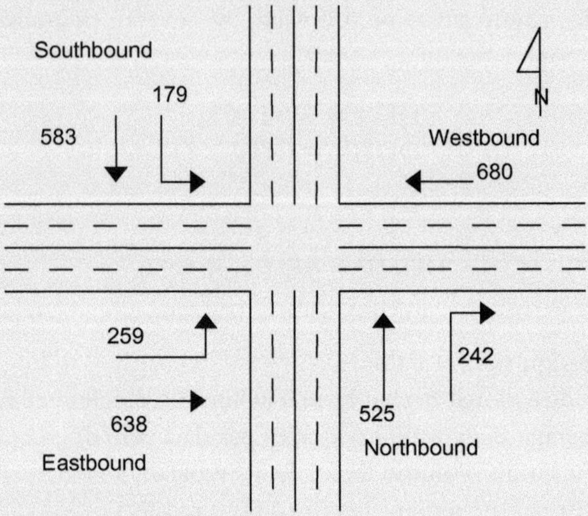

Solution

As directions are given on the drawing, we do not need to multiply the direction factor into the DDHV equation.

$$\text{DDHV} = \text{ADT} \times \text{K}$$

Westbound lane:

$$680 \times 0.10 = 68 \text{ veh/h}$$

Northbound straight:

$$525 \times 0.10 = 52.5 \text{ veh/h}$$
$$= 53 \text{ (because we cannot have 0.5 veh/h)}$$

Northbound right:

$$242 \times 0.10 = 25 \text{ veh/h}$$

Eastbound straight:

$$638 \times 0.10 = 64 \text{ veh/h}$$

Eastbound left:

$$259 \times 0.10 = 26 \text{ veh/h}$$

Southbound straight:

$$583 \times 0.10 = 59 \text{ veh/h}$$

Southbound left:

$$179 \times 0.10 = 18 \text{ veh/h}$$

Peak Hour Volume

When performing traffic analyses, as well as in the design of highways, it is common to use the peak hour volume because this volume places the highest demand on the roadway. The peak hour volume is simply the highest hourly volume on the roadway for a given day. Sometimes separate am and pm (and perhaps noon) peak hour volumes are determined.

Traffic is typically not uniform over the entire peak hour; this means that the rate of flow during the peak 15 minutes (min) of the peak hour will typically be higher than the rate flow based on the entire hour. A Highway Capacity Manual (HCM) analysis therefore applies a peak hour factor (PHF) to convert the maximum 15-min volume into an hourly flow rate. The PHF is found by dividing the peak hour volume by the product of 4 times the peak 15-min volume.

$$\text{PHF} = \frac{\text{peak hour volume}}{4 \times \text{highest 15-min volume}}$$

Dividing the peak hourly volume by the peak hour factor results in an hourly flow rate based on the peak 15-min volume. A high PHF value (1.0 is the highest possible value) represents relatively constant flow throughout the peak hour, while a low PHF (0.25 is the lowest possible value) indicates that most of the traffic occurs during a peak period within the hour. On freeways, typical PHFs range from 0.80 to 0.95. Lower PHFs usually characterize rural freeways, and higher PHFs are typical of urban and suburban peak-hour conditions.

A PHF is only used when design hourly volumes (DHV) are the only available data source. If 15-min traffic volume data are available, the analysis volume can be estimated directly by multiplying the peak 15-min volume by a factor of 4. But the PHF is needed in most planning applications, as 15-min data are generally not available and the analyst thus has to rely on (estimated) DHV values for the traffic analysis.

The following pair of examples illustrate the two extreme values of PHFs.

In summary, the *higher* the PHF is, the more evenly spread out the traffic is over the peak hour; and the *lower* the PHF is, the more "peaked" the traffic is at a certain 15-min time period within the peak hour. Here is another example illustrating how to find the PHF if 15-min counts are provided.

EXAMPLE 2.3

One Hour of Traffic Data

15-min segment	Number of vehicles per hour (vph)
7:00–7:15 am	250
7:15–7:30 am	250
7:30–7:45 am	250
7:45–8:00 am	250

$$\text{Peak hour volume} = 250 + 250 + 250 + 250 = 1000 \text{ vph}$$

$$\text{PHF} = \frac{\text{peak hour volume}}{4 \times \text{highest 15-min volume}} = \frac{\sum \text{15-min vol}}{4 \times \text{highest 15-min vol}}$$

$$= \frac{(250 + 250 + 250 + 250)}{(4 \times 250)} = 1.00$$

Note that if the flow for an entire hour is constant, the PHF is 1.00, which is the highest PHF value possible.

EXAMPLE 2.4

One Hour of Traffic Data

15-min segment	Number of vehicles per hour (vph)
7:00–7:15 am	1000
7:15–7:30 am	0
7:30–7:45 am	0
7:45–8:00 am	0

$$\text{Peak hour volume} = 1000 + 0 + 0 + 0 = 1000 \text{ vph}$$

Note that the peak hour volume is the same as in Example 2.3.

$$\text{PHF} = \frac{\sum \text{15-min vol}}{4 \times \text{highest 15-min vol}} = \frac{(1000 + 0 + 0 + 0)}{(4 \times 1000)} = 0.25$$

Note that if all flow during an hour occurs in one 15-min period, the PHF is 0.25, which is the lowest PHF value possible.

EXAMPLE 2.5

One Hour of Traffic Data

15-min segment	Number of vehicles per hour (vph)
7:00–7:15 am	240
7:15–7:30 am	310
7:30–7:45 am	270
7:45–8:00 am	180

$$\text{Peak hour volume} = 240 + 310 + 270 + 180 = 1000 \text{ vph}$$

Note that the peak hour volume is the same as in Examples 2.3 and 2.4.

$$\text{PHF} = \frac{\sum 15\text{-min vol}}{4 \times \text{highest } 15\text{-min vol}} = \frac{(240 + 310 + 270 + 180)}{(4 \times 310)} = 0.81$$

2.3.1 Planning Definitions and Terms

A list of commonly used transportation planning terms is given in the following, which was taken from FHWA guidance (FHWA, 2007).

Capacity: A transportation facility's ability to accommodate a moving stream of people or vehicles in a given time period.

Capital program funds: Financial assistance from the major capital program of 49 U.S.C. Section 5309. This program enables the Secretary of Transportation to make discretionary capital grants and loans to finance public transportation projects divided among fixed guideway (rail) modernization; construction of new fixed guideway systems and extensions to fixed guideway systems; and replacement, rehabilitation, and purchase of buses and rented equipment, and construction of bus-related facilities.

Congestion management process (CMP): A systematic approach required in transportation management areas (TMAs) that provides for effective management and operation, based on a cooperatively developed and implemented metropolitan-wide strategy of new and existing transportation facilities eligible for funding under title 23 U.S.C. and title 49 U.S.C. through the use of operational management strategies. Provides information on transportation system performance and finds alternative ways to alleviate congestion and enhance the mobility of people and goods, to levels that meet state and local needs.

Department of Transportation (DOT): When used alone, indicates the U.S. Department of Transportation. In conjunction with a place name, indicates state, city, or county transportation agency (e.g., Illinois DOT, Los Angeles DOT).

Federal Highway Administration (FHWA): A branch of the U.S. Department of Transportation that administers the federal-aid highway program, providing financial assistance to states to construct and improve highways, urban and rural roads, and bridges. FHWA also administers the Federal Lands Highway Program, including survey, design, and construction of forest highway system roads, parkways and park roads, Indian reservation roads, defense access roads, and other federal lands roads.

Federal Transit Administration (FTA): A branch of the U.S. Department of Transportation that administers federal funding to transportation authorities, local governments, and states to support a variety of locally planned, constructed, and operated public transportation systems throughout the United States, including buses, subways, light rail, commuter rail, streetcars, monorail, passenger ferry boats, inclined railways, and people movers.

Financial plan: The documentation required to be included with an MTP and TIP (optional for the long-range statewide transportation plan and STIP) that demonstrates the consistency between reasonably available and projected sources of federal, state, local, and private revenues and the costs of implementing the proposed transportation system improvements.

Financial programming: A short-term commitment of funds to specific projects identified in both the regional and the Statewide Transportation Improvement Program.

Fiscal constraint: Making sure that a given program or project can reasonably expect to receive funding within the time allotted for its implementation. The MTP, TIP, and STIP must include sufficient financial information for demonstrating that projects in the MTP, TIP, and STIP can be implemented using committed, available, or reasonably available revenue sources, with reasonable assurance that that the federally supported transportation system is being adequately operated and maintained. For the TIP and the STIP, financial constraint/fiscal constraint applies to each program year. Additionally, projects in air quality nonattainment and maintenance areas can be included in the first two years of the TIP and STIP only if funds are "available" or "committed."

High-occupancy vehicle (HOV): Vehicles carrying two or more people. The number that constitutes an HOV for the purposes of HOV highway lanes may be designated differently by different transportation agencies.

Intelligent transportation systems (ITS): Electronics, photonics, communications, or information processing used singly or in combination to improve the efficiency or safety of a surface transportation system. The National ITS Architecture is a blueprint for the coordinated development of ITS technologies in the United States, providing a systems framework to guide the planning and deployment of ITS infrastructure.

Intermodal: The ability to connect, and connections between, differing modes of transportation.

Interstate Highway System (IHS): The specially designated system of highways, begun in 1956, that connects the principal metropolitan areas, cities, and industrial centers of the United States. Also connects the United States to internationally significant routes in Canada and Mexico.

Land use: Refers to the manner in which portions of land or the structures on them are used (or designated for use in a plan), such as commercial, residential, retail, industrial, etc.

Long-range statewide transportation plan (LRSTP): The official, statewide, multimodal transportation plan covering no less than 20 years developed through the statewide transportation planning processes.

Long-range transportation plan (LRTP): A document resulting from regional or statewide collaboration and consensus on a region's or state's transportation system, and serving as the defining vision for the region's or state's transportation systems and services. In metropolitan areas, this is the official multimodal transportation plan addressing no less than a 20-year planning horizon that is developed, adopted, and updated by the MPO through the metropolitan transportation planning process.

Metropolitan planning area: The geographic area determined by agreement between the metropolitan planning organization (MPO) for the area and the governor, in which the metropolitan transportation planning process is carried out.

Metropolitan planning organization (MPO): The policy board of an organization created and designed to carry out the metropolitan transportation planning process for urbanized areas with populations greater than 50,000, and designated by local officials and the governor of the state.

Metropolitan transportation plan (MTP): The official multimodal transportation plan addressing no less than a 20-year planning horizon that is developed, adopted, and updated by the MPO through the metropolitan transportation planning process.

Mode: A specific form of transportation, such as automobile, subway, bus, rail, air, bicycle, or foot.

Performance measures: Indicators of how well the transportation system is performing with regard to such measures as average speed, reliability of travel, and accident rates. Used as feedback in the decision-making process.

Stakeholders: Individuals and organizations involved in or affected by the transportation planning process. Include federal/state/local officials, MPOs, transit operators, freight companies, shippers, users of the transportation infrastructure, and the general public.

Statewide Transportation Improvement Program (STIP): A statewide prioritized listing/program of transportation projects covering a period of 4 years that is consistent with the long-range statewide transportation plan (LRSTP), metropolitan transportation plans (MTPs), and Transportation Improvement Programs (TIPs), and is required for projects to be eligible for funding under title 23 U.S.C. and title 49 U.S.C. Chapter 53.

Surface Transportation Program (STP): Federal-aid highway funding program that supports a broad range of surface transportation capital needs, including many roads, transit, sea and airport access, vanpool, bike, and pedestrian facilities.

Transportation demand management (TDM): Programs designed to reduce demand for transportation through various means, such as the use of public transit and of alternative work hours.

Transportation Improvement Program (TIP): A prioritized listing/program of transportation projects covering a period of 4 years that is developed by an MPO as part of the metropolitan transportation planning process, consistent with the metropolitan transportation plan (MTP), and required for projects to be eligible for funding under title 23 U.S.C. and title 49 U.S.C. Chapter 53.

Transportation management area (TMA): An urbanized area with a population of 200,000 or more, as defined by the U.S. Bureau of the Census and designated by the Secretary of Transportation, or any additional area where TMA designation is requested by the governor and the MPO and designated by the U.S. Secretary of Transportation.

Unified Planning Work Program (UPWP): A statement of work identifying the planning priorities and activities to be carried out within a metropolitan planning area. At a minimum, a UPWP includes a description of the planning work and resulting products, who will perform the work, time frames for completing the work, the cost of the work, and the source(s) of funds.

2.3 TRIP GENERATION

Trip generation is the first step in the four-step transportation planning process. Trip generation estimates the number of trips generated for a given land use, based on prior data of traffic generators in the same land use category. The number of trips are typically a function of the type of land use, as well as its size. The size of a land use is most commonly defined in terms of a readily measurable dimension, such as square footage for a commercial development, number of units for a residential development, number of fueling stations for a gas station, number of seats for a restaurant, number of spaces for a parking deck, and so on.

Trip generation rates and characteristics in the United States are collected and distributed in the ITE *Trip Generation Manual*, and its various companion documents. The *Trip Generation Manual* is updated every three to five years with new data as applicable (ITE, 2012). Various software implementations of the *Trip Generation Manual* are available to help analysts derive trip generation estimates and integrate them into a planning process.

The *Trip Generation Manual* is primarily focused on uniform land use types. In travel demand modeling applications, each traffic analysis zone or TAZ contains multiple land use categories. These are then aggregated to arrive at an overall trip generation estimate for each zone. The ITE *Trip Generation Manual* further accounts for *internal-capture* effects of land uses, which are especially important for mixed-use developments, where some home-to-work trips, for example, may happen internal to the land use. At the same time, the manual accounts for *pass-by trips*, which are spur-of-the-moment trips attracted to a land use. For example, drivers may stop at a gas station or convenience store while en route for another trip purpose. Both pass-by trips and internal-capture trips represent trips that are subtracted from the overall trip generation rate, as they are no-net additions to the total number of trips on the network.

2.3.1 Land Use Types

The *Trip Generation Manual* contains data for a total of 172 land use types, based on a sample of more than 5500 sites across the United States (ITE, 2012). Common land use types featured in the manual include:

- Port and terminal: 6 land uses, including waterports, airports, transit stations, etc.
- Industrial: 9 land uses, including general light and heavy industrial, manufacturing, warehousing, etc.
- Residential: 19 land uses, including single-family homes, various types of apartments, senior living, mobile home parks, etc.
- Lodging: 5 land uses, including hotels, motels, resorts, etc.
- Recreational: 35 land uses, including various parks, fitness facilities, movie theaters, racetracks, etc.
- Institutional: 17 land uses, including schools, churches, military facilities, museums, libraries, etc.
- Medical: 4 land uses—hospital, nursing home, clinic, and animal hospital
- Office: 11 land uses, including general office, medical offices, government offices, post office, etc.
- Retail: 43 land uses, including various supply stores, convenience stores, supermarkets, sporting goods stores, apparel stores, pet stores, etc.
- Services: 24 land uses, including restaurants, fast food, coffee shops, gas stations, banks, etc.

In recent versions of the *Trip Generation Manual*, new land use codes have been added to reflect the changing needs of the user community, including construction equipment rental store, data center, medical equipment store, mosque, museum, recreational vehicle sales, snow ski area, tractor supply store, truck stop, and variety store.

A key type of land use addressed by the manual is *mixed-used developments*, which are developments that offer a mix of residential, commercial, employment, and retail land uses. Mixed-used developments are also referred to sometimes as *new urban developments*, or *neo-traditional neighborhood designs*. Through the mix of land uses within those developments, a significant portion of trips can be captured internally to the development, as residents can, for example, travel to work or shopping destinations without being on the public road or transit system outside of the development.

2.3.2 Estimating Generated Trips

Trip generation is the first and most important analysis step in travel forecasting, for both short and long term planning analyses. Trip generation estimates how many trips are demanded to or from a site or zone, which depends on the land use(s) in the site or zone. The two main methods for estimating generated trips are to (1) use the ITE *Trip Generation Manual*, or (2) develop a customized equation (e.g., planning model). The first approach is generally preferred, as the development of custom equations is very time and labor intensive.

The ITE manual applies to single land use, homogeneous sites or zones and covers significantly more than 100 land uses, as previously described. Each land use has a compilation of one to several hundred studies conducted in the United States over the past 40 years.

The general process in using the ITE manual begins with choosing your land use, analysis period, and independent variable very carefully. The analyst should take great care as most pages have entering and exiting percentages listed separately. The ITE manual also accounts for estimates for pass-by trips (driver is already on adjacent road and decides to enter site), and assumes minimal transit usage or carpooling. Effects of transit and use of other modes of transportation is part of the later mode-choice-analysis step in the four-step planning process.

For each land use, the ITE manual generally gives a regression equation (R^2) to estimate the number of trips as a function of an independent variable (e.g., square footage of retail). The manual further provides an average of the generated trip rate. The general guidelines for use of the ITE manual are as follows:

- Use equation first, if the model statistical fit defined through the R^2 metric is greater than 0.75. This suggests an acceptable statistical fit to the data used to derive the equation.
- Use average rate next, if standard deviation is less than 1.1 times the average rate.
- Make sure your site or zone is within the calibration limits on the independent variable.

EXAMPLE 2.6

A 500-acre site is being developed to support 400 single-family detached houses and a swimming pool with a clubhouse. Estimate the number of trips (T) exiting the subdivision during a typical am peak hour.

Single-family detached housing (210)

Average vehicle trip ends vs: **Dwelling units**
On a: **Weekday, peak hour of adjacent street traffic, one hour between 7 and 9 am**

Number of studies: 286
Avg. number of dwelling units: 194
Directional distribution: 25% entering, 75% exiting

Trip generation per dwelling unit

Average rate	Range of rates	Standard deviation
0.75	0.33–2.27	0.90

Data plot and equation

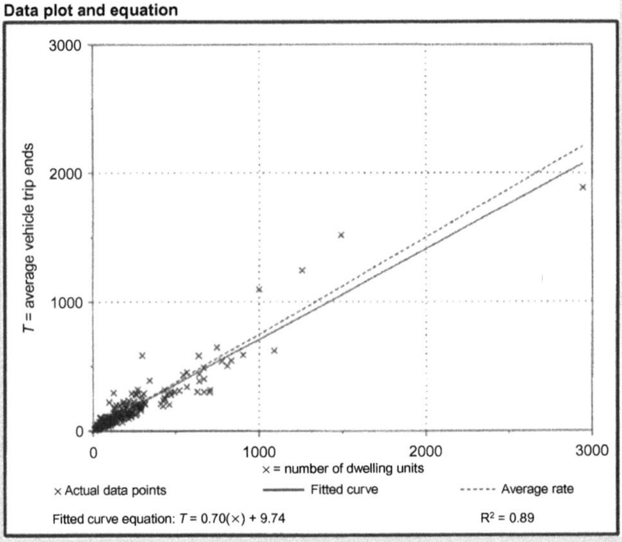

Solution

1. Note that our site with 400 units is within the range of units for the study sites, with the bulk of the study sites having less than 1000 units.
2. $R^2 = 0.89$. Because this is >0.75, we can use the fitted curve equation to solve for the answer versus looking at the average rate.
3. $T = 0.70\ (X) + 9.74$, where X is the number of dwelling units.
   $T = 0.70\ (400) + 9.74$
   $T = 290$ total trips

Now, $T_{exit} = 0.75\ (290) = 218$ trips exiting during the am peak hour. The 0.75 comes from the chart in that 75% are exiting and 25% are entering during the am peak hour.

EXAMPLE 2.7

You have been hired to conduct a TIA for a new fast-food restaurant with a drive-through window. Which of the following statements represents the best choice for estimating the number of trips into and out of the restaurant during the am peak?

a. Can't estimate the trips because there is no equation.

b. Can't estimate the trips because the standard deviation is too high.

c. Can use the average rate because the standard deviation is in acceptable limits.

d. Can use the average rate but should add a factor of safety to it.

Fast-food restaurant with drive-through window (934)

Average vehicle trip ends vs: 1000 sq. feet gross floor area
On a: Weekday, peak hour of adjacent street traffic, one hour between 7 and 9 am

Number of studies: 65
Average 1000 sq. feet GFA: 4
Directional distribution: 51% entering, 49% exiting

Trip generation per 1000 sq. feet gross floor area

Average rate	Range of rates	Standard deviation
49.35	1.02–163.33	28.30

Data plot and equation

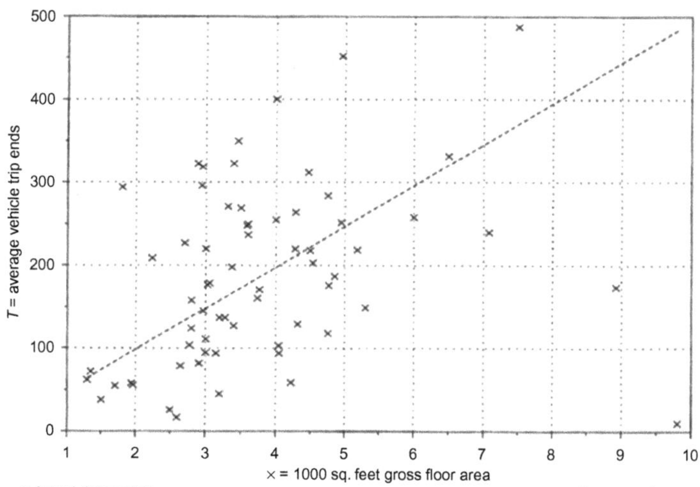

× Actual data points ------ Average rate

Fitted curve equation: Not given R^2 = ••••

(*Continued*)

EXAMPLE 2.7 (Continued)

Solution

As no equation is given for this land use code, it clearly is not an option for estimating trips. The scatter of the data further supports why the development of an equation was not appropriate for this example.

The land use code does contain an average rate of 49.35 trips per 1000 ft of gross floor area, and the standard deviation of the estimate is 28.30 trips. As this is less than 1.1 times the average, it is acceptable to use the average rate. The correct answer is (c).

2.4.1 Sample Network Application of Trip Generation

Regardless of the method used for trip generation, or the level of detail used in the estimation process, the trip generation step results in a list of trips estimated to enter and exit a given land use or TAZ within the analysis period. Figure 2.7 introduces a hypothetical planning network with a series of TAZs and various links, which will be used throughout this section to illustrate the application of the four-step process. The figure shows a sample network with eight internal TAZs (part of the study area) and four external TAZs (not explicitly being studied, but contribute to trips in and out of study area). The figure further shows a roadway network with classification into principal arterials/freeway, principal arterial/surface street, and minor arterials. Collector streets and local roads are omitted from the network in a planning context in this example, but can be included if they carry significant amounts of traffic. As a rule of thumb, the total traffic on nonmodeled links should be less than 10% of the total traffic in the network.

Sample results from the trip generation step are shown in Table 2.2. For each zone, the table gives the total trips produced, and the percentage of exiting trips. These two columns are then used to calculate the estimated entering trips and exiting trips, which will be used as attractions and productions in the trip distribution step, respectively.

2.4 TRIP DISTRIBUTION

Trip distribution is the second step in the travel forecasting process. This step determines where the generated trips are going to and coming from.

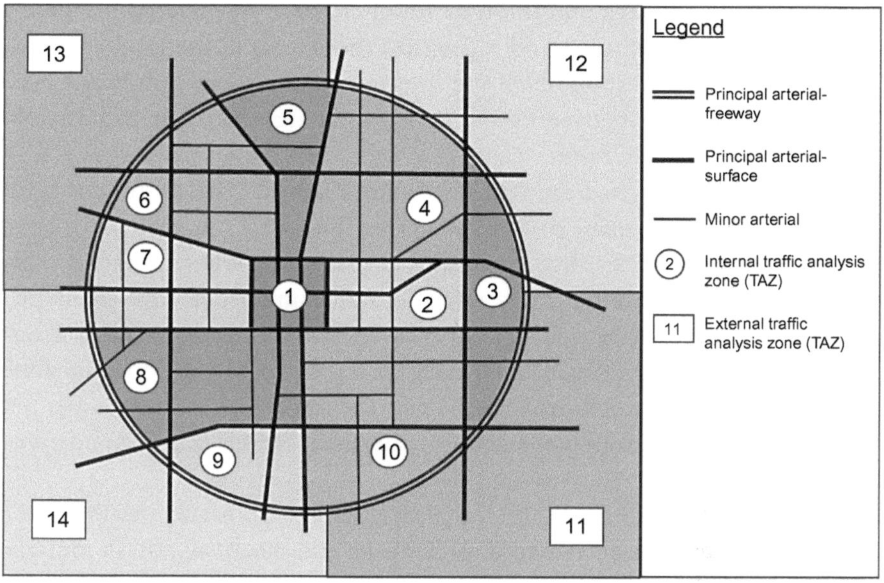

Figure 2.7 Sample network: trip generation step.

Table 2.2 Sample result of trip generation step

Zone number	Type	Total trips	Percent exiting	Entering trips	Exiting trips
1	Internal	1000	75	250	750
2	Internal	750	60	300	450
3	Internal	300	50	150	150
4	Internal	1400	40	840	560
5	Internal	600	40	360	240
6	Internal	900	45	495	405
7	Internal	1100	60	440	660
8	Internal	850	55	382	468
9	Internal	500	30	350	150
10	Internal	950	45	522	428
11	External	1200	30	840	360
12	External	1800	20	1440	360
13	External	3000	15	2550	450
14	External	1600	25	1200	400

Trip generation determines the total number of trips generated from each site (entering and exiting), and aggregates those trips to the level of a traffic analysis zone as applicable. Put another way, trip distribution establishes the *desire lines* between those zones and connects the trips from any one zone to all other zones.

The final product of the trip distribution step is what is known as an *origin–destination matrix*, or *O/D matrix*. The O/D matrix is a large table or spreadsheet that lists all origins as rows and all destinations as columns. The table entries then refer to the number of trips estimated between each zonal pair. It is important to emphasize that trip distribution makes no assignment of how traffic is expected to get from one zone to the other (i.e., which roads do people choose to get from A to B), but rather estimates the total expected flow between A and B, independent of travel path.

In the distribution step, the trips originating from and destined to a zone are generally referred to as *productions* and *attractions*, or *Ps* and *As*. In large travel demand models, the result of the trip distribution step is also referred to as a *trip table*, which essentially contains the O/D matrix, as opposed to a *path table*, which is the result of traffic assignment and physically assigns trips to links and paths on the roadway network.

The trip distribution step is applicable to small planning studies, or traffic impact analyses (TIAs), as well as long-term planning studies using travel demand models (TDMs). For a TIA, trip distribution estimates the trips from the site under study to or from each exit from the study area. For a TDM, trip distributions estimates the trips between each zone and every other zone.

Trip distribution data (O/D patterns) can be estimated from field data collection, synthesized from existing turning movement counts, or estimated through analytical approaches. The most common of those approaches is called the *gravity model*, which is discussed in detail next.

2.4.1 Gravity Model

The gravity model is an analytical methodology to estimate the productions and attractions between each zonal pair in the trip distribution step. It is most applicable for larger studies and networks, and can be quite data intensive to complete. The underlying concept for the gravity model mirrors principles in physics, where two masses attract one another, and where the magnitude of that attraction is proportional to the size of the mass. In other words, larger zones (or more precisely, zones with more activity or higher development density) attract more trips than smaller zones (less activity and a lower development density). That physics-based gravity model is further calibrated by the relative travel time between zones and socioeconomic characteristics.

Specifically, the gravity model estimates the number of trips between each origin zone, *i*, and each destination zone, *j*, as a function of the following characteristics:

- The number of trips produced in each zone, P_i
- The number of trips attracted to each zone, A_j
- A friction factor for each *ij* zonal pair that is a function of the travel time between those zones, F_{ij}
- A socioeconomic factor for each *ij* zonal pair that can be used to calibrate the desire of travelers to travel between those zones, K_{ij}

The general form of the gravity model is shown in the following equation, followed by an example.

$$Q_{ij} = (P_i)\frac{A_j \times F_{ij} \times K_{ij}}{\sum_{1}^{j}(A_j \times F_{ij} \times K_{ij})}$$

where

Q_{ij} = trips from zone *i* to zone *j*
P_i = trips produced
A_j = trips attracted
F_{ij} = friction factor (typically the inverse of distance or travel time)
K_{ij} = socioeconomic factor (calibrated)

The gravity model has to be applied in an iterative fashion for both the total attractions and productions for each zone to match with the entered data. In the first iteration, the productions for each zone will be balanced, but the attractions will likely differ from the entered trip generation data. In repeated iterations, the results of one iteration are fed as input to the next iteration, until a desirable match is obtained.

EXAMPLE 2.8

A new office park is expected to generate 1500 homebound trips in the pm peak hour. Analysts expect the trips to terminate in four residential zones.

Zone number	Travel time from office park (min)	Number of trips attracted during a typical pm peak hour	Socioeconomic factor from office park to zone *i*
1	10	3000	1.2
2	15	2000	0.8
3	25	1800	1.0
4	30	4000	1.5

Using a gravity model, estimate the number of trips to go from the office park to zone 3 during the pm peak hour.

Solution

This problem is best solved by setting up a table to calculate the trips from the office park to each zone. In this example, P_i is equal to 1500 for all rows in the table. The table shows attractions for each destination zone, estimates friction factors as the inverse of travel time, and gives the socioeconomic factor. Those three terms are used to estimate the $A \times F \times K$ product for each destination zone, which are then used to get the estimate of the number of trips for each zonal pair.

Zone j	A_j	F_{ij}	K_{ij}	$A \times F \times K$	$\frac{A \times F \times K}{\sum(A \times F \times K)}$	Q_{ij}
1	3000	1/10 = 0.100	1.2	360	0.4885	733
2	2000	1/15 = 0.067	0.8	107	0.1452	218
3	1800	1/25 = 0.040	1.0	72	0.0977	146
4	4000	1/30 = 0.033	1.5	198	0.2686	403
Σ	—	—	—	737	1.0000	1500

The total number of trips estimated from the office park to zone 3 is 146 trips.

2.4.2 Use of Data and Calibration

Trip distribution data (origin/destination or O/D patterns) can be estimated from field data collection, although the data collection effort can be extensive. Traditional O/D survey methods include license plate matching surveys and market analysis surveys. License-plate matching surveys can be performed manually (noting the last three digits of a plate is often sufficient) or can be automated using automated license plate recognition (ALPR) technology. Modern Bluetooth reader technology can also be used to estimate O/D patterns by recording (anonymous) Bluetooth MAC signatures of mobile devices (cell phones, tablets, etc.) in one location and matching them with readings at other locations.

Bluetooth studies and license plate studies are generally sample based, meaning that only a subset of all traffic is observed to get O/D percentages, which are then extrapolated to the full O/D matrix by scaling the matrix with actual traffic counts. O/D studies are very time consuming and labor intensive, as the data collection units (ALPR, cameras, Bluetooth) need to be deployed simultaneously at all major entry and exit points to the network, as well as at whatever internal zones are desired. In practice, these O/D surveys are therefore often limited to key origin–destination pairs or used as a tool to calibrate an estimated O/D table.

For small networks, the O/D matrix can also be synthesized from existing intersection turning movement counts (% left, % through, % right) by assuming proportionality. For a corridor study, this process may start at a downstream intersection. If we assume an east-to-west corridor, the turning percentages at the eastbound approach at the easternmost intersection can be assumed to be proportionally fed by all trip origins at the next upstream intersection (in this case eastbound through, northbound right, and southbound left). Figure 2.8 illustrates this concept for a one-way street in the eastbound direction.

In part (a), the figure shows turning movement counts. The gray highlighted cells for southbound right (120 veh/h), eastbound through (450 veh/h), and northbound right (60 veh/h) form the combined traffic stream of 630 veh/h traveling toward intersection 2. The contribution of the three streams to that 630 veh/h are 19%, 71%, and 10%, respectively. At intersection 2, the eastbound approach shows left-turn, through, and right-turn volumes of 85, 500, and 45 veh/h, respectively. Because the network is balanced, this also adds up to 630 veh/h.

To estimate the origin–destination pattern, the (destination) turning movement counts at intersection 2 are multiplied by the three origin percentages from intersection 1. For example, the 85 left-turning vehicles at intersection 2 are divided into 19% from intersection 1 north approach (16 veh/h), 71% from intersection 1 west approach (51 veh/h), and 10% from intersection 1 south approach (8 veh/h). The same process is applied for the other movements and the results are shown in part (b) of the figure.

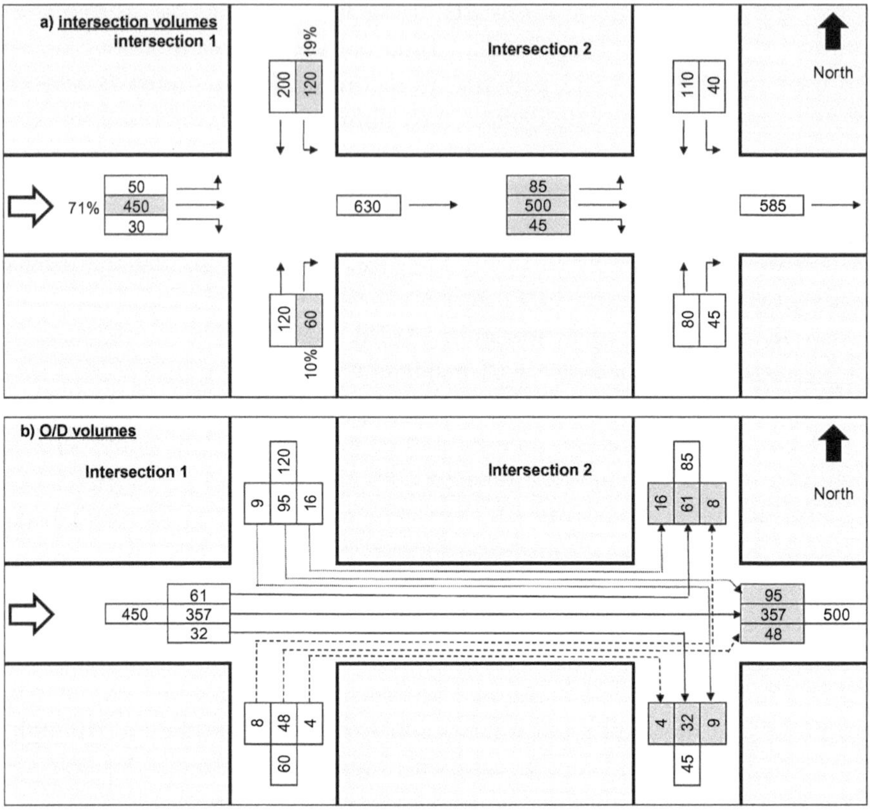

Figure 2.8 O/D synthesis for a corridor study.

Finally, the origins at intersection 1 and the destination 2 are summed internally, to make sure that all the calculations yield consistent volumes with the original turning movement counts. For a longer corridor, this process starts at the most downstream end (e.g., intersection 5 for a five-intersection corridor) and then works its way upstream toward intersection 1.

2.4.3 Sample Network Application of Trip Distribution

Figure 2.9 shows a sample network with eight internal TAZs (part of the study area) and four external TAZs (not explicitly being studied, but contribute to trip in and out of study area). The figure further shows a roadway network with classification into principal arterials/freeway, principal arterial/surface street, and minor arterials. Collector streets and local roads are omitted from the network in a planning context in this example, but

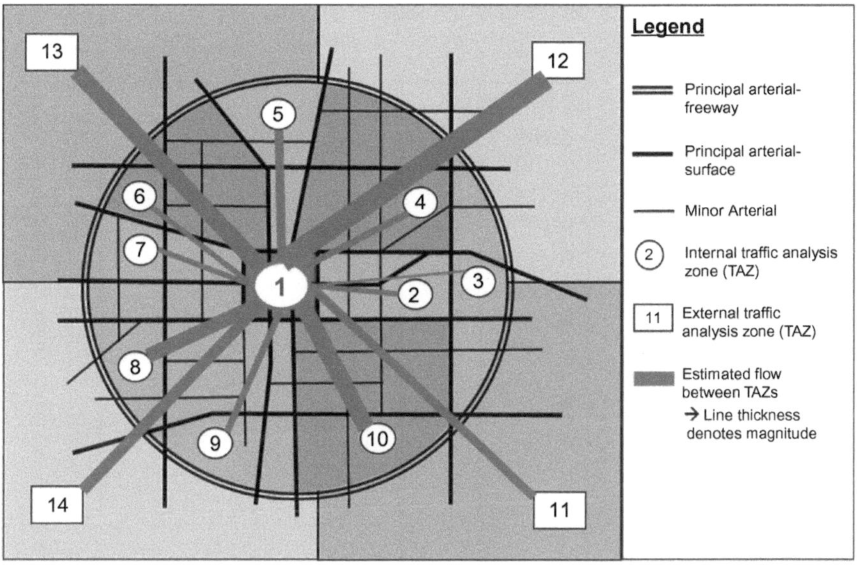

Figure 2.9 Sample network: trip distribution step.

can be included if they carry significant amounts of traffic. As a rule of thumb, the total traffic on nonmodeled links should be less than 10% of the total traffic in the network.

The figure shows estimated trips between zone 1 and all other zones as weighted desire lines, with the weight of each line being proportional to the magnitude of forecasted traffic between each zonal pair.

Sample results from the trip distribution step are shown in Table 2.3 for zone 1. These results would be repeated for all other zones in a planning study. For each zone, the table gives the total trips attracted by the remaining 13 zones from the trip generation step. It then shows the travel times between zones (in minutes), which are converted into friction factors between zones by taking the inverse. The table then shows the estimated socioeconomic factors, assigning larger weights for pairs of zones with more interzonal trips. The table then shows the sum of attractions, friction factors, and socioeconomic factors for each destination zone, which are summed at the bottom of the table. The second to last column divided each "$A \times F \times K$" product by the total sum. In the last column, this product is multiplied by the total trips produced in zone 1 (750 trips, see Table 2.2) to estimate the total trips between zone 1 and the remaining 13 zones.

Table 2.3 Sample result of trip distribution step

Zone 1	A_j	TT_{ij}	F_{ij}	K_{ij}	$A \times F \times K$	$\frac{A \times F \times K}{\sum(A \times F \times K)}$	Q_{ij}
1	—	—	—	—	—	—	—
2	450	5	0.200	1	90	0.072	109
3	150	10	0.100	1	15	0.012	18
4	560	8	0.125	1	70	0.056	84
5	240	6	0.167	2	80	0.064	97
6	405	7	0.143	1	58	0.047	70
7	660	8	0.125	1	83	0.066	100
8	715	5	0.200	2	286	0.230	345
9	240	10	0.100	2	48	0.039	58
10	405	7	0.143	2	116	0.093	140
11	360	18	0.056	3	60	0.048	72
12	360	10	0.100	3	108	0.087	130
13	450	12	0.083	4	150	0.121	181
14	400	15	0.067	3	80	0.064	97
Σ	—	—	—	—	1243	1.000	1500

The results of the trip distribution step are used as a visualization aid in Figure 2.9, where the line weight of the thick bars connecting each zonal pair is proportional to the number of trips between these zones.

The trip distribution step is repeated for the remaining 13 zones, to arrive at a similar table of trips between zones 2 through 14 and the remaining zones. The results of the overall trip distribution step are often displayed in the form of an origin–destination matrix (O/D matrix), which is also referred to as a trip table. The resulting O/D matrix for the example described after one iteration of the gravity model is shown in Table 2.4.

At this point it is again emphasized that multiple iterations of the gravity model are needed to arrive at a balanced O/D matrix or trip table. This is very evident in Table 2.4, in which the row totals from the trip distribution (column "row sum") match the input data from the trip generation step. In other words, each zone produces the correct number of trips as was estimated in that prior step. However, the column totals for each destination zone do not necessarily match, with most destination zone totals ("col sum") being significantly less than the field data.

Consequently, additional iterations of trip distribution are needed, where the results of iteration 1 are used as inputs for iteration 2, and so forth. Sometimes, many iterations are needed for both the row and

Table 2.4 Origin–destination matrix after one iteration of gravity model

Origin	Destination 1	2	3	4	5	6	7	8	9	10	11	12	13	14	Row Sum	Data
1	—	109	18	84	97	70	100	345	58	140	72	130	181	97	1501	1500
2	17	—	9	37	21	12	10	23	28	82	44	71	60	37	451	450
3	3	11	—	8	5	3	2	6	7	21	23	36	16	8	149	150
4	9	38	14	—	39	56	8	20	24	72	76	123	53	28	560	560
5	4	16	6	36	—	5	4	8	10	31	33	53	22	12	240	240
6	7	26	9	58	27	—	6	14	17	49	53	84	36	19	405	405
7	10	39	14	88	40	58	—	20	25	74	79	127	54	29	657	660
8	10	41	15	91	42	60	55	—	26	77	82	131	56	30	716	715
9	3	14	5	31	14	20	19	7	—	26	28	44	19	10	240	240
10	6	25	9	56	26	37	34	13	16	—	50	81	34	19	406	405
11	5	18	6	40	19	27	25	34	33	57	—	58	25	13	360	360
12	7	28	10	62	28	41	38	52	18	52	56	—	38	21	451	450
13	8	31	11	69	32	46	42	57	56	97	87	142	—	23	701	700
14	6	25	9	56	26	37	34	47	16	47	51	90	35	—	479	480
Col Sum	95	421	135	716	416	472	377	646	334	825	734	1170	629	346	—	—
Data	500	300	150	840	360	495	440	585	560	945	840	1350	1300	1120	—	—

column totals to converge, which is especially the case for large networks. Automation of this step in computer models is therefore desirable, to more efficiently conduct these very repetitive computations.

2.5 MODE CHOICE

Mode choice is the third step in the traditional travel-demand forecasting process. This step aims at predicting a distribution of traffic across different modes of transportation. In its most basic form, the mode choice step predicts a diversion rate or percentage to other modes, which results in a reduction in net auto trips. For example, a mode choice model may predict that 15% of trips in a region are expected to occur by transit, reducing the total estimated load of auto trips to 85% of its original estimate. But more complicated (and more realistic) mode choice models predict the relative ratio of different modes directly through advanced modeling techniques, and under consideration of a range of behavioral attributes of travelers and trips.

Fundamentally, mode choice models are divided into *trip end* models and *trip interchange models*, with the latter being the more advanced and more common mode-choice approach in modern travel demand models. Both are discussed in more detail in the following.

A third, highly simplified, mode choice modeling approach is the *direct generation* of a transit, biking, or walking trip. This approach essentially assumes that trips made by car, transit, or other mode are independent and simply a function of land use intensity or other metric. However, standard trip generation practice (and other steps in the four-step process) are not compatible with this approach and there generally doesn't exist much data for predicting the number of nonauto trips. Instead, the common approaches focus on the estimation of person trips, which are then distributed and divided among modes. Therefore, the direct generation approach to mode choice is not discussed in further detail here.

2.5.1 Trip End Models

Trip end models rest on the assumption that the decision to use transit or an alternate mode of transportation (walking, biking, etc.) is primarily a function of socioeconomic ability, or more precisely, a lack thereof. In other words, trip end models assume that individuals take transit because they have to, not because they choose to. Transit riders, for example, are assumed to be *captive* in these models, in that they don't have the means to travel by an alternate type of transportation, for example, they cannot afford to buy and operate a car.

Trip end models may be appropriate in areas and countries in which car ownership is frequent, parking is plentiful, and congestion is at low-enough levels to where there are no clear travel time benefits to using transit. The application of trip end models therefore may hold true for medium-sized U.S. cities without a well-developed transit infrastructure. Trip end models may also be appropriate in developing countries, where transit use is due to necessity, but where individuals are likely switch to cars as soon as they can afford to do so. In that case, it is assumed that once an individual can afford a car, they will use it, and they will not go back to using transit.

A clear advantage of a trip end model is that it is very easy to apply. The percentage of transit trips is simply estimated as a function of land use intensity, or other characteristic of the area under study. For example, the percentage of transit may be a function of the population density or average car ownership in an area, or a combination.

One key consideration in trip end models is that they are applied *prior to trip distribution*, as opposed to trip interchange models, which are applied after trip distribution, following the standard four-step process. As such, trip end models simply deduct a certain percentage of the generated trips and convert them to transit. Because the modal split occurs prior to distribution, the split is not sensitive to any information about travel time between zones or other characteristics (i.e., the socioeconomic calibration factors described in the trip distribution section).

2.5.2 Trip Interchange Models

Trip interchange models are applied after the trip distribution step, once the productions and attractions between zones are known, and once an origin—destination matrix for all zonal pairs has been developed. A trip interchange model is then used to estimate the distribution of trips for each zonal pair across modes. The trip interchange model can take the travel time between zones obtained from the trip distribution step, and a host of other variables or cost functions, into consideration when dividing trips across modes. Trip interchange models take into account the *impedance* of travel between zones by mode.

In modern mode choice application, this impedance is often expressed as the *utility* of travel between zones for each mode, giving a more positive slant on the same concept. The higher the utility for a mode and a given zonal pair (or the lower the impedance), the greater is the likelihood that travelers will make a choice to select that mode for the particular interzonal trip.

The terms *choice* and *likelihood* are critical concepts in the application of these models. *Choice* implies that travelers make a conscious and deliberate decision in favor of a particular mode, which is in stark contrast to the previously mentioned trip end models that assume that travel by transit, for example, is out of necessity (e.g., due to lack of socioeconomic status). *Likelihood* implies that the decision to use a particular mode for a given interzonal trip is not deterministic, but rather a probability that is a function of a host of variables and factors. For example, even if the auto mode is the most desirable way to travel from zone A to B, that doesn't mean that some travelers will not *choose* other modes (transit, walking, or bicycling) for the same trip, or that there is some likelihood of using transit, even if the auto mode is preferred.

The mathematical models used to estimate the likelihood of travel using a specific mode are typically *probit* or *logit* models, with logit models (or more precisely logistic regression) being the most common approach. But before presenting these model forms, the following section first introduces the types of variables and behavioral attributes that may impact mode choice.

Variables Impacting Mode Choice

The attributes that impact mode choice, and thus the independent variables that are used in mode choice models, can be divided into four principal categories:

1. Attributes related to the traveler or person making the trip
2. Attributes related to the mode or transportation alternative
3. Attributes related to the trip context or reason for the trip
4. Attributes describing the interaction of traveler, mode, and trip

In the first category, the attributes related to the person making the trip include age, gender, income, household size, number of vehicles in the household, and others. This category also includes potential combinations of these variables, such as the number of vehicles per member of the household.

In the second category, the attributes related to the specific transportation alternative include travel cost, travel time, travel time spent in-vehicle versus waiting for service, headways between transit vehicles, number of transfers, or variables describing the general level of comfort of a particular mode.

In the third category, attributes related to trip context include the trip purpose (e.g., home-to-work versus recreational or discretionary trips), surrounding area type (suburban, central business district, etc.), employment and population densities at the trip origin and destination, or the availability of parking in the destination zone.

Finally, the fourth category captures interaction effects of these variables, which consider, for example, differences in the value the travelers of varying ages or socioeconomic status place on travel time. A resulting variable may then become travel time by income category or by age group. Another example would be the cost of the trip divided by the household income, where the cost of the trip can include gas prices, transit fees, tolls, parking fees, and vehicle operating costs.

These attributes or variables are typically collected through household travel surveys conducted as part of the census, or a standalone survey instrument. Often, these surveys include travel diaries, which further give important insights into other aspects of the four-step process. The decision of which specific variables are of interest or statistical significance in predicting mode split is a function of the geographic region and community context the model is applied in. As such, modeling agencies often spend significant time and resources collecting customized travel surveys, analyzing the resulting travel data, and calibrating region-specific mode choice models that capture the unique attributes and characteristics of the study area and its residents.

In the application of mode choice models in a travel demand model, the attributes are typically obtained from the underlying transportation network and attribute database through a process called *skimming*. For each zonal pair, the model skims the attributes of two zones, and their relative locations, and thereby makes the skimmed data available for use in the mode choice algorithm. Skimming is also used in the traffic assignment step, and is described in more detail in Section 2.6.

Utility Models

The mathematical formulation used to estimate the mode choice distribution is most commonly in the form of a logistic regression, or logit, model. Just like linear regression, a dependent variable is predicted as a function of one or more independent variables. The difference is that the dependent variable in this case is a discrete choice, for example, between using auto or transit. As such, the more common linear or multilinear regression approach is not statistically correct, and as such the logit function is used. But while the mathematical formulation may seem more difficult (and in fact the algorithms to estimate the logit function are more complex), the underlying concept is the same. A dependent or response variable (e.g., likelihood of using transit) is predicted as a function of various independent variables (e.g., travel time, cost, socioeconomic status, etc.), as listed in the previous section.

The utility of using a particular mode, U_x, can then be expressed as a function of these various attributes, A_i, each multiplied by a regression coefficient b_i.

$$U_x = \sum_{i=1}^{n} b_i A_i$$

where

U_x = utility of transportation mode x (e.g., auto, bus, etc.)

n = number of variables used in mode choice models

A_i = attribute impacting mode choice (e.g., travel time, travel cost, etc.)

b_i = coefficient of attribute A_i determined through regression

In the logit model estimation, the probability of choosing mode x, $P(x)$, can further be expressed by solving the utility expression, which yields the following function:

$$P(x) = \frac{e^{U_x}}{\sum_{i=1}^{n} e^{U_i}}$$

In the case of estimating the simple choice between two modes (auto and transit), the probability of choosing the transit mode then becomes:

$$P(x) = \frac{e^{U_{Transit}}}{e^{U_{Transit}} + e^{U_{Auto}}}$$

Where all variables are as defined previously.

The probability equation is used to estimate the likelihood of using transit as a function of various attributes, which can range between zero and 1. Due to the nature of the logit function, the probability of using transit increases or decreases between the value of 0 and 1, depending on the sign of the coefficient. Figure 2.10 illustrates the probability of using transit as a function of changing independent variables.

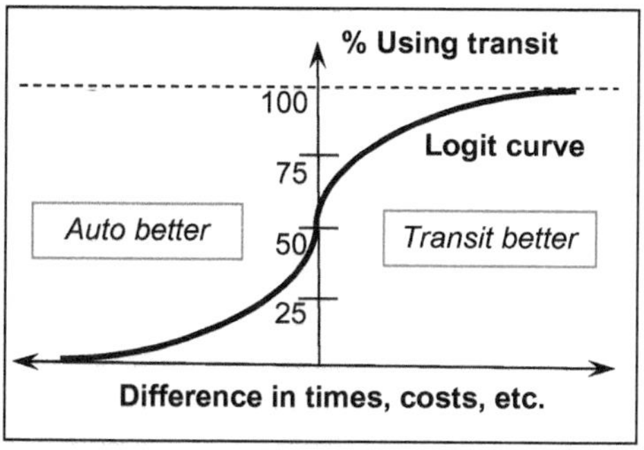

Figure 2.10 Illustration of probability of transit use from logit function. *Source: Urban Mass Transportation Administration, 1977.*

Types of Logit Models

There are several types of logit models; the most common ones for mode choice analysis are the *multinomial logit* and the *nested logit* formulation. More detailed discussion on these logit forms can be found in statistical textbooks or advanced planning texts.

The multinomial logit (MNL, 2006) model assumes a parallel hierarchy of various modal alternative choices. For example, the model may predict the likelihood of using auto, bus, or train, as three parallel choices. The likelihood of using each of these modes may not be the same (auto may be much more likely than others), but the MNL assumes that the decision between these three choices is made at the same time. A special case of the MNL model is the *binary logit* model, where only two alternate decisions are used.

In contrast, a nested logit (NL) model introduces a hierarchy in decision making. In our example, the traveler may first decide on whether to use auto or transit, and then (once the traveler selected the transit mode) whether to use bus or train. In this NL example, the auto-versus-transit decision occurs on the first level of the nesting structure, while the bus-versus-train decision is modeled to occur on the second (lower) level. This basic example and the difference between MNL and NL formulations are shown in Figure 2.11.

In a real-world mode choice model, the nesting structure can be quite a bit more complicated and can include multiple parallel choices in each nesting level. Figure 2.12 shows one example of a mode choice model used by the Capital Area Metropolitan Planning Organization (CAMPO) in Texas.

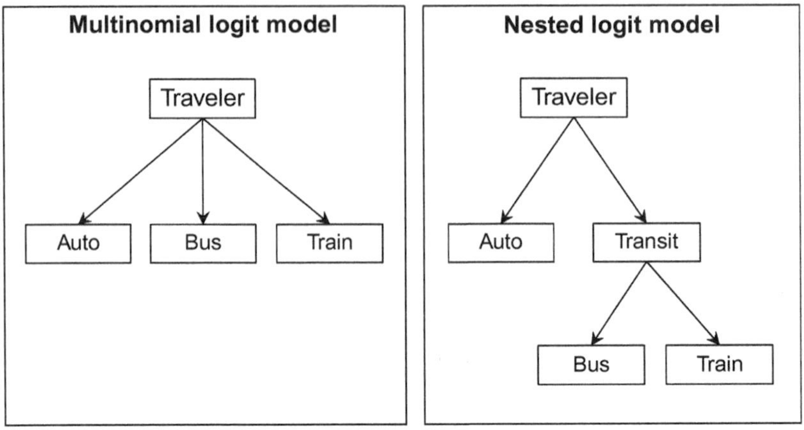

Figure 2.11 Logit model form illustration.

2.6 TRAFFIC ASSIGNMENT

Traffic assignment, the final step in the four-step travel demand forecasting process, estimates the specific routes on the network travelers take to get from one zone to another. This step uses the total trips in the origin–destination (O/D) matrix from the trip distribution step, which may have been divided among modes, and assigns them to physical routes. In a travel demand model application, the input to the traffic assignment step is a *trip table* (essentially the O/D matrix), while the result of the assignment is a *path table.* The path table represents each zone-to-zone trip pair through a chained series of link or segment numbers. Each interzonal pair could have as little as one path assigned, or could have a number of alternative paths, each carrying a fraction of the total trips.

2.6.1 Network, Paths, and Skimming

The previous three steps in the four-step process primarily used the data for different traffic analysis zones (TAZs) in the transportation system to be analyzed. These zones are interconnected by links, which represent the physical characteristics of the road segments and facilities of the transportation network. The traffic assignment step uses the characteristics of all links when determining which paths are most desirable for travelers, and therefore are expected to carry the highest traffic loads. A path is then represented through a series of links, each with its own set of

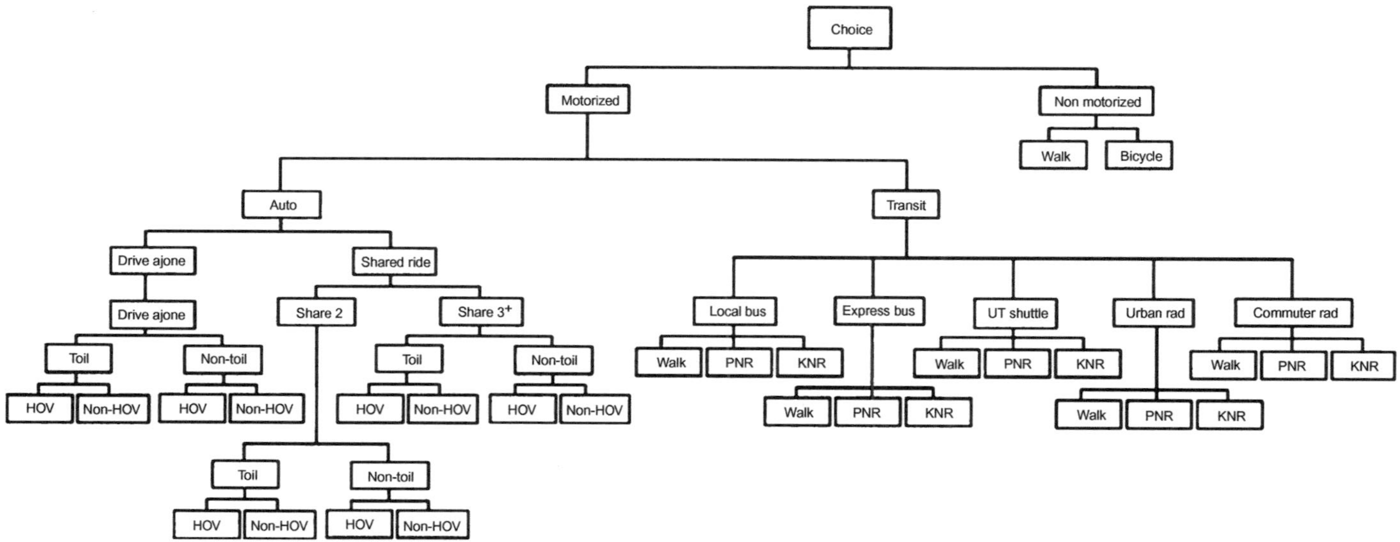

Figure 2.12 Mode choice example nesting structure. PNR, park and ride; KNR, kiss and ride. *http://www.fhwa.dot.gov/planning/tmip/resources/peer_review_program/campo/*

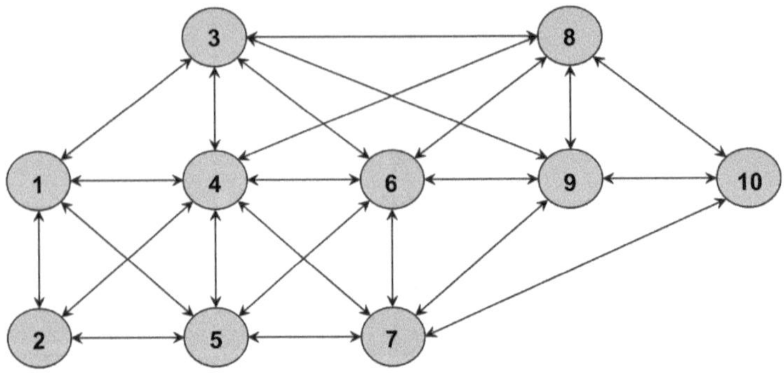

Figure 2.13 Simple network for assignment illustration.

attributes and characteristics. These attributes can include length, speed limit, travel time, capacity, functional classification, and a host of other variables used to describe the characteristics and expected performance of the link.

The link attributes are translated to the path through a process called *skimming*. Conceptually, all the links along a path between two zones are skimmed for their specific characteristics, which are then aggregated to result in path characteristics. Take, for example, the simplified 10-zone network in Figure 2.13. If evaluating a trip with origin in zone 1 and destination in zone 4, the path is represented by link 1–4 that connects the two zones. All attributes of link 1–4 are translated to the path for analysis.

But while link 1–4 has a clear direct path, other zonal pairs have multiple paths. For example, take the O/D pair for links 1 and 7. In this case, one potential path is "1–4→4–7." But several alternate paths exist in this case, including "1–5→5–7," "1–2→2–5→5–7," or "1–4→4–6→6–7." Therefore, the decision of which path drivers will take becomes a function of the attributes of each path, which, in turn, are estimated from the aggregated attributes of each link sequence.

In fact, going back to the earlier example of the path from zone 1 to 4, there technically are other potential paths, including "1–3→3–4," "1–5→5–4," or "1–2→2–4," or even "1–2→2–5→5–4." These alternate paths may seem undesirable compared to the direct path "1–4," but not knowing any characteristics of the network in Figure 2.13, we can't disregard that route. What if link 1–4 is a one-way street that doesn't allow travel in the desired direction? What if it is a neighborhood street with traffic calming? What if it is a drawbridge that opens for 15 minutes every hour

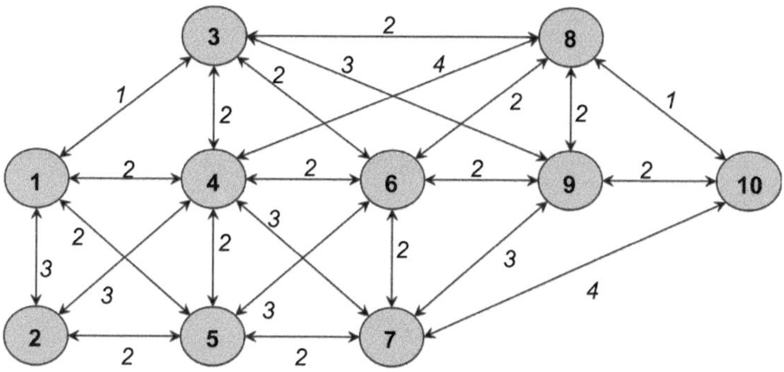

Figure 2.14 Simple assignment network with link travel times.

resulting in a long queue, and so on? In short, we need additional attributes for each path, which are obtained from the attributes of each link along the path through skimming. Figure 2.14 amends the previous example by adding link travel times (in minutes) for each zone-to-zone connection.

Applying our previous example, it appears that link 1−4 is indeed the shortest travel time between zones 1 and 4 at 2 min. The alternate path travel times are 3 min for "1−3→3−4," 4 min for "1−5→5−4," 6 min for "1−2→2−4," and 7 min for "1−2→2−5→5−4." So while several paths exist, path "1→4" clearly has the lowest travel time.

For the second zonal pair example, the path travel times are 5 min for "1−4→4−7," 4 min for "1−5→5−7," 7 min for "1−2→2−5→5−7," and 6 min for "1−4→4−6→6−7." So in this case, path "1−5→5−7" produces the lowest travel time at 4 min.

In this simplified example, the skimming process strictly focused on link travel times. But a host of other variables may be of interest in a traffic assignment step, including cost of travel for each link (through tolls for example) or functional classification with, for example, freeway travel being more desirable and reliable than arterial traffic, even at the same base travel time. These variables can equally be obtained by skimming the path in question. Later in this section, we will briefly introduce *dynamic assignment* approaches, which further take into account time-of-day specific travel time and congestion.

2.6.2 Shortest Path Assignment

The simplest assignment algorithm is one based on the shortest path between zones. Shortest path, in this case, can refer to shortest physical

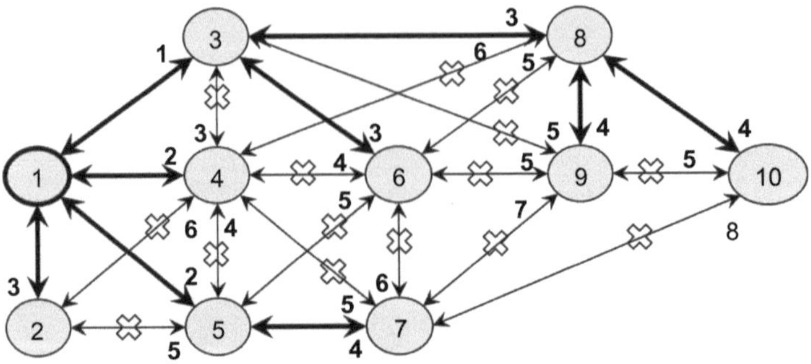

Figure 2.15 Shortest path selection network for zone 1.

distance, but also other variables, such as shortest travel time or lowest cost. Often, only a single variable is used to predict "the best" path or route between two zones based on this method. More importantly even, the assignment at its most basic form is done as an *all-or-nothing* assignment. In other words, all trips between a zonal pair will be assigned to that shortest path. That assumption is further evaluated and refined in advanced assignment algorithms.

Referring back to the simple network in Figure 2.14, the provided link travel times can be used to determine the shortest path assignment. The assignment process evaluates the cumulative link travel time for each path between zones and eliminates those route options with higher travel time. In demand modeling language, each path is *skimmed* for the total travel time and compared to other paths for that zone-to-zone connection. Of course, this skimming process can involve other variables or multiple factors.

Figure 2.15 shows the shortest path selection process for the simple example network we have been using. The link travel times have been replaced with the cumulative travel times to get to a particular zone. The example focuses on trips from zone 1 to all other zones, but would eventually be repeated for all other origin zones as well.

Starting with link 1–4, the cumulative travel time from zone 1 to zone 4 is a simple 2 min. For an alternate path from zone 1 to 4, the analysis sums the travel time from zone 1 to 3 (1 min) and the travel time from zone 3 to 4 (2 min) for a cumulative travel time of 3 min. Because that cumulative travel time of 3 min is greater than the link 1–4 connection, link 3–4 is eliminated from the shortest path analysis for zone 1

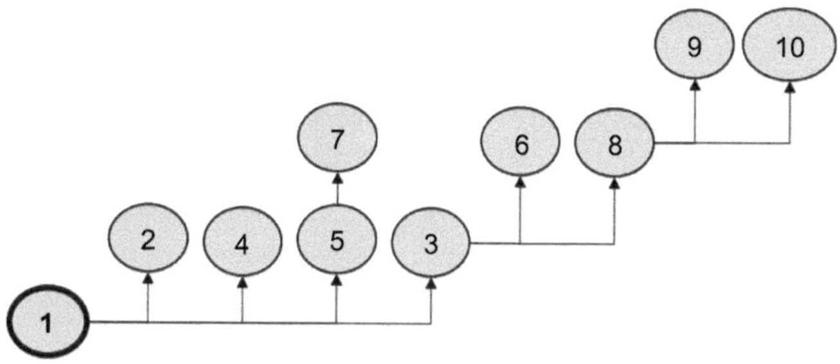

Figure 2.16 Minimum path tree for zone 1.

(to be sure, one could check the possible connections between zones 1 and 3 and arrive at the conclusion that the trip on link 1−3 with 1 min travel time outperforms the trip on link sequence 1−4 plus 4−3 at a total travel time of 3 min). Using the same procedure, link sequence 1−2 to 2−4 (cumulative travel time of 6 min) and sequence 1−5 to 5−4 (4 min) are eliminated, leaving link 1−4 as the shortest path connection between zones 1 and 4.

This process is repeated for all subsequent paths to additional zones, but is simplified because links that were previously eliminated no longer need to be considered. As such, link 1−3 directly gives the shortest path between zones 1 and 3, for example. Using similar processes, link 1−2 is the shortest zone 1 to 2 connection, and link 1−5 is the shortest path from zone 1 to zone 5. Moving on to zone 6, it is evident that the path "1−3→3−6" (3 min) outperforms the path "1−4→4−6" (4 min) and the path "1−5→5−6" (5 min). As a result, links 4−6 and 5−6 are eliminated from further consideration.

This process is repeated until all zones in the network have been connected through a shortest path from zone 1. Figure 2.16 shows the resulting shortest path diagram from zone 1 to all other zones.

From the minimum path tree in Figure 2.16 the total load on each link can be estimated by assigning the forecasted trips between zones from the trip distribution step, which is illustrated in Example 2.9. Note that we have only covered trips originating from zone 1 at this point. In a full travel demand modeling application, this series of steps would have to be repeated for the remaining nine origin zones. Even for this simple network, it is evident that this process can be quite cumbersome and challenging to complete.

EXAMPLE 2.9

The traffic assignment for the sample network in Figure 2.14 resulted in the minimum path tree for trips from zone 1 as shown in Figure 2.16. The following table lists the number of trips (from the trip distribution step) for all destination zones, as well as a listing of impacted links.

Origin zone	Destination zone	Number of trips estimated	Sequence of links
1	2	30	1−2
	3	50	1−3
	4	40	1−4
	5	30	1−5
	6	30	1−3, 3−6
	7	20	1−5, 5−7
	8	60	1−3, 3−8
	9	10	1−3, 3−8, 8−9
	10	100	1−3, 3−8, 8−10

Using this shortest path and all-or-nothing assignment, estimate the total number of zone 1 origin trips on each of the links in the network.

Solution

This problem is solved by summing up the trips for each link using all zonal combinations from zone 1. Note that links 2−4, 2−5, 3−4, 3−9, 4−5, 4−6, 4−7, 5−6, 6−7, 6−8, 6−9, 7−9, 7−10, and 9−10 were eliminated from the shortest path assignment due to high cumulative travel times. These are not shown in the table, as the resulting number of trips (originating in zone 1) is zero. The result is shown in the following table.

Link	Estimated trips	Total trips
1−2	30	30
1−3	50 + 30 + 60 + 10 + 100	250
1−4	40	40
1−5	30 + 20	50
3−6	30	30
3−8	60 + 10 + 100	170
5−7	20	20
8−9	10	10
9−10	100	100

2.7.1 Capacity-Constrained Assignment

There are several problems with the shortest path and all-or-nothing assignment presented in Figure 2.15 and Figure 2.16. For example, the assignment doesn't take into account that the total traffic demand assigned to a particular link may result in increases in travel time on that link, thereby increasing link travel time and decreasing its desirability in the assignment. In extreme cases, of course, the demand may actually exceed the capacity of a link, making it impossible to process the full demand from all zones. A capacity-constrained assignment takes these travel time increases and link capacity constraints into consideration.

A capacity constrained assignment applies a relationship between the volume-to-capacity ratio (v/c) on a link and the resulting travel time to each link. The higher the v/c ratio estimated, the higher is the resulting effect on link travel time. The most commonly used relationship was developed by the U.S. Bureau of Public Roads and is referred to as the *BPR function*. The BPR formula for estimating link travel time under capacity constraint is as follows:

$$t_{CR,i} = t_{0,i} \times \left[1 + 0.15 \times \left(\frac{v_i}{c_i}\right)^4\right]$$

where

$t_{CR,i}$ = capacity restrained travel time of link i (min)
$t_{0,i}$ = free-flow travel time of link i (min)
v_i = demand volume forecast for link i (veh/h)
c_i = capacity of time of link i (veh/h)

The form of the equation results in an increase in the estimated link travel time as a function of link capacity and v/c ratio. The link travel time is equal to 1.15 times the free-flow travel time at capacity (v/c = 1) and increases exponentially past the capacity. A sample application for a link with free-flow travel time of 5 min and capacity of 2000 veh/h is shown in Figure 2.17.

Note that there are ways to calibrate the BPR function. For example, changing the 1.15 coefficient to a higher value increases the initial slope of the travel time curve for low v/c ratios. Similarly, increasing the v/c exponent of 4 to a higher value will drastically increase the slope beyond the capacity value.

It is important to point out that this assignment method will continue to assign flows to a link even if the demand exceeds capacity. In a travel

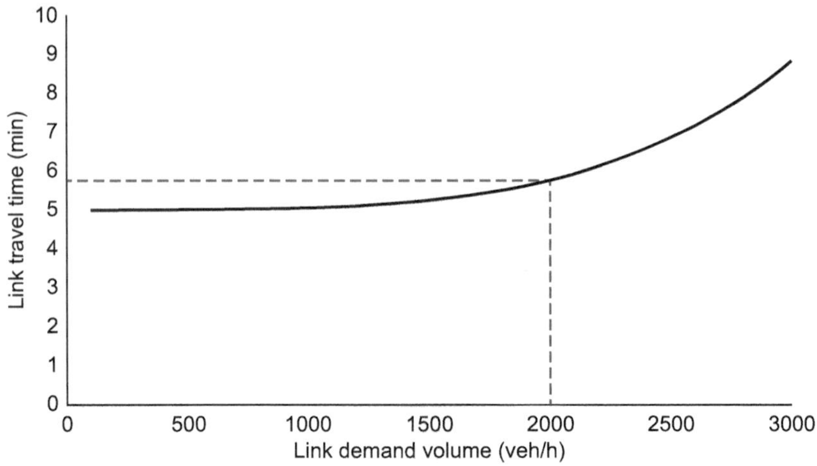

Figure 2.17 Sample application of BPR function for capacity-constrained assignment.

demand application this is important to assure that the model continues to be able to perform calculations and doesn't "lock up." In other words, the BPR function creates additional link impedance at high v/c ratios, but doesn't actually cap the link volume to the capacity.

The application of the capacity-constrained assignment is iterative. After an initial (shortest path) assignment, the link travel times are updated based on the BPR function or another impedance curve. The assignment is then repeated with the updated link travel times until an equilibrium is reached. Equilibrium in this case means that the assigned flows and link sequences no longer change, and travel times on all links are thus stable between successive assignments.

2.7.2 Dynamic Traffic Assignment and Other Advanced Algorithms

The previous assignment methods are useful to illustrate the basic concepts of traffic assignment, and form the basis for more advanced assignment algorithms. But in practice, an all-or-nothing shortest path assignment, or even a capacity-restrained assignment, are often not representative of the assignment algorithms used in travel demand models.

These algorithms generally use more advanced functions that are applied in an automated fashion to search for potential assignment solutions in a software implementation. A common challenge, for example, with the simpler algorithms is that they may not result in an equilibrium solution. For example, in a large network with several thousand zones, an all-or-nothing assignment may result in overloading of key links to a point, where even repeated iterations cannot find sufficient alternate paths. Travel demand model–based assignments therefore sometimes begin with a reduced trip distribution table (e.g., 10% of trips), run an initial assignment, and then gradually add traffic in successive assignment iterations until the full O/D table is assigned and equilibrium is reached. A detailed discussion of these advanced assignment algorithms is beyond the scope of this text, and is available in customized planning references.

A special category of assignment algorithms is referred to as *dynamic traffic assignment* (DTA) algorithms (Patriksson, 2015; Meyer, 2015; Weiner, 2012). These are commonly found in simulation models that are applied to a transportation network, and may interact with the travel demand model. Simulation tools are further classified as *microscopic* or *mesoscopic simulation*, which are concepts discussed in more detail in Part 5 of this book. These simulation tools differ in the level of complexity of the algorithms used to model individual vehicles, and both types can contain DTA logic. Both tools model the movement of individual vehicles between zones on the network and are able to estimate congestion and bottleneck points in the network as a function of the actual traffic loads and the underlying capacity (more concepts discussed in Part 5).

A dynamic assignment uses the actual measured travel times from these vehicles to inform successive assignment iterations. So, rather than approximating travel times based on BPR functions and the *skimming* process, DTA algorithms use the actual vehicle travel times. DTA models are therefore often believed to be more accurate, and are increasingly used to estimate network-wide diversion impacts as a result of major work zones, bridge closures, or emergency evacuation events. A more detailed discussion of DTA tools is found in the literature (DTA, 2011).

2.7.3 Sample Network Application for Traffic Assignment

Turning back to the sample network introduced in the trip generation and trip distribution steps, the traffic assignment now assigns the origin–destination matrix to the physical paths taken by the trips between zones.

Figure 2.18 shows a sample network with eight internal TAZs (part of the study area) and four external TAZs (not explicitly being studied, but contribute to trips in and out of study area). The figure further shows a roadway network with classification into principal arterials/freeway,

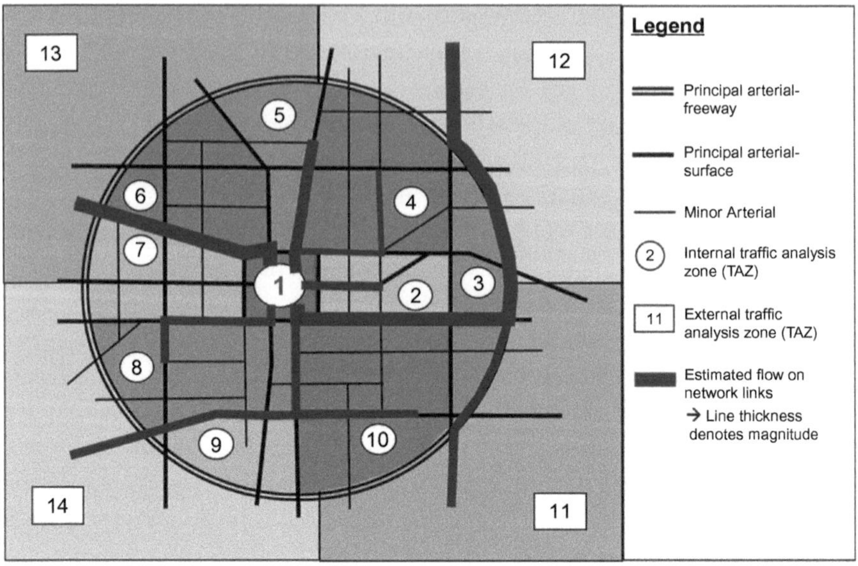

Figure 2.18 Sample network: traffic assignment step.

principal arterial/surface street, and minor arterials. Collector streets and local roads are omitted from the network in a planning context in this example, but can be included if they carry significant amounts of traffic. As before, we will focus on the trips originating from zone 1.

The figure shows the estimated flow between zone 1 and the 13 remaining zones, with the line thickness. Notice that in this example several zone-to-zone trips use the same links, such as for trips to zones 6, 7, and 13. Trips originating from other zones are not shown, but those would ultimately all add to the estimated load of trips on each link in the network by superimposing them on the zone 1 trips.

If this example illustrates the results of an all-or-nothing assignment, the analyst may now introduce a capacity constraint, which may further distribute those trips. Other advanced assignment algorithms can also help to distribute the trips for each zonal pair across multiple links.

2.7 TRAVEL DEMAND MODEL APPLICATIONS

This section provides an overview of some additional planning topics, including scoping a planning study, an overview of planning software, and a high-level discussion of specialized planning applications, including activity-based models, forecasting nonmotorized travel, and freight models. References to additional reading on these topics are provided throughout.

2.7.1 Scoping a Planning Study

The planning concepts introduced in this part are intended to provide an overview of the basic four-step process, and the basic analysis principles involved in each step. Clearly, many more details are available to fully understand each of these steps, and transportation planners (or teams of planners) often work on a planning study for several weeks or months. Furthermore, large regional travel demand networks often are maintained continuously by a highly trained staff involving multiple people and close interaction of the various stakeholders in the planning region. A report by the National Cooperative Highway Research Program (NCHRP, 2012) provides a good overview of modern demand modeling applications and techniques.

The scope of a planning study depends on the problem statement at hand, the geographic coverage of the planning network, the size of the network (in terms of number of links, number of zones, number of trips, etc.), the planning horizon (3 years vs. 10 years, vs. 20 years, etc.), data availability, and various other factors.

As such, a three- to five-year traffic forecast of trips generated by a new gas station is a problem of a very different magnitude from forecasting 20-year traffic for a large regional network with a population of several million people. While the first may be completed by a single analyst within a few days (depending on how much data collection is needed), the latter involves a team of analysts who work on a regional travel demand model continuously and full time. Such a regional travel demand model may cover multiple municipalities, metropolitan planning organizations, transit agencies, and state agencies that all collaborate on developing, calibrating, and continuously updating one integrated model.

2.7.2 Travel Demand Modeling Software

Travel demand modeling software is often an essential part of the planning process, and numerous such software tools exist on the market today. Each tool has its own unique approach to graphical interface, automation of data entry, and so on. But at the heart, all of these tools integrate the basic four-step process, as well as potentially many advanced features.

The goals and benefits of such software are clear. With the computations for trip generation, trip distribution, mode split, and traffic assignment being complex and time consuming, software provides a way to introduce great efficiency in the planning process and automate many calculations. In the examples in this part, even those using simplified networks with few zones and even though examples were often condensed to select representative zones, several pages of calculations and tables were needed to illustrate the concepts. For larger networks, any calculations by hand are futile.

But software also brings limitations and challenges. For example, the analyst has to rely on the validity and computational accuracy of the algorithms that were implemented by the developers, and often has little control over modifying these algorithms. Clearly, software developers will do their best to assure the validity of their product, but it is still outside of the control of the analyst. Moreover, the analyst has to be trained in the particular software product at hand, which requires time and resources. As such, the analyst needs to have a deep understanding of planning methods and concepts, but also needs to learn software interface and the way these algorithms are implemented.

Another potential danger of software is relying on automated results without including the close scrutiny and assessment that should be applied. Engineering and professional judgment are needed in all planning steps, and the analyst needs to closely check assumptions, interim results, and the final planning estimates based on his or her professional expertise.

2.7.3 Activity-Based Models

The traditional four-step process is limited by the assumption that the desire of a traveler to complete a trip is fixed and that the main objectives of a planning model are to figure out where that trip will end up, what mode the traveler chooses, and what route is ultimately selected.

An alternate approach to travel forecasting is given by *activity-based models*. These models center around predicting the types of activities individuals choose to perform (e.g., go to work, go shopping, etc.) and what trips are necessary to complete those activities. The trip thereby becomes a means to an end to support other activities, as opposed to being an end in itself. The focus of an activity-based forecasting model is to predict when and where individuals want to perform what activities, and travel then becomes just one part of the broader activity decision model for that person.

Activity-based models are generally believed to be able to better model nonwork and nonpeak period trips, which are often more discretionary. Decisions about which household member will do the shopping, whether that shopping trip happens before or after work, and what other activities can be combined with the shopping trip are difficult to account for in the traditional four-step process. An activity-based model can take these decisions into account, consider tradeoffs of different decisions, and take into account nontraditional performance measures that go beyond travel time and cost. For example, activity-based models can weigh the environmental impacts of different transportation choices or health effects in mode choice decisions. As such, the models can more directly illustrate these nontraditional factors, as well as individual responses to changing conditions.

Activity-based models are generally very complex and are integrated in travel demand modeling software. They take into account that decisions about trips can vary in how constrained they are in space and time. Some decisions are constrained in space (e.g., work location) while others are more flexible (e.g., where to shop). Similarly, decisions can be constrained in time (e.g., school start time) or can have flexibility (e.g., flexible work hours).

Additional details on activity-based modeling are given in other, planning-specific works in the literature (Meyer, 2015; Weiner, 2012).

2.7.4 Forecasting Nonmotorized Travel

Most travel demand models do not explicitly forecast nonmotorized travel, but focus instead on auto and, to some extent, transit modes. But the fraction of nonmotorized travel can be significant in urban and downtown areas, university campus environments, or even technology campuses. For some cities in the United States, and even more so internationally, nonmotorized commuting options can represent a significant portion of home-to-work trips, for example.

One of the biggest challenges in predicting nonmotorized travel is that travel choices for pedestrians and cyclists are highly sensitive to factors not generally part of the four-step process. Specifically, weather and climate, as well as the quality of service and safety of various links in the transportation network, must be considered. Nonmotorized travel can be incorporated in various stages of the four-step process, or become an independent prediction process in an activity-based model.

In trip generation, exclusive pedestrian and bicycle trips for recreational or exercise purposes can be estimated. These trips are intrinsically generated as a nonmotorized trip and would not have existed otherwise. Alternatively, mode choice models can incorporate nonmotorized options (in addition to auto and transit) to estimate, for example, the share of home-to-work trips by bicycle (i.e., bike commuting). This second category represents an existing trip diverted to an alternate mode, while recreational bicycling trips would not have occurred otherwise, and are net additions to the total trips in the network.

Additional information about modeling nonmotorized transportation can be found in the literature (FHWA, 1999). In addition, multimodal quality of service methodologies developed under NCHRP Report 616 (Dowling and Bonneson, 2008), and included in the *Highway Capacity Manual*, can provide methods for estimating performance measures for nonmotorized road users, which can then be incorporated in utility functions to predict mode choices.

2.7.5 Forecasting Freight and Goods Movements

The prediction of the movement of freights and goods are of increasing importance to national policy and to assure regional and global economic competitiveness. Trucks on the surface transportation network carried 64% of freight by value and 67% by weight in 2012 according to the Bureau of Transportation Statistics (Strocko et al., 2014). The transportation network, therefore, is the backbone of commerce domestically and internationally.

Freight forecasting models are generally standalone models that are separate from the metropolitan or regional planning models. Freight movements further tend to cover very large distances, and freight models are therefore generally much larger in geographic extent than urban planning models. At the same time, freight movements are generally coarser in their spatial resolution and are often limited to the freeway network, strategic highway corridors, and principal arterials in urban networks. As such, the level of spatial detail in freight models is coarser, with most local roads, collectors, and minor arterials excluded from the model.

Freight models generally follow a similar structure to urban planning models, incorporating steps of trip generation, trip distribution, and traffic assignment, with mode choice being generally not applicable.

The Freight Analysis Framework (FAF) represents a national freight forecasting model. The FAF contains data on freight trip generators and an inventory of the nation's freight transportation network. The FAF further estimates the distribution of trips between zones and then assigns trips to the network. Similar to urban models, the FAF is calibrated from extensive national data on truck and freight movements. Figure 2.19 shows examples of the FAF network and trip distribution between select

Average daily long-haul freight truck traffic on the national highway system: 2035

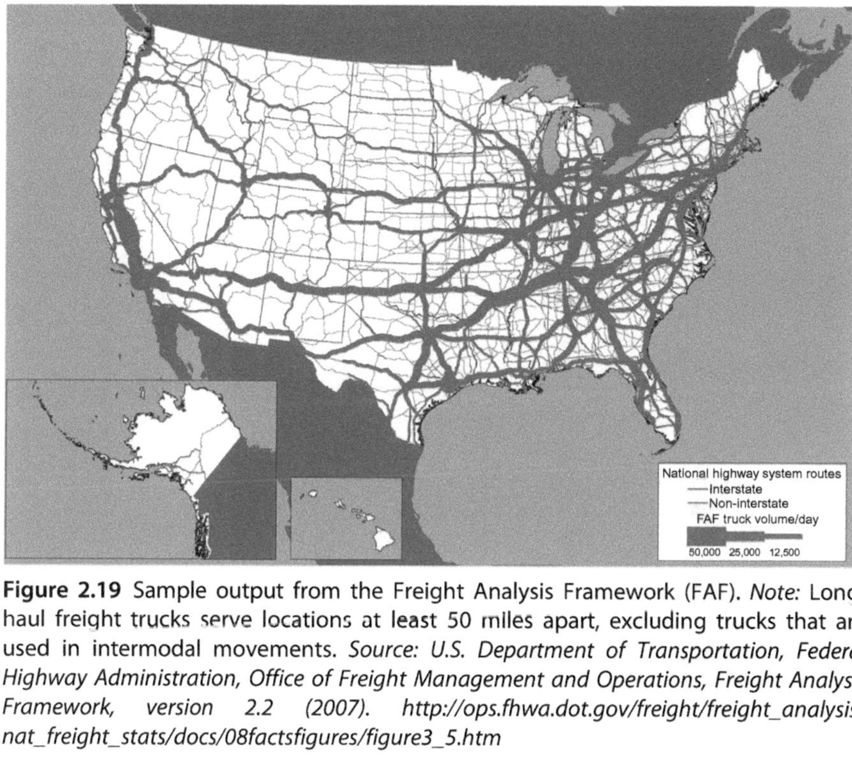

Figure 2.19 Sample output from the Freight Analysis Framework (FAF). *Note:* Long-haul freight trucks serve locations at least 50 miles apart, excluding trucks that are used in intermodal movements. *Source: U.S. Department of Transportation, Federal Highway Administration, Office of Freight Management and Operations, Freight Analysis Framework, version 2.2 (2007). http://ops.fhwa.dot.gov/freight/freight_analysis/nat_freight_stats/docs/08factsfigures/figure3_5.htm*

zones, as well as a completed assignment; network links with more assigned traffic are drawn as thicker lines. This type of visualization is common for all travel demand models and allows for an easy visual assessment of traffic patterns and strategic network links.

2.8 PRACTICE PROBLEMS

Problem 2.1

For an urban arterial street, 24-hour traffic counts were conducted for 10 days. Based on the data collected determine the ADT.

Data: 1500, 2750, 1325, 1427, 1800, 1329, 1132, 1527, 1200, 2000

Problem 2.2

For the following counts taken over a period of 5 days, determine the ADT, DHV, and DDHV. State all assumptions!

Data: 1100, 1352, 4421, 3421, 987

Problem 2.3

Determine the DDHV based on the ADT for the following interchange.

K = 10%

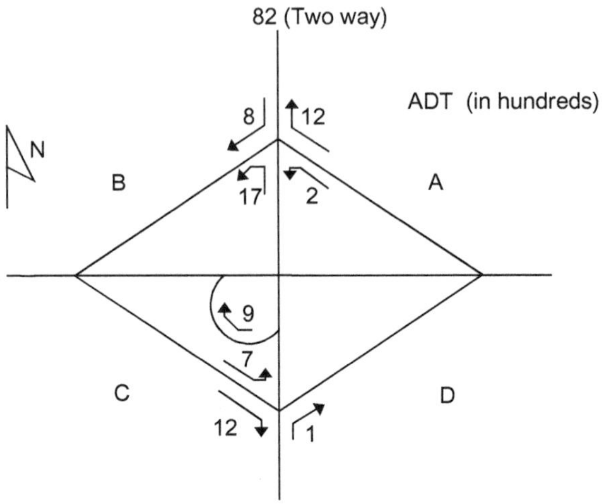

Problem 2.4

Given the following one hour of traffic data, determine the PHF.

One hour of traffic data

15-min segment	Number of vehicles per hour (vph)
1st	120
2nd	154
3rd	187
4th	113

Problem 2.5

Given an hourly volume of 352 and a typical PHF (rural road) of 0.88, find the peak 15-min volume.

$$PHF = \frac{\text{peak hour volume}}{4 \times \text{highest 15-min volume}}$$

Problem 2.6

A research and development center will employ about 500 people after 3 years of operation. You are analyzing the impact of traffic from this center on the adjacent street network. Estimate the number of trips entering the site during the am peak hour when the center is fully employed.

Research and development center (760)

Average vehicle trip ends vs: Employees
On a: weekday, am peak hour

Number of studies: 28
Avg. number of employees: 1,038
Directional distribution: 86% entering, 14% exiting

Trip generation per employee

Average rate	Range of rates	Standard deviation
0.43	0.20–1.39	0.67

Data plot and equation

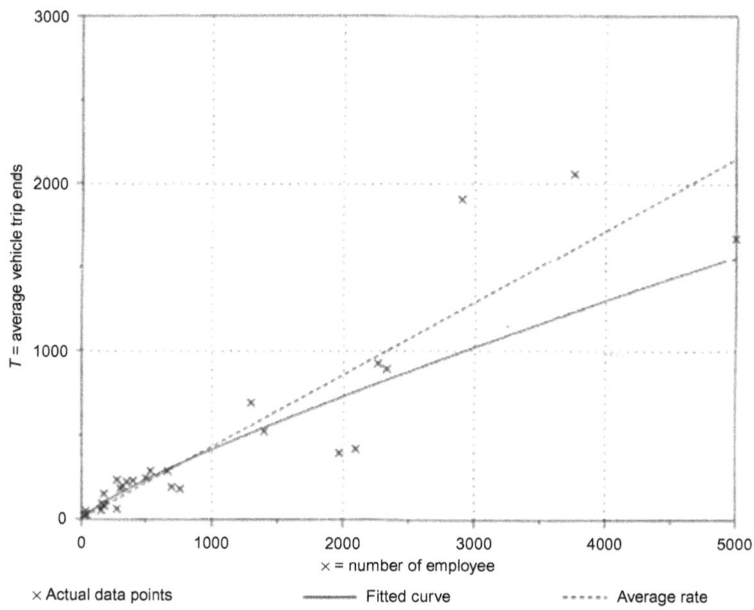

Fitted curve equation: $\text{Ln}(T) = 0.82\ \text{Ln}(x) + 0.33$ $R^2 = 0.90$

REFERENCES

Codina, E., Barceló, J., 1995. Dynamic traffic assignment: considerations on some deterministic modelling approaches. Ann. Oper. Res. 60 (1), 1–58, <http://link.springer.com/search?facet-author=%22E.+Codina%22>, <http://link.springer.com/search?facet-author=%22J.+Barcel%C3%B3%22>, <http://link.springer.com/journal/10479>, <http://link.springer.com/journal/10479/60/1/page/1>.

Dowling, R.G., Bonneson, J., 2008. NCHRP Report 616: Multimodal Level of Service Analysis for Urban Streets. Transportation Research Board of the National Academies, Washington, DC.

DTA, 2011. Dynamic Traffic Assignment – A Primer (June 2011), <http://onlinepubs.trb.org/onlinepubs/circulars/ec153.pdf>.

FHWA, 1999. Guidebook on Methods to Estimate Non-Motorized Travel, PUBLICATION NO. FHWA-RD-98-165 JULY 1999, <http://safety.fhwa.dot.gov/ped_bike/docs/guidebook1.pdf>.

FHWA. The Transportation Planning Process: Key Issues - A Briefing Book for Transportation Decisionmakers, Officials, and Staff. FHWA-HEP-07-039. https://www.planning.dot.gov/documents/briefingbook/bbook_07.pdf. Washington, DC 2007.

FHWA, 2013. Federal Highway Administration. Traffic Monitoring Guide, 2013.

FHWA, 2014. FHWA AIIG Summer 2014, <http://safety.fhwa.dot.gov/intersection/alter_design/pdf/fhwasa14067_ddi_infoguide.pdf>, <http://safety.fhwa.dot.gov/intersection/alter_design/pdf/fhwasa14068_dlt_infoguide.pdf>, <http://safety.fhwa.dot.gov/intersection/alter_design/pdf/fhwasa14069_mut_infoguide.pdf>, <http://safety.fhwa.dot.gov/intersection/alter_design/pdf/fhwasa14070_rcut_infoguide.pdf>.

ITE, 2012. Trip Generation Manual 9th Edition. Institute of Transportation Engineers. Washington, D.C. 2012, REF: <http://www.ite.org/tripgeneration/trippubs.asp>.

Meyer, M.D., 2015. Transportation Planning Handbook. ITE, Wiley, <https://www.google.com/search?tbo=p&tbm = bks&q = inauthor:%22ITE%22>, <https://www.google.com/search?tbo=p&tbm = bks&q = inauthor:%22Michael + D.+Meyer%22>.

MNL, 2006. A Self Instructing Course in Mode Choice Modeling: Multinomial and Nested Logit Models, <http://www.caee.utexas.edu/prof/bhat/COURSES/LM_Draft_060131Final-060630.pdf>.

National Cooperative Highway Research Program (NCHRP), 2012. Travel Demand Forecasting: Parameters and Techniques, vol. 716. Transportation Research Board.

Patriksson, M., 2015. The Traffic Assignment Problem: Models and Methods. Courier Dover Publications.

Strocko, E., Sprung, M., Nguyen, L., Rick, C., Sedor, J., 2014. Freight Facts and Figures (No. FHWA-HOP-14-004).

TRB, 2010. Highway Capacity Manual 2010. Transportation Research Board. Washington, D.C.

Urban Mass Transportation Administration, 1977. An Introduction to Urban Travel Demand Forecasting – A Self Instructional Text. U.S. Department of Transportation – Federal Highway Administration.

Weiner, E., 2012. Urban Transportation Planning in the Unites States: History, Policy, and Practice. Springer Science & Business Media.

PART 3

Horizontal and Vertical Alignment

Daniel J. Findley, PhD, PE
Institute for Transportation Research and Education, North Carolina State University

Contents

3.1 INTRODUCTION

Highway alignment involves the decisions related to choosing an optimal position, given substantial environmental and design considerations. Corridor selection is comprised of the broad task of choosing a highway location through decisions relating to minimizing costs and impacts to the human and natural environment. The engineering computations of such corridors are derived from consideration of the highway segments as a traverse. Traverse calculations are based on geometric and trigonometric principles. The horizontal (elements in the plan view) and vertical (elements in the profile view) components each affect the highway location and require an iterative process to balance the various quantitative measures and tradeoffs of a particular alternative, as well as include feedback gathered from stakeholders in the public involvement process. At any point along a highway, the driver should be able to perceive an obstruction or change in alignment and react by changing his or her speed, direction, or path. The distance required to perform this maneuver—the sight distance—is an integral part of highway alignment.

3.2 CORRIDOR SELECTION

Corridor selection follows the planning process with the identification of a specific route that will meet the demand recognized during planning. A consistent right-of way width can be applied to the corridor during this process, depending on the functional classification and setting of the highway. Engineering and environmental (human and natural) factors are considered during the corridor selection process. Corridor selection elements can have competing objectives and must be balanced to result in a final project that is acceptable to a wide variety of stakeholders. A full environmental analysis is typically not completed during corridor selection, but key elements are considered when determining which corridors provide the needed benefits with a minimal impact. The gathering of pertinent data for a thorough corridor selection process can be resource and time intensive. The justification for a highway involves increased access or mobility, which must be balanced with other factors. These factors can include elements that primarily affect human or natural, or a combination of both, environments, as shown in Figure 3.1.

Because highway projects can have significant negative effects on communities and the environment, governments have enacted numerous

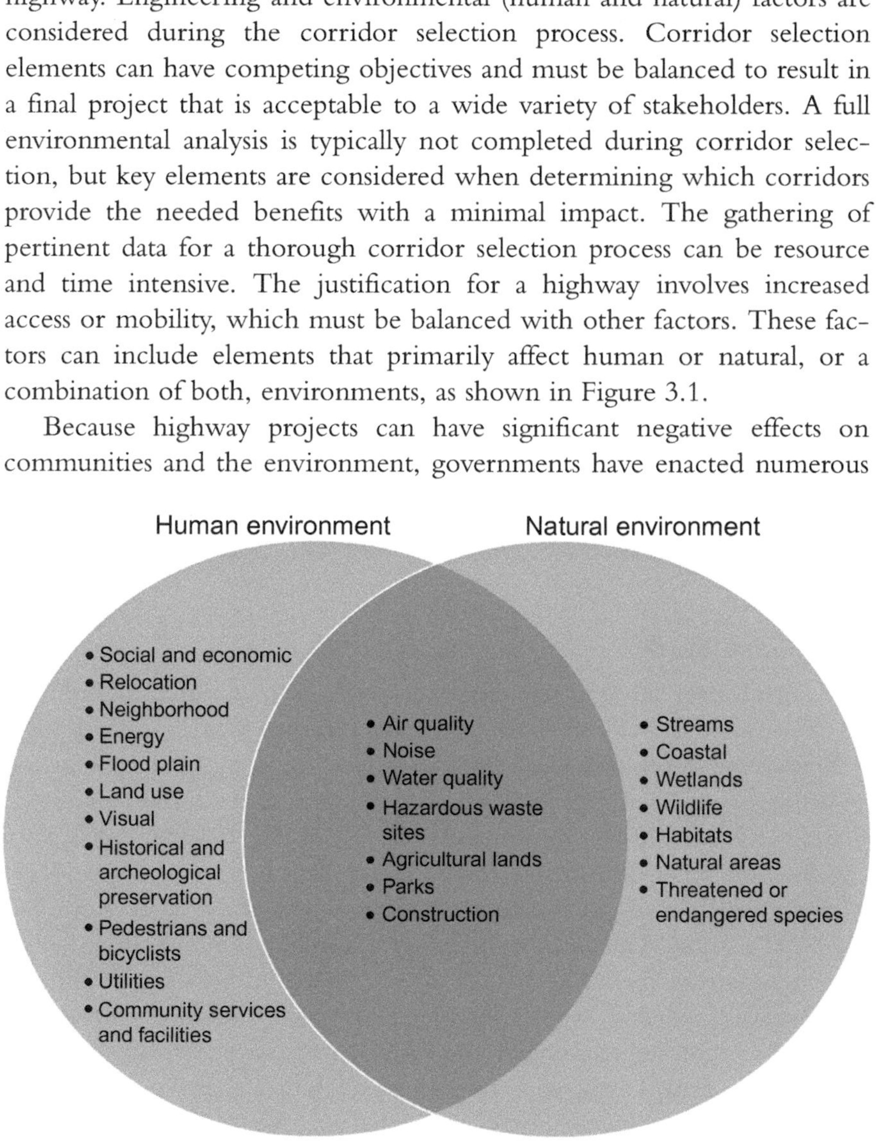

Figure 3.1 Potential human and natural environmental impacts of a highway.

laws, regulations, and policies. One such policy is the National Environmental Policy Act of 1969 (NEPA), which directs how the government should protect and enhance the environment while making decisions. NEPA ensures that information about the environmental impacts of any federally funded action is available to public officials and citizens before decisions are made and actions are taken.

In addition to environmental impacts, costs are an important consideration during corridor selection. These will include capital costs from the construction of the highway and the acquisition of necessary rights of way, and may also include safety, environmental mitigation, travel time, vehicle operation, and maintenance costs. An initial and thorough list of possible corridor alternatives can be reduced to a few of the most feasible options for further detailed analysis. Many factors will be monetized and compared with economic analysis tools (benefit/cost, net present value, internal rate of return, or other). Other nonquantifiable factors, such as neighborhood impacts and historical preservation, can be evaluated qualitatively and compared to the impact levels from the competing alternatives. Sustainability, or the activities involved with accommodating present needs without compromising the ability for meeting future needs, is another concern for highway projects. Aspects of highway projects that can be scrutinized for potentially unsustainable conditions are vehicular emissions, energy insecurity, congestion, and ecological impacts.

3.2.1 Case Study

The corridor for the final segment of an outer loop around the southeastern portion of the city of Raleigh, North Carolina must be determined. Officials have developed 17 alternative routes consisting of a combination of 10 segments identified by colors in the map and diagram shown in Figures 3.2 (NCDOT 2013b) and 3.3 (NCDOT 2013c). Preliminary community and environmental data show the length of each alternative, in addition to human and natural environmental impacts (Figure 3.4, NCDOT 2013d). These factors include the number of homes or businesses and parks within each 300-ft right of way, in addition to the quantity of streams, wetlands, and flood plains affected by each. Extensive efforts were required to thoroughly compare the various alternatives through data collection and analysis, as well as gathering feedback from stakeholders through three public meetings conducted as open houses. The most heavily debated route was the Red Corridor, which was developed as an alternative that would pass on the north side of Lake Benson and result in a shorter overall project, thereby reducing some of the natural environmental impacts (NCDOT, Complete 540 Frequently Asked Questions, 2013a). However, the Red Corridor would have severe impacts on the communities in and near the town of Garner. The Red Corridor is a prime example of the difficult decisions and analysis necessary to compare impacts to the natural and human environment. The following process, from public input on alternatives to the opening of the highway for traffic, is expected to take almost 10 years to complete (NCDOT, Complete 540 Frequently Asked Questions, 2013a). The dates are subject to change depending on approval of key documents and availability of funding, but the timeline illustrates the duration of events that occur prior to construction. Approximately half of the project timeline occurs between the public workshops and construction, including the development and studying of alternatives, completion of necessary technical and environmental impact studies, public outreach, and final corridor selection.

- Public workshops on alternatives (fall 2013)
- Finalize detailed study alternatives (winter 2013)
- Complete required technical studies (fall 2014)
- Receive approval of the draft environmental impact statement (EIS; spring 2015)
- Draft EIS review period and hold public hearing (summer 2015)
- Selection of the preferred alternative (fall 2015)
- Approval of final environmental impact statement (spring 2016)
- Publication of the Record of Decision (summer 2016)
- Complete financial feasibility (spring 2017)
- Begin right-of-way acquisition (summer 2017)
- Begin construction (spring 2018)
- Open to traffic (spring 2022)

REFERENCES

North Carolina Department of Transportation (NCDOT). Frequently Asked Questions. Complete 540. Triangle Expressway Southeast Extension. September 2013a. Raleigh, NC.

North Carolina Department of Transportation (NCDOT). Recommended Detailed Study Alternatives. Complete 540. Triangle Expressway Southeast Extension. October 2013b. Raleigh, NC.

North Carolina Department of Transportation (NCDOT). Alternatives by Segment. Complete 540. Triangle Expressway Southeast Extension. October 2013c. Raleigh, NC.

North Carolina Department of Transportation (NCDOT). Community and Environmental Data. Complete 540. Triangle Expressway Southeast Extension. October 2013d. Raleigh, NC.

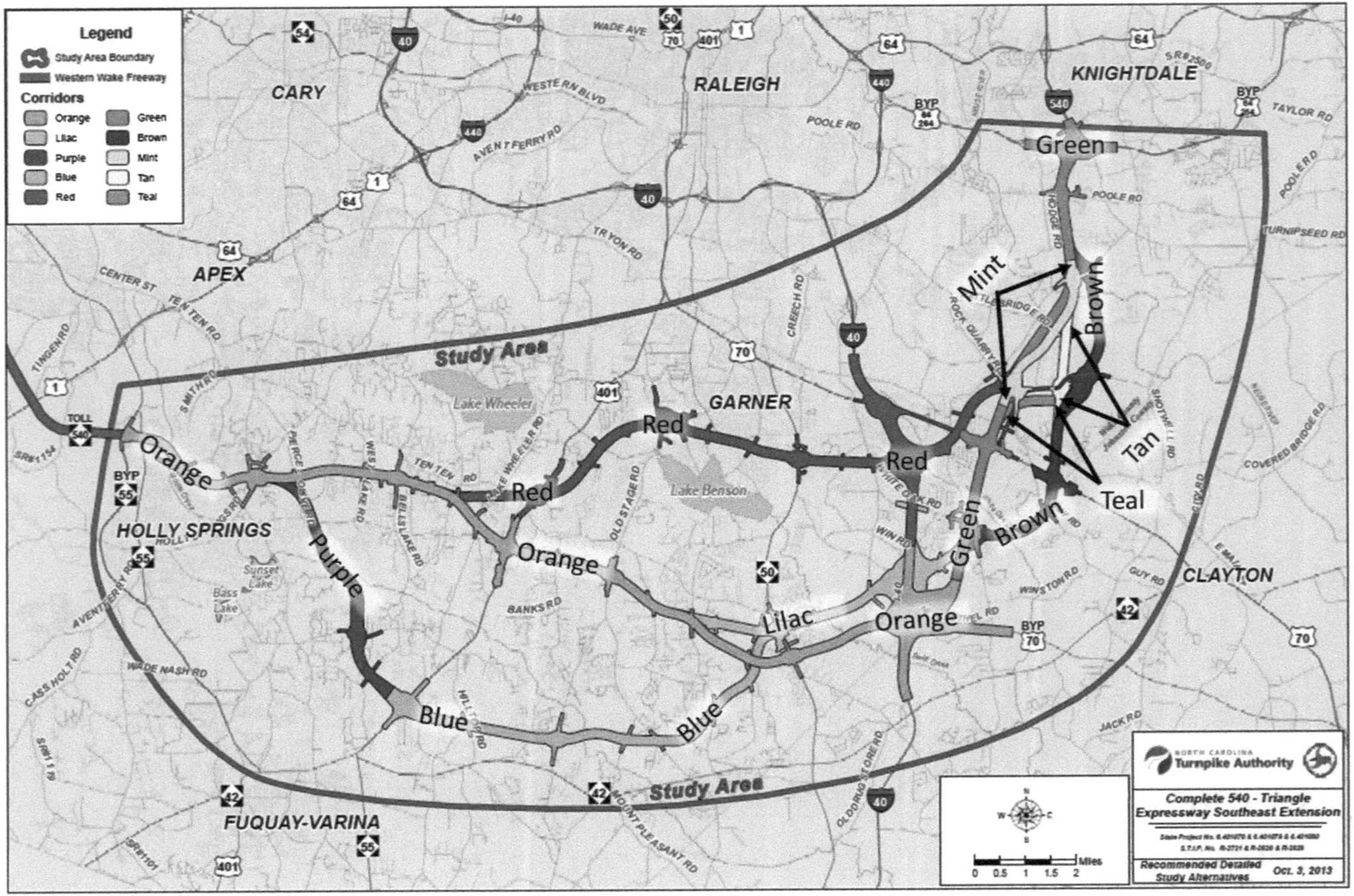

Figure 3.2 Case study: recommended detailed study alternatives (NCDOT 2013b).

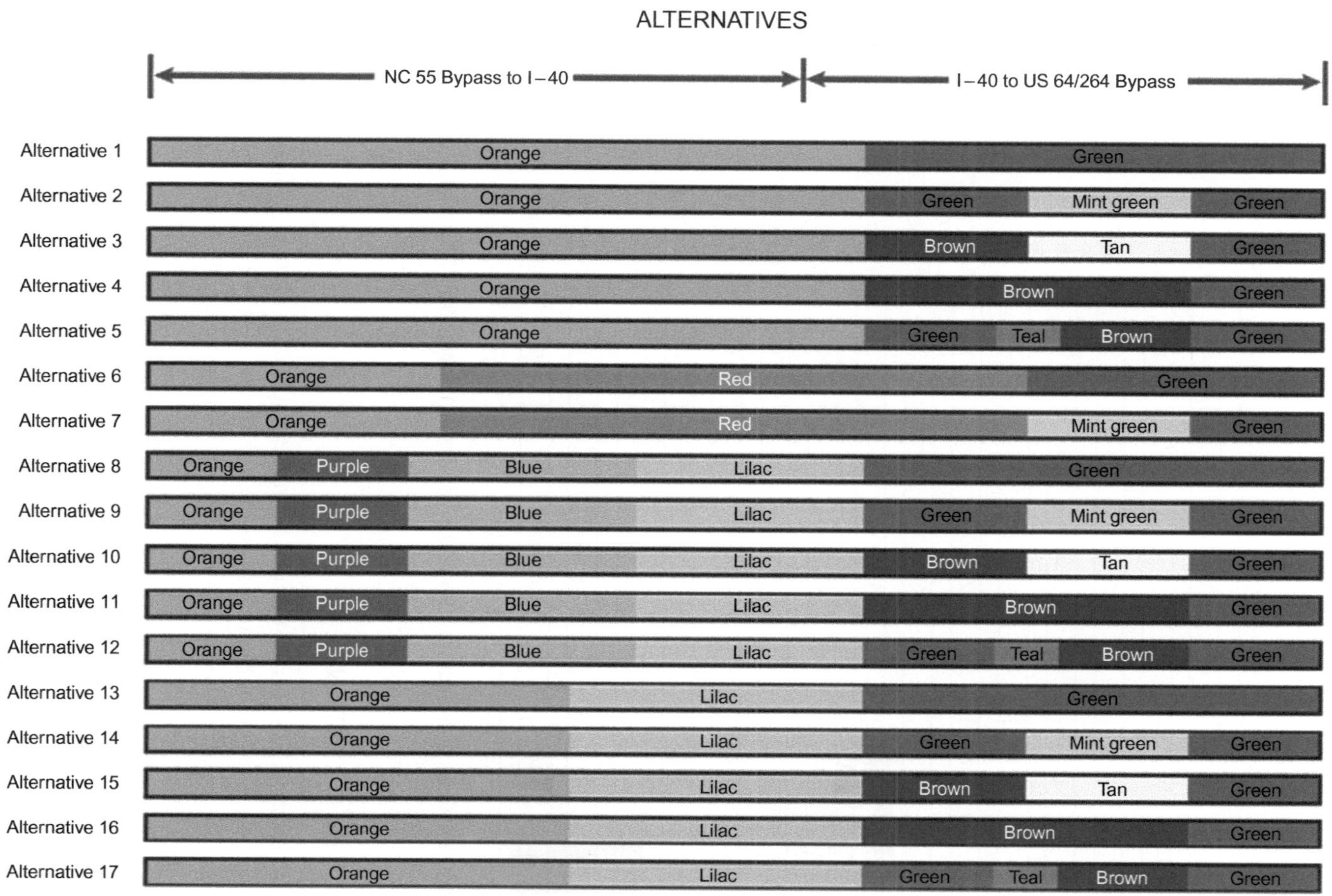

Figure 3.3 Case study: alternatives by segment (NCDOT 2013c).

Community and environmental data

Length (miles)	Homes or businesses	Parks (existing or planned)	Streams (linear ft.)	Wetlands (acres)	100-year flood plain (acres)	
27.4	217	0	36,110	88	140	Alternative 1
27.5	220	0	37,140	93	147	Alternative 2
27.9	204	1	37,280	91	146	Alternative 3
28.2	187	1	36,410	89	114	Alternative 4
28.4	209	0	36,270	89	114	Alternative 5
23.9	404	2	24,520	44	129	Alternative 6
24.0	407	2	25,350	48	136	Alternative 7
29.8	353	1	36,840	47	137	Alternative 8
29.9	356	1	37,870	51	145	Alternative 9
30.5	338	2	38,260	50	144	Alternative 10
30.9	321	2	37,390	48	111	Alternative 11
30.9	345	1	37,000	48	111	Alternative 12
26.4	366	0	33,140	56	104	Alternative 13
26.4	369	0	34,160	60	111	Alternative 14
27.0	351	0	34,550	59	110	Alternative 15
27.4	334	0	33,690	57	78	Alternative 16
27.4	358	0	33,290	57	78	Alternative 17

Figure 3.4 Case study: community and environmental data (NCDOT 2013d).

3.3 SIGHT DISTANCE

Sight distance is a fundamental consideration for highway alignment. Throughout the alignment process, the ability of a driver to perceive and react to changes in the direction of the highway or the presence of potential hazards is essential for the safe operation of a highway. Areas of a highway that require complex maneuvers may warrant more sight distance to allow a driver more time to recognize the need for the operation and act on that decision. The amount of sight distance provided to the driver is influenced by the horizontal, vertical, and cross-section design of the highway. The minimum sight distance for every point along a highway is the stopping sight distance, which is affected by the operating speed of the roadway, the time to perceive the need to stop, and the distance required to complete the stopping maneuver. On two-lane highways another sight distance, known as passing sight distance, can be used to design or identify sections that can be used for safe passing movements. At locations that demand complex decision making, decision sight distance can be used to provide more sight distance for drivers.

3.3.1 Stopping and Decision Sight Distance

Stopping sight distance is comprised of the distance required for a driver to perceive a need to stop and complete the braking process to bring their vehicle to a stop. These two components can be computed separately: $d_{reaction}$, the reaction distance, and $d_{braking}$, the braking distance. The typical assumed reaction time and deceleration rate are based on research findings that encompass 90% or greater of a driver's and passenger vehicle's capabilities (AASHTO, 2011, p. 3–3). These assumptions include conservative estimates of the properties present for wet pavements, poorly maintained vehicles, and elderly drivers.

$$\text{Stopping Sight Distance} = d_{reaction} + d_{braking}$$

where

$$d_{reaction} = 1.47\ Vt \qquad \text{(AASHTO, 2011, equation 3 − 2)}$$

$$d_{braking} = \frac{V^2}{30\left(\frac{a}{32.2}\right) \pm G} \qquad \text{(AASHTO, 2011, equation 3 − 3)}$$

V = design speed in miles per hour
t = brake reaction time in seconds (typically assumed as 2.5 s)
a = deceleration rate in ft/s^2 (typically assumed as 11.2 ft/s^2)
G = longitudinal grade of the highway in ft/ft (input 0 for a flat grade)

Decision sight distance utilizes the principles of stopping sight distance and applies the equations to provide additional sight distance for complex situations. Decision sight distances are larger than stopping sight distances to compensate for additional time needed to perceive an obstruction and determine the appropriate action to take to avoid a collision. These areas may have merging traffic streams or multiple actions competing for visual attention, which require the driver to make multiple successive decisions or to undertake other complex driving tasks. Typical locations that might benefit from the extended sight distance provided by the decision sight distance process include:

- Unconventional interchange design or ramp locations on freeways
- Unexpected reductions in the number of through lanes
- Unexpected geometry or traffic control features
- Heavily developed commercial corridors

The AASHTO[1] Green Book presents five scenarios (A, B, C, D, and E) that address decision sight distance with varying assumptions for the brake reaction time. Avoidance maneuver A considers a scenario on a rural highway with a brake reaction time of three seconds, which results in the vehicle stopping. Avoidance maneuver B accounts for a stopping vehicle on an urban highway with a significantly increased brake reaction time of 9.1 seconds. The increased reaction time accounts for the likely delay in perceiving the need to stop in an urbanized area, which has the potential for multiple simultaneous activities that compete for the driver's attention. Avoidance maneuvers C, D, and E result in a decision by the driver to change his or her speed, path, or direction to avoid the conflict, with brake reaction time ranges of increasing duration for rural, suburban, and urban highways (10.2–11.2 s, 12.1–12.9 s, and 14.0–14.5 s, respectively). Table 3.1 presents stopping and decision sight distances by design speed and maneuver type recommended by the AASHTO Green Book (Tables 3.1 and 3.3).

Reference

American Association of State Highway and Transportation Officials (AASHTO, 2011). A Policy on Geometric Design of Highways and Streets, 6th ed.

Stopping Sight Distance for Horizontal Curves

A common area of sight distance concerns is along the inside of horizontal curves as drivers attempt to see far enough in front of their vehicle to stop, if necessary. Vegetation, including trees, shrubs, and grasses, can present issues, particularly as it grows and creates an obstruction long after the curve is constructed. The backslope of a ditch, a retaining wall, signs, fences, or buildings can also reduce the available sight distance for drivers. An equation that relates the curvature, stopping sight distance, and distance to the obstruction can be used to design for adequate sight distance (AASHTO, 2011, eq. 3–36). The middle of the inside lane is the key location for measurements when analyzing sight distance on horizontal curves to assume the most conservative scenario of the position of the driver's field of vision; if adequate sight distance is provided in the inside lane, drivers in the outside lane will also have appropriate sight distance.

[1] This publication is commonly known as the AASHTO "Green Book."

Table 3.1 Stopping and decision sight distances

Design speed (mph)	Stopping sight distance (ft)	Decision sight distance (ft)				
		Avoidance maneuver				
		A	B	C	D	E
15	80	N/A	N/A	N/A	N/A	N/A
20	115	N/A	N/A	N/A	N/A	N/A
25	155	N/A	N/A	N/A	N/A	N/A
30	200	220	490	450	535	620
35	250	275	590	525	625	720
40	305	330	690	600	715	825
45	360	395	800	675	800	930
50	425	465	910	750	890	1030
55	495	535	1030	865	980	1135
60	570	610	1150	990	1125	1280
65	645	695	1275	1050	1220	1365
70	730	780	1410	1105	1275	1445
75	820	875	1545	1180	1365	1545
80	910	970	1685	1260	1455	1650

Note: Avoidance maneuvers: A = stop on rural highway (3.0 s), B = stop on rural highway (9.1 s), C = speed, path, or direction change on rural highway (10.2–11.2 s), D = speed, path, or direction change on suburban highway (12.1–12.9 s), E = speed, path, or direction change on urban highway (14.0–14.5 seconds).
Adapted from American Association of State Highway and Transportation Officials (AASHTO), 2011. A Policy on Geometric Design of Highways and Streets, 6th ed. Tables 3–1 and 3–3.

$$M = r\left[1 - \cos\left(\frac{28.65\ S}{r}\right)\right]$$

where

M = middle ordinate in feet, as measured from the middle of the inside lane to the edge of the sight obstruction

S = stopping sight distance in feet, as measured along the middle of the inside lane

r = radial distance in feet, as measured from the origin of the circle to the center of the inside lane (for a two-lane highway, this distance is half of a lane width shorter than the radius of the curve, as shown in Figure 3.5)

The equation can also be rearranged to provide the sight distance needed given a radial distance and distance to the obstruction.

$$S = \frac{r}{28.65}\left[\cos^{-1}\left(\frac{r - M}{r}\right)\right]$$

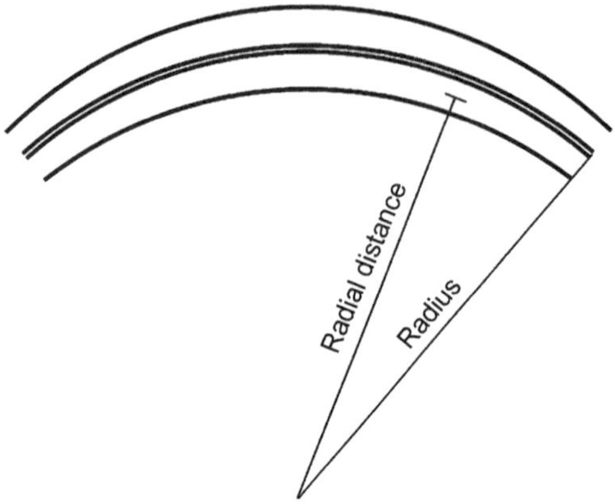

Figure 3.5 Comparison of radius and radial distance.

Table 3.2 Passing sight distances for two-lane highways

Design speed (mph)	Assumed speeds (mph)		Passing sight distance (ft)
	Passed vehicle	Passing vehicle	
20	8	20	400
25	13	25	450
30	18	30	500
35	23	35	550
40	28	40	600
45	33	45	700
50	38	50	800
55	43	55	900
60	48	60	1000
65	53	65	1100
70	58	70	1200
75	63	75	1300
80	68	80	1400

From American Association of State Highway and Transportation Officials (AASHTO), 2011. A Policy on Geometric Design of Highways and Streets, 6th ed. Table 3–4.

3.3.2 Passing Sight Distance

On two-lane, two-way highways, it is desirable to provide faster drivers with the capability to pass slower drivers in a safe and designated setting. To determine the required sight distance needed to provide safe passing areas, the concepts of passing sight distance can be applied. A driver wishing to pass another vehicle must be able to clearly see the approaching travel lane, accelerate past the other vehicle, and return to the travel lane before encountering an opposing vehicle. Table 3.2 presents passing sight distance design values in addition to assumed speeds for passing vehicles and vehicles being passed. Passing sight distance only applies to two-lane, two-way highways because highways with additional lanes are not constrained by the risk posed by opposing traffic.

3.4 HIGHWAY ALIGNMENT

Horizontal and vertical alignment provide the basis for locating the centerline of the highway at a defined point: horizontal alignment in the plan view and vertical alignment in the profile view. Cross sections provide the third element, which then give the highway its volumetric proportions. Figure 3.6 shows an example of four highway perspectives. The driver's perspective shows the highway as a driver would observe it. In the plan view, as seen from above the highway, the highway is horizontally straight until the end of the visible section, when a curve to the right occurs. In the profile view, showing the vertical alignment, a curve is present at the beginning of the segment as a downgrade transitions to an upgrade in the direction the driver is proceeding along the highway. The cross-section view displays the elements that give the highway its width (lanes, shoulders, and ditches).

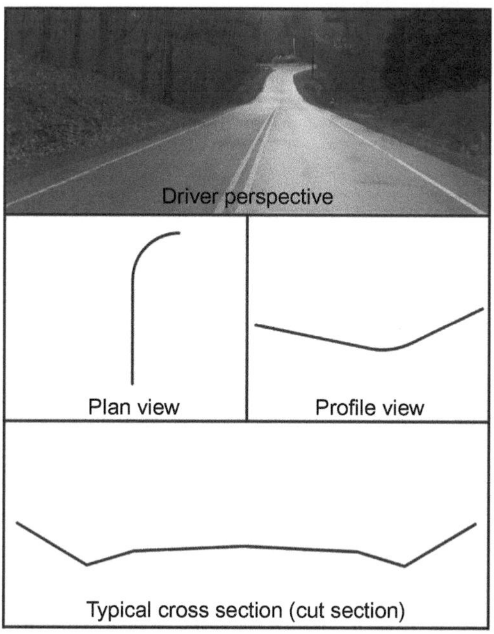

Figure 3.6 Highway perspectives.

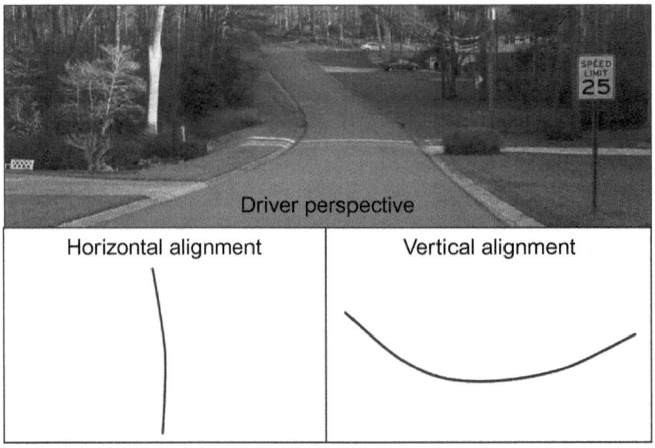

Figure 3.7 Highway perspectives.

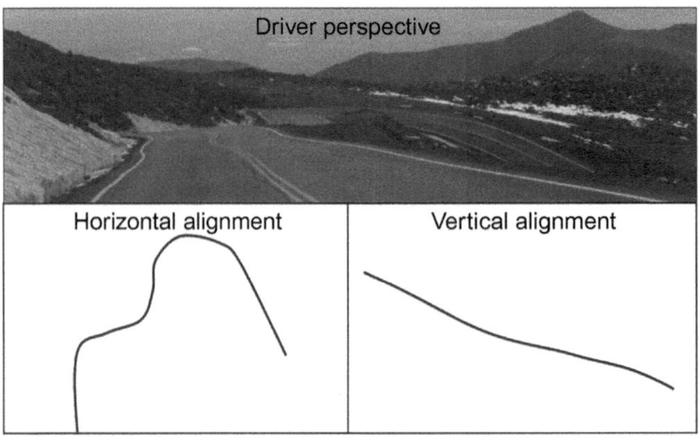

Figure 3.8 Highway perspectives.

Figure 3.7 shows another perspective of the relationship between the driver's perspective and the horizontal and vertical alignment of the highway. In the plan view, the horizontal alignment has one curve near the middle of the segment bounded by tangents. In the profile view, the vertical alignment has a sag curve, with a concave up shape, near the middle of the segment with a downgrade going into the curve and an upgrade coming out of the curve.

In Figure 3.8, the horizontal alignment is complex, with multiple horizontal curves turning to the left and right, and the driver progresses along the segment, which eventually results in a switchback where the final direction is approximately 180° different from the initial direction. The vertical alignment has a fairly constant downgrade with minor vertical curves along the segment.

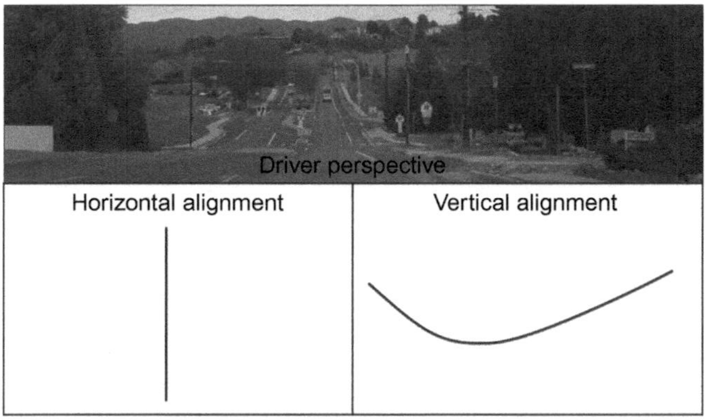

Figure 3.9 Highway perspectives.

In Figure 3.9, the horizontal alignment is tangent with no horizontal curves, meaning that the driver should not need to make a significant steering movements to keep a vehicle in the travel lane. The vertical alignment has one sag curve with a downgrade going into the curve and an upgrade coming out of the curve.

Developing the alignment of a highway involves the fundamental process of locating an unknown point in space based on a known point. The three elements necessary to locate an unknown point are the distance, elevation, and angle from a known point. To locate unknown points, surveying techniques can be deployed for site investigations and highway alignment, including plane surveying or geodetic surveying. Plane surveying assumes measurements are taken on a flat plane and that the curvature of the earth does not affect these measurements over relatively short distances, while geodetic surveying accounts for the curvature of the earth. The accuracy of a survey should be considered and should be within the threshold accuracy established by the jurisdiction, typically at a minimum accuracy of 1:10,000 (i.e., 1 ft of error in a 10,000-ft measurement is acceptable).

Angles can be described as azimuths or bearings. There are 360 degrees in a circle with 60 minutes in a degree and 60 seconds in a minute. The azimuth of a line is the direction measured clockwise from a standard direction (north is the common standard direction). Azimuths must have a positive value and be less than 360°. The bearing of a line is the direction within a quadrant with reference to a meridian (north or south) line. The reference is selected based on the closest meridian; therefore, bearing angles must be less than 90°. While the azimuth is described by a single numerical value, bearings must reference the leading latitude direction (north or south), the angle turned from the meridian, and the direction of the turn from the meridian (east or west). For example, the SW line (the line that bisects the SW quadrant), has an azimuth of 225° and a bearing of S45°W. The azimuth of 225° is the angular measurement from north and the bearing first lists the reference axis, south, then the angle of turn toward the horizon, west.

EXAMPLE 3.1 Azimuths and Bearings

Determine the azimuth and bearing of each line, AB, AC, and AD.

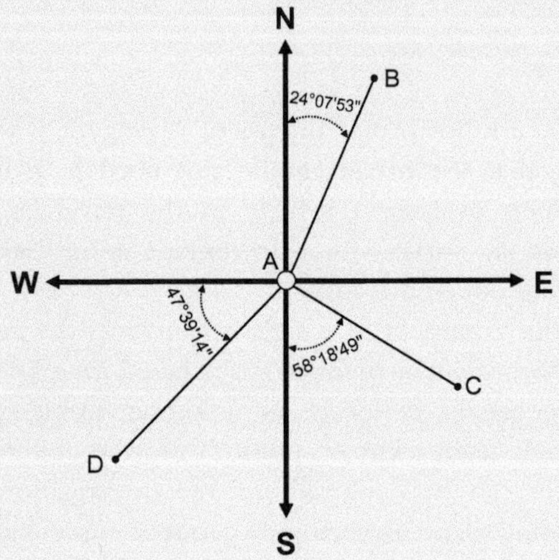

Solution

- Line AB: The given angle NAB is the angle measured from north of 24°07′53″. Therefore, the azimuth of line AB is directly known as 24° 07′53″. The bearing of line AB is also directly known as N24°07′53″E.
- Line AC: The given angle SAC is the angle measured from south of 58° 18′49″. Therefore, the azimuth can be calculated by subtracting this angle from 180°; the resulting azimuth of line AC is 121°41′11″. The bearing of line AC is referenced from south toward the east as S58°18′49″E.
- Line AD: The given angle WAD is the angle measured from west of 47°39′14″. Therefore, the azimuth can be calculated by subtracting this angle from 270°; the resulting azimuth of line AD is 222°20′46″. The bearing of line AD is referenced from south toward the west, so the given angle must be subtracted from 90°, which results in a bearing of S42°20′46″W.

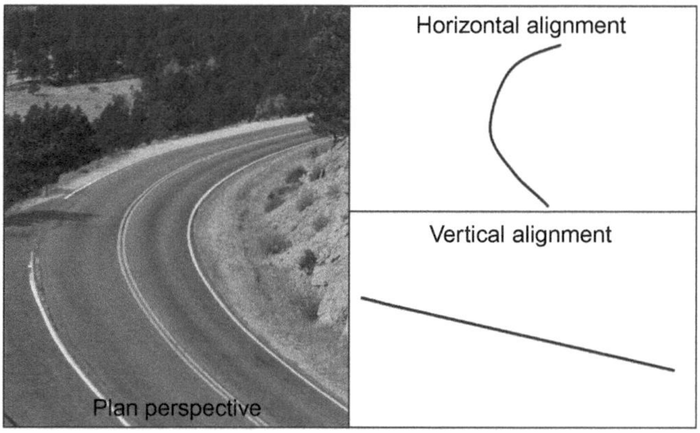

Figure 3.10 Highway plan perspective.

3.4.1 Horizontal Alignment

Horizontal alignment is comprised of tangent sections that are smoothly connected by curves. Horizontal alignment is viewed in the plan perspective (from an overhead view) as shown in Figure 3.10, which shows a horizontal curve that occurs along a consistent upgrade in the vertical alignment.

Traverses

The initial process of developing a highway's alignment is through a series of straight lines, or tangents, that form a traverse. The process of measuring and assigning stations to a traverse is an essential component of horizontal alignment. A traverse should be checked for positional accuracy through comparison to multiple known points or the amount of error present in a traverse that returns to the start. One station is equal to 100 ft and takes the form of $1^{+}00$. The overall beginning station of a project should not be set to $0^{+}00$, to avoid negative values if revisions to the project require an advanced starting point. The stations established during the development of the traverse will be revised when horizontal curves are included, as the stationing will follow the centerline of the highway through the curve. The key elements of interest are where the tangents sections meet, called the points of intersection (PI), which are used throughout horizontal alignment. A line segment can be separated into two components: (1) *latitude* is used to describe the north/south components and north is typically assigned a positive value, and (2) *departure* describes the east/west components and east is typically assigned

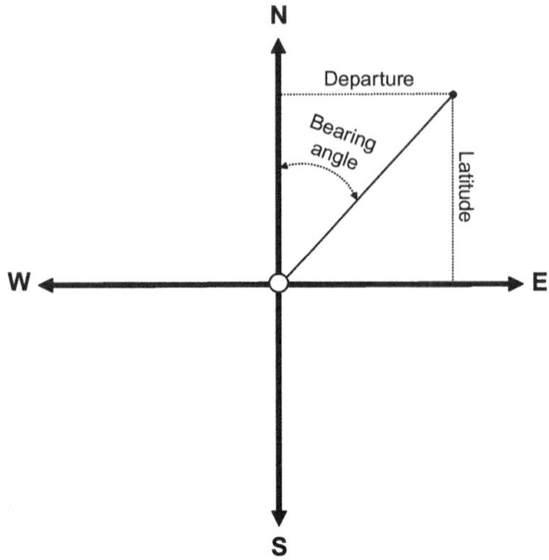

Figure 3.11 Latitude and departure.

a positive value (see Figure 3.11). For a traverse with known beginning and end points, the following steps can be used to evaluate the accuracy of the measurements:

1. List the line segments, bearings, and distances for the traverse in a table.
2. Compute the latitude and departure for each line.
3. Compute the coordinates for each point based on the latitude and departure for each line.
4. Compare the computed coordinate of the end point with the given end point coordinates. If the given and computed coordinates match, the traverse can be considered closed and no adjustment is necessary. However, perfect measurements are uncommon and when the coordinates do not match, the traverse must be adjusted.
5. Compute the accuracy of the survey based on the computed and given end points. The accuracy must meet the requirements of the jurisdiction or new measurements may be required.
6. Adjust the latitudes and departures of each line segment by proportioning the measurement errors to each line based on their length, under the assumption that all distances and bearings were measured with equal accuracy. The latitude and departure correction for each line must be opposite in sign from the error in each component to

remove the closure error (i.e., if the latitude error is off in the positive direction, the correction must be negative). The corrected values can be computed by adding the corrections to each latitude and departure value. After this process, the computed end point should match the given end point coordinates.

7. Compute corrected distances and bearings based on the corrected latitude and departure values.

$$\text{Latitude} = \text{length} \times \cos(\text{bearing angle})$$
$$\text{Depature} = \text{length} \times \sin(\text{bearing angle})$$
$$\text{Latitude error} = \text{Latitude}_{\text{given}} - \text{Latitude}_{\text{computed}}$$
$$\text{Departure error} = \text{Departure}_{\text{given}} - \text{Departure}_{\text{computed}}$$
$$\text{Traverse error} = \sqrt{(\text{Departure error})^2 + (\text{Latitude error})^2}$$
$$\text{Precision} = \frac{\text{Traverse error}}{\text{Traverse length}}$$
$$\text{Latitude correction} = \text{Latitude error}\left(\frac{\text{Line length}}{\text{Traverse length}}\right)$$
$$\text{Departure correction} = \text{Departure error}\left(\frac{\text{Line length}}{\text{Traverse length}}\right)$$
$$\text{Corrected length} = \sqrt{(\text{Corrected latitude})^2 + (\text{Corrected departure})^2}$$
$$\text{Corrected bearing} = \tan^{-1}\left(\frac{\text{Corrected departure}}{\text{Corrected latitude}}\right)$$

EXAMPLE 3.2 Traverses

In the following figure, a highway connection is needed between the two existing highways. The new highway should connect at Points 1 and 2 and avoid the water, forests, and buildings to the extent possible. A traverse can be established to lay out the horizontal alignment of the highway. Determine the precision of the traverse and revise the traverse based on the error present. Using a local coordinate system, Points 1 and 2 are known from existing control networks on each existing highway. Point 1 has a known position of 13,422.516, 7,661.229 and Point 2 has a known position of 15,434.186, 7,683.880. The local regulations require an accuracy of 1:10,000 or better.

(Continued)

EXAMPLE 3.2 Traverses—(Continued)

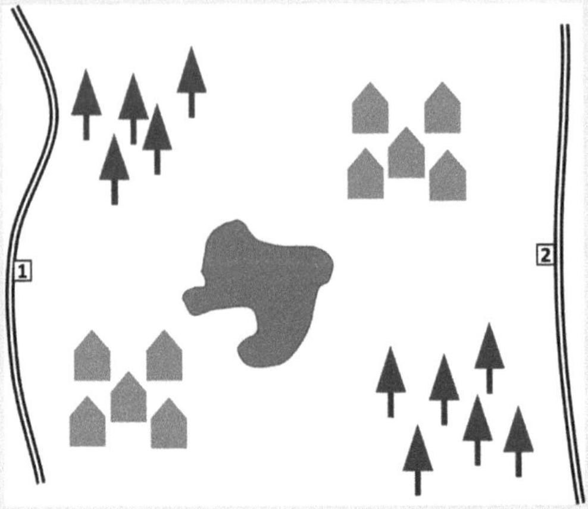

Solution

One option for completing the project is shown in the following figure. Points 1 and 2 are connected by three tangent highway sections that intersect at Points A and B. The bearing, from the perspective of traveling from Point 1 to Point 2, of each line is displayed on the figure. The length of line 1A is 807.6 ft, line AB is 645.3 ft, and line B2 is 613.2 ft.

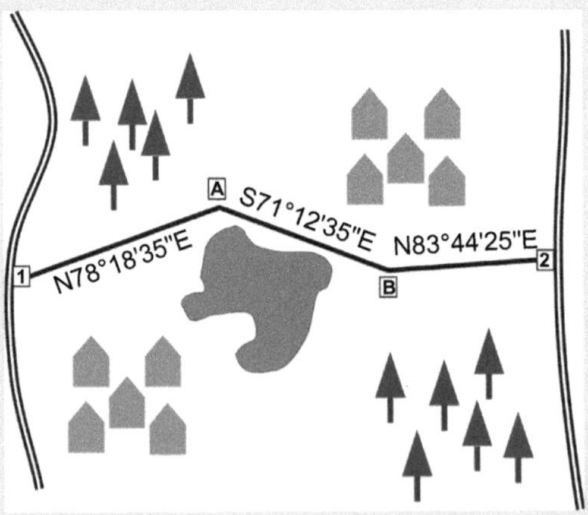

The control points from each of the existing highways can be used to check the accuracy of measurements made along the new highway.

(Continued)

EXAMPLE 3.2 Traverses—(Continued)

Course	Bearing	Length (ft)	Latitude (ft)	Departure (ft)
1A	N78°18′35″E	807.6	163.636	790.848
AB	S71°12′35″E	645.3	−207.854	610.908
B2	N83°44′25″E	613.2	66.861	609.544

Point	X (east/west)	Y (north/south)
1 (given)	13,422.516	7,661.229
A	14,213.364	7,824.865
B	14,824.272	7,616.011
2 (computed)	15,433.816	7,683.872
2 (given)	15,433.896	7,683.880

$$\text{Latitude error} = 7,683.880 - 7,683.872 = 0.008 \text{ ft}$$
$$\text{Departure error} = 15,433.896 - 15,433.816 = 0.08 \text{ ft}$$
$$\text{Traverse error} = \sqrt{(0.08)^2 + (0.008)^2} = 0.0804 \text{ ft}$$
$$\text{Accuracy} = \frac{0.0804 \text{ ft}}{2,066.1 \text{ ft}} \approx \frac{1 \text{ ft}}{25,698 \text{ ft}}$$

After comparing the computed coordinates with the given coordinates at Point 2, the measurements were determined to have an accuracy of 1 ft of error for every 25,698 ft. Therefore, these measurements meet the required accuracy threshold for this project of 1:10,000 ft. The corrected dimensions can now be determined. The sum of the latitude/departure correction factors for each line segment must total the latitude/departure error, with an opposite sign to eliminate the error. Therefore, the latitude correction for this example must be negative and the departure correction must be positive.

$$\text{Latitude correction}_{\text{line 1A}} = 0.008 \text{ ft}\left(\frac{807.6 \text{ ft}}{2066.1 \text{ ft}}\right) = -0.003$$

$$\text{Latitude correction}_{\text{line AB}} = 0.008 \text{ ft}\left(\frac{645.3 \text{ ft}}{2066.1 \text{ ft}}\right) = -0.003$$

$$\text{Latitude correction}_{\text{line B2}} = 0.008 \text{ ft}\left(\frac{613.2 \text{ ft}}{2066.1 \text{ ft}}\right) = -0.002$$

$$\text{Departure correction}_{\text{line 1A}} = 0.08 \text{ ft}\left(\frac{807.6 \text{ ft}}{2066.1 \text{ ft}}\right) = -0.03$$

$$\text{Departure correction}_{\text{line AB}} = 0.08 \text{ ft}\left(\frac{645.3 \text{ ft}}{2066.1 \text{ ft}}\right) = -0.03$$

$$\text{Departure correction}_{\text{line 2B}} = 0.08 \text{ ft}\left(\frac{613.2 \text{ ft}}{2066.1 \text{ ft}}\right) = -0.02$$

(Continued)

EXAMPLE 3.2 Traverses—(Continued)

$$\text{Corrected length}_{\text{line 1A}} = \sqrt{(163.633\ \text{ft})^2 + (790.818\ \text{ft})^2} = 807.570\ \text{ft}$$

$$\text{Corrected length}_{\text{line AB}} = \sqrt{(-207.857\ \text{ft})^2 + (610.878\ \text{ft})^2} = 645.272\ \text{ft}$$

$$\text{Corrected length}_{\text{line B2}} = \sqrt{(66.859\ \text{ft})^2 + (609.524\ \text{ft})^2} = 613.180\ \text{ft}$$

$$\text{Corrected bearing}_{\text{line 1A}} = \tan^{-1}\left(\frac{791.105\ \text{ft}}{162.633\ \text{ft}}\right) = \text{N}78^\circ 18' 34'' \text{E}$$

$$\text{Corrected bearing}_{\text{line AB}} = \tan^{-1}\left(\frac{610.948\ \text{ft}}{-207.851\ \text{ft}}\right) = \text{S}71^\circ 12' 21'' \text{E}$$

$$\text{Corrected bearing}_{\text{line B2}} = \tan^{-1}\left(\frac{609.581\ \text{ft}}{66.859\ \text{ft}}\right) = \text{N}83^\circ 44' 25'' \text{E}$$

Course	Corrected Latitude (ft)	Corrected Departure (ft)	Corrected Length (ft)	Corrected Bearing
1A	163.633	790.818	807.570	N78°18′34″E
AB	−207.857	610.878	645.272	S71°12′21″E
B2	66.859	609.524	613.180	N83°44′25″E

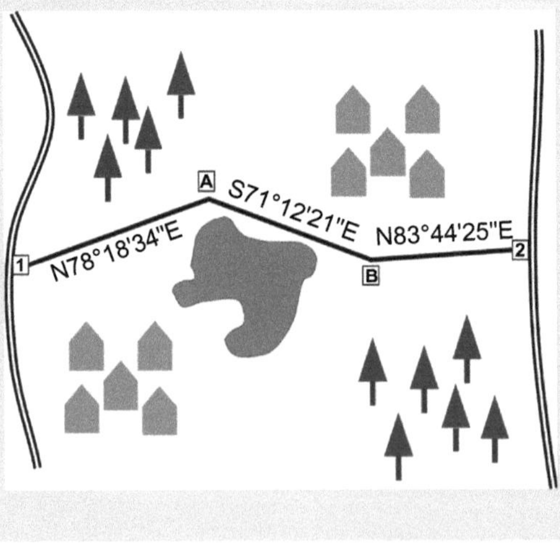

Horizontal Curve Fundamentals

Horizontal curves provide a smooth transition from one tangent to the next. From a safety perspective, horizontal curves are a particular focus because of the prevalence of collisions at curves relative to tangent sections. Proper design of a horizontal curve, including elements within a single curve and consistency of curvature along a highway, can reduce the likelihood of collisions. Figure 3.12 shows the four types of horizontal curves: (1) simple, (2) reverse, (3) compound, and (4) spiral. The radius is a term used to describe the sharpness or flatness of a curve; a large radius denotes a flat curve, while a small radius signifies a sharp curve. The simple curve, an arc of circle with a constant radius, plays an integral role in each of the other curve types. The spiral curve consists of a simple curve bounded by spiral transitions on each end. The spiral transition provides a smoother transition from the tangent segment, with an infinite radius, to the simple curve, with a fixed radius. The reverse curve system contains two simple curves in opposite directions that facilitate the transition from

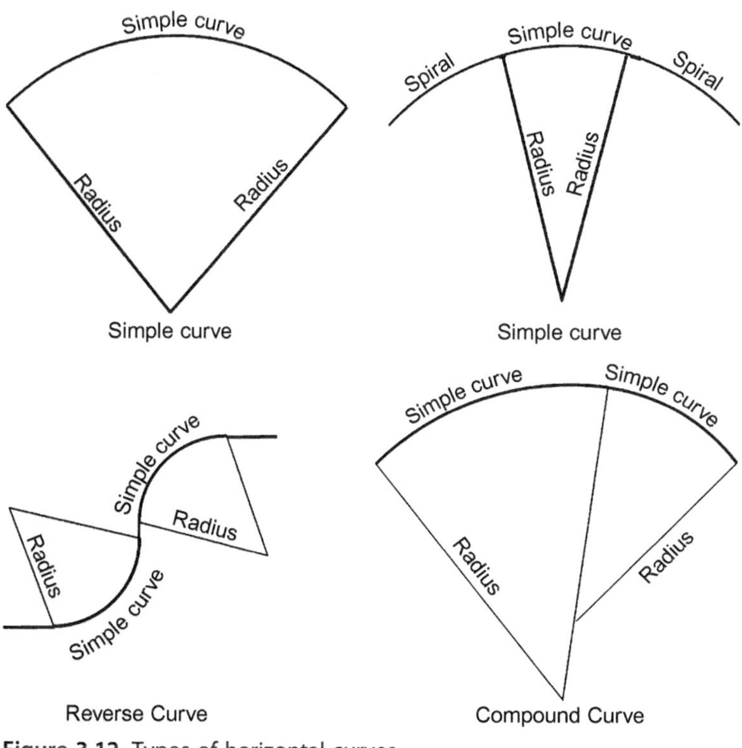

Figure 3.12 Types of horizontal curves.

two parallel highway segments and allows a designer to change the direction of the highway over a short distance. A compound curve also contains two curves, but each turns in the same direction and the radii are not equal. A compound curve is often used on loops at interchanges to reduce the amount of right of way required for the project; the larger radius curve will be placed adjacent to the higher speed highway.

Simple Horizontal Curves

In the English unit measurement system, the degree of curve is commonly used to describe a horizontal curve. There is a direct relationship between the degree of curve and radius based on the circumference of the unit circle (2π), degrees in a circle (360°), and the curvature present in arc with a length of 100 ft. In railroad engineering, a chord of 100 ft is used instead of an arc length of 100 ft.

$$\text{Degree of curve} = D_c = \left(\frac{360^\circ \times 100\ \text{ft}}{2\pi \times \text{Radius}}\right) = \frac{18{,}000}{\pi\ \text{Radius}}$$

Figure 3.13 shows the primary elements of a simple curve. These include:

- *Point of intersection* (*PI*): The location where the back and forward tangents intersect. The point of intersection is one of the stations on the preliminary traverse.
- *Intersecting angle* (*I*): The deflection angle at the PI. The intersecting angle can be computed from the preliminary traverse.
- *Radius* (*R*): The length of the line from the center of the circle to the perimeter, which describes the amount of curvature of the curve.

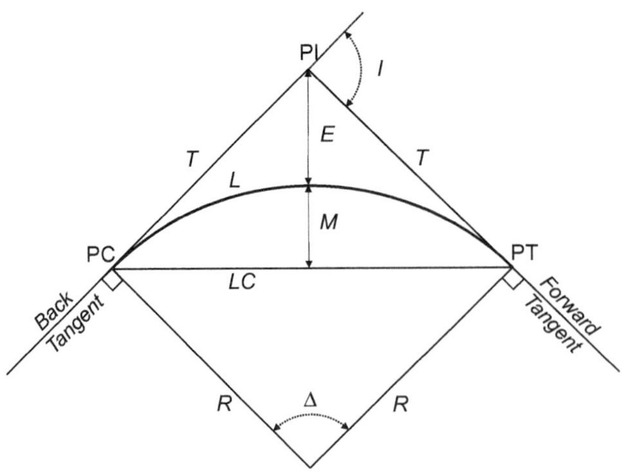

Figure 3.13 Simple horizontal curve.

- *Point of curvature (PC)*: The point where the circular curve begins and the highway leaves the tangent. The back tangent is at a right angle to the curve at this point.
- *Point of tangency (PT)*: The point where the circular curve ends and the highway returns to the tangent. The forward tangent is at a right angle to the curve at this point.
- *Curve length (L)*: The distance along the arc of the curve from the PC to the PT.
- *Tangent distance (T)*: The distance along the tangents from the PI to the PC or PT. Each tangent distance is equal for a simple curve.
- *Central angle (Δ)*: The angle formed by the two radii drawn from the center of the circle to the PC and PT. The central angle is equal in value to the intersecting angle.
- *Long chord (LC)*: The chord length, straight line segment, from the PC to the PT.
- *External distance (E)*: The distance from the PI to the midpoint of the curve.
- *Middle ordinate (M)*: The distance from the midpoint of the curve to the midpoint of the long chord.

$$R = \frac{18,000}{\pi D_c}$$

$$T = R \tan\left(\frac{\Delta}{2}\right)$$

$$L = 100\frac{\Delta}{D_c}$$

$$LC = 2R \sin\left(\frac{\Delta}{2}\right)$$

$$M = R\left(1 - \cos\frac{\Delta}{2}\right)$$

$$E = T \tan\left(\frac{\Delta}{4}\right)$$

$$\mathrm{PC} = \mathrm{PI} - T$$

$$\mathrm{PT} = \mathrm{PC} + L$$

Deflection angles can assist with the layout of a horizontal curve. This process is useful because it allows a curve to be developed incrementally with chords without locating the center of the circle. From the perspective of the PC, deflection angles can be measured relative to the back tangent, as shown in Figure 3.14. These deflection angles are equal to half of the

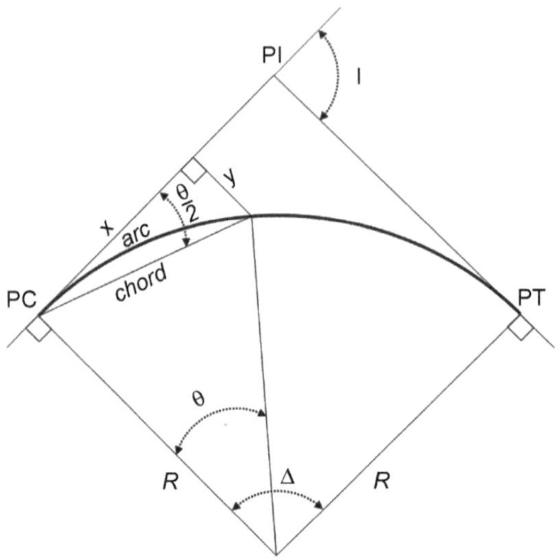

Figure 3.14 Deflection on simple horizontal curve.

central angle of the subtended arc. To compute the deflection to a point on a simple horizontal curve, the following additional elements are needed:

- *Arc length* (*l*): The length along the curve to the point of interest.
- *Chord*: Connects any two points on the circumference of a circle. For deflection measurements, the chord starts at the PC and connects in a straight line to the point of interest.
- *Deflection angle* ($\theta/2$): The angle enclosed by the tangent and the chord with a vertex at the PC. Each incremental deflection angle is equal to half of the central angle of the subtended arc. The deflection angle defined by the chord connecting the PC and the PT is half of the overall intersecting angle of the curve.
- *Central angle* (θ): The angle subtended by the arc.
- *Tangent length* (x): The distance along the tangent from the PC to the point on the tangent that is perpendicular to the point of interest.
- *Tangent offset* (y): The perpendicular distance from the tangent to the point on the curve.

$$1 = 100\left(\frac{\theta}{D_c}\right)$$

$$y = R - \sqrt{R^2 - x^2}$$
$$y = R(1 - \cos\theta)$$
$$x = R\sin\theta$$

EXAMPLE 3.3 Simple Horizontal Curve

Determine the values and stationing for a 6° curve to be placed on the following preliminary traverse. The length of line 1A is 807.6 ft and line AB is 645.3 ft. The beginning station of the project is at Point 1 and is assigned as 10 + 00.

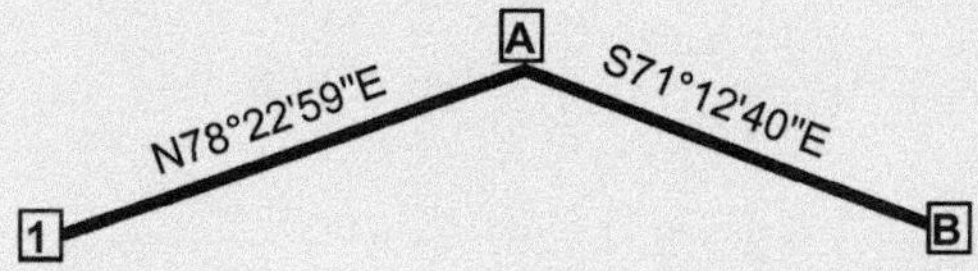

Solution

The intersecting angle—the amount of deflection between the back tangent and the forward tangent—can be found using the bearing angles from the given traverse:

$$\Delta = 180^\circ - 78^\circ 22' 59'' - 71^\circ 12' 40'' = 30^\circ 24' 21''$$

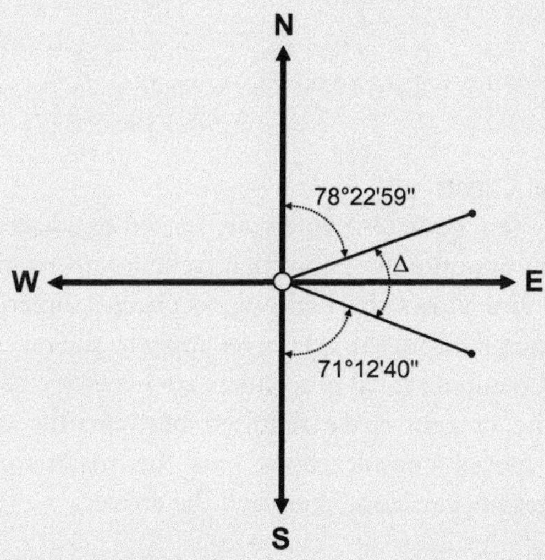

The station of the PI, at Point A, can be found using the project beginning station at Point 1 and the length of line 1A:

$PI_{station} = 10^{+}00 + 8^{+}07.60 = 18^{+}07.60$

The other curve elements of interest can be determined from the standard curve equations (additional decimal places are presented for each value for illustration purposes):

(Continued)

EXAMPLE 3.3 Simple Horizontal Curve—(Continued)

$$R = \frac{18,000}{\pi D_c} = \frac{18,000}{\pi 6^\circ} = 954.9297 \text{ ft}$$

$$T = R \tan\left(\frac{\Delta}{2}\right) = 954.9297 \text{ ft} \tan\left(\frac{30^\circ 24' 21''}{2}\right) = 259.501 \text{ ft}$$

$$L = 100\frac{\Delta}{D_c} = 100\left(\frac{30^\circ 24' 21''}{6^\circ}\right) = 506.7639 \text{ ft}$$

$$LC = 2R \sin\left(\frac{\Delta}{2}\right) = 2 \times 954.9297 \text{ ft} \sin\left(\frac{30^\circ 24' 21''}{2}\right) = 500.8383 \text{ ft}$$

$$M = R\left(1 - \cos\frac{\Delta}{2}\right) = 954.9297 \text{ ft}\left(1 - \cos\frac{30^\circ 24' 21''}{2}\right) = 33.42 \text{ ft}$$

$$E = T \tan\left(\frac{\Delta}{4}\right) = 259.501 \text{ ft} \tan\left(\frac{30^\circ 24' 21''}{4}\right) = 34.63 \text{ ft}$$

The stationing of the curve can be computed based on the preceding calculations and given information.

$$\text{PC} = \text{PI} - T = 18^{+}07.60 - 2^{+}59.50 = 15^{+}48.10$$
$$\text{PT} = \text{PC} + L = 15^{+}48.10 + 5^{+}06.76 = 20^{+}54.86$$

Reverse Horizontal Curves

A reverse curve system contains two closely spaced or adjacent simple curves with deflections in opposite directions that facilitates the transition from two parallel segments and allows the highway to change direction over a short distance. The principles from simple curves apply to the reverse curve system and no additional computational procedures are necessary to design a reverse curve system. The tangent space required between the curves varies by jurisdiction and special considerations must be made for superelevation transitions (changes in cross slope) between the curves.

Compound Horizontal Curves

A compound curve is composed of two simple curves of unequal radii that are adjacent to each other and turn in the same direction. A compound curve is also referred to as a two-centered compound curve because the differing radius values create two circular centers. A common application of a compound curve is on a loop at interchanges to reduce the amount of right of way required for the project as compared to one large radius. On an interchange exit and entrance loop, the larger radius curve will be placed adjacent to the freeway to facilitate higher speeds. Compound curves can also be used to tie two simple curves together in an instance of unusual site obstructions that must be avoided or when extending a highway at the point of an existing curve.

Because two simple curves are present, most of the terminology is similar to simple curves and the computations for each individual curve remain the same. As shown in Figure 3.15, there are two possibilities for compound curves: (1) the larger radius curve is adjacent to the PC or (2) the smaller radius curve is adjacent to the PC. The PI, PC, and PT refer to the locations created by the composite curve system, while each individual curve has these points, which can be differentiated by subscripts. The parameters of the larger radius curve are always denoted by the subscript 1 and the smaller radius parameters take the subscript 2. Because the radii are not equal, the tangent distance between the PI and PC and the PI and PT are no longer equal and the proper attribution of the tangent lengths, T_a or T_b, is essential. T_a is tangent length for side of the curve with the larger radius and T_b is the tangent length for the side of the curve with the smaller radius. The overall deflection angle for the compound curve is comprised of the deflection angle of each individual simple curve. The point of compound curvature (PCC) denotes the location where the two curves meet, that is, the PT of the first curve intersects with the PC of the second individual curve. The line that is

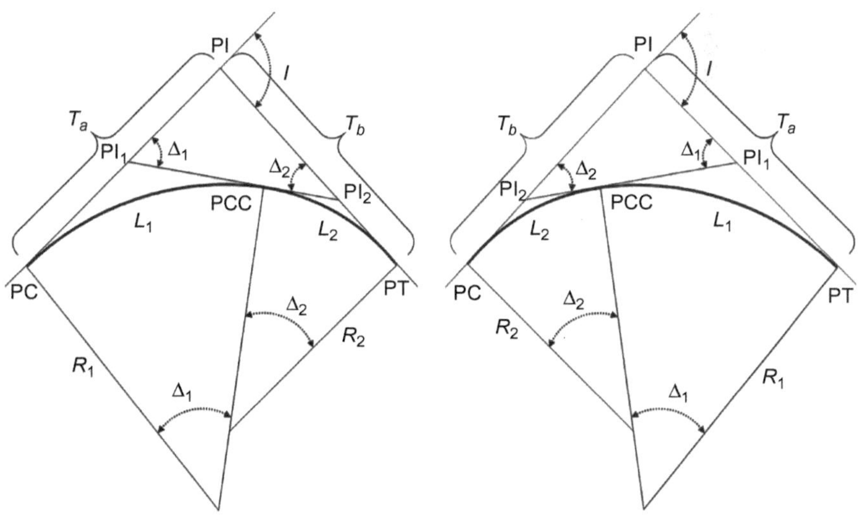

Figure 3.15 Types of compound curves.

tangent to the curve, perpendicular to the radius, at the PCC is an extension of the tangent for each of the individual curves (PI_1 to PCC to PI_2 and vice versa).

$$T_a = \frac{R_2 - (R_1 \cos I) + [(R_1 - R_2) \cos \Delta_2]}{\sin I}$$

$$T_b = \frac{R_1 - (R_2 \cos I) + [(R_1 - R_2) \cos \Delta_1]}{\sin I}$$

EXAMPLE 3.4 Compound Horizontal Curve

Determine the values and stationing for a compound curve to be placed on an exit loop of a freeway. The stationing of the loop construction will start at the point the taper departs the freeway with a station of 10 + 00. The PI of the compound curve will begin 2700 ft after the taper starts. The first curve has a radius of 2400 ft and a deflection angle of 33°45′ and the second curve has a radius of 1900 ft and a deflection angle of 25°15′.

Solution

The first curve in the direction of stationing is the larger curve; therefore, the following notations will be used for calculations.

$R_1 = 2400$ ft
$\Delta_1 = 33°45'$
$R_2 = 1900$ ft
$\Delta_2 = 25°15'$
$I = \Delta_1 + \Delta_2 = 33°45' + 25°15' = 59°00'$

Because the larger curve is adjacent to the PC, T_a is needed to determine the PC station based on the given PI station information. The length of each individual curve will also be needed to determine the stationing of the curve.

$$T_a = \frac{R_2 - (R_1 \cos I) + [(R_1 - R_2)\cos \Delta_2]}{\sin I}$$

$$T_a = \frac{1900' - (2400' \cos 59°00') + [(2400' - 1900')\cos 25°15']}{\sin 59°00'}$$

$$= 1302.12'$$

$$L_1 = \frac{\pi R_1 \Delta_1}{180} = \frac{\pi 2400' \times 33°45'}{180} = 1413.72'$$

$$L_2 = \frac{\pi R_2 \Delta_2}{180} = \frac{\pi 1900' \times 25°15'}{180} = 837.32'$$

(Continued)

EXAMPLE 3.4 Compound Horizontal Curve—(Continued)

$PI_{station} = 10^{+}00 + 27^{+}00 = 37^{+}00$

$PC_{station} = PI_{station} - T_a = 37^{+}00 - 13^{+}02.12 = 23^{+}97.88$

$PCC_{station} = PC_{station} + L_1 = 23^{+}97.88 + 14^{+}13.72 = 38^{+}11.60$

$PT_{station} = PCC_{station} + L_2 = 38^{+}11.60 + 8^{+}37.32 = 46^{+}48.92$

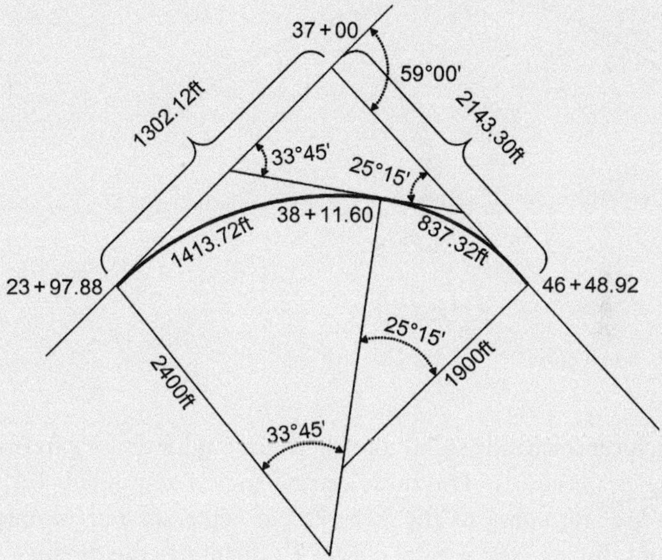

Spiral Horizontal Curves

A spiral curve consists of a simple curve bounded by spiral transitions on each end. The spiral transitions provide a transition from the tangent segment, which allows for the equilibrium of vehicles to be maintained throughout the curve in a designed manner (on simple curves, drivers tend to drive a spiral transition even if one is not designed into the curve). Geometrically, the spiral has a constantly changing radius and the transition connects the tangent section, which has an infinite radius, to the simple curve, which has a fixed radius. Figure 3.16 shows the basic components of a spiral curve. The point of intersection (PI) remains as the intersection between the back and forward tangents. However, four points of interest are present along the curve, due to the presence of the spiral transitions. The central angle of the spiral curve is comprised of three components, Δ_c, which represents the angle inscribed by the simple curve, and two angles

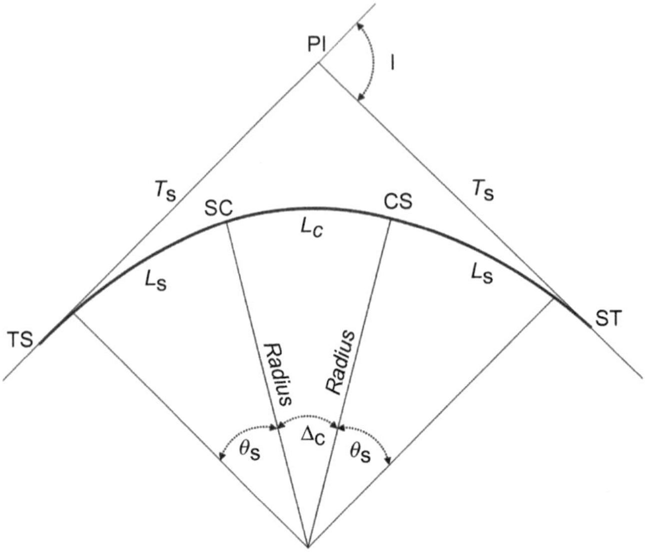

Figure 3.16 Spiral curve.

of θ_s, which represent the spiral transitions (in a symmetric spiral curve, the two θ_s angles are equal). The three central angle components (Δ_c and two θ_s angles) are still equal to the intersecting angle, as with simple curves. The tangent distance for a spiral curve, T_s, quantifies the distance from the PI to the start of the spiral curve (and the distance from the PI to the end of the curve). From the perspective of increasing stationing, going from left to right along in the figure, these points include:

- *Tangent to spiral point* (*TS*): The point where the tangent ends and the first spiral transition begins. At the TS point, the spiral transition has an infinite radius (matching the radius of the tangent), which then constantly decreases along the transition toward the simple curve.
- *Spiral to curve point* (*SC*): The point where the spiral transition ends and the simple curve begins. The distance from the TS to SC point is the length of the spiral, L_S. At the SC point, the spiral transition has a radius equal to the simple curve, thereby completing the transition from the infinite radius to the fixed radius. The curve has a consistent, fixed radius until the next spiral transition is reached.
- *Curve to spiral point* (*CS*): The point where the simple curve ends and the second spiral transition begins. The distance from the SC to CS point is the length of the simple curve, L_C. At the SC point, the spiral transition has a radius equal to the simple curve, which then constantly increases along the transition toward the tangent.

- *Spiral to tangent point* (*ST*): The point where the second spiral ends and the tangent begins. The distance from the CS to ST point is the length of the spiral, L_S. At the ST point, the spiral transition has an infinite radius (matching the radius of the tangent).

The design of the spiral curves begins with the selection of an appropriate radius for a simple horizontal curve. To accommodate the spiral transitions, the length of the curve is shorter, which shrinks the inscribed angle, (i.e., Δ_c must be less than the intersecting angle). A shift of the simple curve, known as the offset, toward the center of the circle is also necessary to allow for the spiral transitions. The lateral offset is a key element of spiral curves and is measured as the perpendicular distance from the tangent at the point of the PC (if a simple horizontal curve is assumed) to the point of the shifted simple curve. A driver naturally traverses a spiraled path, which is offset from his or her original path by 0.66 ft (0.20 m); this value serves as the minimum offset for designing a spiral curve (AASHTO, 2011, p. 3–72). If the computations for a spiral curve result in an offset that is less than the minimum value, a spiral is not needed for the given curve, because the driver's natural path is sufficient for making the transition from the tangent section to the simple curve. The spiral transition is also often used as the transition of superelevation from normal crown to design superelevation. Figure 3.17 is a detailed diagram of spiral transition elements from the TS to SC; equivalent relationships exist between the CS and ST for a symmetric spiral curve.

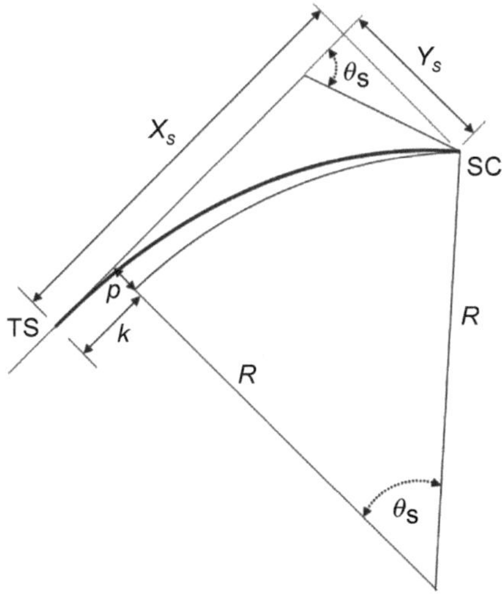

Figure 3.17 Spiral curve transition elements (TS to SC).

$$\theta_s = \frac{L_s D_c}{200}$$
$$\Delta = \Delta_c + 2\theta_s$$
$$L_c = 100\frac{\Delta_c}{D_c}$$
$$Y_s = \frac{L_s^2}{6\,R}$$
$$X_s = L_s - \left(\frac{Y_s^2}{2\,L_s}\right)$$
$$p = Y_s - R(1 - \cos\theta_s)$$
$$k = X_s - R\sin\theta_s$$
$$T_s = (R + p)\tan\left(\frac{\Delta}{2}\right) + k$$

where

Δ = central angle, which is equal in value to the intersecting angle

Δ_c = simple curve angle, which is the portion of the central angle assigned to the simple curve

θ_s = spiral angle, which is the portion of the central angle assigned to the spiral transition (for a symmetric spiral curve, both θ_s values are equal)

k = distance from the beginning of the spiral (TS or ST) to the location where the PC or PT would be located if a simple curve was used

p = lateral offset, measured as the perpendicular distance from the tangent at the PC or PT (if a simple horizontal curve is assumed) to the point of the shifted simple curve

X_s = tangent length to the end of the spiral

Y_s = tangent offset, lateral distance, from the tangent to the simple curve at the end of the spiral

For metric system measurements, the following equations should be used in place of the previous equations, which utilize the degree of curvature. Lengths and radii in these equations should be measured in meters.

$$\theta_s = \frac{90L_s}{\pi R}$$
$$L_c = \frac{\pi R\Delta_c}{180}$$

EXAMPLE 3.5 Spiral Horizontal Curve

Determine the stationing of key points for a spiral curve with an intersecting angle of 40°, a 4° curve, and a length of spiral of 250 ft. The curve has a PI at station $45^{+}89.22$.

Solution

The individual components of the spiral curve must be computed first.

$$\theta_s = \frac{L_s D_c}{200} = \frac{250\text{ ft} \times 4^\circ}{200} = 5^\circ$$

$$\Delta_c = \Delta - 2\,\theta_s = 40^\circ - 2(5^\circ) = 30^\circ$$

$$L_c = 100\frac{30^\circ}{4^\circ} = 750\text{ ft}$$

$$R = \frac{18,000}{\pi\, D_C} = \frac{18,000}{\pi\, 4^\circ} = 1432.39\text{ ft}$$

$$Y_s = \frac{L_s^2}{6R} = \frac{250\text{ ft}^2}{6 \times 1432.39\text{ ft}} = 7.272\text{ ft}$$

$$X_s = L_s - \left(\frac{Y_s^2}{2L_s}\right) = 250 - \left(\frac{7.272\text{ ft}^2}{2 \times 250\text{ ft}}\right) = 249.957\text{ ft}$$

$$p = Y_s - R(1 - \cos\theta_s) = 7.272\text{ ft} - 1432.39\text{ ft}\,(1 - \cos 5^\circ) = 1.82\text{ ft}$$

$p > 0.66$, therefore, the effect of constructing a spiral curve is justified in the impact it will have on vehicles.

$$k = X_s - R\sin\theta_s = 249.957\text{ ft} - 1432.39\text{ ft}\,(\sin 5^\circ) = 125.12\text{ ft}$$

$$T_s = (R + p)\tan\left(\frac{\Delta}{2}\right) + k = (1432.39\text{ ft} + 1.82\text{ ft})\tan\left(\frac{40^\circ}{2}\right) + 125.12\text{ ft}$$

$$= 647.13\text{ ft}$$

The stationing can be developed using the basic parameters of this spiral curve.

PI station $= 45^{+}89.22$

TS station $=$ PI station $- \; T_s = 45^{+}89.22 - 6^{+}47.13 = 39^{+}42.09$

SC station $=$ TS station $+ L_s = 39^{+}42.09 + 2^{+}50 = 41^{+}92.09$

CS station $=$ SC station $+ L_c = 41^{+}92.09 + 7^{+}50 = 49^{+}42.09$

ST station $=$ CS station $+ L_s = 49^{+}42.09 + 2^{+}50 = 51^{+}92.09$

(Continued)

EXAMPLE 3.5 Spiral Horizontal Curve—(Continued)

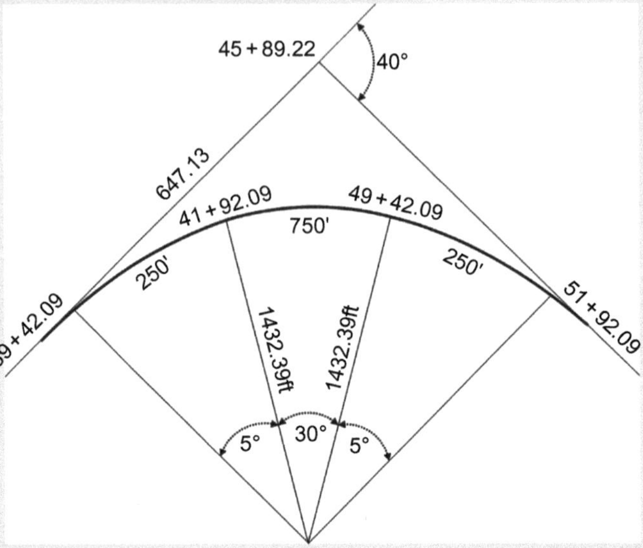

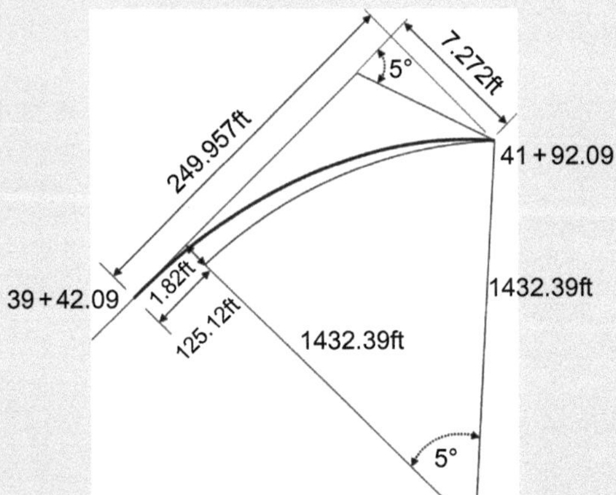

Superelevation

A straight highway segment is typically designed with a normal crown for the purpose of providing sufficient drainage of water off the surface of the highway. The term *normal crown* is used to describe the type of cross slope and the percentage of the slope. Cross slope can be defined in percent (%) or decimal form (ft/ft). For example, a 2% cross slope represents a change

Figure 3.18 (a) Normal crown, (b) superelevation.

in transverse elevation of 0.02 ft for each 1.0 ft of pavement width. Normal crown has a rooftop shape that peaks in the center of the roadway and falls away from the centerline at a typical rate of 2%. This means that for a highway with 12-ft lanes, the edge of the roadway will be 0.24 ft below the elevation of the centerline. Most drivers do not notice this amount of cross slope and this cross slope does not typically affect the ride quality of the vehicle. However, normal crown is essential for reducing the likelihood of hydroplaning by directing water off of the traveled way. Figure 3.18 presents the typical implementation of normal crown in the cross-section view as would be experienced by a driver traveling along the highway.

Horizontal curves can be also superelevated, with an elevated cross slope along the width of the pavement, to allow vehicles to travel through the curve at higher speeds, as shown in Figure 3.18. The cross section of the pavement must be rotated, about the centerline or edgeline of a highway, to create the superelevation. Rotation about the edgeline of a highway can be necessary when underground utilities or water features could be problematic; rotating the pavement about the edgeline ensures that no part of the pavement will go lower in elevation than a section in normal crown, unlike centerline rotation, which can push the inside edge lower than the normal crown section. The superelevation rate must be selected so that equilibrium is maintained for the vehicles based on the design speed and radius of the curve. If a design superelevation of 6% is selected, on a two-lane highway with 12-ft lanes, the outside edge of pavement will be 1.44-ft higher in elevation than the inside edge of pavement. The length of the spiral transition of a spiral curve is often used to accomplish the transition from normal crown to the designed superelevation. The transition distance consists of two lengths: tangent runout and superelevation runoff, as shown in Figure 3.19 with a horizontal curve

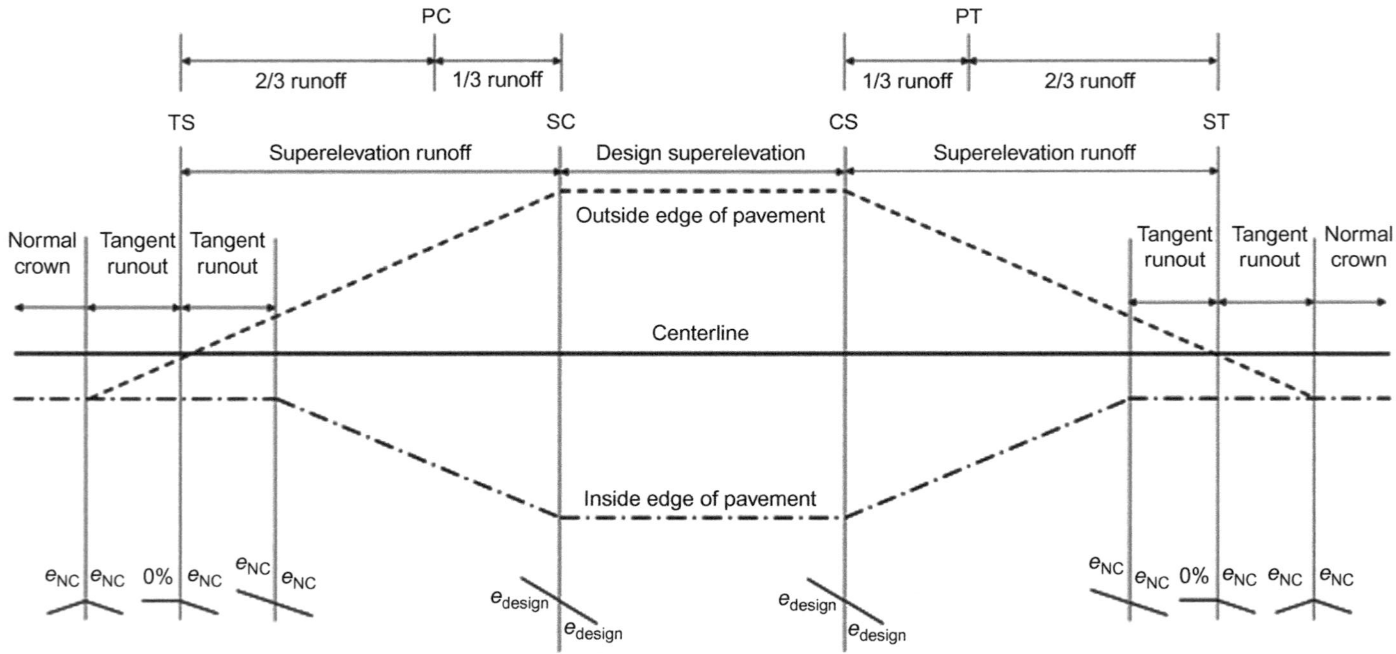

Figure 3.19 Relative gradient profile and cross-section view of centerline rotation.

that turns to the right and moves from Point A through Point H. If a simple curve is used, the transition from normal crown to the design superelevation must still be provided to allow for the safe travel of drivers. The application of the superelevation transition should apply to the majority of the transition prior to reaching the point of curvature; a common practice is to apply two-thirds of the runoff prior to the curve and one-third on the curve, as shown in the figure. The figure presents the relative gradient profile, and corresponding key locations in the cross-section view (with the outside travel way on the left portion and inside travel way on the right portion) are displayed beneath the relative gradient profile. Figure 3.19 also includes the following points of interest when transitioning from normal crown to design superelevation and back to normal crown:

- Point A signifies the beginning of the transition from normal crown toward design superelevation.
- Point B is the point where the adverse crown is removed; the outside edge of pavement is at the same elevation as the centerline. The distance from Point A to Point B is the tangent runout length. Point B represents the TS point for a spiral curve or the beginning of the superelevation runoff for a simple and spiral curve.
- Point C is the point of reverse crown; the entire pavement width has a cross slope equal to the normal crown slope. The distance from Point B to Point C is the tangent runout length. At Point C, the inside begins to rotate for the first time and will rotate at the same rate and magnitude relative to the centerline as this outside edge of pavement.
- Point D is the point at which the design superelevation has been reached, and the entire highway width has a cross slope equal to the design superelevation. Point D represents the SC point as the spiral is now complete and the simple curve begins. The distance from Point B to Point D is the superelevation runoff.
- Point E is the final point for the simple curve and the location where the design superelevation will begin the transition back toward normal crown. The entire pavement width will rotate simultaneously about the centerline. Point E is the CS point signifying the end of the simple curve and beginning of the spiral transition.

- Point F is the reverse crown point; beyond this point, no more rotation of the inside edge of pavement will occur.
- Point G is the adverse crown removed point and represents the ST point for a spiral curve and the end of the superelevation runoff for a simple curve and spiral curve. The distance from Point F to Point G is the tangent runout length.
- Point H is end of the transition back to normal crown. The distance from Point G to Point H is the tangent runout length.

Forces on a Vehicle in a Horizontal Curve

On a straight highway section, the lateral forces on a vehicle are negligible. On a horizontal curve, the vehicle is subjected to a centripetal acceleration that acts toward the center of the curve. This acceleration is sustained by the friction between the tires and the pavement, as well as the weight of the vehicle. The friction force acts along the cross slope of the roadway in a perpendicular direction from the normal force. The weight of the vehicle is a function of the mass of the vehicle and gravity. The superelevation counteracts the tendency of a vehicle traveling at high speeds to slide out of the curve. All forces acting on the vehicle must be in equilibrium for the vehicle to resist the tendency to slide up or down the pavement while traveling through the curve. The components include the weight (W), the side frictional resistance (F), and the normal force (N). The friction force is equal to the side friction factor, f, multiplied by the normal force (Figure 3.20).

There is only one nonzero acceleration present, which is $\frac{V^2}{R}$ toward the center of the curve. Therefore, the forces in the horizontal direction equal mass $\times$ acceleration, while the vertical forces equal 0. These relationships

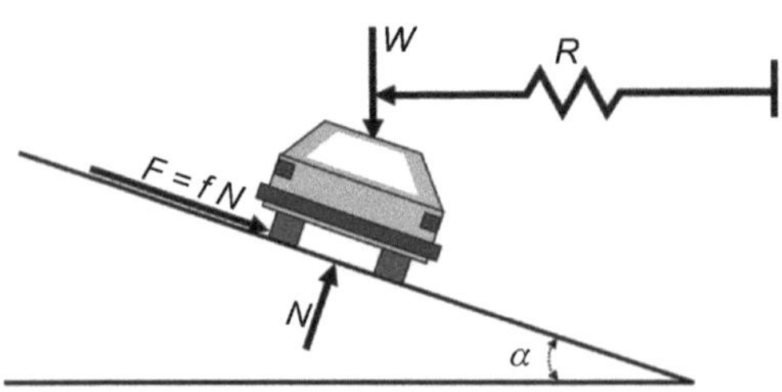

Figure 3.20 Vehicular forces in a horizontal curve. W = weight (mg); N = normal force (all 4 tires); F = friction force (all 4 tires) = fN (at maximum grip); f = side friction factor; α = amount of banking; R = radius of curve.

for horizontal and vertical forces can be used to develop two equations that will be used to solve for the radius of the curve, as shown in the following.

Horizontal forces ($\sum F_{\text{Horizontal}} = ma$):

$$F\cos\alpha + N\sin\alpha = \text{ma}$$
$$(fN)\cos\alpha + N\sin\alpha = \frac{W \times V^2}{g \times R}$$
$$N(f\cos\alpha + \sin\alpha) = \frac{W \times V^2}{g \times R}$$

Vertical forces ($\sum F_{\text{vertical}} = 0$):

$$-F\sin\alpha + N\cos\alpha - W = 0$$
$$-(fN)\sin\alpha + N\cos\alpha = W$$
$$N(-f\sin\alpha + \cos\alpha) = W$$
$$N = \frac{W}{\cos\alpha - f\sin\alpha}$$

The equation for the radius can be determined by substituting the equation from the vertical forces into the equation for the horizontal forces. The equation is presented in terms of English units, with the radius in feet and speed in miles per hour.

$$\frac{W}{\cos\alpha - f\sin\alpha}(f\cos\alpha + \sin\alpha) = \frac{W \times V^2}{g \times R}$$
$$\frac{f\cos\alpha + \sin\alpha}{\cos\alpha - f\sin\alpha} = \frac{V^2}{g \times R}$$
$$\frac{f + \tan\alpha}{1 - f\tan\alpha} = \frac{V^2}{g \times R}(\text{Let } e = \tan\alpha)$$
$$\frac{f + e}{1 - f\,e} = \frac{V^2}{g \times R}$$

For highway design purposes, it is common to solve this relationship for the radius, so it can be directly applied to highway design.

$$R = \frac{V^2}{g \times R}\left(\frac{1 - f\,e}{f + e}\right)$$

For English units (R in ft, V in mph):

$$R = \frac{V^2}{15}\left(\frac{1 - f\,e}{f + e}\right)$$

The previous equation includes the term of $e \times f$. However, $e \times f$ is very small and can be assumed as 0 for highway design purposes. Therefore, the final equation relating the radius, speed, superelevation, and side friction factor is as follows. Due to the wide variations of vehicle speeds on a curve, the side friction factor applied in design is usually substantially less than the coefficient of friction at the point where a vehicle would begin to skid. Many factors within the variables of this equation play into this overall relationship including the weather and the type and condition of the pavement and tires.

$$R = \frac{V^2}{15\,(e+f)}$$

where

R = radius of curve, ft
V = vehicle speed, mph
f = side friction factor
e = superelevation rate (ft/ft)

Horizontal Curve Design Process

The fundamentals of horizontal curves are important to explore for the purposes of implementing and computing necessary components of horizontal curves. However, it is equally important to be able to generate horizontal curve designs that conform to standard engineering practice. The standard professional reference for horizontal curve design parameters is *A Policy on Geometric Design of Highways and Streets*, commonly referred to as the "Green Book," which is published by the American Association of State Highway and Transportation Officials (AASHTO). Although deviations from the recommended values presented in the Green Book may be necessary given site conditions or restrictions, designs should be consistent with AASHTO recommendations.

To meet the goal of providing a highway that matches driver's expectations, the design of horizontal curves on a road should be consistent with curve parameters to the extent possible; similar curve radii and lengths should be used. Additionally, the safety performance of a highway is generally better when evenly spaced, consistently designed horizontal curves are present, as opposed to highways that have occasional, but sharp, horizontal curves. Two simple curves in the same direction that are separated by a short tangent section, known as broken-back curves, should be avoided due to the potential for driver confusion and design challenges.

The horizontal alignment of a roadway must also consider impacts based on the associated vertical alignment and cross-sectional values at any given point. An iterative process may be needed to prevent unnecessary negative impacts on nearby features and to properly align the horizontal and vertical components. For instance, a vertical curve that is followed immediately by a horizontal curve might pose a safety risk to drivers.

The general design process for a horizontal curve involves the following steps:

1. Apply relevant design controls, particularly maximum superelevation rates for the highway's jurisdiction and any other horizontal curve–specific elements identified by the controlling agency. Maximum superelevation rates, e_{max}, are typically between 4% and 12%, with higher superelevation rates accepted on higher mobility highways. However, highways that are susceptible to icy conditions may need to be limited to 8% or less to reduce the likelihood of a vehicle sliding (toward the inside of the curve) into adjacent lanes or off the highway in those instances. Additionally, an intersection in a horizontal curve may need to have limited superelevation rates to accommodate the crossing and turning traffic volumes.
2. Conduct traverse activities by choosing tangents (angles and distances), measuring the angle of intersection between successive tangents, and developing stationing along the points of intersection of the traverse. The fundamentals of these calculations are built into contemporary highway design software packages and are computed for the user.
3. Determine an appropriate minimum radius based on the design speed and e_{max} for the type of highway using Table 3.3 (AASHTO, 2011). However, a larger radius value will provide additional safety and should be used if possible.
4. Determine whether a simple curve or spiral curve should be used based on the impact of introducing a spiral transition. Table 3.4 (AASHTO, 2011) shows the maximum radius that would necessitate the use of a spiral transition. Horizontal curves with radius values smaller than the maximum would benefit from a spiral transition, while the benefit of the transition is negligible for larger radius curves.
5. Determine the design superelevation, e_{design}, based on the selected radius, if a value larger than the minimum radius is chosen. The values shown in Table 3.3 incorporate a balancing of forces on the vehicle to maintain equilibrium, but when a larger radius is selected, the curve is flatter and the pavement does not require as much superelevation.

Table 3.5 (AASHTO, 2011) presents the minimum radii and corresponding e_{design} with an e_{max} of 6%, and Table 3.6 (AASHTO, 2011) presents the same data for an e_{max} of 8%. Refer to the AASHTO Green Book for additional e_{max} values from 2% to 12% and for metric system units.

6. Determine the minimum superelevation runoff length, L_r, based on the design speed, the number of lanes rotated, and the design superelevation from the following equation and corresponding reference data. The minimum values presented in Table 3.7 are appropriate for a two-lane highway that is rotated about the centerline, corresponding to one lane being rotated. The superelevation runoff length provides for the transition of the cross section from adverse crown removed (outside edge of pavement is flat) to design superelevation.

$$L_r = \frac{(LW)(n)(n_{factor})e_{design}}{\Delta} \quad \text{(AASHTO Equation 3 − 23)}$$

where

LW = lane width of one lane of traffic, in feet (typically 12 ft)

e_{design} = design superelevation rate

Δ = maximum relative gradient, in percent, as specified in Table 3.8, representing the rate of transition of the cross slope of the pavement

n = number of lanes rotated (1 for a two-lane highway rotated about the centerline, 2 for a four-lane highway rotated about the centerline, etc.)

n_{factor} = adjustment factor for the number of lanes rotated, as computed from the following equation (1 for 1 lane rotated, 0.83 for 1.5 lanes, 0.75 for 2 lanes, 0.70 for 2.5 lanes, 0.67 for 3 lanes, and 0.64 for 3.5 lanes), adjusts for the number of lanes to reduce the potential for excessive runoff lengths (AASHTO, 2011, Table 3.16)

$$n_{factor} = \frac{1 + 0.5\,(n-1)}{n} \quad \text{(AASHTO Table 3 − 16)}$$

7. Compute the runout length, L_t, based on the normal crown, design superelevation, and superelevation runoff. The runout length provides for the transition from normal crown, e_{NC}, to adverse crown removed. Therefore, the sum of the runout length and runoff length complete the transition of the cross section from normal crown to e_{design}.

Equation 3-24 from the AASHTO Green Book defines the equation for L_t in the following equation with cross slopes in percent and lengths in feet.

$$L_t = \frac{e_{\text{NC}}}{e_{\text{design}}} L_r$$

8. Determine necessary curve features for the layout of the horizontal curve, as appropriate for a simple curve or spiral curve, as determined previously. These elements should include the stationing of key points, curve length, tangent lengths, etc.

   a. For a spiral curve, the design process is complete; employ the curve parameters and spiral transitions for a typical spiral curve. Figure 3.19 shows the method for accomplishing the superelevation transition.

   b. For a simple curve, the transition from normal crown to design superelevation must still be accounted for in the design process. As with spiral curves, the runoff and runout lengths are used to accomplish the transition. Table 3.9 (AASHTO, 2011, Table 3-18) presents the portion of superelevation runoff that should be located prior to the curve, based on design speed and number of lanes rotated. Another common practice is to apply two-thirds of the runoff prior to the curve and one-third on the curve. The majority of the runoff should be applied to the highway prior to the point of curvature. Figure 3.19 shows the method for accomplishing the superelevation transition.

Table 3.3 Minimum radii

Design speed (mph)	Maximum *e* (%)	Maximum *f*	Radius (ft)
10	4	0.38	16
15	4	0.32	42
20	4	0.27	86
25	4	0.23	154
30	4	0.20	250
35	4	0.18	371
40	4	0.16	533
45	4	0.15	711
50	4	0.15	926
55	4	0.13	1190
60	4	0.12	1500

(*Continued*)

Table 3.3 (Continued)

Design speed (mph)	Maximum *e* (%)	Maximum *f*	Radius (ft)
10	6	0.38	15
15	6	0.32	39
20	6	0.27	81
25	6	0.23	144
30	6	0.20	231
35	6	0.18	340
40	6	0.16	485
45	6	0.15	643
50	6	0.15	833
55	6	0.13	1060
60	6	0.12	1330
65	6	0.11	1660
70	6	0.10	2040
75	6	0.09	2500
80	6	0.08	3050
10	8	0.38	14
15	8	0.32	38
20	8	0.27	76
25	8	0.23	134
30	8	0.20	214
35	8	0.18	314
40	8	0.16	444
45	8	0.15	587
50	8	0.15	758
55	8	0.13	960
60	8	0.12	1200
65	8	0.11	1480
70	8	0.10	1810
75	8	0.09	2210
80	8	0.08	2670
10	10	0.38	14
15	10	0.32	36
20	10	0.27	72
25	10	0.23	126
30	10	0.20	200
35	10	0.18	292
40	10	0.16	410
45	10	0.15	540
50	10	0.15	694
55	10	0.13	877
60	10	0.12	1090
65	10	0.11	1340

(Continued)

Table 3.3 (Continued)

Design speed (mph)	Maximum *e* (%)	Maximum *f*	Radius (ft)
70	10	0.10	1630
75	10	0.09	1970
80	10	0.08	2370
10	12	0.38	13
15	12	0.32	34
20	12	0.27	68
25	12	0.23	119
30	12	0.20	188
35	12	0.18	272
40	12	0.16	381
45	12	0.15	500
50	12	0.15	641
55	12	0.13	807
60	12	0.12	1000
65	12	0.11	1220
70	12	0.10	1480
75	12	0.09	1790
80	12	0.08	2130

From American Association of State Highway and Transportation Officials (AASHTO), 2011. A Policy on Geometric Design of Highways and Streets, 6th ed. Table 3–7.

Table 3.4 Maximum radius for spiral curves

Design speed (mph)	Maximum radius (ft)
15	114
20	203
25	317
30	456
35	620
40	810
45	1025
50	1265
55	1531
60	1822
65	2138
70	2479
75	2846
80	3238

From American Association of State Highway and Transportation Officials (AASHTO), 2011. A Policy on Geometric Design of Highways and Streets, 6th ed. Table 3–20.

Table 3.5 Radii values for 6% e_{max}

e (%)	Design speed (V_{design})													
	15 mph	**20 mph**	**25 mph**	**30 mph**	**35 mph**	**40 mph**	**45 mph**	**50 mph**	**55 mph**	**60 mph**	**65 mph**	**70 mph**	**75 mph**	**80 mph**
NC	868	1580	2290	3130	4100	5230	6480	7870	9410	11,100	12,600	14,100	15,700	17,400
RC	614	1120	1630	2240	2950	3770	4680	5700	6820	8060	9130	10,300	11,500	12,900
2.2	543	991	1450	2000	2630	3370	4190	5100	6110	7230	8200	9240	10,400	11,600
2.4	482	884	1300	1790	2360	3030	3770	4600	5520	6540	7430	8380	9420	10,600
2.6	430	791	1170	1610	2130	2740	3420	4170	5020	5950	6770	7660	8620	9670
2.8	384	709	1050	1460	1930	2490	3110	3800	4580	5440	6200	7030	7930	8910
3.0	341	635	944	1320	1760	2270	2840	3480	4200	4990	5710	6490	7330	8260
3.2	300	566	850	1200	1600	2080	2600	3200	3860	4600	5280	6010	6810	7680
3.4	256	498	761	1080	1460	1900	2390	2940	3560	4250	4890	5580	6340	7180
3.6	209	422	673	972	1320	1740	2190	2710	3290	3940	4540	5210	5930	6720
3.8	176	358	583	864	1190	1590	2010	2490	3040	3650	4230	4860	5560	6320
4.0	151	309	511	766	1070	1440	1840	2300	2810	3390	3950	4550	5220	5950
4.2	131	270	452	684	960	1310	1680	2110	2590	3140	3680	4270	4910	5620
4.4	116	238	402	615	868	1190	1540	1940	2400	2920	3440	4010	4630	5320
4.6	102	212	360	555	788	1090	1410	1780	2210	2710	3220	3770	4380	5040
4.8	91	189	324	502	718	995	1300	1640	2050	2510	3000	3550	4140	4790
5.0	82	169	292	456	654	911	1190	1510	1890	2330	2800	3330	3910	4550
5.2	73	152	264	413	595	833	1090	1390	1750	2160	2610	3120	3690	4320
5.4	65	136	237	373	540	759	995	1280	1610	1990	2420	2910	3460	4090
5.6	58	121	212	335	487	687	903	1160	1470	1830	2230	2700	3230	3840
5.8	51	106	186	296	431	611	806	1040	1320	1650	2020	2460	2970	3560
6.0	39	81	144	231	340	485	643	833	1060	1330	1660	2040	2500	3050

NC = normal crown, RC = reverse crown.
From American Association of State Highway and Transportation Officials (AASHTO), 2011. A Policy on Geometric Design of Highways and Streets, 6th ed. Table 3–9.

Table 3.6 Radii values for 8% e_{max}

e (%)	Design speed (V_{design})													
	15 mph	**20 mph**	**25 mph**	**30 mph**	**35 mph**	**40 mph**	**45 mph**	**50 mph**	**55 mph**	**60 mph**	**65 mph**	**70 mph**	**75 mph**	**80 mph**
NC	932	1640	2370	3240	4260	5410	6710	8150	9720	11,500	12,900	14,500	16,100	17,800
RC	676	1190	1720	2370	3120	3970	4930	5990	7150	8440	9510	10,700	12,000	13,300
2.2	605	1070	1550	2130	2800	3570	4440	5400	6450	7620	8600	9660	10,800	12,000
2.4	546	959	1400	1930	2540	3240	4030	4910	5870	6930	7830	8810	9850	11,000
2.6	496	872	1280	1760	2320	2960	3690	4490	5370	6350	7180	8090	9050	10,100
2.8	453	796	1170	1610	2130	2720	3390	4130	4950	5850	6630	7470	8370	9340
3.0	415	730	1070	1480	1960	2510	3130	3820	4580	5420	6140	6930	7780	8700
3.2	382	672	985	1370	1820	2330	2900	3550	4250	5040	5720	6460	7260	8130
3.4	352	620	911	1270	1690	2170	2700	3300	3970	4700	5350	6050	6800	7620
3.6	324	572	845	1180	1570	2020	2520	3090	3710	4400	5010	5680	6400	7180
3.8	300	530	784	1100	1470	1890	2360	2890	3480	4140	4710	5350	6030	6780
4.0	277	490	729	1030	1370	1770	2220	2720	3270	3890	4450	5050	5710	6420
4.2	255	453	678	955	1280	1660	2080	2560	3080	3670	4200	4780	5410	6090
4.4	235	418	630	893	1200	1560	1960	2410	2910	3470	3980	4540	5140	5800
4.6	215	384	585	834	1130	1470	1850	2280	2750	3290	3770	4310	4890	5530
4.8	193	349	542	779	1060	1390	1750	2160	2610	3120	3590	4100	4670	5280
5.0	172	314	499	727	991	1310	1650	2040	2470	2960	3410	3910	4460	5050
5.2	154	284	457	676	929	1230	1560	1930	2350	2820	3250	3740	4260	4840

(*Continued*)

Table 3.6 (Continued)

e (%)	Design speed (V_{design})													
	15 mph	20 mph	25 mph	30 mph	35 mph	40 mph	45 mph	50 mph	55 mph	60 mph	65 mph	70 mph	75 mph	80 mph
5.4	139	258	420	627	870	1160	1480	1830	2230	2680	3110	3570	4090	4640
5.6	126	236	387	582	813	1090	1390	1740	2120	2550	2970	3420	3920	4460
5.8	115	216	358	542	761	1030	1320	1650	2010	2430	2840	3280	3760	4290
6.0	105	199	332	506	713	965	1250	1560	1920	2320	2710	3150	3620	4140
6.2	97	184	308	472	669	909	1180	1480	1820	2210	2600	3020	3480	3990
6.4	89	170	287	442	628	857	1110	1400	1730	2110	2490	2910	3360	3850
6.6	82	157	267	413	590	808	1050	1330	1650	2010	2380	2790	3240	3720
6.8	76	146	248	386	553	761	990	1260	1560	1910	2280	2690	3120	3600
7.0	70	135	231	360	518	716	933	1190	1480	1820	2180	2580	3010	3480
7.2	64	125	214	336	485	672	878	1120	1400	1720	2070	2470	2900	3370
7.4	59	115	198	312	451	628	822	1060	1320	1630	1970	2350	2780	3250
7.6	54	105	182	287	417	583	765	980	1230	1530	1850	2230	2650	3120
7.8	48	94	164	261	380	533	701	901	1140	1410	1720	2090	2500	2970
8.0	38	76	134	214	314	444	587	758	960	1200	1480	1810	2210	2670

NC = normal crown, RC = reverse crown.
From American Association of State Highway and Transportation Officials (AASHTO), 2011. A Policy on Geometric Design of Highways and Streets, 6th ed. Table 3–10b.

Table 3.7 Superelevation runoff, L_r, for one-lane rotation (two-lane highway)

e (%)	Design speed (V_{design})													
	15 mph	20 mph	25 mph	30 mph	35 mph	40 mph	45 mph	50 mph	55 mph	60 mph	65 mph	70 mph	75 mph	80 mph
1.5	23	24	26	27	29	31	33	36	38	40	42	45	47	52
2.0	31	32	34	36	39	41	44	48	51	53	56	60	636	69
2.2	34	36	38	40	43	46	49	53	56	59	61	66	69	75
2.4	37	39	41	44	46	50	53	58	61	64	67	72	76	82
2.6	40	42	45	47	50	54	58	62	66	69	73	78	82	89
2.8	43	45	48	51	54	58	62	67	71	75	78	84	88	96
3.0	46	49	51	55	58	62	67	72	77	80	84	90	95	103
3.2	49	52	55	58	62	66	71	77	82	85	89	96	101	110
3.4	52	55	58	62	66	70	76	82	87	91	95	102	107	117
3.6	55	58	62	65	70	74	80	86	92	96	100	108	114	123
3.8	58	58	65	69	74	79	84	91	97	101	106	114	120	130
4.0	62	62	69	73	77	83	89	96	102	107	112	120	126	137
4.2	65	68	72	76	81	87	93	101	107	112	117	126	133	144
4.4	68	71	75	80	85	91	98	106	112	117	123	132	139	151
4.6	71	75	79	84	89	95	102	110	117	123	128	138	145	158
4.8	74	78	82	87	93	99	107	115	123	128	134	144	152	165
5.0	77	81	86	91	97	103	111	120	128	133	140	150	158	171
5.2	80	84	89	95	101	108	116	125	133	139	145	156	164	178
5.4	83	88	93	98	105	112	120	130	138	144	151	162	171	185
5.6	86	91	96	102	108	116	124	134	143	149	156	168	177	192
5.8	89	94	99	105	112	120	129	139	148	155	162	174	183	199
6.0	92	97	103	109	116	124	133	144	153	160	167	180	189	206
6.2	95	101	106	113	120	128	138	149	158	165	173	186	196	213
6.4	98	104	110	116	124	132	142	154	163	171	179	192	202	219
6.6	102	107	113	120	128	137	147	158	169	176	184	198	208	226
6.8	105	110	117	124	132	141	151	163	174	181	190	204	215	233
7.0	108	114	120	127	135	145	156	168	179	187	195	210	221	240

(Continued)

Table 3.7 (Continued)

e (%)	Design speed (V_{design})													
	15 mph	20 mph	25 mph	30 mph	35 mph	40 mph	45 mph	50 mph	55 mph	60 mph	65 mph	70 mph	75 mph	80 mph
7.2	111	117	123	131	139	149	160	173	184	192	201	216	227	247
7.4	114	120	127	135	143	153	164	178	189	197	207	222	234	254
7.6	117	123	130	138	147	157	169	182	194	203	212	228	240	261
7.8	120	126	134	142	151	161	173	187	199	208	218	234	246	267
8.0	123	130	137	145	155	166	178	192	204	213	223	240	253	274
8.2	126	133	141	149	159	170	182	197	209	219	229	246	259	281
8.4	129	136	144	153	163	174	187	202	214	224	234	252	265	288
8.6	132	139	147	156	166	178	191	206	220	229	240	258	272	295
8.8	135	143	151	160	170	182	196	211	225	235	246	264	278	302
9.0	138	146	154	164	174	186	200	216	230	240	251	270	284	309
9.2	142	149	158	167	178	190	204	221	235	245	257	276	291	315
9.4	145	152	161	171	182	194	209	226	240	251	262	282	297	322
9.6	148	156	165	175	186	199	213	230	245	256	268	288	303	329
9.8	151	159	168	178	190	203	218	235	250	261	273	294	309	336
10.0	154	162	171	182	194	207	222	240	255	267	279	300	316	343
10.2	157	165	175	185	197	211	227	245	260	272	285	306	322	350
10.4	160	169	178	189	201	215	231	250	266	277	290	312	328	357
10.6	163	172	182	193	205	219	236	254	271	283	296	318	335	363
10.8	166	175	185	196	209	223	240	259	276	288	301	324	341	370
11.0	169	178	189	200	213	228	244	264	281	293	307	330	347	377
11.2	172	182	192	204	217	232	249	269	286	299	313	336	354	384
11.4	175	185	195	207	221	236	253	274	291	304	318	342	360	391
11.6	178	188	199	211	225	240	258	278	296	309	324	348	366	398
11.8	182	191	202	215	228	244	262	283	301	315	329	354	373	405
12.0	185	195	206	218	232	248	267	288	306	320	335	360	379	411

Adapted from American Association of State Highway and Transportation Officials (AASHTO), 2011. A Policy on Geometric Design of Highways and Streets, 6th ed. Table 3–17b.

Table 3.8 Maximum relative gradient

Design speed (mph)	Maximum relative gradient (%)
15	0.78
20	0.74
25	0.70
30	0.66
35	0.62
40	0.58
45	0.54
50	0.50
55	0.47
60	0.45
65	0.43
70	0.40
75	0.38
80	0.35

Adapted from American Association of State Highway and Transportation Officials (AASHTO), 2011. A Policy on Geometric Design of Highways and Streets, 6th ed. Table 3–15.

Table 3.9 Recommended runoff locations for simple curves—portion located prior to curve

Design speed (mph)	Number of lanes rotated			
	1.0	1.5	2.0–2.5	3.0–3.5
15–45	0.80	0.85	0.90	0.90
50–80	0.70	0.75	0.80	0.85

Adapted from American Association of State Highway and Transportation Officials (AASHTO), 2011. A Policy on Geometric Design of Highways and Streets, 6th ed. Table 3–18.

EXAMPLE 3.6 Horizontal Curve Design

Determine key elements for a horizontal curve with a 30° intersecting angle on a two-lane highway with 12-ft lanes and a 60-mph design speed with a designated e_{max} of 8%. The radius for the horizontal curve has been set to 1720 ft. The PI is located at station 85 + 29.14 and the highway has a constant 1.3% longitudinal grade throughout this section, and the centerline of the highway has an elevation of 245.78 ft at the beginning of the curve.

(Continued)

EXAMPLE 3.6 Horizontal Curve Design—(Continued)

Solution

Using Table 3.3, the selected radius of 1720 ft corresponds to a design superelevation, e_{design}, of 7.2%. Based on Table 3.4, a radius of 1720 ft for a 60-mph design speed will be positively impacted with the design of spiral curve; therefore, a spiral curve will be applied for this design. From Table 3.7, the superelevation runoff length, L_r, is 192 ft, which provides for the transition of the cross section from adverse crown removed (outside edge of pavement is flat) to design superelevation. The superelevation runoff length will be used for the spiral length, $L_r = L_s$.

The runout length, L_t, is

$$L_t = \frac{e_{\text{NC}}}{e_{\text{design}}} L_r = \frac{2\%}{7.2\%} 192 \text{ ft} = 53.33 \text{ ft}$$

$$D_c = \frac{18,000}{R\pi} = \frac{18,000}{1720\pi} = 3.33^\circ$$

$$\theta_s = \frac{L_s D_c}{200} = \frac{192 \text{ ft} \times 3.33^\circ}{200} = 3.20^\circ$$

$$\Delta_c = \Delta - 2\theta_s = 30^\circ - 2(3.20^\circ) = 23.60^\circ$$

$$L_c = 100 \frac{23.60^\circ}{3.33^\circ} = 708.7 \text{ ft}$$

$$Y_s = \frac{L_s^2}{6\,R} = \frac{192 \text{ ft}^2}{6 \times 1720 \text{ ft}} = 3.57 \text{ ft}$$

$$X_s = L_s - \left(\frac{Y_s^2}{2\,L_s}\right) = 192 - \left(\frac{3.57 \text{ ft}^2}{2 \times 192 \text{ ft}}\right) = 191.967 \text{ ft}$$

$$p = Y_s - R(1 - \cos\theta_s) = 3.57 \text{ ft} - 1720 \text{ ft}(1 - \cos 3.20^\circ) = 0.89 \text{ ft}$$

$p > 0.66$, therefore the effect of constructing a spiral curve is justified in the impact it will have on vehicles.

$$k = X_s - R\sin\theta_s = 191.967 \text{ ft} - 1720 \text{ ft}\,(\sin 3.20^\circ) = 95.95 \text{ ft}$$

$$T_s = (R + p)\tan\left(\frac{\Delta}{2}\right) + k = (1720 \text{ ft} + 0.89 \text{ ft})\tan\left(\frac{30^\circ}{2}\right) + 95.95 \text{ ft} = 557.06 \text{ ft}$$

The stationing can be developed using the basic parameters of this spiral curve.

PI station $= 85^{+}29.14$

TS station = PI station $- T_s = 85^{+}29.14 - 5^{+}57.06 = 79^{+}72.06$

SC station = TS station $+ L_s = 79^{+}72.06 + 1^{+}92 = 81^{+}64.08$

CS station = SC station $+ L_c = 81^{+}64.08 + 7^{+}08.7 = 88^{+}72.78$

ST station = CS station $+ L_s = 88^{+}72.78 + 1^{+}92 = 90^{+}64.78$

(Continued)

EXAMPLE 3.6 Horizontal Curve Design—(Continued)

The relevant centerline and edgeline elevations can also be determined (based on points described earlier and presented in Figure 3.19). Determining the centerline elevation along the highway can be completed first.

CL elevation_B (given) = 245.78 ft
CL elevation_A = 245.78 ft − 1.3% (53.33 ft) = 245.087 ft
CL elevation_C = 245.78 ft + 1.3% (53.33 ft) = 246.473 ft
CL elevation_D = 245.78 ft + 1.3% (192 ft) = 248.276 ft
CL elevation_E = 248.276 ft + 1.3% (708.7 ft) = 257.489 ft
CL elevation_F = 257.489 ft + 1.3% (192 ft − 53.33 ft) = 259.292 ft
CL elevation_G = 257.489 ft + 1.3% (192 ft) = 259.985 ft
CL elevation_H = 259.985 ft + 1.3% (53.33 ft) = 260.678 ft

Based on the centerline elevations, the outside edge of pavement (EOP) and inside edge of pavement elevations can be computed.

Outside EOP elevation_A (normal crown) = 245.087 ft − 12 ft (2%) = 244.847
Outside EOP elevation_B (adverse crown removed) = 245.78 ft
Outside EOP elevation_C (reverse crown) = 246.473 ft + 12 ft (2%) = 246.713 ft
Outside EOP elevation_D (design superelevation) = 248.276 ft + 12 ft (7.2%) = 249.14 ft
Outside EOP elevation_E (design superelevation) = 257.489 ft + 12 ft (7.2%) = 258.353 ft
Outside EOP elevation_F (reverse crown) = 259.292 ft + 12 ft (2%) = 259.532 ft
Outside EOP elevation_G (adverse crown removed) = 259.985 ft
Outside EOP elevation_H (normal crown) = 260.678 ft − 12 ft (2%) = 260.438 ft

Inside EOP elevation_A (normal crown) = 245.087 ft − 12 ft (2%) = 244.847
Inside EOP elevation_B (adverse crown removed) = 245.78 ft − 12 ft (2%) = 245.54 ft
Inside EOP elevation_C (reverse crown) = 246.473 ft − 12 ft (2%) = 246.233 ft
Inside EOP elevation_D (design superelevation) = 248.276 ft − 12 ft (7.2%) = 247.412 ft
Inside EOP elevation_E (design superelevation) = 257.489 ft − 12 ft (7.2%) = 256.625 ft
Inside EOP elevation_F (reverse crown) = 259.292 ft − 12 ft (2%) = 259.052 ft
Inside EOP elevation_G (adverse crown removed) = 259.985 ft − 12 ft (2%) = 259.745 ft
Inside EOP Elevation_H (normal crown) = 260.678 ft − 12 ft (2%) = 260.438 ft

(Continued)

EXAMPLE 3.6 Horizontal Curve Design—(Continued)

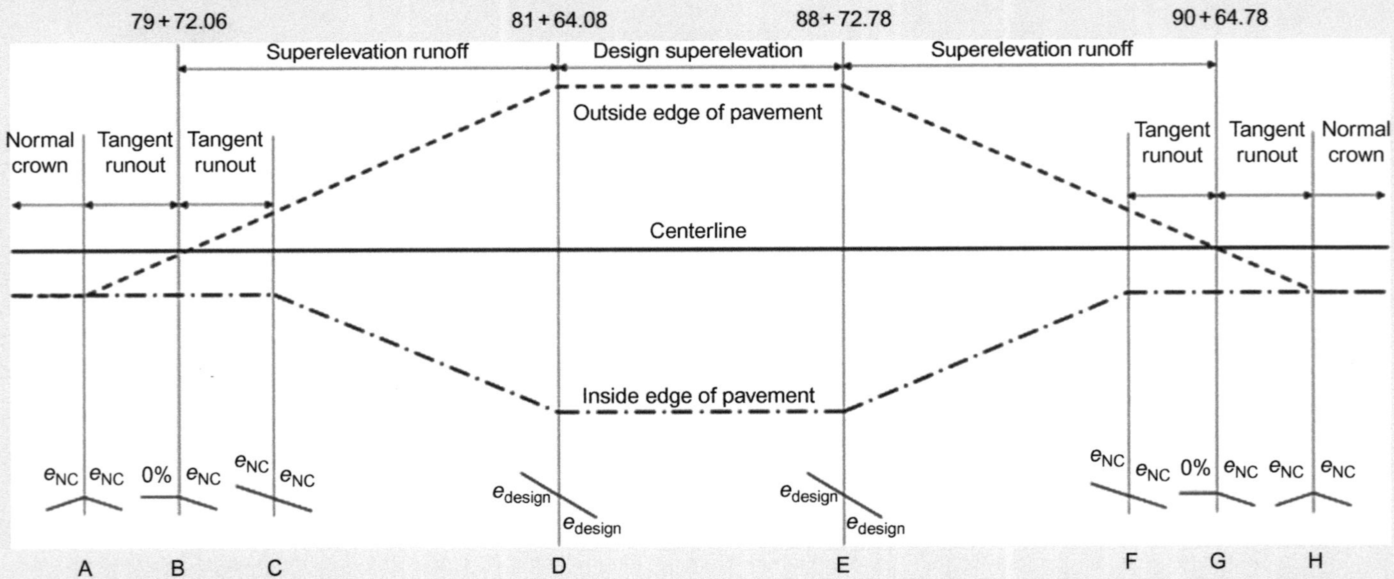

3.4.2 Vertical Alignment

Similar to horizontal alignment, the core of vertical alignment is comprised of tangents that are smoothly connected by curves. However, vertical alignment is displayed in the profile view and the curvature has a parabolic shape instead of a constant radius. The tangents represent grades that can either be flat, uphill, or downhill. Minimizing the amount of earthwork and balancing the amount of cut and fill within a project is a major consideration for vertical alignment. Figure 3.21 shows the profile perspective of a highway with a complex horizontal alignment and a vertical alignment with a relatively constant upgrade with minor vertical curves along the segment.

Vertical Curve Fundamentals

The typical vertical curve is a symmetric, parabolic curve with a shape defined by the parabolic equation. The information required to fully define a vertical curve is the elevation of the beginning of the curve, the grades of the two tangents that are connected, and the length of the curve. Figure 3.22 depicts a typical vertical curve showing the profile view of the centerline of the highway. For vertical alignment, distance measurements are made on the horizontal plane, based on stationing from horizontal alignment; this process is necessary to avoid creating an additional station for the purpose of vertical alignment. There are two general types of vertical curves: (1) a crest curve, which has a concave down shape and (2) a sag curve, which has a concave up shape. For a symmetric, parabolic curve, the length of the curve is evenly divided on each

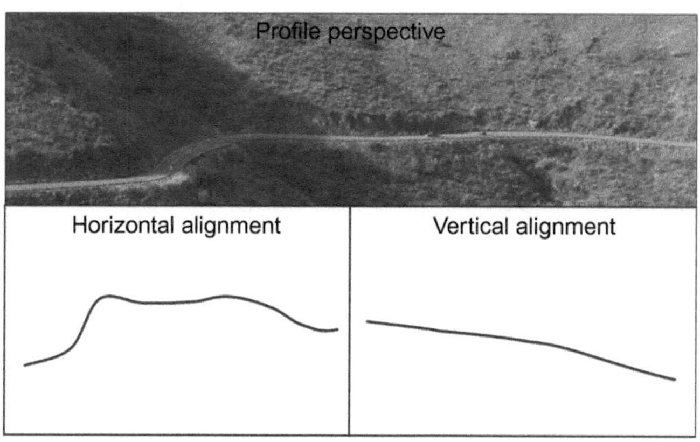

Figure 3.21 Highway profile perspective.

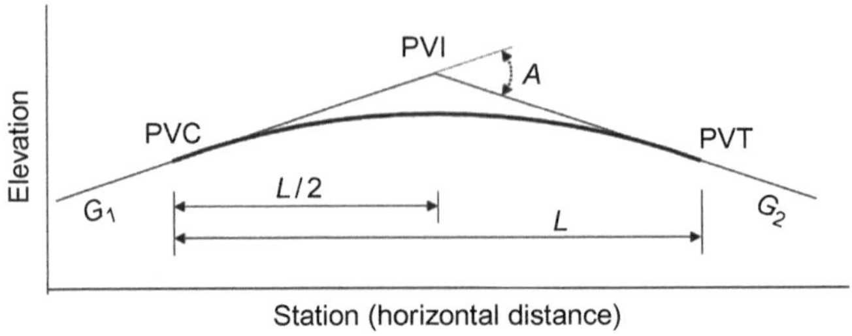

Figure 3.22 Vertical curve.

side of the point of vertical intersection (PVI). The distance from the point of vertical curvature (PVC) to the PVI and from the PVI to the point of vertical tangency (PVT) is half of the length of the curve.

The primary elements of a symmetric, parabolic vertical curve include:

- Point of vertical intersection (PVI): The location where the entrance grade, G_1, and the exit grade, G_2, intersect.
- Algebraic difference in grades (A): This measures the difference in grades between G_1 and G_2.
- Point of vertical curvature (PVC): The point where the vertical curve begins and the highway leaves the tangent.
- Point of vertical tangency (PVT): The point where the vertical curve ends and the highway returns to the tangent.
- Curve length (L): The horizontal distance (not the distance along the arc) between the PVC and PVT. With a symmetric, parabolic vertical curve, the length is bisected by the PVI. The curve length is expressed in stations.
- Entrance grade (G_1): The grade of the tangent leading into the vertical curve. The grade is expressed as a percent.
- Exit grade (G_2): The tangent leading out of the vertical curve. The grade is expressed as a percent.

The general equation for a second-degree parabola takes the form of: $Y = ax^2 + bx + c$. This general equation can be modified for each individual curve to describe the relationship between the distance, x, and the elevation, Y, at any point along the vertical curve. The term x is used to denote any horizontal distance along the curve and is bounded by 0, which would designate the PVC point, and the length of the curve, which would designate the PVT point. The term a becomes the tangent offset, the change in elevation from the tangent to the curve. The grade of G_1 replaces the term b, and c becomes the elevation of the PVC. The parabolic equation, when adapted specifically in terms of vertical curves, takes the following form:

$$Y = \frac{(G_2 - G_1)}{2L}x^2 + G_1 x + \text{PVC}_{\text{elevation}}$$

where

Y = elevation on the curve at a distance x from the PVC, ft

x = distance from the PVC to the point of interest on the curve, stations

G_1 = entrance grade of the curve, percent (− for downgrades and + for upgrades)

G_2 = exit grade of the curve, percent (− for downgrades and + for upgrades)

L = length of the vertical curve measured horizontally, stations

$\text{PVC}_{\text{elevation}}$ = elevation of the PVC

High/Low Point

The high or low point, also known as the turning point, of a vertical curve occurs when the slope of the highway is equal to zero. Therefore, using basic calculus principles, the derivative of the parabolic equation can be used to find the location of the high/low point. The derivative of the parabolic equation, $Y = ax^2 + bx + c$, is $\frac{DY}{dx} = 2ax + b$. By setting this equation equal to zero and substituting the vertical curve specific components for a and b, the equation for the location of the high or low point can be determined. The parabolic equation can be used to determine the elevation at this point. The high point can be identified for a crest curve and the low point can be found for a sag curve. These points are never above (for a crest curve) or below (for a sag curve) the PVI. Identifying and locating the high or low point of a vertical curve is important for drainage considerations and for determining where the appropriate clearance is provided underneath an overhead structure or on top of underground utilities. The following process develops the equation for x that can be used to find the location of the high or low point based on the point's distance from the PVC.

$$Y = ax^2 + bx + c$$

$$\frac{DY}{dx} = 2ax + b = 0$$

$$x = \frac{-b}{2a} = \frac{-G_1}{2\left(\frac{(G_2 - G_1)}{2L}\right)} = \frac{-G_1 L}{G_2 - G_1}$$

EXAMPLE 3.7 Horizontal Vertical Curve

Determine the elevation of the centerline of a highway at 50-ft increments from the PVC of an 800-ft vertical curve that connects an entrance grade of −3.7% and an exit grade of 1.9%. The PVI of the curve is located at station $146^+17.18$ with an elevation of 314.22 ft. Also determine the location and elevation of the low point.

Solution

$G_1 = -3.7\%$
$G_2 = 1.9\%$
$L = 8.00$ stations
$\mathrm{PVC}_{\mathrm{elevation}} = \mathrm{PVI}_{\mathrm{elevation}} - G_1\ (L/2) = 314.22\ \mathrm{ft} + 3.7\%\ (400\ \mathrm{ft})$
$= 329.02$ ft
$\mathrm{PVC}_{\mathrm{station}} = \mathrm{PVI}_{\mathrm{station}} - L/2 = 146^+17.18 - (800/2\ \mathrm{ft}) = 142^+17.18$
$\mathrm{PVT}_{\mathrm{station}} = \mathrm{PVI}_{\mathrm{station}} + L/2 = 146^+17.18 + (800/2\ \mathrm{ft}) = 150^+17.18$

The parabolic equation that defines this vertical curve can be determined using the basic information that described the curve:

$$Y = \frac{(G_2 - G_1)}{2L}x^2 + G_1 x + \mathrm{PVC}_{\mathrm{elevation}}$$

$$Y = \frac{(1.9\% - -3.7\%)}{2(8\ \mathrm{stations})}x^2 + -3.7\%\,x + 329.02$$

$$Y = 0.35x^2 + -3.7x + 329.02$$

Using the parabolic equation, the elevation Y, can be computed at every 50-ft increment along the centerline of the roadway.

x (stations)	***Y* (ft)**
0	329.02
0.5	327.26
1	325.67
1.5	324.26
2	323.02
2.5	321.96
3	321.07
3.5	320.36

(*Continued*)

EXAMPLE 3.7 Horizontal Vertical Curve—(Continued)

x (stations)	*Y* (ft)
4	319.82
4.5	319.46
5	319.27
5.5	319.26
6	319.42
6.5	319.76
7	320.27
7.5	320.96
8	321.82

The elevation of the PVT can also be checked geometrically to confirm that the parabolic equation was developed accurately.

$$\text{PVT}_{\text{elevation}} = \text{PVI}_{\text{elevation}} + G_2(L/2) = 314.22 \text{ ft} + 1.9\%(400 \text{ ft}) = 321.82$$

The location of the low point can be computed using the following equation.

$$x = \frac{-G_1\ L}{G_2 - G_1} = \frac{--3.7\% \times 8 \text{ stations}}{1.9 - -3.7} = 5.29 \text{ stations}$$

Therefore, the low point occurs 529 ft or 5.29 stations from the PVC. The low point (or high point) will always be located on the side of the PVI with the flatter grade. Using the parabolic equation, the elevation of the low point can also be determined.

$$Y = 0.35\ (5.29)^2 + -3.7\ (5.29) + 329.02 = 319.24 \text{ ft}$$

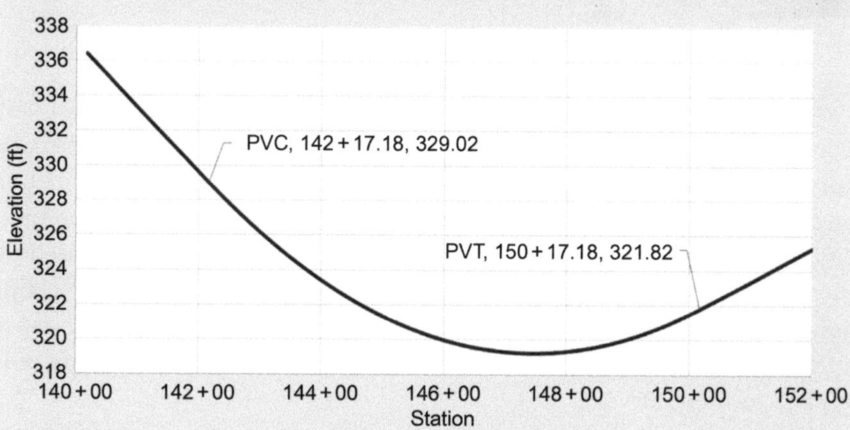

EXAMPLE 3.8 Vertical Curve through Fixed Point

Determine the minimum length of a vertical curve with an entrance grade of 4.2% and an exit grade of −3.2% that will provide 5 ft of cover for a 54-in diameter concrete pipe that has a 3-in wall thickness. The invert elevation of the concrete pipe is 874.51 ft and is located at station $70^{+}12.43$. The PVC is located at station $65^{+}29.94$ at an elevation of 871.73 ft.

Solution

$G_1 = 4.2\%$

$G_2 = -3.2\%$

$PVC_{elevation} = 898.73$ ft

$x = 70^{+}12.43 - 65^{+}29.94 = 4^{+}82.49$

The desired elevation at station $70^{+}12.43$, including the diameter of the pipe, thickness of the pipe, and a necessary cover over the pipe is:

$$\text{Desired elevation}_{70+12.43} = 874.51 \text{ ft} + 54'' + 3'' + 5 \text{ ft} = 884.26 \text{ ft}$$

The parabolic equation that defines this vertical curve can be determined using the basic information that describes the curve:

$$Y = \frac{(G_2 - G_1)}{2L}x^2 + G_1 x + PVC_{elevation}$$

$$884.26 \text{ ft} = \frac{(-3.2 - 4.2)}{2L}(4.8249)^2 + 4.2(4.8249) + 871.73 \text{ ft}$$

$$884.26 \text{ ft} = \frac{(-172.269)}{2L} + 20.26458 + 898.73 \text{ ft}$$

$L = 11.1363$ stations $= 1113.63$ ft

Vertical Curve Design Process

Vertical curve design is based on sight distance considerations. Vertical curve design consistency along a roadway is desirable and coordination with horizontal alignment is necessary for an optimal design. The computations to determine the minimum length of a vertical curve are based on the product of the algebraic difference in grades, A, of the intersecting tangents and the rate of vertical curvature, K. Stopping sight distance applies to crest and sag vertical curves, but passing sight distance is only applied to crest curves.

For crest vertical curves, a driver's ability to see upcoming obstructions in the highway can become blocked by the highway itself.

Standard assumptions from the AASHTO Green Book for the driver's eye height (3.5 ft) and object height (2 ft), such as the taillight of a stopped or slowly moving car, allow for design values for the rate of vertical curvature to be utilized directly to provide for a sufficient minimum stopping sight distance. For the passing sight distance criterion for two-lane, two-way highways, the design rate of vertical curvature assumes an object height of 3.5 ft for the average height of a passenger vehicle and larger sight distance is required to allow for the passing maneuver to safely occur.

For sag vertical curves, darkness is the constraining factor for sight distance. When traversing a sag curve, the headlights of the vehicle limit the driver's ability to perceive obstructions on the highway in front of the vehicle. Computations for an appropriate sag curve length include the assumption of a 1° angle above the headlight direction until the light strikes the surface of the highway.

The general design process for a vertical curve involves the following steps:

1. Apply relevant design controls, particularly maximum grades for the highway's jurisdiction and any other vertical curve specific elements identified by the controlling agency. Maximum grades are based on the highway type and performance characteristics of design vehicles.
2. Choose tangents and associated grades based on design controls, terrain, intersecting roadways, bridge clearances, drainage, water features, or utility cover requirements. The algebraic difference in grades, A, is important for further vertical curve calculations. Grades impact the operational characteristics of heavy trucks more than passenger vehicles. Passenger vehicle speeds are not significantly impacted until longitudinal grades exceed 5%; while heavy trucks may increase speeds by up to 5% on downgrades and reduce speeds by up to 7% on upgrades, compared to level terrain (AASHTO, 2011). The specific impact of grades on a heavy truck's operation depends on the length and steepness of grade, power-to-weight ratio, entering speed, aerodynamic resistance, and skill of the driver. Reasonable guidelines for maximum grades that account for heavy truck performance are typically 5% for a 70-mph design speed, 7–12% for a 30-mph design speed (7–8% for primary highways), and grades in-between these values for design speeds of

40–60 mph. A minimum grade of 0.5% is also needed in most instances to provide for adequate drainage of water on the roadway. When the speed of heavy trucks is 10 mph or more below the average operating speed of other vehicles, the speed differential can pose a safety risk on the highway, which can occur with a 5% upgrade on a 1000-ft-long highway section (AASHTO, 2011, Figure 3-28).

3. Determine an appropriate minimum vertical curve length based on the controlling sight distance (stopping or passing). However, a larger curve length will provide additional safety and should be used if possible. Compute the minimum vertical curve length based on rate of vertical curvature, K, multiplied by the algebraic difference in grades, A. The rate of vertical curvature is determined by the type of curve (sag or crest), type of sight distance (stopping or passing), and design speed as presented in Table 3.10. If the value for A is small, it is possible to obtain small values for the length of curve. Therefore, the AASHTO Green Book recommends that the minimum length of vertical curve should be three times the design speed. Additionally, long vertical curves and large K values can create the need for extensive drainage work to facilitate the movement of water off of the surface of the highway.
4. Determine necessary curve features for the layout of the vertical curve, particularly the parabolic equation, which defines the elevation at any point along the curve.

EXAMPLE 3.9 Vertical Curve

Determine the minimum length for a vertical curve that is needed to connect a −4.3% grade and a 3.1% grade on a 60-mph multilane highway.

Solution

From Table 3.10, the rate of vertical curvature is 151. The algebraic difference in grades, A, can be computed as $|-4.3\% - 3.1\%| = |-7.4\%| = 7.4\%$.

Therefore, the length, L, can be computed as $L = KA = 151 \times 7.4\% = 1117.4\ \text{ft} \approx 1120\ \text{ft}$. This value is acceptable because it exceeds the minimum vertical curve length of 180 ft ($3 \times$ design speed).

Table 3.10 Rate of vertical curvature design values

Design speed (mph)	Design rate of vertical curvature, *K*		
	Sag	Crest	
	Stopping sight distance	Stopping sight distance	Passing sight distance
15	10	3	N/A
20	17	7	57
25	26	12	72
30	37	19	89
35	49	29	108
40	64	44	129
45	79	61	175
50	96	84	229
55	115	114	289
60	136	151	357
65	157	193	432
70	181	247	514
75	206	312	604
80	231	384	700

Adapted from American Association of State Highway and Transportation Officials (AASHTO), 2011. A Policy on Geometric Design of Highways and Streets, 6th ed. Tables 3–34, 3–35, 3–36.

3.5 PRACTICE PROBLEMS

Problem 3.1

Determine the appropriate sight distance for designing the following highway elements.

A. An urban freeway with a 65-mph design speed approaching a cloverleaf interchange
B. A rural freeway with a 75-mph design speed
C. An urban arterial with a 50-mph design speed
D. A rural two-lane arterial with a 60-mph design speed
E. An urban two-lane collector with a 35-mph design speed

Solution

A. Evaluate the approach to the interchange using decision sight distance, to provide for additional distance for drivers to react to the complex situation. Based on Case E for an urban environment in Table 3.1, provide a minimum of 1365 ft.

B. Evaluate the highway for stopping sight distance. Based on Table 3.1, provide a minimum of 820 ft.

C. Evaluate the arterial for stopping sight distance. Based on Table 3.1, provide a minimum of 425 ft.

D. Evaluate the arterial for stopping sight distance and passing sight distance, for approach areas to allow for passing maneuvers. Based on Table 3.1, provide a minimum of 570 ft at all points along the highway. For locations suitable for passing maneuvers, provide a minimum of 1000 ft.

E. Evaluate the collector for stopping sight distance. Based on Table 3.1, provide a minimum of 250 ft.

Problem 3.2

Determine the azimuth and bearing for lines AB, AC, AD, and AE.

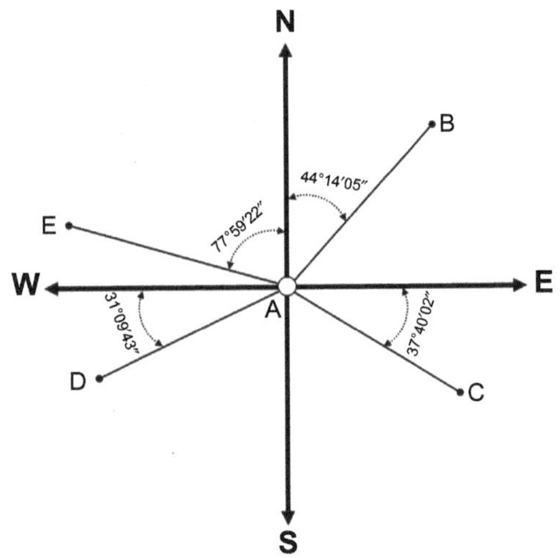

Solution

Line AB:

Bearing = N44°14′05″E
Azimuth = 44°14′05″

Line AC:

Bearing = S52°19′58″E
Azimuth = 127°40′02″

Line AD:

Bearing = S58°50′17″W

Azimuth = 238°50′17″

Line AE:

Bearing = N77°59′22″W

Azimuth = 282°0′38″

Problem 3.3

For the following driver perspectives, sketch the approximate horizontal and vertical alignments that correspond to the highway alignment.

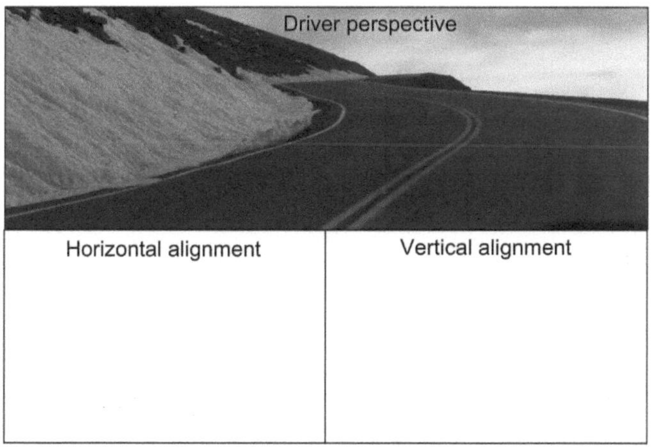

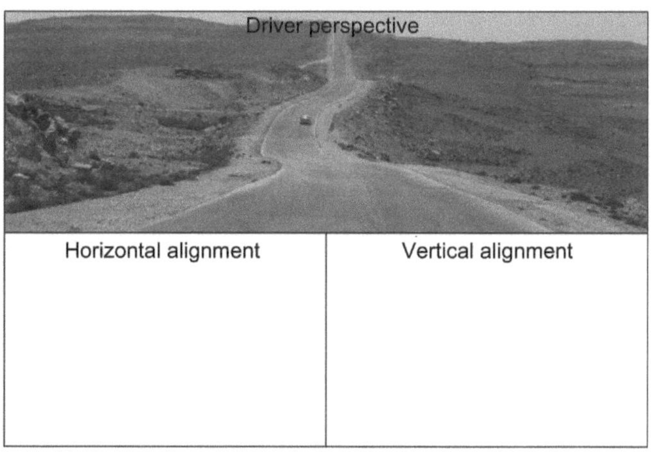

Solution

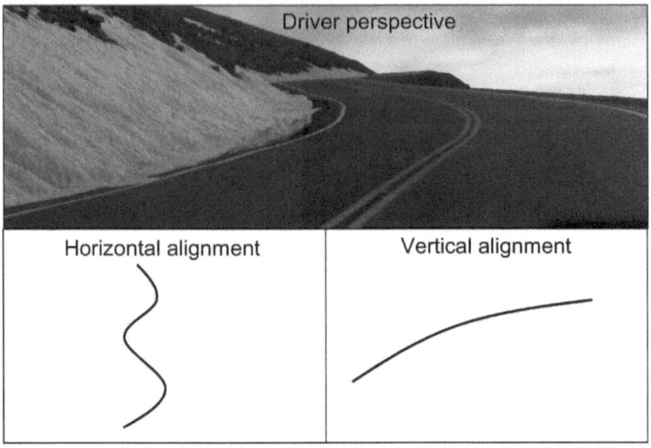

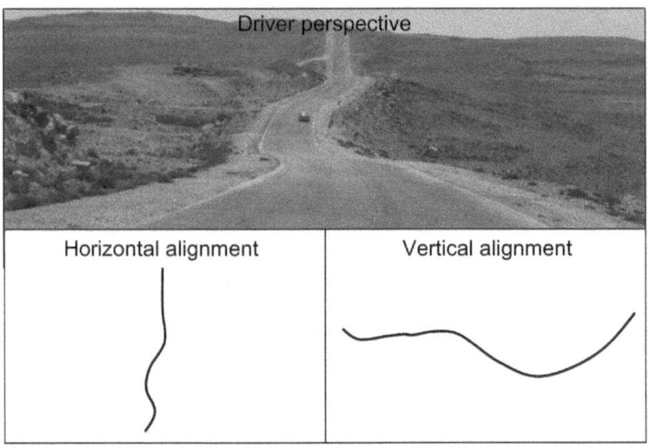

Problem 3.4

A sign is located 7 ft from the edge of the pavement on a 3° horizontal curve. Determine if the sign should be relocated further from the edge of the highway to provide the necessary stopping sight distance. The two-lane highway has a design speed of 50 mph with 12-ft wide lanes.

Solution

From Table 3.1, the minimum stopping sight distance is 425 ft.

$$R = \frac{18,000}{D_C \pi} = \frac{18,000}{3^\circ \pi} = 1909.86 \text{ ft}$$

r (radial distance to the middle of the inside lane) = 1909.86 ft − 6 ft = 1903.86 ft

$$M = r\left[1 - \cos\left(\frac{28.65\ S}{r}\right)\right] = 1903.86\ \text{ft}\left[1 - \cos\left(\frac{28.65(425\ \text{ft})}{1903.86\ \text{ft}}\right)\right]$$
$$= 11.85\ \text{ft}$$

The middle ordinate of 11.85 ft is measured from the middle of the inside lane to the edge of the sight obstruction. The minimum distance for this case from the edge of pavement is 5.85 ft (11.85 ft − 6 ft). Because the provided distance of 7 ft is greater than the minimum distance of 5.85 ft, the location of the sign does not restrict the recommended minimum stopping sight distance on the curve.

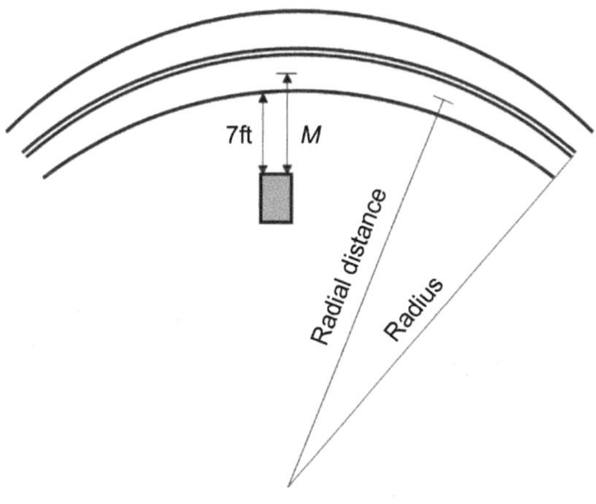

Problem 3.5

Determine the stationing for the PC and PT of a 4.5° curve to be placed on tangents that intersect at station $33^{+}49.86$ with an intersecting angle of 47°.

Solution

$$R = \frac{18,000}{\pi D_C} = \frac{18,000}{\pi 4.5^\circ} = 1273.24\ \text{ft}$$

$$T = R\tan\left(\frac{\Delta}{2}\right) = 1273.24\ \text{ft}\tan\left(\frac{47^\circ}{2}\right) = 553.62\ \text{ft}$$

$$L = 100\frac{\Delta}{D_C} = 100\left(\frac{47^\circ}{4.5^\circ}\right) = 1044.44\ \text{ft}$$

$$PC_{station} = PI - T = 33^{+}49.86 - 5^{+}53.62 = 27^{+}96.24$$
$$PT_{station} = PC + L = 27^{+}96.24 + 10^{+}44.44 = 38^{+}40.68$$

Problem 3.6

Determine the stationing for a compound curve to be placed on an exit loop of a freeway. The PI of the compound curve is $103^{+}52.11$ with an intersecting angle of 112°. The radii of the curves are 1400 ft and 3000 ft, and the larger radius has a central angle of 54°.

Solution

The stationing will increase in the direction of travel and the larger radius curve should be placed first on the exit loop to allow vehicles to decelerate before reaching the smaller radius curve. Therefore, the following notations will be used for calculations:

$R_1 = 3000\text{ ft}\ \Delta_1 = 54°$

$R_2 = 1400\text{ ft}\ \Delta_2 = \Delta - \Delta_1 = 112° - 54° = 58°$

Because the larger curve is adjacent to the PC, T_a is needed to determine the PC station based on the given PI station information. The length of each individual curve will also be needed to determine the stationing of the curve.

$$T_a = \frac{R_2 - (R_1 \cos I) + [(R_1 - R_2)\cos \Delta_2]}{\sin I}$$

$$T_a = \frac{1400\text{ ft} - (3000\text{ ft}\cos 112°) + [(3000\text{ ft} - 1400\text{ ft})\cos 58°]}{\sin 112°}$$

$$= 3,636.49\text{ ft}$$

$$L_1 = \frac{\pi R_1 \Delta_1}{180} = \frac{\pi\ 3000\text{ ft} \times 54°}{180} = 2,827.43\text{ ft}$$

$$L_2 = \frac{\pi R_2 \Delta_2}{180} = \frac{\pi 1400\text{ ft} \times 58°}{180} = 1417.21\text{ ft}$$

$$PC_{station} = PI_{station} - T_a = 103^{+}52.11 - 36^{+}36.49 = 67^{+}15.62$$
$$PCC_{station} = PC_{station} + L_1 = 67^{+}15.62 + 28^{+}27.43 = 95^{+}43.05$$
$$PT_{station} = PCC_{station} + L_2 = 95^{+}43.05 + 14^{+}17.21 = 109^{+}60.26$$

Problem 3.7

Determine the stationing of key points for a spiral curve with an intersecting angle of 29°, a 6° curve, and a length of spiral of 200 ft. The curve has a PI at station $32^{+}29.87$.

Solution

The individual components of the spiral curve must be computed first.

$$\theta_s = \frac{L_S D_C}{200} = \frac{200\text{ ft} \times 6^\circ}{200} = 6^\circ$$

$$\Delta_c = \Delta - 2\,\theta_S = 29^\circ - 2(6^\circ) = 17^\circ$$

$$L_c = 100\frac{17^\circ}{6^\circ} = 283.33\text{ ft}$$

$$R = \frac{18{,}000}{\pi D_c} = \frac{18{,}000}{\pi 6^\circ} = 954.93\text{ ft}$$

$$Y_s = \frac{L_s^2}{6\text{ R}} = \frac{(200\text{ ft})^2}{6 \times 954.93\text{ ft}} = 6.981\text{ ft}$$

$$X_s = L_s - \left(\frac{Y_s^2}{2\,L_s}\right) = 200 - \left(\frac{6.981\text{ ft}^2}{2 \times 200\text{ ft}}\right) = 199.878\text{ ft}$$

$$p = Y_s - R(1 - \cos\theta_s) = 6.981\text{ ft} - 954.93\text{ ft}\,(1 - \cos 6^\circ) = 1.75\text{ ft}$$

$p > 0.66$, therefore, the effect of constructing a spiral curve is justified in the impact it will have on vehicles.

$$k = X_s - R\sin\theta_s = 199.878\text{ ft} - 954.93\text{ ft}\,(\sin 6^\circ) = 100.06\text{ ft}$$

$$T_s = (R + p)\tan\left(\frac{\Delta}{2}\right) + k = (954.93\text{ ft} + 1.75\text{ ft})\tan\left(\frac{29^\circ}{2}\right) + 100.06\text{ ft} = 347.47\text{ ft}$$

The stationing can be developed using the basic parameters of this spiral curve.

PI station $= 32^{+}29.87$

TS station = PI station $- \ T_s = 32^{+}29.87 - 3^{+}47.47 = 28^{+}82.40$

SC station = TS station $+ L_s = 28^{+}82.40 + 2^{+}00 = 30^{+}82.40$

CS station = SC station $+ L_c = 30^{+}82.40 + 2^{+}83.33 = 33^{+}65.73$

ST station = CS station $+ L_s = 33^{+}65.73 + 2^{+}00 = 35^{+}65.73$

Problem 3.8

Determine key elements (stationing and pavement elevations for critical locations) for a horizontal curve with a 37° intersecting angle on a two-lane highway with 12-ft lanes and a 55-mph design speed with a designated e_{max} of 6%. The radius for the horizontal curve has been set to 3040 ft. The PI is located at station $152^{+}03.44$ and the highway has a constant −2.1% longitudinal grade throughout this section, and the centerline of the highway has an elevation of 384.69 ft at the beginning of the curve.

Solution

Using Table 3.5, the selected radius of 3040 ft corresponds to a design superelevation, e_{design}, of 3.8%. Based on Table 3.4, a radius of 3040 ft for a 55-mph design speed will not be positively impacted with the design of spiral curve; therefore, a spiral curve will not be applied for this design. The simple curve elements of interest include:

$$D_c = \frac{18,000}{R\pi} = \frac{18,000}{3040\pi} = 1.885^\circ$$

$$T = R\tan\left(\frac{\Delta}{2}\right) = 3040\text{ ft}\tan\left(\frac{37^\circ}{2}\right) = 1017.17\text{ ft}$$

$$L = 100\frac{\Delta}{D_c} = 100\left(\frac{37^\circ}{1.885^\circ}\right) = 1962.87\text{ ft}$$

The stationing of the curve can be computed based on the preceding calculations and given information.

$$\text{PC} = \text{PI} - T = 152^{+}03.44 - 10^{+}17.17 = 141^{+}86.27$$
$$\text{PT} = \text{PC} + L = 141^{+}86.27 + 19^{+}92.87 = 161^{+}79.14$$

From Table 3.7, the superelevation runoff length, L_r, is 77 ft, which provides for the transition of the cross section from adverse crown removed (outside edge of pavement is flat) to design superelevation.

The runout length, L_t, is

$$L_t = \frac{e_{NC}}{e_{design}}L_r = \frac{2\%}{3.8\%}77\text{ ft} = 40.53\text{ ft}$$

The relevant centerline and edgeline elevations can also be determined (based on points described earlier and presented in Figure 3.19). Determining the centerline elevation along the highway can be completed first.

CL $elevation_{PC}$ (given) = 384.69 ft

CL $elevation_A$ = 384.69 ft + 2.1% (2/3 × 77 ft + 40.53 ft) = 386.619 ft

CL $elevation_B$ = 384.69 ft + 2.1% (2/3 × 77 ft) = 385.768 ft

CL $elevation_C$ = 385.768 ft − 2.1% (40.53 ft) = 384.917 ft

CL $elevation_D$ = 384.69 ft − 2.1% (1/3 × 77 ft) = 384.151 ft

CL $elevation_E$ = 384.151 ft − 2.1% (1962.87 ft − 1/3 × 77 ft − 1/3 × 77 ft) = 344.009 ft

CL $elevation_F$ = 344.009 ft − 2.1% (77 ft − 40.53 ft) = 343.243 ft

CL $elevation_G$ = 343.243 ft − 2.1% (40.53 ft) = 342.392 ft

CL $elevation_H$ = 343.392 ft − 2.1% (40.53 ft) = 341.541 ft

Based on the centerline elevations, the outside edge of pavement (EOP) and inside edge of pavement elevations can be computed.

Outside EOP $elevation_A$ (normal crown) = 386.619 ft − 12 ft (2%) = 386.379 ft

Outside EOP $elevation_B$ (adverse crown removed) = 385.768 ft

Outside EOP $elevation_C$ (reverse crown) = 384.917 ft + 12 ft (2%) = 385.157 ft

Outside EOP $elevation_D$ (design superelevation) = 384.151 ft + 12 ft (3.8%) = 384.607 ft

Outside EOP $elevation_E$ (design superelevation) = 344.009 ft + 12 ft (3.8%) = 344.465 ft

Outside EOP $elevation_F$ (reverse crown) = 343.243 ft + 12 ft (2%) = 343.483 ft

Outside EOP $elevation_G$ (adverse crown removed) = 342.392 ft

Outside EOP $elevation_H$ (normal crown) = 341.541 ft − 12 ft (2%) = 341.301 ft

Inside EOP $elevation_A$ (normal crown) = 386.619 ft − 12 ft (2%) = 386.379 ft

Inside EOP $elevation_B$ (adverse crown removed) = 385.768 ft − 12 ft (2%) = 385.528 ft

Inside EOP $elevation_C$ (reverse crown) = 384.917 ft − 12 ft (2%) = 384.677 ft

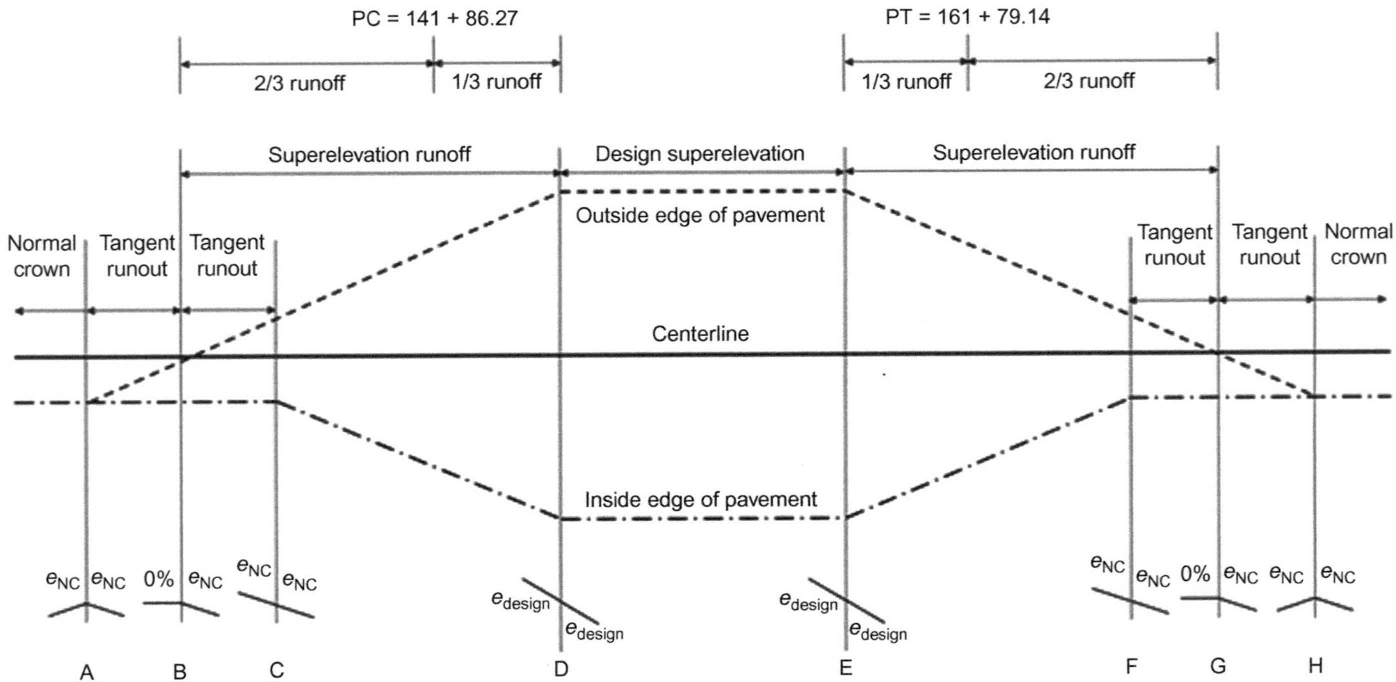

Relative gradient profile and cross-section view for Problem 8.

Inside EOP elevation_D (design superelevation) = 384.151 ft − 12 ft (3.8%) = 383.695 ft

Inside EOP elevation_E (design superelevation) = 344.009 ft − 12 ft (3.8%) = 343.553 ft

Inside EOP elevation_F (reverse crown) = 343.243 ft − 12 ft (2%) = 343.003 ft

Inside EOP elevation_G (adverse crown removed) = 342.392 ft − 12 ft (2%) = 342.152 ft

Inside EOP elevation_H (normal crown) = 341.541 ft − 12 ft (2%) = 341.301 ft

Problem 3.9

Determine the elevation of the centerline of a highway at 50-ft increments from the PVC of a 1300-ft vertical curve that connects an entrance grade of 2.3% and an exit grade of 5.4%. The PVI of the curve has an elevation of 874.32 ft.

Solution

$G_1 = 2.3\%$

$G_2 = 5.4\%$

$L = 13.00$ stations

$\text{PVC}_{\text{elevation}} = \text{PVI}_{\text{elevation}} - G_1\ (L/2) = 874.32\text{ ft} - 2.3\%\ (1300/2\text{ ft}) = 859.37\text{ ft}$

The parabolic equation that defines this vertical curve can be determined using the basic information that described the curve:

$$Y = \frac{(G_2 - G_1)}{2L}x^2 + G_1 x + \text{PVC}_{\text{elevation}}$$

$$Y = \frac{(5.4\% - 2.3\%)}{2\ (13\text{ stations})}x^2 + 2.3\%\ x + 859.37\text{ ft}$$

$$Y = 0.119231\ x^2 + 2.3\ x + 859.37$$

Using the parabolic equation, the elevation Y, can be computed at every 50-ft increment along the centerline of the roadway.

x (stations)	*Y* (ft)
0	859.37
0.5	860.55
1	861.79
1.5	863.09
2	864.45
2.5	865.87
3	867.34
3.5	868.88
4	870.48
4.5	872.13
5	873.85
5.5	875.63
6	877.46
6.5	879.36
7	881.31
7.5	883.33
8	885.40
8.5	887.53
9	889.73
9.5	891.98
10	894.29
10.5	896.67
11	899.10
11.5	901.59
12	904.14
12.5	906.75
13	909.42

The elevation of the PVT can be checked geometrically to confirm that the parabolic equation was developed accurately.

$$PVT_{elevation} = PVI_{elevation} + G_2(L/2) = 874.32\text{ft} + 5.4\%\ (1300/2\ \text{ft})$$
$$= 909.42\ \text{ft}$$

Problem 3.10

Determine the minimum length of a vertical curve with an entrance grade of 4.2% and an exit grade of −3.2% that will provide 5 ft of cover for a 54-in diameter concrete pipe that has a 3-in wall thickness. The invert elevation of the concrete pipe is 874.51 ft and is located at station $70^{+}12.43$. The concrete pipe is 74.32 ft before the PVI. The PVI has an elevation of 895.12 ft.

Solution

$G_1 = 4.2\%$

$G_2 = -3.2\%$

The desired elevation at station $70^{+}12.43$, including the diameter of the pipe, thickness of the pipe, and necessary cover over the pipe is:

$\text{Desired elevation}_{70+12.43} = 874.51\text{ ft} + 54\text{ in} + 3\text{ in} + 5\text{ ft} = 884.26\text{ ft}$

The parabolic equation that defines this vertical curve can be determined using the basic information that describes the curve:

$$Y = \frac{(G_2 - G_1)}{2L}x^2 + G_1x + \text{PVC}_{\text{elevation}}$$

$$x = \frac{L}{2} - 0.7432\text{ stations}$$

$$\text{PVC}_{\text{elevation}} = \text{PVI}_{\text{elevation}} - G_1\left(\frac{L}{2}\right) = 895.12\text{ ft} - 4.2\left(\frac{L}{2}\right)$$

$$884.26\text{ ft} = \frac{(-3.2 - 4.2)}{2L}\left(\frac{L}{2} - 0.7432\right)^2 + 4.2\left(\frac{L}{2} - 0.7432\right) + 895.12\text{ ft} - 4.2\left(\frac{L}{2}\right)$$

$$0 = \frac{-3.7}{L}\left(\frac{L}{2} - 0.7432\right)^2 + 2.1\,L - 3.1214 + 895.12\text{ ft} - 2.1\,L - 884.26\text{ ft}$$

$$0 = \frac{-3.7}{L}\left(\frac{L^2}{4} - 0.7432L + 0.5523\right) + 7.7386\text{ ft}$$

$$0 = -0.925L + 2.7498 - \frac{2.0435}{L} + 7.7386\text{ ft}$$

$$0 = -0.925L + 10.4884\text{ ft} - \frac{2.0435}{L}$$

$$0 = -0.925L^2 + 10.4884\text{ ft}L - 2.0435$$

Solve using the quadratic equation:

$$L = \frac{-b \pm \sqrt{b^2 - 4ac}}{2a}$$

Where, $a = -0.925$, $b = 10.884$, $c = -2.0435$

$$L = \frac{-10.884 \pm \sqrt{10.884^2 - 4(-0.925)(-2.0435)}}{2(-0.925)}$$

$L = 11.1405$ stations or 0.1983 stations

Select $L = 11.14$ stations (0.1983 or 19.83 ft results in a curve length that is too short).

PART 4

Highway Geometric Design

Daniel J. Findley, PhD, PE
Institute for Transportation Research and Education, North Carolina State University

Contents

4.1 INTRODUCTION

Highway geometric design comprises the processes necessary for choosing appropriate geometric features for a highway. Design controls govern key aspects of highway design and are essential for safety and efficiency. The geometric features considered in this section include the basic components that guide horizontal and vertical alignment, including curvature and grades, and elements that form the cross section of the highway, including lanes, shoulders, and medians. Intersections and interchanges are important parts of highway design due to their significant impact on safety performance and operational efficiency.

4.2 DESIGN CONTROLS

The design of a highway must fit within the confines of the broader highway network and is guided by policy-level parameters known as design controls. Design controls guide the design of the highway and serve to provide desired performance levels, consistency among similar functional highway classes, and implementation of proven safety and efficiency elements. The design controls consist of dimensional values, which directly affect the dimensions of the highway, and indirect factors, which influence the dimensional values, but are not explicitly apparent on the highway. A variety of sources provide recommendations for appropriate ranges of design elements, including federal, state, and local guidance. The primary general reference is the American Association of State Highway and Transportation Officials (AASHTO, 2011a) Green Book; however, other references may be more appropriate for specific highway types. Design controls do not include attributes that are not critical for safety and efficiency, and are better suited to local decisions made within the context of the area; these include considerations of cost, right of way, adjacent land uses, and so on. For example, the presence, width, and type of median is not a design control.

Although engineering judgment should be applied to every design, the following discussion of design controls sets the framework within which a designer must complete a design to sufficiently meet the needs of the project.

4.2.1 Indirect Design Factors

Indirect design factors are essential components that guide the selection of dimensional design values, but are not directly implemented in construction of the highway. The indirect design factors include:

- Design year
- Functional class
- Access control
- Design vehicle
- Design driver/user
- Capacity
- Design speed

Design Year

The design year has important implications on the design of a highway, particularly in the anticipation of future conditions and the ability of the highway to efficiently and safely process the demand. For large projects, the design year is commonly 20–30 years into the future. Predicting future demand is a difficult process and a well-developed future demand can inform the design process to reduce the risk of under- or overdesigning the highway. If predictions are unreliable or future demands are expected to remain constant in the near term and increase rapidly in the long term, design elements can be incorporated that will decrease future costs and construction time. These elements may include initial widening of bridges to accommodate additional lanes or acquiring necessary additional right of way for a wider facility.

Functional Class

The functional classification of a highway is typically identified during planning processes and is not at the discretion of the designer. The function of a highway has a considerable impact on the design elements, especially in the framework of providing the appropriate balance of mobility and accessibility. Arterials are designed with a goal of higher mobility, while local roads are designed for higher accessibility.

Access Control

Related to the functional classification, the type of access on a roadway affects mobility and accessibility. Interstate highways have full control of the access to the highway, providing access to drivers only through ramps at interchanges. While local and collector streets typically allow each parcel owner to directly access the road, higher-speed highways may use frontage roads to connect properties to other highways. The density of access points directly impacts the safety and efficiency of the road.

Design Vehicle

The design vehicle is critical for turning movements and is most influential at intersections. Large vehicles require larger turning radii to accommodate turns (particularly right turns), higher vertical clearances, and longer segments for acceleration and deceleration. Basic dimensions of common design vehicles are presented in Table 4.1 and additional dimensional data are available from AASHTO (2011a, Table 2-1b), including the front and rear overhang and the wheelbase (WB) lengths.

Design Driver/User

The design driver is inherent in the design of a highway when attempting to meet drivers' expectations and provide actionable information. A typical design driver is usually assumed to have slower than average information processing, reacting, and decision-making abilities. These attributes are applied through reaction times, signing, marking, and other human factor elements. Similarly, nonmotorized users, such as pedestrians and

Table 4.1 Design vehicle dimensions

Design vehicle type		**Height (ft)**	**Width (ft)**	**Length (ft)**
Passenger car/single unit trucks	Passenger car (P)	4.3	7.0	19.0
	Single-unit truck (SU-30)	11.0–13.5	8.0	30.0
	Single-unit truck (three-axle) (SU-40)	11.0–13.5	8.0	39.5
Buses	Intercity bus (motor coaches) (BUS-40)	12.0	8.5	40.5
	Intercity bus (motor coaches) (BUS-45)	12.0	8.5	45.5
	City transit bus (CITY-BUS)	10.5	8.5	40.0
	Conventional school bus (65 passengers) (S-BUS 36)	10.5	8.0	35.8
	Large school bus (84 passengers) (S-BUS 40)	10.5	8.0	40.0
	Articulated bus (A-BUS)	11.0	8.5	60.0
Combination trucks	Intermediate semitrailer (WB-40)	13.5	8.0	45.5
	Interstate semitrailer (WB-62)	13.5	8.5	69.0
	Interstate semitrailer (WB-67)	13.5	8.5	73.5
	"Double-bottom" semitrailer/trailer (WB-67D)	13.5	8.5	72.3
	Rocky mountain double-semitrailer/trailer (WB-92D)	13.5	8.5	97.3
	Triple-semitrailer/trailers (WB-100T)	13.5	8.5	104.8
	Turnpike double-semitrailer/trailer (WB-109D)	13.5	8.5	114.0
Recreational vehicles	Motor home (MH)	12.0	8.0	30.0
	Car and camper trailer (P/T)	10.0	8.0	48.7
	Car and boat trailer (P/B)	—	8.0	42.0
	Motor home and boat trailer (MH/B)	12.0	8.0	53.0

Adapted from AASHTO (2011a). Table 2-1b.

bicyclists, may require special considerations in the design of a roadway, particularly at crossing locations or intersections. These design elements may include sidewalks, bicycle lanes, median refuge islands, accessibility components for persons with disabilities, and so on.

Table 4.2 Recommended level of service

Functional highway classification	Rural: level and rolling terrain	Rural: mountainous terrain	Urban and suburban
Freeway	B	C	C or D
Arterial	B	C	C or D
Collector	C	D	D
Local	D	D	D

Adapted from AASHTO (2011a). Table 2-5.

Capacity

The capacity of the highway and other related operational measures are essential to consider when developing design characteristics for a specific highway. The capacity will affect critical design features including the number of through lanes and turning lanes, intersection configuration, and traffic control. The level of service associated with the capacity varies depending on the setting and functional class. Table 4.2 presents recommended design levels of service by functional highway classification, setting, and type of terrain. Level of service is rated on a scale of A through F, where A corresponds to free-flowing conditions and F represents breakdown conditions.

Design Speed

The design speed is an impact factor in the selection of dimensional values for many highway components, particularly the elements related to horizontal and vertical alignment. The design speed is the maximum safe operating speed under favorable vehicle, highway, and weather conditions. The speed limit is typically 5 or 10 mph lower than the design speed to provide an appropriate factor of safety for drivers. Consistency in design speed along a highway is desirable for safe and efficient operations. Table 4.3 presents recommended design speeds.

4.2.2 Dimensional Design Values

The dimensional design values are features that directly affect the size and shape of the highway. The dimensional design values include:

- Lane width
- Shoulder width
- Bridge width
- Horizontal alignment

Table 4.3 Recommended design speeds by roadway type and terrain

Roadway type	Terrain type	Rural design speed (mph)	Urban design speed (mph)
Freeway	Level	70	50+
	Rolling	70	50+
	Mountainous	50–60	50+
Arterial	Level	60–75	30–60
	Rolling	50–60	30–60
	Mountainous	40–50	30–60
Collector	Level	40–60	30+
	Rolling	30–50	30+
	Mountainous	20–40	30+
Local	Level	30–50	20–30
	Rolling	20–40	20–30
	Mountainous	20–30	20–30

Adapted from AASHTO (2011a).

- Superelevation
- Vertical alignment
- Grade
- Stopping sight distance
- Cross slope
- Vertical clearance
- Lateral offset to obstruction
- Structural capacity

Lane Width

Travel lanes provide the primary throughput capacity for a roadway; the width of travel lanes are an important value. Typical lane widths are between 9 and 12 ft. Twelve-feet-wide lanes are common on higher-speed roadways, while 9-ft-wide lanes may be used for low-volume, rural roads. Lane widths directly affect the required right of way necessary to construct the road, the cost of paving and repaving, crossing distances for pedestrians, operating speeds, likelihood of collisions resulting from leaving the travel lane, and available maneuvering and turning space for large vehicles. Table 4.4 presents recommended lane widths by roadway type.

Shoulder Width

Shoulders can be constructed with a variety of materials including pavement, gravel, turf, or earth. Shoulders are typically 2–12 ft wide

Table 4.4 Recommended lane width by roadway type

Roadway type	Lane width (ft)
Freeway	12
Ramp (1-lane)	12–30
Ramp (2-lane)	12–23
Arterial	10–12 (urban), 11–12 (rural)
Collector	10–12
Local	9–12

Adapted from AASHTO (2011a).

Table 4.5 Recommended shoulder width by roadway type

Roadway type	Shoulder width (ft)
Freeway	4–12
Ramp	2–12
Arterial	2–8
Collector	2–8
Local	2–8

Adapted from AASHTO (2011a).

(see Table 4.5). Adequate shoulder widths that are properly maintained can provide space for collision avoidance or errant vehicle recovery, disabled vehicle storage, maintenance activities, enforcement, additional space for bicyclists or oversized vehicles, improved sight distance, and support for travel lane pavement. To maximize the benefits available from shoulders, it is important for shoulders to be properly maintained (particularly unpaved shoulders) and continuous along the route, particularly on or under bridges.

Bridge Width

Insufficient bridge width has similar effects as reduced lane and shoulder widths. The reduced cross-section width decreases the amount of space available for motorists to correct any driving mistakes and increases the likelihood of collisions with other vehicles or the bridge railings.

Horizontal Alignment

Horizontal curves are a significant source of collisions on roadways due to the required cognitive and physical requirements of drivers. Adverse weather and roadway conditions can contribute to safety concerns at horizontal curves. A minimum radius, dictating the amount of curvature, can be determined by a maximum superelevation rate and design speed using AASHTO recommended values (AASHTO, 2011a). Longitudinal grades and design consistency of other curves along the roadway can also influence the safety performance. For more information on horizontal alignment, refer to Part 3.

Superelevation

Horizontal curves can be superelevated, with an elevated cross slope along the width of the pavement, to allow vehicles to travel through the curve at higher speeds (Figure 4.1). The cross section of the pavement must be rotated to create the superelevation. The superelevation rate must be selected so that equilibrium is maintained for the vehicles based on the design speed and radius of the curve. For more information on superelevation, refer to Part 3.

Vertical Alignment

Vertical alignment considerations as a design control include vertical curves (sag and crest) and grades. The following descriptions of grade

Figure 4.1 Superelevation in horizontal curve.

and vertical curve stopping sight distance provide additional details about considerations for vertical alignment. For more information on vertical alignment, refer to Part 3.

Grade

Grade, the longitudinal slope of a roadway, has significant operational and safety impacts on vehicles. Operationally, vehicles (particularly heavy vehicles) will lose speed on sustained upgrades of significant length and may need to descend at reduced speeds to alleviate braking issues. Reduced speeds by a portion of the vehicles can create a safety concern for other vehicles who may be traveling faster. Additionally, if heavy vehicles are unable to control their speed on steep descents, lane departure collisions may result. For flat or nearly flat grades, hydroplaning can be a serious concern. Table 4.6 presents the recommended maximum grades by design speed, roadway type, and type of terrain.

Stopping Sight Distance

Sight distance is a fundamental consideration for horizontal and vertical alignment. The ability of a driver to perceive and react to changes in the direction of the highway or the presence of potential hazards is essential for the safe operation of a highway. The amount of sight distance provided to the driver is influenced by the horizontal, vertical, and cross-section design of the highway. The minimum sight distance for every point along a highway is the stopping sight distance, which is affected by the operating speed of the roadway, the time to perceive and react to the need to stop, and the distance required to complete the stopping maneuver. For horizontal alignment, available stopping sight distance on a horizontal curve can be limited by vegetation, backslopes, walls, signs, fences, or buildings. Vertical curves affect stopping sight distance based on the limitations from the illumination of headlights at nighttime for sag curves and from physical restrictions caused by grade changes for crest curves. Inadequate stopping sight distance can result in collisions with hazards in the roadway, stopped or slow-moving vehicles, or vehicles turning onto the roadway. For more information on stopping sight distance, refer to Part 3.

Cross Slope

A straight highway segment is designed with a normal crown for the purpose of providing sufficient drainage of water off the surface of the

Table 4.6 Recommended maximum grade by roadway type

Terrain type	Freeway design speed (mph)[1]						
	50	**55**	**60**	**65**	**70**	**75**	**80**
Level	4	4	3	3	3	3	3
Rolling	5	5	4	4	4	4	4
Mountainous	6	6	6	5	5		

Terrain type	Urban arterial road design speed						
	30	**35**	**40**	**45**	**50**	**55**	**60**
Level	8	7	7	6	6	5	5
Rolling	9	8	8	7	7	6	6
Mountainous	11	10	10	9	9	8	8

Terrain type	Rural arterial road design speed								
	40	**45**	**50**	**55**	**60**	**65**	**70**	**75**	**80**
Level	5	5	4	4	3	3	3	3	3
Rolling	6	6	5	5	4	4	4	4	4
Mountainous	8	7	7	6	6	5	5	5	5

Terrain type	Urban collector road design speed								
	20	**25**	**30**	**35**	**40**	**45**	**50**	**55**	**60**
Level	9	9	9	9	9	8	7	7	6
Rolling	12	12	11	10	10	9	8	8	7
Mountainous	14	13	12	12	12	11	10	10	9

Terrain type	Rural collector road design speed								
	20	**25**	**30**	**35**	**40**	**45**	**50**	**55**	**60**
Level	7	7	7	7	7	7	6	6	5
Rolling	10	10	9	9	8	8	7	7	6
Mountainous	12	11	10	10	10	10	9	9	8

Terrain type	Local rural road design speed[2]								
	15	**20**	**25**	**30**	**40**	**45**	**50**	**55**	**60**
Level	9	8	7	7	7	7	6	6	5
Rolling	12	11	11	10	10	9	8	7	6
Mountainous	17	16	15	14	13	12	10	10	

[1]Freeways in urban areas (with right-of-way constraints) or in mountainous areas may use grades that are 1% steeper.
[2]Local urban roads should be limited to a maximum of 15% and 8% in commercial/industrial areas.
Adapted from AASHTO (2011a).

highway. Normal crown has a rooftop shape that peaks in the center of the roadway and falls away from the centerline at a typical rate of 1.5–2% (steeper cross slopes may be used in situations where water ponding is likely to occur, such as with curb and gutter). Insufficient cross slope can create drainage and hydroplaning concerns on the roadway, while excessive cross slope can result in unexpected lateral forces on vehicles that can destabilize the vehicle during braking maneuvers or when crossing the peak of the cross slope. Rollover, the algebraic difference in cross slopes between adjacent roadway features, should not exceed 4% between lanes or 8% at the edge of the shoulder. For more information on cross slope, refer to Part 3.

Vertical Clearance

The amount of vertical clearance between the surface of the roadway and any obstructions crossing the roadway is dictated to allow for safe and efficient movement of vehicles. Interstate minimum vertical clearance standards were initially developed to support logistics for national defense needs. The basis for the minimum value is to provide for a buffer of 1 ft between the bottom of the vertical structure and roadway surface when accounting for the height of the tallest legal vehicle. The value is the minimum allowable vertical clearance across the lanes and shoulders and should include considerations of the impact of future resurfacing that might raise the height of the pavement surface or the accumulation of snow and ice. Insufficient vertical clearance can limit the usefulness of a route for large vehicles and has the potential for collisions with the overhead structure for drivers that disobey or are unaware of the limited vertical clearance. The recommended minimum vertical clearance is 14.5 ft.

Lateral Offset to Obstruction

As a design control, the lateral offset to obstruction is concerned with the distance from the edge of the travel lane to a roadside obstruction and its influence on roadway operations. Obstructions are elements that have a vertical dimension and influence vehicle operations; the impact on safety is not considered in this design control. Safety impacts are considered when designing an appropriate clear zone, which is an important measure, but is not included as a design control due to the flexibility required for specific local conditions. Common obstructions include barriers, curbs, retaining walls, signs, trees, and utility poles. If an inadequate lateral offset is applied, the obstruction may influence speed and lane position of vehicles, as well as vehicle access when on-street parking is allowed. The minimum offset to diminish the impact of the obstruction on vehicles operations is 18 in, while 4–6 ft is recommended (AASHTO, 2011b).

Structural Capacity

The structural capacity of a bridge must be considered when determining the volume and types of vehicles that will use the roadway. This load-carrying capacity should be designed using appropriate structural engineering resources.

4.2.3 Design Exceptions

The Federal Highway Administration (FHWA) requires a formal exception if the criteria for design controls are not met on the National Highway System.[1] The design controls evaluated by FHWA include all the dimensional values described previously, as well as design speed. FHWA considers these attributes to be significant in the safe and efficient operations of highways. This formal allowance process for design exceptions is an integral part of the design process that requires professional judgment to alleviate or minimize any potential negative impact to the motoring public. A design exception may be needed to reduce impacts to the natural or human environment or to reduce construction or right-of-way costs. The acceptance of a design exception may be supported by satisfactory performance of the same design exception on a similar facility, collision modeling, input from public involvement, or natural or human environmental impacts. The design exception process includes the following steps[1]:

1. Determine the costs and impacts to meet design criteria
2. Develop and evaluate multiple alternatives
3. Evaluate risk
4. Evaluate mitigation measures
5. Document, review, and approve
6. Monitor and evaluate in-service performance

For each design element, Table 4.7 presents potential mitigation strategies based on the objective of the measure. For example, if the design

[1] FHWA. *Mitigation Strategies for Design Exceptions*. Federal Highway Administration. United States Department of Transportation. July 2007.

Table 4.7 Mitigation strategies for design exceptions

Design element	Objective	Mitigation strategy
Design speed	Reduce operating speeds to the design speed	Cross-sectional elements to manage speed
Lane width and shoulder width	Optimize safety and operations by distributing available cross-sectional width	Select optimal combination of lane and shoulder width for given site characteristics
	Provide advance warning of lane width reduction	Signing
	Improve ability to stay within the lane	Wide pavement markings Recessed or raised pavement markings Delineators Lighting Centerline or shoulder rumble strips
	Improve ability to recover if driver leaves the lane	Paved shoulders Safety edge
	Reduce crash severity if driver leaves the roadway	Remove or relocate fixed objects Traversable slopes Breakaway safety hardware Shield fixed objects and steep slopes
	Provide space for enforcement and disabled vehicles	Pull-off areas
Bridge width	Provide advance warning and delineation of narrow bridge and improve visibility of narrow bridge, bridge rail, and lane lines	Signing Reflectors on approach guardrail and bridge rail Post-mounted delineators Object markers High-visibility bridge rail Bridge lighting Enhanced pavement markings
	Maintain pavement on bridge that will provide safe driving conditions	Skid-resistant pavement Antiicing systems
	Reduce crash severity if driver leaves the roadway	Crashworthy bridge rail and approach guardrail

(*Continued*)

Table 4.7 (Continued)

Design element	Objective	Mitigation strategy
	Provide space for disabled vehicles or emergencies on long bridges	Pull-off areas
	Provide quick response to disabled vehicles or emergencies on long bridges	Real-time video monitoring
Horizontal alignment and superelevation	Provide advance warning	Signing Pavement marking messages Dynamic curve warning systems
	Provide delineation	Chevrons Post-mounted delineators Reflectors on barrier
	Improve ability to stay within the lane	Widen the roadway Skid-resistant pavement Enhanced pavement markings Lighting Centerline or shoulder rumble strips Painted edgeline rumble strips
	Improve ability to recover if driver leaves the lane	Paved or partially paved shoulders Safety edge
	Reduce crash severity if driver leaves the roadway	Remove or relocate fixed objects Traversable slopes Breakaway safety hardware Shield fixed objects and steep slopes
Vertical alignment	Refer to mitigation strategies for grade and stopping sight distance	
Grade	Provide advanced warning	Signing
	Improve safety and operations for vehicles ascending or descending steep grades	Climbing or descending lanes

(*Continued*)

Table 4.7 (Continued)

Design element	Objective	Mitigation strategy
	Capture out-of-control vehicles descending steep grades	Escape ramps
	Improve ability to stay within the lane	Enhanced pavement markings Delineators Centerline or shoulder rumble strips
	Improve ability to recover if driver leaves the lane	Paved shoulders Safety edge
	Reduce crash severity if driver leaves the roadway	Remove or relocate fixed objects Traversable slopes Breakaway safety hardware Shield fixed objects and steep slopes
	Address drainage on flat grades	Adjusting gutter profile on curbed cross sections Continuous drains
Stopping sight distance	Mitigate sight distance restrictions	Signing and speed advisory plaques (crest vertical curves) Lighting (sag vertical curves) Adjust placement of lane within the roadway cross section (horizontal) Cross-sectional elements to manage speed
	Improve ability to avoid crashes	Wide shoulders Wider clear recovery area
	Improve driver awareness on approach to intersections	Advanced warning signs Dynamic warning signs Larger or additional stop/yield signs Intersection lighting
Cross slope	Provide warning of slick pavement	Signing

(Continued)

Table 4.7 (Continued)

Design element	Objective	Mitigation strategy
	Improve surface friction	Pavement grooving (concrete pavement) Open-graded friction courses (asphalt pavement)
	Improve drainage	Transverse pavement grooving (concrete pavement) Open-graded friction courses (asphalt pavement) Pavement edge drains
	Mitigate cross-slope break on the high side of superelevated curves	Modified shoulder cross slope
Vertical clearance	Advance warning	Signing
	Preventing impacts with low structures	Alternate routes Large vehicle restrictions
Lateral offset to obstruction	Improve visibility of objects near the roadway	Delineate objects Lighting
	Optimize operations by distributing available cross-sectional width	Provide full outside lane width and/or additional offset
	Improve visibility of the lane lines	Enhanced pavement markings
Structural capacity	Refer to relevant structural design sources for detailed alternatives	

Adapted from FHWA. *Mitigation Strategies for Design Exceptions*. Federal Highway Administration. United States Department of Transportation. July 2007.

speed needed to be reduced below the prevailing design speed of the roadway, cross-sectional elements, such as lane width or shoulder width could be reduced to encourage drivers to decrease their operating speeds.

4.3 BASIC HIGHWAY SEGMENTS

Basic highway segments are nonintersection locations and can be considered the typical or desired design features for the road. The layout consists

Figure 4.2 Typical urban/suburban freeway section with concrete median barrier.

of a collection of relevant design controls, covered in the previous section, and design elements that should be selected based on the context and function of the road.

4.3.1 Arterials

Arterials focus primarily on mobility with an emphasis on providing high-speed, uninterrupted flow. Freeways, a subclassification of arterials, are the highest functional classification of highways and carry a significant portion of traffic volumes, based on lane miles of road. In the United States, although the majority of the interstate system is complete, new sections and loops are under development or in planning stages. Additionally, as the design life of many freeways are approached or exceeded, there is a need for extensive rehabilitation. Freeways are an essential part of the highway network, particularly for travel that occurs between cities, regions, and states. Well-designed freeways have the ability to support economic development through the safe and efficient travel of goods and people. Figure 4.2 shows a typical freeway segment in an urban/suburban setting with a concrete median barrier to minimize

Figure 4.3 Typical rural freeway section with wide median.

the necessary right of way and Figure 4.3 presents a typical freeway segment in a rural setting with a wide median to enhance safety and minimize barrier costs.

A focal point of freeway design is the management of the convergence and divergence of routes and lanes. Interstates and primary highway routes regularly run concurrently before splitting, requiring the need for communicating the convergence and divergence of the routes to drivers. Applying the concepts of route and lane continuity can reduce collisions and improve operations. Lane or route continuity violations can result in drivers who make late lane merging and diverging movements.

Route continuity concerns facility-level decisions about prioritizing merge and diverge points of concurrent routes. The mainline should be dedicated to the principle route and ramps should be used for the secondary route; this is consistent with the expectation of drivers, who presume that they will not need to take an exit ramp to stay on the primary route. Implementing the concept of route continuity can reduce the total miles driven and sudden lane- and/or speed-changing behavior. Typical prioritization of routes starts with two-digit interstate routes and is followed by three-digit interstate loops or bypasses, U.S.-designated routes, state routes, and finally, other secondary routes. Occasionally, routes of significant local importance may be considered a higher priority than a concurrent route of a higher designation, if traffic volumes and driver expectations suggest that the lower designated route would provide more benefits.

Similar to route continuity, applying the concepts of lane continuity can improve the safety performance and operations of a highway. The primary consideration of lane continuity is that a consistent set of lanes does not end. Drivers expect that while traveling in the main through lanes of a highway, the lanes will not end or exit and that no lane changing should be required to remain on the primary route. For additional capacity, which commonly occurs as highways transition from rural to urban settings, auxiliary lanes should be added and terminated on the same side of the roadway. When through lanes are terminated or used as exit-only ramps, lane and speed changing is a common action of drivers, which can disrupt traffic.

4.3.2 Collector

Functionally, arterials and collectors may have considerable differences, as arterials typically focus on mobility, while collectors have a blended objective of maintaining mobility and access. Although arterials and collectors are differentiated primarily by their operations, from a geometric design perspective arterials (without full control of access, such as freeways) and collectors may have very similar features with only slight differences to favor mobility or access more heavily. Collectors cover a wide spectrum of needs as the connecting roadway types between arterials and local streets. Therefore, careful consideration should be given to the design elements of these roadways, particularly the type and quantity of access that is provided to adjacent land uses and potential future land uses (Figure 4.4).

Access Management

A primary design decision on arterials and collectors is how to manage access to developments. Restricting turns (particularly left turns) and using frontage roads are common techniques in access management for high-volume roadways.

Figure 4.4 Collector roadway with pedestrian facilities and frequent access points.

Turns can be restricted from the major roadway and from side streets and developments. For traffic turning along a roadway, medians with selected openings are a common physical restriction, while signs may communicate prohibited turning movements for all or a portion of the day. For traffic turning onto a roadway, diverters and islands (along with appropriate signage) are common treatments to allow traffic to the right only. Concentrating left turns at designated locations (intersections and other median openings) can improve the operations and safety performance along a corridor.

Frontage roads, which are roads adjacent and parallel to an arterial, can be used to accumulate traffic from developments and concentrate turning movements on the arterial at major intersections. Frontage roads can significantly improve the efficiency and safety of the arterial, but require additional right of way and special considerations for the closely spaced intersections.

4.3.3 Local

Local streets provide direct connectivity to businesses, residences, and other land uses (Figure 4.5). Local streets can be designed to provide

Figure 4.5 Local street.

access while minimizing speeds. A prevalence of turning movements and nonmotorized usage on local streets accentuates the need for controlling speeds. Techniques for controlling speeds, known as traffic calming, can be implemented during construction or as a retrofit to streets experiencing high speeds.

Traffic Calming

Traffic calming techniques are used to provide more balance between motorized and nonmotorized roadway users on local and neighborhood roads. Many roadways are automobile focused and agencies strive to minimize travel time and delay for motorists, while on other roadways, agencies may seek to encourage heavy pedestrian volumes. Figure 4.6 shows an example of a significant change along a roadway and techniques used to communicate to drivers that the function of the roadway has changed. In the foreground of the exhibit, the roadway environment is heavily pedestrian-focused, while the background is more automobile-focused.

Figure 4.6 Traffic calming measures between automobile-focused and pedestrian-focused zones.

The roadway features change significantly between these two zones, including the use of auditory and vibratory signals through the pavement texture (using bricks) to communicate the change in the roadway function. Several visual elements also serve to reduce speeds and convey the change in user priority toward nonmotorized users, including pavement

Figure 4.7 Horizontal feature for transition to low speed residential area.

color, the lack of lane markings for vehicles, and the availability of on-street parking.

Traffic calming elements typically have a horizontal, vertical, or informative component. Horizontal elements modify the cross section of the road, including realignment of the roadway, changes in the width, or roadside elements that create friction for drivers. Figure 4.7 shows horizontal elements used to restrict the available roadway width to communicate a change from a higher speed area to a residential zone and to discourage use by trucks. Figure 4.8 shows a horizontal feature in the center of a roadway to provide a refuge area for pedestrians and to reduce vehicular speeds near the crosswalk. Vertical elements affect speeds directly through a vertical deflection in the road. Figure 4.9 presents an example of a vertical element, a speed table, used to reduce speeds and, in this instance, provide a raised crosswalk for pedestrians. Information directed toward drivers may also serve to calm traffic, including traffic control devices (signs and markings) and pavement changes. Law enforcement efforts and driver education and outreach can also serve to modify drivers' behavior. Figure 4.10 shows signage that alerts drivers of the upcoming traffic calming measures in addition to vertical elements.

Figure 4.8 Horizontal feature for pedestrian refuge and speed reduction at crosswalk.

Figure 4.9 Vertical feature for pedestrian crossing and speed reduction at crosswalk.

Traffic calming measures should be deployed with a consideration for the local roadway network. Although the measures may have the intended outcomes on the road with the traffic calming implementation, adjacent roads may begin to serve the diverted traffic, resulting in higher speeds and volumes on other nearby roads. From a network-level perspective, efficient arterial and collector roads with acceptable levels of delay can reduce the desire for through traffic to utilize local streets.

Figure 4.10 Informational signage coordinated with vertical feature for traffic calming.

4.4 CROSS-SECTION ELEMENTS

Cross sections provide the third element (along with horizontal and vertical alignment) that gives the highway its volumetric proportions. Elements included in the cross section give the roadway its width (lanes, shoulders, ditches, etc.). The cross section of a roadway extends from the edge of the right of way on one side to the edge of the right of way on the other side of the roadway. Cross sections provide the details needed to connect the roadway features with the existing elevations on the natural ground and are essential for estimating earthwork quantities necessary for construction cost estimates. The width included in the right of way may be significantly larger than the width of the travel lanes to incorporate adequate drainage features, allow for future expansion of roadway capacity, and to provide enough lateral distance alongside the roadway for errant drivers to regain control of their vehicles. Some cross-section elements have minimum values dictated by design controls, but many of the elements that significantly affect the width of cross section should be selected based on the context of the road and specific design objectives (see Figures 4.11–4.16).

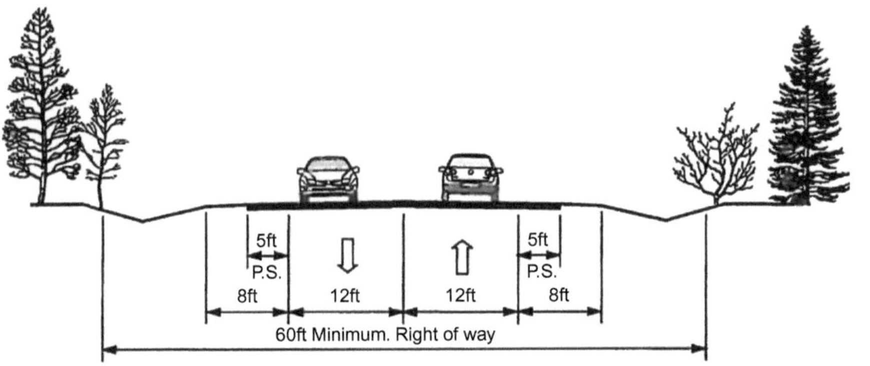

Figure 4.11 Typical cross section for two-lane undivided roadway with paved shoulders (PS) with 55 mph posted speed limit. *Source: NCDOT (2014).*

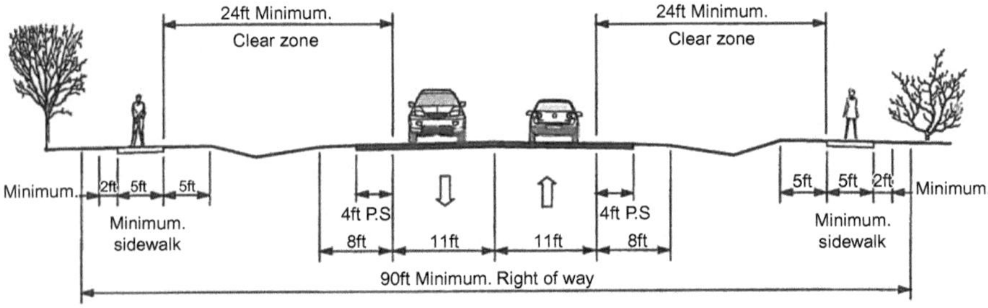

Figure 4.12 Typical cross section for two-lane undivided roadway with paved shoulders and sidewalks with 25–45 mph posted speed limit. *Source: NCDOT (2014).*

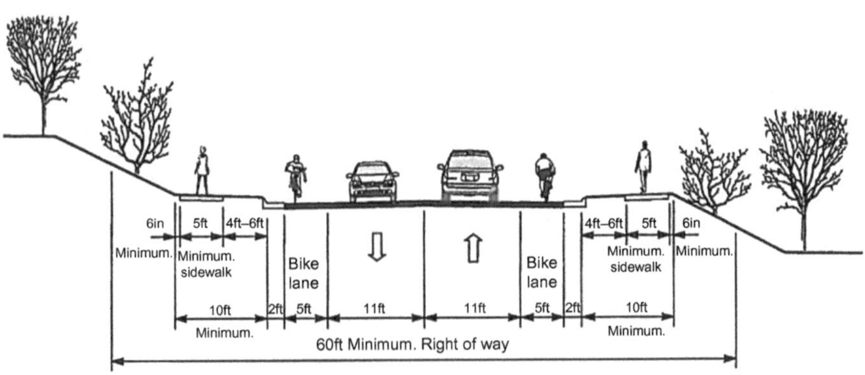

Figure 4.13 Typical cross section for two-lane undivided roadway with curb and gutter, bicycle lanes, and sidewalks with 25–45 mph posted speed limit. *Source: NCDOT (2014).*

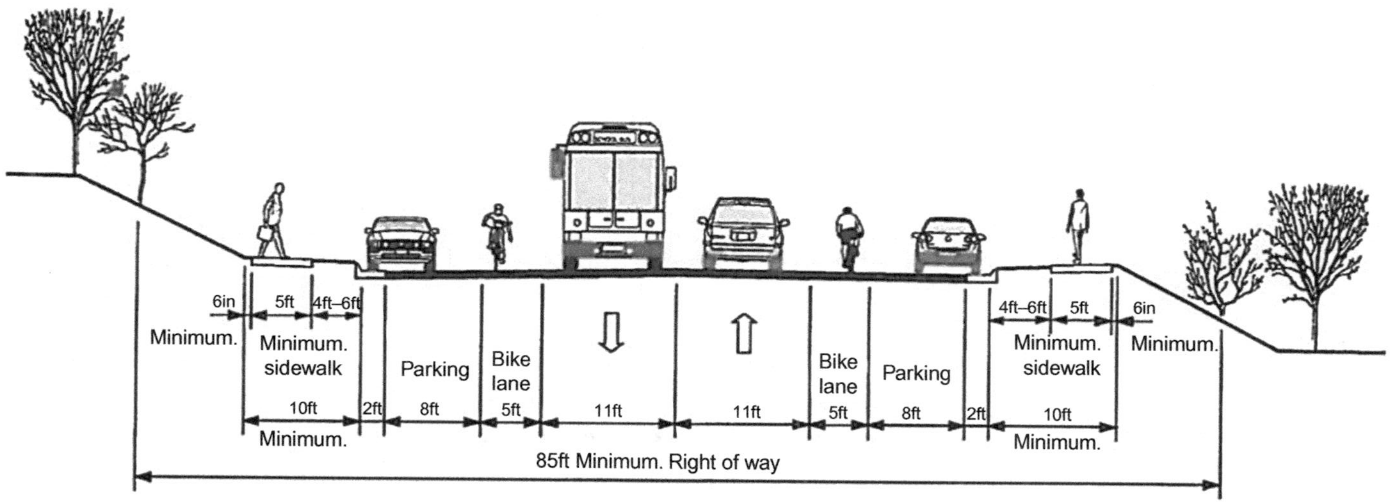

Figure 4.14 Typical cross section for two-lane undivided roadway with curb and gutter, bicycle lanes, on-street parking, and sidewalks with 25–45 mph posted speed limit. *Source: NCDOT (2014).*

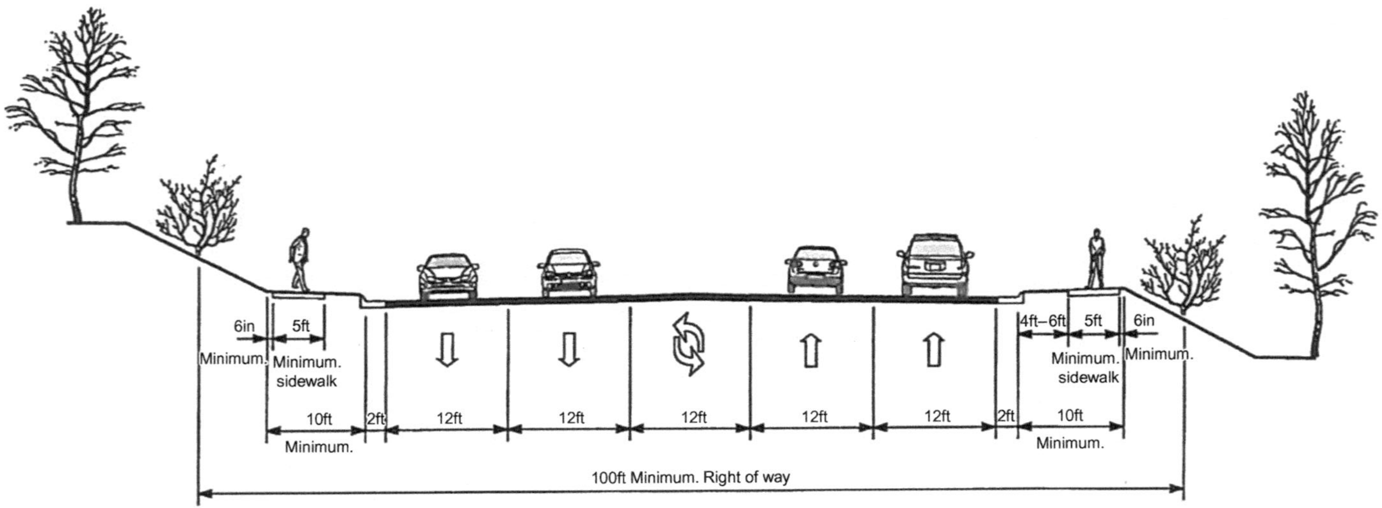

Figure 4.15 Typical cross section for four-lane roadway with a two-way left-turn lane, curb and gutter, and sidewalks with 35–45 mph posted speed limit. *Source: NCDOT (2014).*

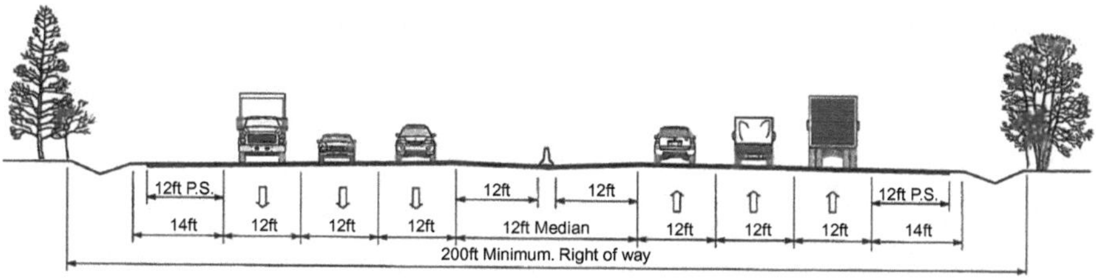

Figure 4.16 Typical cross section for six-lane divided roadway with paved shoulders and concrete median barrier with a 55–70 mph posted speed limit. *Source: NCDOT (2014).*

4.4.1 Roadway Design

Several cross-section elements and factors affecting the roadway cross section were covered in detail in the design controls section:

- Capacity
- Lane width
- Shoulder width
- Bridge width
- Superelevation
- Cross slope
- Lateral offset to obstruction

Additional cross-section elements, which are important to the overall safety and efficiency of a roadway but are not explicit design controls, include the following elements:

- Number of lanes
- Passing lanes
- Parking lanes
- Median treatment
- Bicycle and pedestrian facilities
- Curb and gutter

Number of Lanes

The necessary minimum capacity of the roadway is typically determined during planning processes, which may specify a minimum number of through travel lanes. However, the configuration and need for additional lanes may be determined after detailed analyses for the specific project. Additional travel way width is usually a positive feature for vehicular traffic, but the resulting higher speeds, longer crossing distances, and potentially larger volumes can negatively affect pedestrians and bicyclists.

At intersections, larger than necessary travel way widths can decrease the efficiency of the movements from additional delay and result in unnecessary construction and maintenance costs.

Passing Lanes

Passing lanes are supplemental lanes utilized in strategic locations to allow for passing maneuvers without requiring vehicles to cross into the opposing traffic lane. Passing lanes can significantly improve travel times, particularly where steep grades limit the operations of heavy vehicles and sharp horizontal curves constrain the opportunities for passing. Passing lanes installed in key locations can improve the operations and safety of sections of roadway beyond the end of the lane.

Parking Lanes

On-street parking lanes are a common feature of collector and local streets. Parking spaces can be designed as parallel, angled (back-in or pull-in), or perpendicular to the direction of traffic. Parking lanes can provide a buffer between moving vehicles and pedestrians; parking for retail businesses, residences, and restaurants; and additional roadway width for bus lanes and right-turn lanes at intersections. However, on-street parking lanes consume a portion of the right of way that could otherwise be used for different purposes, and parking maneuvers can create safety issues for bicyclists and motorists. A typical minimum width for parking lanes is 8 ft, while common widths range from 10–12 ft. Low-volume, low-speed local streets may allow parking on both sides of the street, which reduces the travel way to one lane.

Median Treatment

The specific configuration of the travel way to accommodate the projected future demand can also be selected during the design process. The inclusion of a center, two-way left-turning lane can serve to store and process left-turning traffic outside of the through lanes, or a median can be used to restrict left-turning movements to designated median openings (Figure 4.17). Two-way left-turning lanes, which can improve safety and efficiency relative to configurations with the same number of through lanes without a center turning lane, are popular in a three-lane configuration and also common in a five-lane configuration. Medians offer similar, but typically improved, safety and operational advantages as

Figure 4.17 Three-lane configuration with two-way left-turn lane.

two-way left-turning lanes through the consolidation of turns at designated locations.

Medians can be flush with the travel lanes or elevated about the travel lanes with a curb. Flush medians offer constructability and maintenance advantages, but might allow some vehicles to make prohibited movements. Raised medians provide effective access control, can be paved or landscaped, but can also be a fixed object hazard for errant vehicles. The width of the median can vary from a minimal amount to provide separation between the directions of traffic and a pedestrian refuge, to a large median meant to reduce the likelihood of crossover head-on collisions. Similar to the advantages provided by adequate shoulder widths, medians that are properly maintained can provide space for collision avoidance or errant vehicle recovery, disabled vehicle storage, aesthetics, drainage, maintenance activities, enforcement, improved sight distance, and increased turning radii. To maximize the benefits available from shoulders, it is important for medians to be properly maintained (particularly unpaved medians) and continuous along the route.

Bicycle and Pedestrian Facilities

Sidewalks, multiuse paths, and bicycle lanes provide separate facilities for nonmotorized users. Separation of nonmotorized users from moving motor vehicles, through parked vehicles, landscaping, curbs, markings, or other methods, can improve comfort levels. To maximize the effectiveness of nonmotorized networks, bicycle and pedestrian facilities should be connected and consistent.

Curb and Gutter

The curb and gutter controls water runoff from the roadway and directs it to drainage features. Curb, the vertical component, and gutter, the horizontal component connected to the base of the curb, can also assist with access management. The gutter is commonly 1- or 2-ft wide and has a slope steeper than the traveled way to facilitate drainage. Curbs are commonly 6-in tall and the slope of the curb face can vary from nearly vertical to traversable.

4.4.2 Roadside Design

Roadside design emphasizes a premise that the roadside environment should be forgiving for errant vehicles. Cost effectiveness is an important consideration for efficiently implementing roadside safety features. The roadside environment usually varies by functional category, operating speeds, traffic volume, available right of way, and terrain. The AASHTO *Roadside Design Guide* (2011b) is a primary reference for roadside design; it is intended to direct the design of roadsides on new facilities and can also be used to inform safety improvements for reconstructed roadways.

Barriers

Barriers can be a useful tool for shielding errant motorists from roadside hazards. Barriers are intended to contain and redirect a vehicle without exerting excessive forces on occupants. The AASHTO *Manual for Assessing Safety Hardware* (*MASH*) contains the current recommendations for testing and evaluating the safety performance of highway features and hardware (AASHTO, 2009). A general hierarchy of options (listed in order of priority) can be applied to evaluate possible alternatives.

1. Remove the obstacle.
2. Redesign the obstacle to include traversable features.

3. Relocate the obstacle to a location further offset from the edge of the roadway.
4. Reduce the severity of impact with inclusion of an appropriate breakaway device.
5. Shield the obstacle with a longitudinal traffic barrier designed for vehicle redirection or use a crash cushion.
6. Delineate the obstacle if the preceding alternatives are not appropriate.

Potential hazards for motorists include obstacles and nontraversable terrain, as presented in the following list and identified by AASHTO in its *Roadside Design Guide* (2011b). The characteristics and site conditions of each potential hazard should guide the decision of whether a barrier is necessary and what type of barrier is appropriate. Installing a barrier itself presents a potential hazard to errant motorists and can increase the quantity of less-severe collisions; therefore, barriers should only be installed when shielding would improve the overall safety performance of the roadside environment. Potential hazards include:

- Bridge piers, abutments, and railing ends[2]
- Boulders[3]
- Culverts, pipes, and headwalls[3]
- Foreslopes and backslopes (smooth)[4]
- Foreslopes and backslopes (rough)[3]
- Ditches (parallel)[3]
- Ditches (transverse)[2]
- Embankment[3]
- Retaining walls[3]
- Sign/luminaire supports (breakaway)[4]
- Sign/luminaire supports (nonbreakaway)[2]
- Traffic signal supports[3]
- Trees[3]
- Utility poles[2]
- Permanent bodies of water[3]

[2] Denotes elements that generally need shielding.
[3] Denotes elements that typically only need shielding when warranted.
[4] Denotes elements that typically do not require shielding.

EXAMPLE 4.1 Potential Hazard

A raised concrete drop inlet is present in the median of a freeway (Figure 4.18). A collision with the structure could result in severe injuries and significant vehicle damage. The hierarchy of options can be used to select a potential solution. The first and ideal option is to remove the hazard, but this feature is needed to collect storm-water runoff from the road and median. Next, a redesign of the obstacle for traversable features should be considered. Because a type of traversable drop inlet is available, the recommended option is to reconstruct the drop inlet to be traversable.

Figure 4.18 Raised concrete drop inlet in highway median.

If a barrier is determined to be the recommended solution to reduce collision severity, a variety of considerations, listed as follows, should be evaluated to select an appropriate barrier for the location:

- Appropriate structural strength (consult *MASH* recommendations for specific values)
- Cost and cost-effectiveness
- Acceptable deflection
- Height of the barrier

- Design vehicle weight and dimensions
- Design speed of the facility
- Roadside slopes
- Angle and direction of impact
- Climate
- Installation
- Replacement
- Maintenance (particularly for collision repair)
- Material storage
- Traffic volume
- Aesthetics
- Sight distance
- Site conditions
- Performance in similar conditions
- Compatibility with other barrier systems and terminal units

Three general categories of barriers can be used to shield motorists from hazardous obstacles: flexible, semirigid, and rigid barriers (see Figures 4.19–4.22). Generally, more flexible barriers are able to absorb more of the impact of the collisions; thereby transferring lower levels of the force of the impact to the occupants. However, flexible barriers require wider areas to allow for maximum deflection. Barriers should be installed as far from the edge of the travel lanes as possible to realize the advantages of shoulders and minimize the effect on traffic operations from a fixed object near traveling vehicles. End treatments of guardrails must be specially designed to absorb impact forces from vehicles that strike the

Figure 4.19 Flexible barrier (three-strand cable).

Figure 4.20 Flexible barrier (W-beam with weak steel post).

Figure 4.21 Semirigid barrier (W-beam with steel post and wood block).

Figure 4.22 Rigid barrier (concrete).

Figure 4.23 Earth berm.

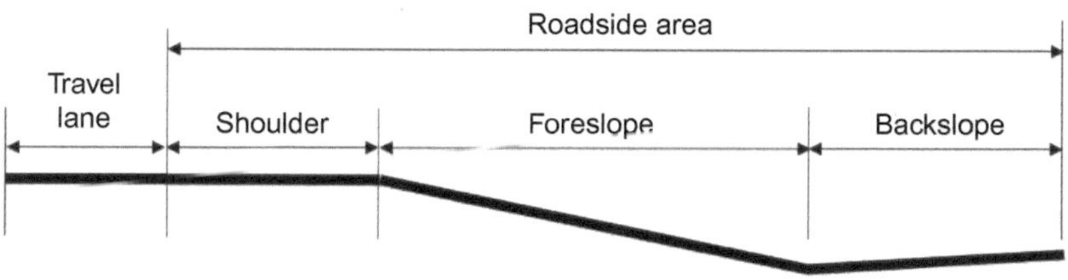

Figure 4.24 Roadside features.

end of the guardrail. Properly designed end treatments can reduce the occurrence of the guardrail spearing, launching, or puncturing the vehicle.

Earth berms are another useful barrier treatment to protect errant motorists and critical infrastructure (Figure 4.23).

Clear Zone

The clear zone is an important feature of the roadside environment that improves a driver's ability to control and recover their errant vehicle. The clear zone is an area adjacent to the edge of pavement that is traversable and unobstructed and varies in width depending on the design speed, slope, and risk of collisions. The *Roadside Design Guide* (AASHTO, 2011b) provides suggested design dimensions for clear zones. The width of the shoulder, foreslope, and backslope can contribute to the provision of the clear zone (Figure 4.24). The design speed and traffic volumes are

Table 4.8 Suggested clear zone distance

Design speed (mph)	Design Average Daily Traffic (ADT)	Suggested clear zone distance (ft)				
		Foreslopes		Backslopes		
		1V:6H (or flatter)	1V:5H–1V:4H	1V:3H	1V:5H–1V:4H	1V:6H (or flatter)
≤40	<750	7–10	7–10	7–10	7–10	7–10
	750–1500	10–12	12–14	12–14	12–14	12–14
	1500–6000	12–14	14–16	14–16	14–16	14–16
	>6000	14–16	16–18	16–18	16–18	16–18
45–50	<750	10–12	12–14	8–10	8–10	10–12
	750–1500	14–16	16–20	10–12	12–14	14–16
	1500–6000	16–18	20–26	12–14	14–16	16–18
	>6000	20–22	24–28	14–16	18–20	20–22
55	<750	12–14	14–18	8–10	10–12	10–12
	750–1500	16–18	20–24	10–12	14–16	16–18
	1500–6000	20–22	24–30	14–16	16–18	20–22
	>6000	22–24	26–32	16–18	20–22	22–24
60	<750	16–18	20–24	10–12	12–14	14–16
	750–1500	20–24	26–32	12–14	16–18	20–22
	1500–6000	26–30	32–40	14–18	18–22	24–26
	>6000	30–32	36–44	20–22	24–26	26–28
65–70	<750	18–20	20–26	10–12	14–16	14–16
	750–1500	24–26	28–36	12–16	18–20	20–22
	1500–6000	28–32	34–42	16–20	22–24	26–28
	>6000	30–34	38–46	22–24	26–30	28–30

Adapted from AASHTO (2011a). Table 3-1.

necessary to determine the dimensions of an appropriate clear zone, as shown in Table 4.8.

In addition to the width of the shoulder, the suggested clear zone distance can be accomplished on the foreslope or a combination of the foreslope and backslope. For designs that use the foreslope and backslope, the more conservative (larger) value should be selected from Table 4.8. Three general categories of roadside slopes include 1. recoverable slopes (1Vertical:4Horizontal or flatter), 2. critical slopes (1V:3H or steeper), and 3. slopes in-between, nonrecoverable slopes. Recoverable slopes are preferred, and the driver of an errant vehicle can be expected to reduce his or her speed and/or return to the roadway. In general, flatter slopes are more desirable than steeper slopes. Nonrecoverable slopes do not add to the width of the clear zone and any nonrecoverable slope distance that occurs within the suggested clear zone width should be added to the end of the slope, known as the clear runout area, to provide the total suggested distance. The clear runout area should have a minimum width of 10 ft. For steeper nonrecoverable slopes (approaching 1V:3H), fixed objects should not be present near the toe of the slope because of the high likelihood of vehicles to reach the end of the slope. A critical slope terminates the clear zone (i.e., a critical slope within the clear zone distance would result in a roadside that does not meet the suggested clear zone width). Site-specific experience with collisions may necessitate larger clear zones to provide additional safety for vehicles that leave the roadway. Conversely, if similar roadways have a history of acceptable safety performance, clear zones can be limited to 30 ft. Low-volume roadways may also require additional attention to determine the appropriate distance needed.

EXAMPLE 4.2 Clear Zone

Determine the minimum foreslope width to provide the recommended clear zone distance on a roadway with a 55-mph design speed, a design year annual average daily traffic (AADT) of 4000 vehicles per day, and other parameters as shown in the figure.

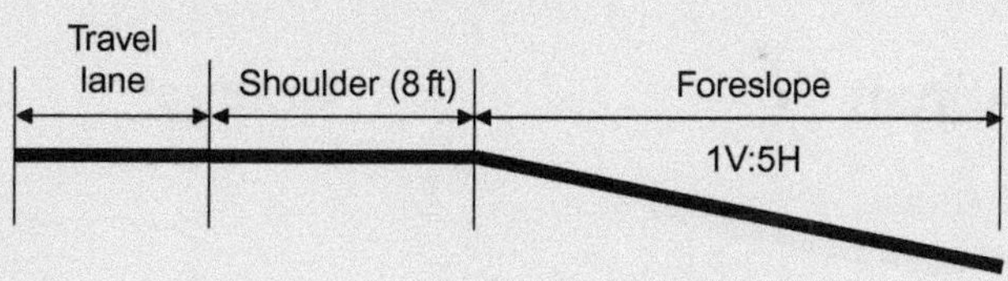

Solution

From Table 4.8 (55 mph design speed, 4000 vehicles per day, and 1V:5H slope), the recommended clear zone distance is 24–30 ft. The shoulder counts in the determination of the clear zone width, so the foreslope should be 16–22 ft wide.

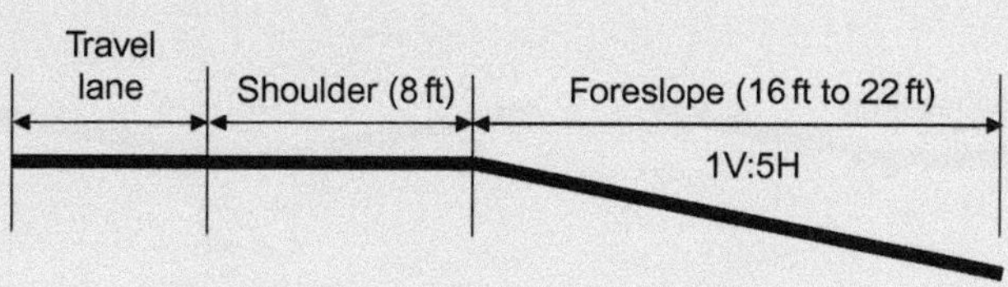

Drainage Channels

The connection of the foreslope and backslope forms a drainage channel. Drainage channels are typically offset and parallel to the roadway and collect storm-water runoff from the roadway and roadside regions for delivery to appropriate drainage systems. Although drainage channels serve the important function of storm-water removal, they should also be designed in a manner that does not present a hazard to motorists. An excessive change between the foreslope and backslope poses a risk to motorists who leave the roadway at a sharp angle (higher probability of sharp angles on the outside of horizontal curves), which may need a barrier or other design features for adequate safety performance.

Two general categories of drainage channels exist (AASHTO, 2011b): (1) abrupt and (2) gradual slope changes, as shown in Figure 4.25. Abrupt slope changes take the form of a V-shape, a rounded-bottom channel with a width of less than 8 ft, or a trapezoidal-bottom channel with a width of less than 4 ft. Gradual slope channels have a rounded-bottom width of greater than or equal to 8 ft or a trapezoidal-bottom channel with a width of greater than or equal to 4 ft. Table 4.9 shows the maximum recommended backslope for commonly used foreslopes.

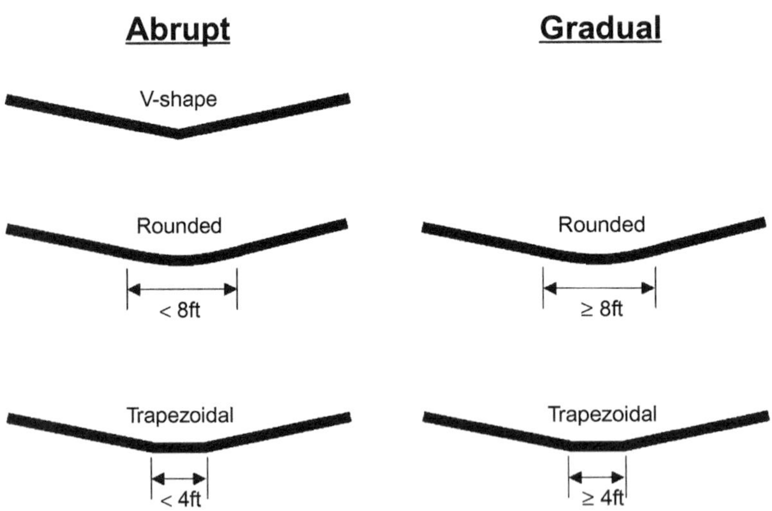

Figure 4.25 Drainage channel categories.

Table 4.9 Drainage channel dimensions by category

Foreslope (V:H)	Maximum steepness of backslope for abrupt slope changes (V:H)	Maximum steepness of backslope for gradual slope changes (V:H)
1:10	1:3.25	1:2.4
1:8	1:3.5	1:2.6
1:6	1:3.75	1:2.8
1:5	1:4.5	1:3
1:4	1:6	1:4
1:3	N/A	1:7
1:2	N/A	N/A

Adapted from AASHTO (2011b).

EXAMPLE 4.3 Drainage Channel

Determine the maximum backslope steepness for the drainage channel configuration shown in the following image.

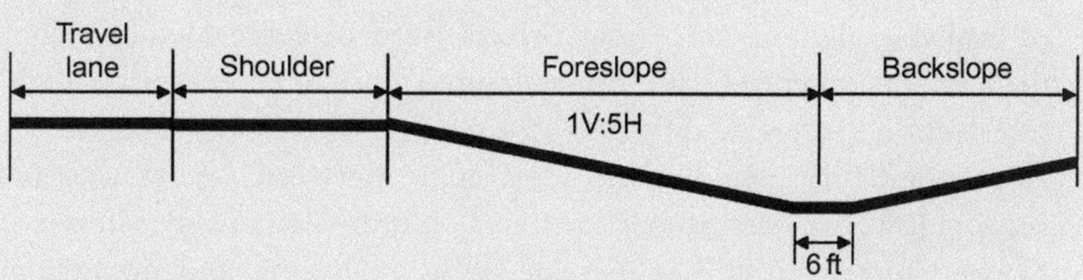

Solution

This drainage channel is defined as a gradual slope change because it has a trapezoidal shape with a base that is greater than or equal to 4 ft wide. From Table 4.9, the recommended maximum steepness of the backslope is 1V:3H.

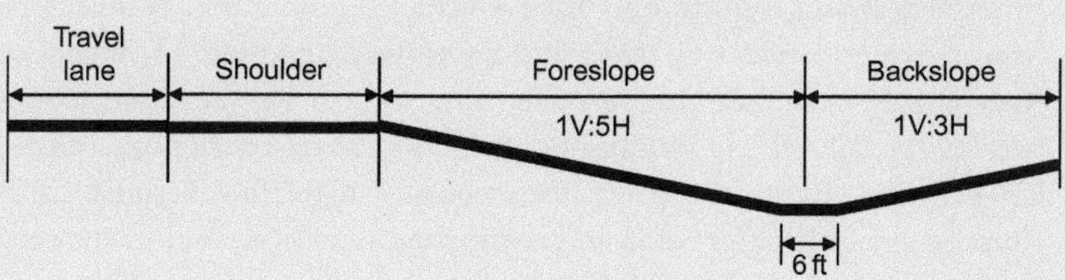

4.5 INTERSECTION DESIGN

Intersections are locations where roads cross each other; they are a primary focus in highway engineering. Intersections can be signalized or unsignalized and require special consideration from designers to appropriately manage conflicting, through, and turning vehicles. Intersections are leading locations for the occurrence of collisions and limiting factors in the capacity of arterials and collectors. Similar to other roadway facilities, intersections can balance motorized and nonmotorized users by considering the effects to all users. In addition to the features of individual intersections, corridor-level characteristics also affect the operational and safety performance of the road. Some general guidance for intersection design includes:

- The distance, or spacing, between consecutive intersections can be used to facilitate better quality of service for nonmotorized users with shorter distances. Alternatively, longer distances with better progression for coordinated signal systems benefit motorized users.
- The location of an intersection in relation to the horizontal and vertical alignment should target straight and flat areas to the extent possible. Horizontal and vertical curves near intersections can restrict the amount of available sight distance that drivers need to react to traffic control devices and other vehicles. Superelevated horizontal curves through an intersection can create difficulties for turning or crossing vehicles.
- A perpendicular angle is the ideal angle between intersecting roadways. However, site restrictions may require a skewed intersection. Skewed intersections can increase delay, collisions, and infrastructure costs. When possible, angles should differ by less than 30° from perpendicular. For intersections of excessive skew, it is possible to use reconstruction, as shown in Figure 4.26, to improve the angle.
- The number and directional restrictions of intersection approaches affect the safety and efficiency of the intersection, primarily through the reduced number of conflict points and signal phases. A standard three-leg intersection and roundabouts have 8 vehicle conflict points; an intersection between a two-way street and a one-way street has 13 vehicle conflict points; and a standard four-leg intersection between two two-way streets has 32 vehicle conflict points. Directional restrictions (one-way streets) can positively affect intersections with fewer signal phases, improved signal progression in coordinated networks, and fewer vehicle and pedestrian conflicts. However, directional restrictions on streets can also have negative impacts including increased travel distance and time, the likelihood of wrong-way movements, and increased speeds.

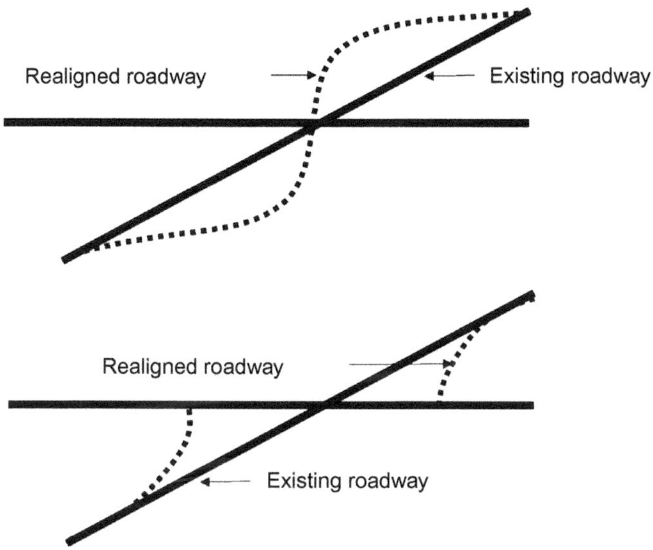

Figure 4.26 Realignment options for skewed intersections.

- Access points, driveways, or minor streets should be located as far from an intersection as possible. Closely spaced access points increase collisions and decrease efficiency due to turning vehicles.
- Pedestrians benefit from shorter crossing distances and median or island refuge areas to make multistage crossings. Design features should provide adequate accommodations for users of all abilities.
- Auxiliary through lanes—lanes added in advance of and removed after an intersection—can provide a boost to capacity and reduce overall intersection delay. However, the specific design of auxiliary through lanes must be carefully planned to ensure that vehicle utilization will make the additions cost effective.

4.5.1 Turn Lanes

Turn lanes for left- or right-turning vehicles allow space for vehicles to decelerate prior to turning and for vehicles to queue outside the through lanes. Separating turning and through traffic can reduce delay and improve safety. However, turn lanes also increase the width of the roadway, increasing the necessary right of way, making the crossing distance longer for opposing traffic and pedestrians, and increasing the likelihood for needing additional protected signal phases. The length of the full width of the turn lane should be adequate to allow for deceleration and more vehicles than the average queue produces.

4.5.2 Dimensions for Turning Movements

The radius for right turns, defined by the curb line or connection of two pavement edgelines at an intersection, has vehicular and pedestrian impacts. Vehicles, particularly heavy vehicles, benefit from larger turning radii for higher turning speeds and easier turning movements, while pedestrians encounter increased crossing distances with larger curb radii. Three options exist for the design of the turn: (1) a constant radius (Figure 4.27), (2) a constant radius connected to each edge of the travel lane by a taper (Figure 4.28), and (3) a three-center compound curve (Figure 4.29).

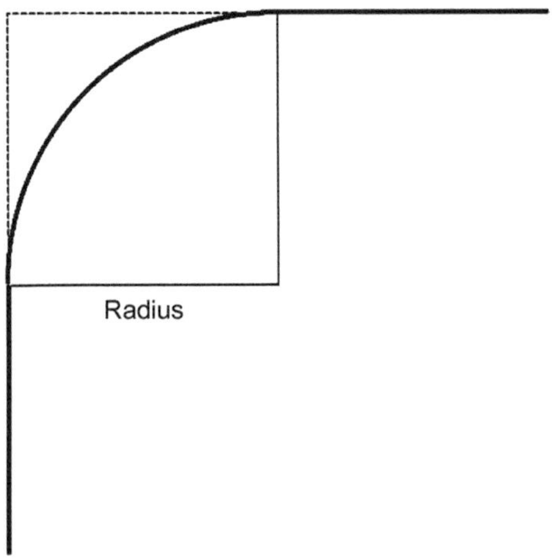

Figure 4.27 Simple curve radius.

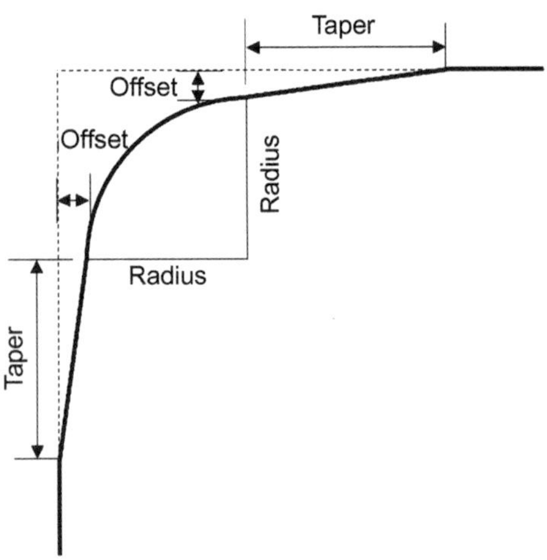

Figure 4.28 Simple curve radius with taper.

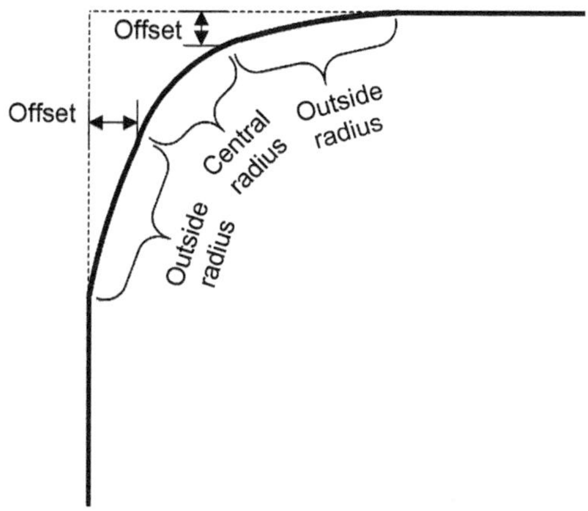

Figure 4.29 Three-center compound curve.

The constant radius is an economical and simple option, while the three-center compound curve is more complex, but provides a more natural turning movement for drivers. The constant radius with tapers is a reasonable compromise between the other options, with better constructability than the three-center compound curve and a better turning path than the constant radius. Table 4.10 presents recommended dimensions for each turning radius option. For the three-center compound curve, only the symmetric alternative is presented, but an asymmetric curve is also possible.

4.5.3 Channelization

Islands can be used to channelize, or separate, traffic (Figure 4.30). Channelization improves safety and operations by separating conflicting vehicles, indicating appropriate turning movements, and providing a refuge area for pedestrians. Islands are commonly located at an intersection to restrict left turns or to separate vehicles with different turning destinations. Islands may also serve to decrease the potential for wrong-way movements by communicating permitted movements to drivers, which can be especially useful for unconventional designs. Islands that restrict movements to only allow vehicles to make a right turn into and out of the road are often effective left-turn restrictions from driveways or side streets. However, the presence of islands is an additional roadway feature that errant vehicles may strike, and they also add to the size of an intersection. To maximize effectiveness, islands should be large enough to be prominently noticed by drivers and help guide them through the

Table 4.10 Turning radius values

Angle of turn	Design vehicle	Simple curve radius (ft)	Simple curve radius with taper			Three-center compound curve	
			Radius (ft)	Offset (ft)	Taper (ft) (length: width)	Curve radii (ft)	Symmetric offset (ft)
30	P	60	N/A	N/A	N/A	N/A	N/A
	SU-30	100	N/A	N/A	N/A	N/A	N/A
	SU-40	140	N/A	N/A	N/A	N/A	N/A
	WB-40	150	N/A	N/A	N/A	N/A	N/A
	WB-62	360	220	3.0	15:1	N/A	N/A
	WB-67	380	220	3.0	15:1	460–175–460	4.0
	WB-92D	365	190	3.0	15:1	550–155–550	4.0
	WB-100T	260	125	3.0	15:1	220–80–220	4.5
	WB-109D	475	260	3.5	20:1	550–250–550	5.0
45	P	50	N/A	N/A	N/A	N/A	N/A
	SU-30	75	N/A	N/A	N/A	N/A	N/A
	SU-40	115	N/A	N/A	N/A	N/A	N/A
	WB-40	120	N/A	N/A	N/A	N/A	N/A
	WB-62	230	145	4.0	15:1	460–240–460	2.0
	WB-67	250	145	4.5	15:1	460–175–460	2.0
	WB-92D	270	145	4.0	15:1	525–155–525	5.0
	WB-100T	200	115	2.5	15:1	250–80–250	4.5
	WB-109D	N/A	200	4.5	20:1	550–200–550	5.0
60	P	40	N/A	N/A	N/A	N/A	N/A
	SU-30	60	N/A	N/A	N/A	N/A	N/A
	SU-40	100	N/A	N/A	N/A	N/A	N/A
	WB-40	90	N/A	N/A	N/A	N/A	N/A
	WB-62	170	140	4.0	15:1	400–100–400	15.0
	WB-67	200	140	4.5	15:1	400–100–400	8.0
	WB-92D	230	120	5.0	15:1	480–110–480	6.0
	WB-100T	150	95	2.5	15:1	250–80–250	4.5
	WB-109D	N/A	180	4.5	20:1	650–150–650	5.5
75	P	35	25	2.0	10:1	100–25–100	2.0
	SU-30	55	45	2.0	10:1	120–45–120	2.0
	SU-40	90	60	2.0	10:1	200–35–200	5.0
	WB-40	N/A	60	2.0	15:1	120–45–120	5.0
	WB-62	N/A	145	4.0	20:1	440–75–440	15.0
	WB-67	N/A	145	4.5	20:1	420–75–420	10.0
	WB-92D	N/A	110	5.0	15:1	500–95–500	7.0
	WB-100T	N/A	85	3.0	15:1	250–80–250	4.5
	WB-109D	N/A	140	5.5	20:1	700–125–700	6.5
90	P	30	20	2.5	10:1	100–20–100	2.5
	SU-30	50	40	2.0	10:1	120–40–120	2.0
	SU-40	80	45	4.0	10:1	200–30–200	7.0

(*Continued*)

Table 4.10 (Continued)

Angle of turn	Design vehicle	Simple curve radius (ft)	Simple curve radius with taper			Three-center compound curve	
			Radius (ft)	Offset (ft)	Taper (ft) (length: width)	Curve radii (ft)	Symmetric offset (ft)
	WB-40	N/A	45	4.0	10:1	120–40–120	5.0
	WB-62	N/A	120	4.5	30:1	400–70–400	10.0
	WB-67	N/A	125	4.5	30:1	440–65–440	10.0
	WB-92D	N/A	95	6.0	10:1	470–75–470	10.0
	WB-100T	N/A	85	2.5	15:1	250–70–250	4.5
	WB-109D	N/A	115	2.9	15:1	700–110–700	6.5
105	P	N/A	20	2.5	8:1	100–20–100	2.5
	SU-30	N/A	35	3.0	10:1	100–35–100	3.0
	SU-40	N/A	45	4.0	10:1	200–35–200	6.0
	WB-40	N/A	40	4.0	10:1	100–35–100	5.0
	WB-62	N/A	115	3.0	15:1	520–50–520	15.0
	WB-67	N/A	115	3.0	15:1	500–50–500	13.0
	WB-92D	N/A	80	8.0	10:1	500–80–500	8.0
	WB-100T	N/A	75	3.0	15:1	250–60–250	5.0
	WB-109D	N/A	90	9.2	20:1	700–95–700	8.0
120	P	N/A	20	2.0	10:1	100–20–100	2.0
	SU-30	N/A	30	3.0	10:1	100–30–100	3.0
	SU-40	N/A	35	6.0	8:1	200–35–200	6.0
	WB-40	N/A	35	5.0	8:1	120–30–120	6.0
	WB-62	N/A	100	5.0	15:1	520–70–520	10.0
	WB-67	N/A	105	5.2	15:1	550–45–550	15.0
	WB-92D	N/A	80	7.0	10:1	500–70–500	10.0
	WB-100T	N/A	65	3.5	15:1	250–60–250	5.0
	WB-109D	N/A	85	9.2	20:1	700–85–700	9.0
135	P	N/A	20	1.5	10:1	100–20–100	1.5
	SU-30	N/A	30	4.0	10:1	100–30–100	4.0
	SU-40	N/A	40	4.0	8:1	200–40–200	4.0
	WB-40	N/A	30	8.0	15:1	120–30–120	6.5
	WB-62	N/A	80	5.0	20:1	600–60–600	12.0
	WB-67	N/A	85	5.2	20:1	550–45–550	16.0
	WB-92D	N/A	75	7.3	10:1	450–70–450	9.0
	WB-100T	N/A	65	5.5	15:1	250–60–250	5.5
	WB-109D	N/A	85	8.5	20:1	700–70–700	12.5
150	P	N/A	18	2.0	10:1	75–20–75	2.0
	SU-30	N/A	30	4.0	8:1	100–30–100	4.0
	SU-40	N/A	35	7.0	8:1	200–35–200	6.5
	WB-40	N/A	30	6.0	8:1	100–30–100	6.0
	WB-62	N/A	60	10.0	10:1	480–55–480	15.0

(Continued)

Table 4.10 (Continued)

Angle of turn	Design vehicle	Simple curve radius (ft)	Simple curve radius with taper			Three-center compound curve	
			Radius (ft)	Offset (ft)	Taper (ft) (length: width)	Curve radii (ft)	Symmetric offset (ft)
	WB-67	N/A	65	10.2	10:1	550–45–550	19.0
	WB-92D	N/A	65	11.0	10:1	350–60–350	15.0
	WB-100T	N/A	65	7.3	10:1	250–60–250	7.0
	WB-109D	N/A	65	15.1	10:1	700–65–700	15.0
180	P	N/A	15	0.5	20:1	50–15–50	0.5
	SU-30	N/A	30	1.5	10:1	100–30–100	1.5
	SU-40	N/A	35	6.4	10:1	150–35–150	6.2
	WB-40	N/A	20	9.5	5:1	100–20–100	9.5
	WB-62	N/A	55	10.0	15:1	800–45–800	20.0
	WB-67	N/A	55	13.8	10:1	600–45–600	20.5
	WB-92D	N/A	55	16.8	10:1	400–55–400	16.8
	WB-100T	N/A	55	10.2	10:1	250–55–250	9.5
	WB-109D	N/A	55	20.0	10:1	700–55–700	20.0

Adapted from AASHTO (2011a). Tables 9-15 and 9-16.

Figure 4.30 Channelization to restrict left turns.

intersection with rounded edges that match turning paths. A variety of considerations should guide the selection of an island's surface, shape, and offset from travel paths, including cost, aesthetics, maintenance requirements, safety impacts if struck, and visibility.

Figure 4.31 Multilane roundabout.

4.5.4 Roundabouts

The modern roundabout features low design speeds, yielding by entering vehicles, and deflection for every movement through the intersection (Figure 4.31). Roundabouts function well as the controlling device for moderate traffic volumes on two intersecting roads. When operating below capacity, roundabouts typically provide a delay savings relative to signalized intersections. Resulting from fewer conflict points and lower speeds, especially the elimination of conflicting traffic streams that can result in severe collisions, roundabouts provide improved safety performance over other intersection types.

4.5.5 Intersection Sight Distance

Sight distance is essential for the safe operation of intersections. Drivers approaching and maneuvering at an intersection must be able to adequately observe traffic control devices and conflicting traffic to take the appropriate action. Vertical, horizontal, and cross-sectional elements of each intersecting roadway may obscure a driver's sight lines necessary to perceive and react. Common visual impediments in the sight triangle

(Figures 4.32–4.35) include terrain, landscaping/vegetation, parked cars, and other roadside objects. For sight distance considerations at unsignalized intersections, any movements that conflict with a traffic stream that has the legal right of way should be examined for appropriate sight distance values. At signalized intersections, specific approaches that allow permitted left turns or right turns during the red interval should be the focus of sight distance evaluations. Additionally, if a signalized intersection is planned for operations in flashing mode during off-peak periods, all appropriate movements should be studied. The amount of sight distance provided to the driver at each relevant position should equal or exceed recommended minimum values. AASHTO (2011a) defines six cases for sight distance:

- Case A—Intersections with no traffic control
- Case B—Stop control on the minor road
- Case C—Yield control on the minor road
- Case D—Traffic signal control
- Case E—All-way stop control
- Case F—Left turns from the major road

Case A—Intersections with No Traffic Control

Uncontrolled intersections, with no traffic control signs or signals, rely on drivers to observe conflicting vehicles and stop before entering the intersection. The arrival of multiple vehicles at the same time is infrequent because uncontrolled intersections typically handle very low traffic volumes. The length of the approach legs defines the intersection sight distance for uncontrolled intersections (Figure 4.32 for left view and Figure 4.33 for right view). The sight triangle bounded by approach lengths should be clear of obstructions that could impeded the driver's sight of a conflicting vehicle. An obstruction is any object that impedes the line of sight between any two points between opposite legs within the sight triangle between points that are 3.5 ft above the surface of the roadway. The length of the legs for the approach sight triangle is based on driver behavior at intersections with no traffic control and stopping sight distance actions. If the grade of the approach to the intersection exceeds ±3%, the value selected from Table 4.11 should be adjusted by multiplying by the appropriate value in Table 4.12.

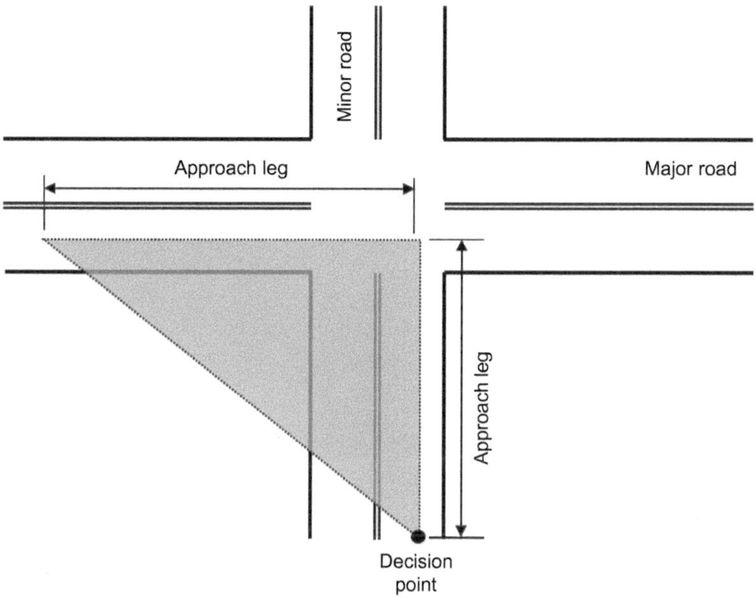

Figure 4.32 Left-view sight triangle for minor road approach to uncontrolled intersection. *Source: Adapted from AASHTO (2011a), Figure 9-15A.*

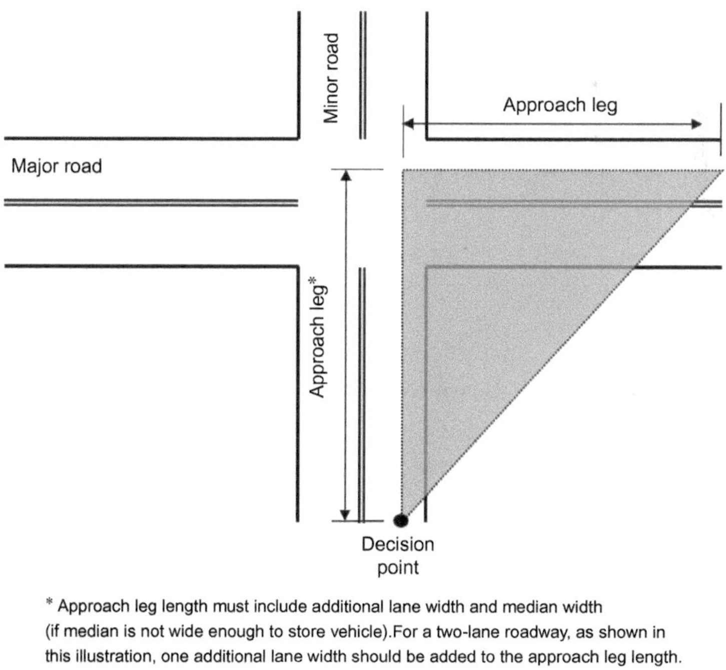

Figure 4.33 Right-view sight triangle for minor road approach to uncontrolled intersection. *Source: Adapted from AASHTO (2011a), Figure 9-15A.*

Table 4.11 Sight triangle approach leg length for uncontrolled intersections

Design speed (mph)	Length of approach leg (ft)
15	70
20	90
25	115
30	140
35	165
40	195
45	220
50	245
55	285
60	325
65	365
70	405
75	445
80	485

Adapted from AASHTO (2011a). Table 9-3.

Table 4.12 Multiplicative sight distance adjustment factors by approach grade

Approach grade (%)	Design speed													
	15	20	25	30	35	40	45	50	55	60	65	70	75	80
−6	1.1	1.1	1.1	1.1	1.1	1.1	1.1	1.2	1.2	1.2	1.2	1.2	1.2	1.2
−5	1.0	1.0	1.1	1.1	1.1	1.1	1.1	1.1	1.1	1.1	1.2	1.2	1.2	1.2
−4	1.0	1.0	1.0	1.1	1.1	1.1	1.1	1.1	1.1	1.1	1.1	1.1	1.1	1.1
−3 to +3	1.0	1.0	1.0	1.0	1.0	1.0	1.0	1.0	1.0	1.0	1.0	1.0	1.0	1.0
+4	1.0	1.0	1.0	1.0	0.9	0.9	0.9	0.9	0.9	0.9	0.9	0.9	0.9	0.9
+5	1.0	1.0	1.0	0.9	0.9	0.9	0.9	0.9	0.9	0.9	0.9	0.9	0.9	0.9
+6	1.0	1.0	0.9	0.9	0.9	0.9	0.9	0.9	0.9	0.9	0.9	0.9	0.9	0.9

Adapted from AASHTO (2011a). Table 9-4.

Case B—Stop Control on the Minor Road

Case B applies to vehicles that arrive at a stop-controlled approach of a minor road with an uncontrolled major road. This case applies to intersections commonly known as two-way stop-controlled intersections. At the stop-controlled approach on the minor road, three movements are possible: left (Case B1), right (Case B2), and through (Case B3). These movements apply to vehicles that have stopped in accordance with the stop sign, and the sight distance is necessary to safely accommodate their departure from the stop.

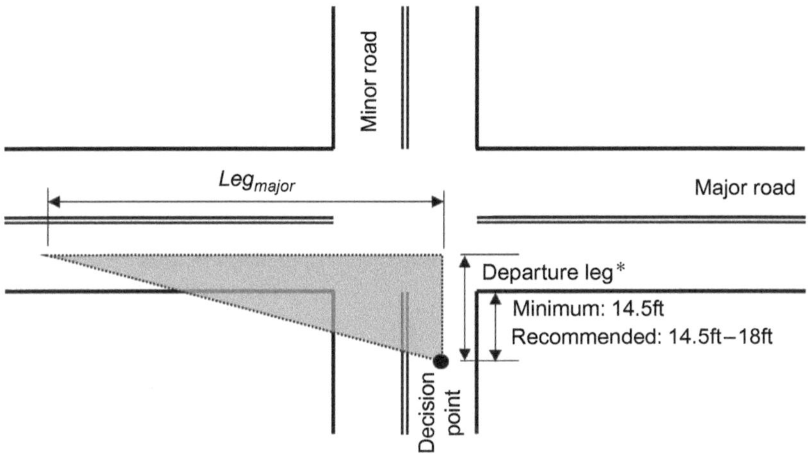

Figure 4.34 Left-view sight triangle for minor road departure from stop-controlled intersection. *Source: Adapted from AASHTO (2011a), Figure 9-15B.*

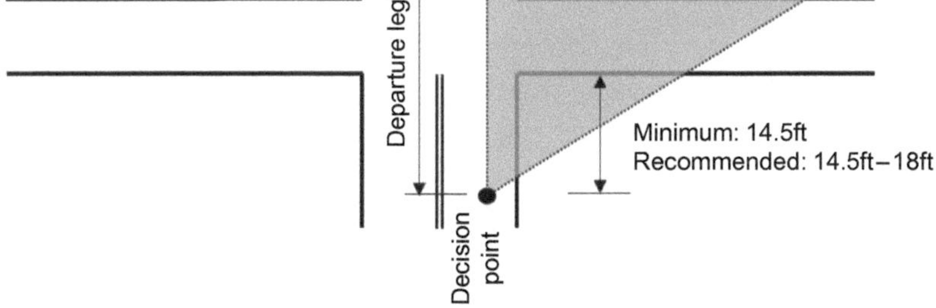

Figure 4.35 Right-view sight triangle for minor road departure from stop-controlled intersection. *Source: Adapted from AASHTO (2011a), Figure 9-15B.*

Table 4.13 Sight triangle approach leg time gap for stop-controlled minor road left movement

Design vehicle	Base time gap (s)	Supplement to time gap	
		Extra lanes to cross on major road	Approach grade
Passenger car	7.5	0.5 s for each additional lane	0.2 s for each additional percent beyond 3% grade
Single-unit truck	9.5	0.7 s for each additional lane	0.2 s for each additional percent beyond 3% grade
Combination truck	11.5	0.7 s for each additional lane	0.2 s for each additional percent beyond 3% grade

Adapted from AASHTO (2011a). Table 9-5.

Case B1—Left from Minor Road

For left-turning movements from the minor road onto the major road, the following equation and associated table can be used to determine the recommended minimum sight distance along the major road. Table 4.13 contains the base time gap by design vehicle that can be used in the analysis of a two-way, two-lane road with approach grades less than 3%. Adjustments should be made to increase the base time gap when additional lanes must be crossed and/or when the approach grade exceeds an upgrade of 3%.

$$Leg_{major} = 1.47\,V_{major}t_g \quad \text{(Adapted from AASHTO, 2011a, Equation 9-1)}$$

where

Leg_{major} = length of the leg of the sight triangle along the major road in feet

V_{major} = design speed of the major road in miles per hour

t_g = time gap necessary for the vehicle stopped on the minor road to turn left onto the major road in seconds (see Table 4.13 for details to determine time gap for specific site conditions)

Case B2—Right from Minor Road

For right-turning movements from the minor road onto the major road, the following equation and associated table can be used to determine the recommended minimum sight distance along the major road. Table 4.14 contains the base time gap by design vehicle. Adjustments should be made to increase the base time gap when the approach grade exceeds an upgrade of 3%.

$$Leg_{major} = 1.47\,V_{major}t_g \quad \text{(Adapted from AASHTO, 2011a, Equation 9-1)}$$

Table 4.14 Sight triangle approach leg time gap for stop-controlled minor road right or through movement

Design vehicle	Base time gap (s)	Supplement to time gap for approach grade
Passenger car	6.5	0.1 s for each additional percent beyond 3% grade
Single-unit truck	8.5	0.1 s for each additional percent beyond 3% grade
Combination truck	10.5	0.1 s for each additional percent beyond 3% grade

Adapted from AASHTO (2011a). Table 9-7.

where

Leg_{major} = length of the leg of the sight triangle along the major road in feet

V_{major} = design speed of the major road in miles per hour

t_g = time gap necessary for the vehicle stopped on the minor road to turn right onto the major road in seconds (see Table 4.14 for details to determine time gap for specific site conditions)

Case B3—Minor Road Crossing

For through movements from the minor road across the major road, the following equation and associated table can be used to determine the recommended minimum sight distance along the major road. Table 4.15 contains the base time gap by design vehicle that can be used in the analysis for a two-way, two-lane road with approach grades less than 3%. Adjustments should be made to increase the base time gap when additional lanes must be crossed, the median is not wide enough to allow for vehicle storage in a multistage crossing, and/or when the approach grade exceeds an upgrade of 3%.

$$Leg_{major} = 1.47\, V_{major} t_g \quad \text{(Adapted from AASHTO, 2011a, Equation 9-1)}$$

where

Leg_{major} = length of the leg of the sight triangle along the major road in feet

V_{major} = design speed of the major road in miles per hour

t_g = time gap necessary for the vehicle stopped on the minor road to cross the major road in seconds (see Table 4.15 for details to determine time gap for specific site conditions)

Table 4.15 Sight triangle approach leg length for stop-controlled minor road crossing movement

Design vehicle	Base time gap (s)	Supplement to time gap	
		Extra lanes to cross on major road	Approach grade
Passenger car	6.5	0.5 s for each additional lane and if median is narrow	0.1 s for each additional percent beyond 3% grade
Single-unit truck	8.5	0.7 s for each additional lane and if median is narrow	0.1 s for each additional percent beyond 3% grade
Combination truck	10.5	0.7 s for each additional lane and if median is narrow	0.1 s for each additional percent beyond 3% grade

Adapted from AASHTO (2011a). Table 9-7.

EXAMPLE 4.4 Sight Distance for Stop-Controlled Minor Street Approach

Determine the minimum sight distance for a minor street approach to a stop-controlled intersection on a two-way, four-lane divided roadway with a design speed of 55 mph and 12-ft-wide lanes. The roadway has a median that is 24-ft wide. The minor street approach grade is 2% and the design vehicle is a single-unit truck.

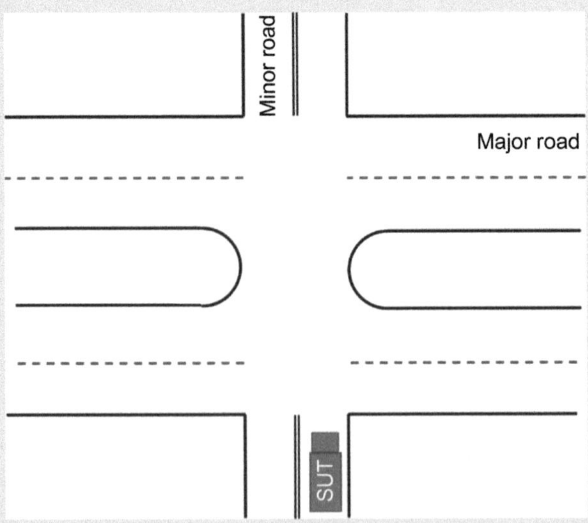

(Continued)

EXAMPLE 4.4 Sight Distance for Stop-Controlled Minor Street Approach—(Continued)

Solution

Case B1—Left from Minor Road

The time gap and design speed are needed to calculate the length of the leg of the sight triangle along the major road (for the left turn from the minor road, this distance will be needed in both directions along the major road). The base time gap for a single-unit truck is 9.5 s, and 0.7 s is needed for each additional lane or equivalent median width (if the median is too narrow to store the design vehicle). For the left-turn movement, the vehicle will have to cross one extra lane of traffic and the median (with a width equivalent to two lanes of traffic). No additional time gap is needed for the grade of the approach because it is less than 3%. Therefore, the additional time gap is 2.1 s (3 × 0.7 s).

$$Leg_{major} = 1.47 V_{major} t_g$$
$$Leg_{major} = 1.47(55 \text{ mph})(9.5 \text{ s} + 2.1 \text{ s})$$
$$Leg_{major} = 938 \text{ ft} \approx 940 \text{ ft}$$

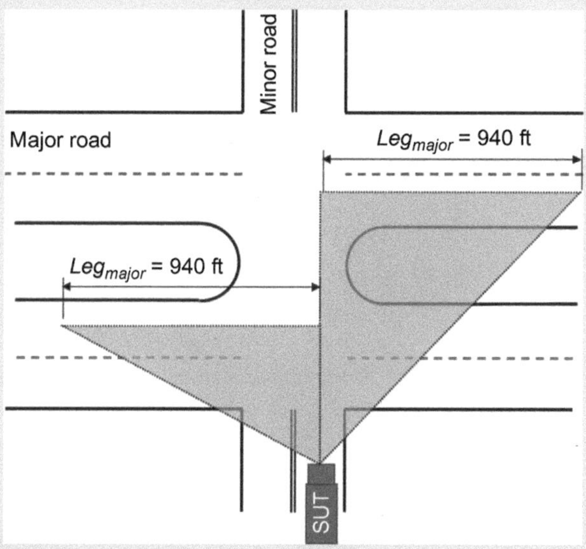

Case B2—Right from Minor Road

The time gap and design speed are needed to calculate the length of the leg of the sight triangle along the major road (for the right turn from the minor road, this distance will be needed to the left of the vehicle along the major road). The base time gap for a single-unit truck is 8.5 s, and no adjustments for lanes are needed because the nearest lane on the major

(Continued)

EXAMPLE 4.4 Sight Distance for Stop-Controlled Minor Street Approach—(Continued)

roadway will be used for the right-turn maneuver. No additional time gap is needed for the grade of the approach because it is less than 3%.

$$Leg_{major} = 1.47\,V_{major}t_g$$
$$Leg_{major} = 1.47(55\text{ mph})(8.5\text{ s})$$
$$Leg_{major} = 687\text{ ft} \approx 690\text{ ft}$$

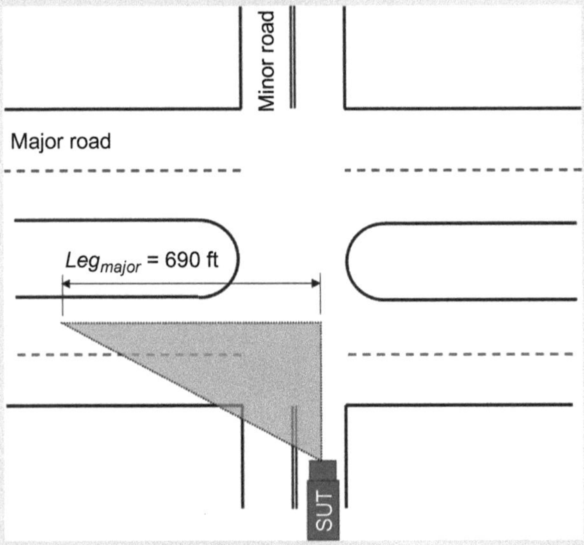

Case B3—Minor Road Crossing

The time gap and design speed are needed to calculate the length of the leg of the sight triangle along the major road (for the crossing of the major road from the minor road, this distance will be needed in both directions along the major road). The base time gap for a single-unit truck is 8.5 s, and 0.7 s is needed for each additional lane or equivalent median width (if the median is too narrow to store the design vehicle). For the crossing movement, the vehicle will have to cross two extra lanes of traffic and the median (with a width equivalent to two lanes of traffic). No additional time gap is needed for the grade of the approach because it is less than 3%. Therefore, the additional time gap is 2.8 s (4 × 0.7 s).

$$Leg_{major} = 1.47\,V_{major}t_g$$
$$Leg_{major} = 1.47(55\text{ mph})(8.5\text{ s} + 2.8\text{ s})$$
$$Leg_{major} = 914\text{ ft} \approx 915\text{ ft}$$

(Continued)

EXAMPLE 4.4 Sight Distance for Stop-Controlled Minor Street Approach—(Continued)

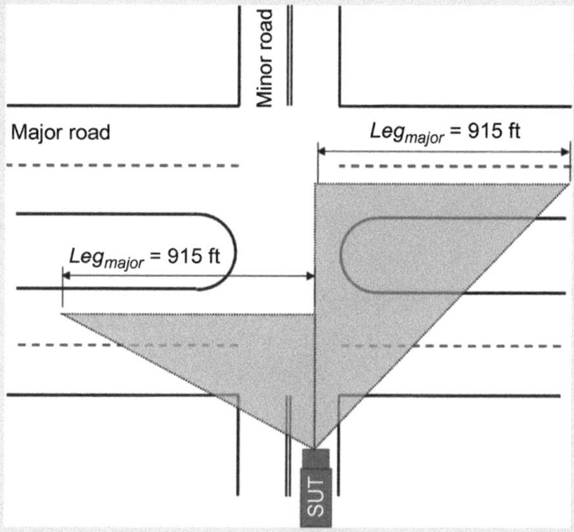

Summary of Case B—Minor Road Stop-Controlled Intersection

The maximum length of the leg of the sight triangle along the major road should be used for each sight triangle (to the left and to the right of the vehicle). In this example, Case B1 (left from the minor road) resulted in the largest values (940 ft) in each direction that define the length of the sight triangle along the major road.

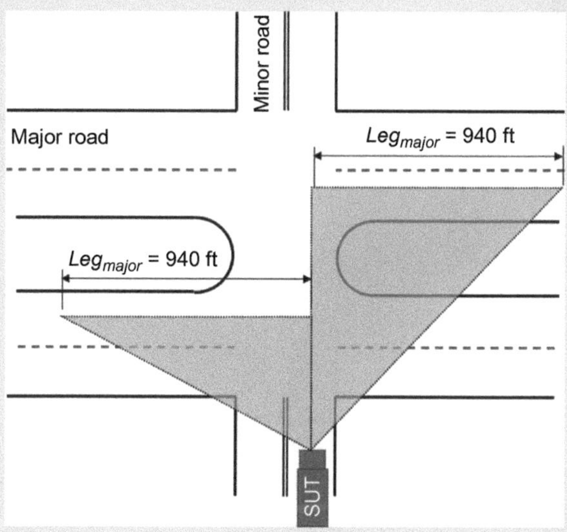

Case C—Yield Control on the Minor Road

Case C applies to vehicles that arrive at a yield-controlled approach of a minor road with an uncontrolled major road. At the yield-controlled approach on the minor road, three movements are possible: through (Case C1), left (Case C2), and right (Case C2). The length of the approach legs defines the intersection sight distance for yield-controlled intersections (Figure 4.36 for left view and Figure 4.37 for right view).

Case C1—Crossing from Minor Road

For through movements from the minor road across the major road, the following equation and associated table can be used to determine the recommended minimum sight distance along the major road. Table 4.15 contains the base time gap by design vehicle that can be used in the analysis for a two-way, two-lane road with approach grades of less than 3%. Adjustments should be made to increase the base time gap when additional lanes must be crossed, the median is not wide enough to allow for vehicle storage in a multistage crossing, and/or when the

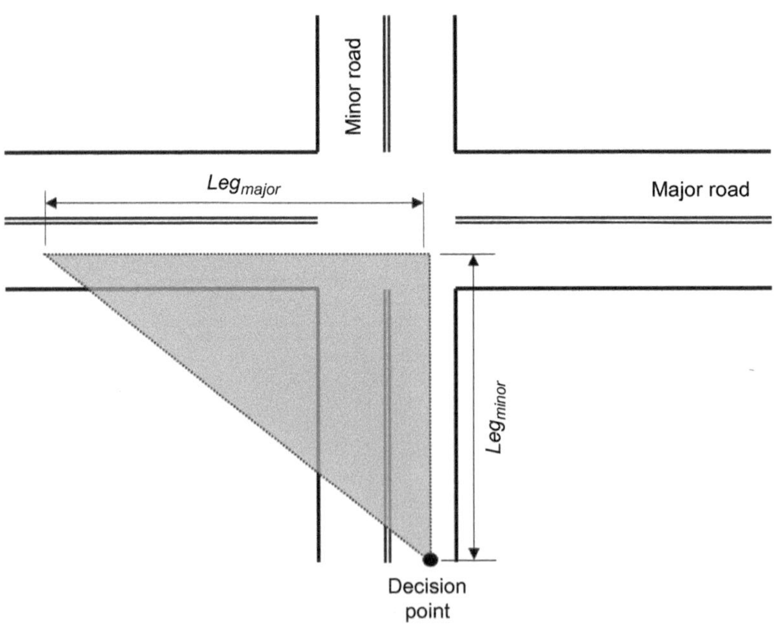

Figure 4.36 Left-view sight triangle for minor road approach to yield-controlled intersection for through movements. *Source: Adapted from AASHTO (2011a), Figure 9-15A.*

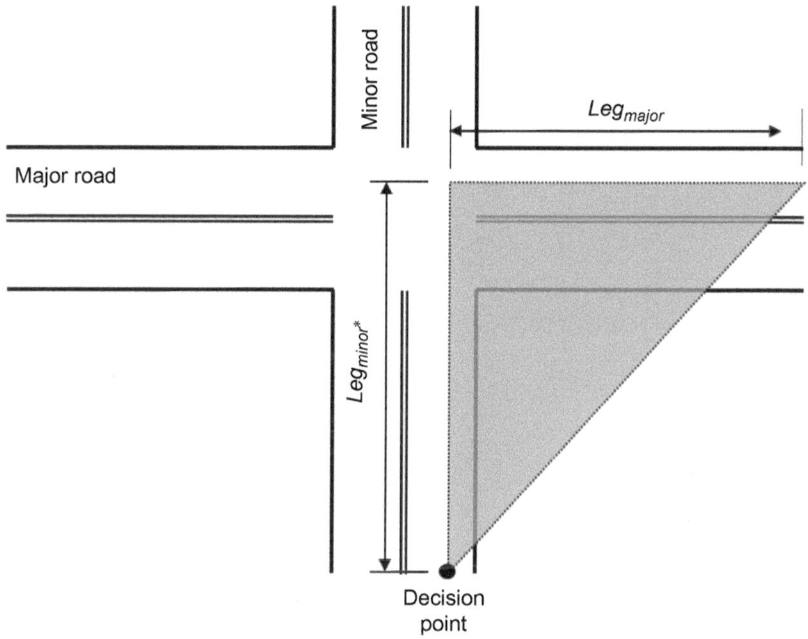

Figure 4.37 Right-view sight triangle for minor road approach to yield-controlled intersection for through movements. *Source: Adapted from AASHTO (2011a), Figure 9-15A.*

approach grade exceeds an upgrade of 3% (grade adjustment factors are available in Table 4.12).

$$Leg_{major} = 1.47\,V_{major}t_g \text{(Adapted from AASHTO, 2011a, Equation 9-1)}$$

where

Leg_{major} = length of the leg of the sight triangle along the major road in feet

V_{major} = design speed of the major road in miles per hour

t_g = travel time to reach and clear the major road in seconds (see following equation and accompanying table to determine time gap for specific site conditions)

$$t_g = t_a + \frac{w + L_a}{0.88\,V_{minor}} \text{(AASHTO, 2011a, Equation 9-2)}$$

Table 4.16 Sight triangle approach leg length for yield-controlled minor road crossing movement

Design speed (mph)	**Length of leg for sight triangle along minor road (Leg_{minor}) (ft)**	**Travel time to reach major road (t_a) (s)**
15	75	3.4
20	100	3.7
25	130	4.0
30	160	4.3
35	195	4.6
40	235	4.9
45	275	5.2
50	320	5.5
55	370	5.8
60	420	6.1
65	470	6.4
70	530	6.7
75	590	7.0
80	660	7.3

Adapted from AASHTO (2011a). Table 9-9.

where

t_g = travel time to reach and clear the major road in seconds

t_a = travel time to reach the major road from decision point for a vehicle that does not stop (see Table 4.16 to determine travel time for specific site conditions)

w = width of the intersection to be crossed in feet

L_a = length of the design vehicle in feet (design vehicle dimensions are presented in Table 4.1)

V_{minor} = design speed of the minor road in miles per hour

The calculated values for t_g should be greater than or equal to the values for stop-controlled intersections shown in Table 4.14.

Case C2—Right from Minor Road

For right-turning movements from the minor road onto the major road, the following equation and associated table can be used to determine the recommended minimum sight distance along the major road based on the approach to the intersection. Table 4.17 contains the base time gap by design vehicle that can be used in the analysis for a two-way, two-lane road. For the sight triangle, as shown in Figure 4.38, the distance along the minor road, 82 ft, from the center of the near lane of the major road is based on the minor road driver slowing to 10 mph for the turn (AASHTO, 2011a).

Table 4.17 Sight triangle approach leg time gap for yield-controlled minor road right movement

Design vehicle	Base time gap (s)
Passenger car	8.0
Single-unit truck	10.0
Combination truck	12.0

Adapted from AASHTO (2011a). Table 9-11.

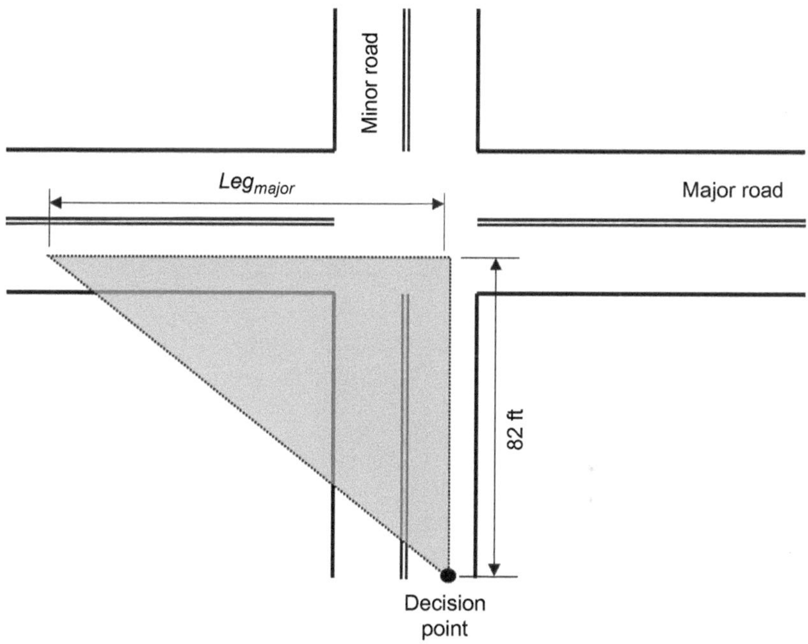

Figure 4.38 Left-view sight triangle for minor road approach to yield-controlled intersection for right-turning movements. *Source: Adapted from AASHTO (2011a), Figure 9-15A.*

$$Leg_{major} = 1.47\, V_{major} t_g \text{ (Adapted from AASHTO, 2011a, Equation 9-1)}$$

where

Leg_{major} = length of the leg of the sight triangle along the major road in feet

V_{major} = design speed of the major road in miles per hour

t_g = time gap necessary for the vehicle stopped on the minor road to turn right onto the major road in seconds (see Table 4.17 for details to determine time gap for specific site conditions)

Table 4.18 Sight triangle approach leg time gap for yield-controlled minor road left movement

Design vehicle	Base time gap (s)	Supplement to base time gap: extra lanes to cross on major road
Passenger car	8.0	0.5 s for each additional lane
Single-unit truck	10.0	0.7 s for each additional lane
Combination truck	12.0	0.7 s for each additional lane

Adapted from AASHTO (2011a). Table 9-11.

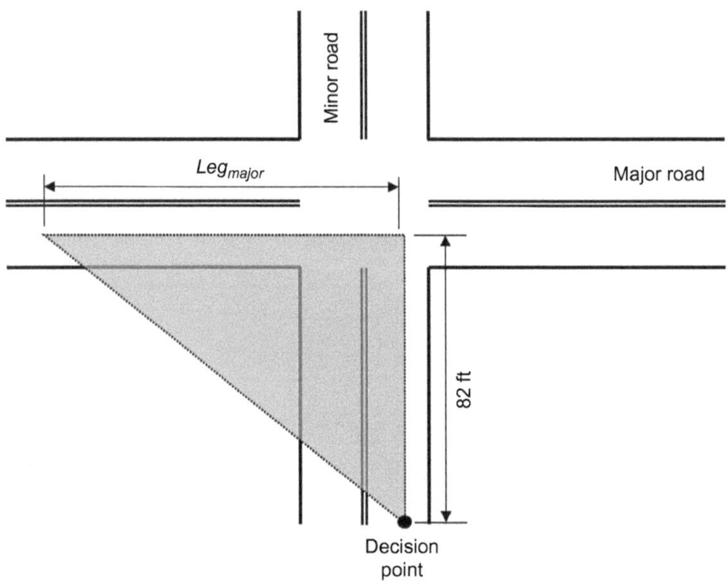

Figure 4.39 Left-view sight triangle for minor road approach to yield-controlled intersection for left-turning movements *Source: Adapted from AASHTO (2011a), Figure 9-15A.*

Case C2—Left from Minor Road

For left-turning movements from the minor road onto the major road, the following equation and associated table can be used to determine the recommended minimum sight distance along the major road based on the approach to the intersection. Table 4.18 contains the base time gap by design vehicle that can be used in the analysis for a two-way, two-lane road. Adjustments should be made to increase the base time gap when additional lanes must be crossed and/or the median is not wide enough to allow for vehicle storage in a multistage crossing. For the left-view sight triangle, as shown in Figure 4.39, the distance along the

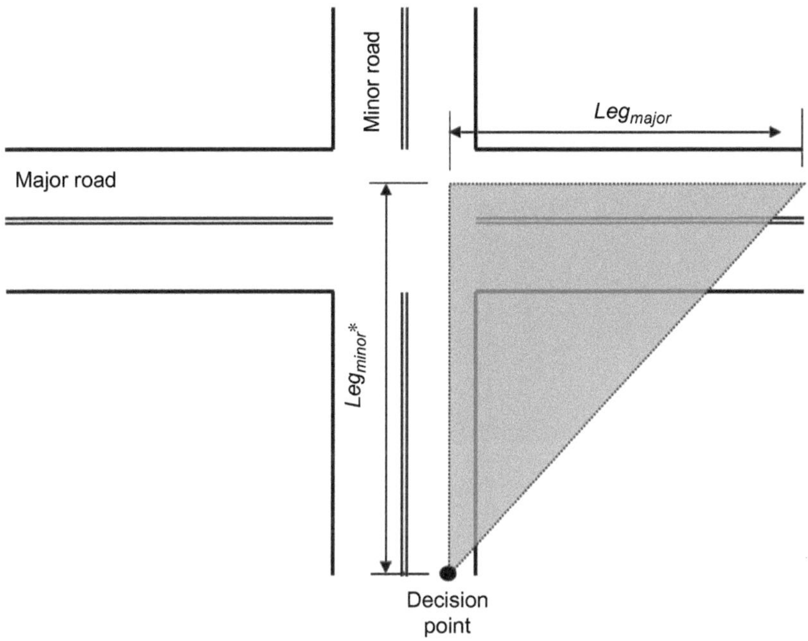

∗ Approach leg length must include additional lane width and median width (if median is not wide enough to store vehicle). For a two-lane roadway, as shown in this illustration, one additional lane width should be added to the approach leg length of 82 feet.

Figure 4.40 Right-view sight triangle for minor road approach to yield-controlled intersection for left-turning movements *Source: Adapted from AASHTO (2011a), Figure 9-15A.*

minor road, 82 ft, from the center of the near lane of the major road is based on the minor road driver slowing to 10 mph for the turn (AASHTO, 2011a). The right-view sight triangle includes the additional distance necessary to reach the center of the nearest travel lane on the major road that the left-turning vehicle will merge into (Figure 4.40).

$$Leg_{major} = 1.47\, V_{major} t_g \text{ (Adapted from AASHTO, 2011a, Equation 9-1)}$$

where

Leg_{major} = length of the leg of the sight triangle along the major road in feet

V_{major} = design speed of the major road in miles per hour

t_g = time gap necessary for the vehicle stopped on the minor road to turn right onto the major road in seconds (see Table 4.18 for details to determine time gap for specific site conditions)

EXAMPLE 4.5 Sight Distance for Yield-Controlled Minor Street Approach

Determine the minimum sight distance for a minor street approach with a 40-mph design speed to a yield-controlled intersection on a two-way, four-lane divided roadway with a design speed of 55 mph and 12-ft-wide lanes. The roadway has a median that is 24-ft wide. The minor street approach grade is 2% and the design vehicle is a single-unit truck.

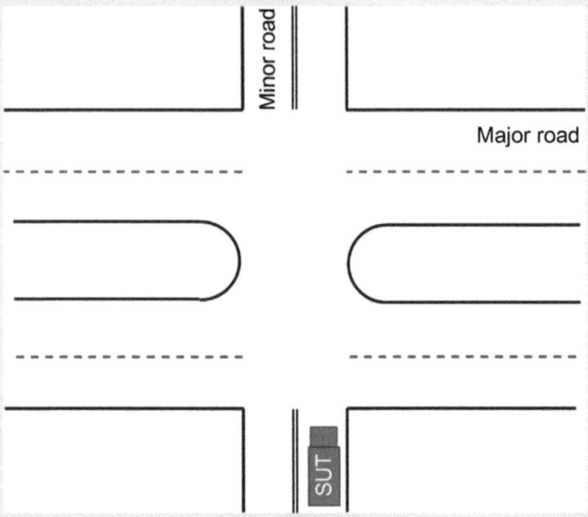

Solution

Case C1—Crossing from Minor Road

The time gap and design speeds are needed to calculate the length of the leg of the sight triangle along the major road. The width of the intersection is 72 ft and the length of the design vehicle is 30 ft.

$$t_g = t_a + \frac{w + L_a}{0.88 V_{minor}}$$

$$t_g = 4.9 \text{ s} + \frac{72 \text{ ft} + 30 \text{ ft}}{0.88(40 \text{ mph})} = 7.7978 \text{ s} \approx 7.8 \text{ s}$$

A check of the stop-controlled crossing maneuver time gap values shows that the single-unit truck time gap is 8.5 s. Because the calculated

(*Continued*)

EXAMPLE 4.5 Sight Distance for Yield-Controlled Minor Street Approach—(Continued)

value of 7.8 s is less than the stop-controlled values, 8.5 s will be used for this analysis.

$$Leg_{major} = 1.47\,V_{major}t_g$$

$$Leg_{major} = 1.47(55\text{ mph})(8.5\text{ s})$$

$$Leg_{major} = 687\text{ ft} \approx 690\text{ ft}$$

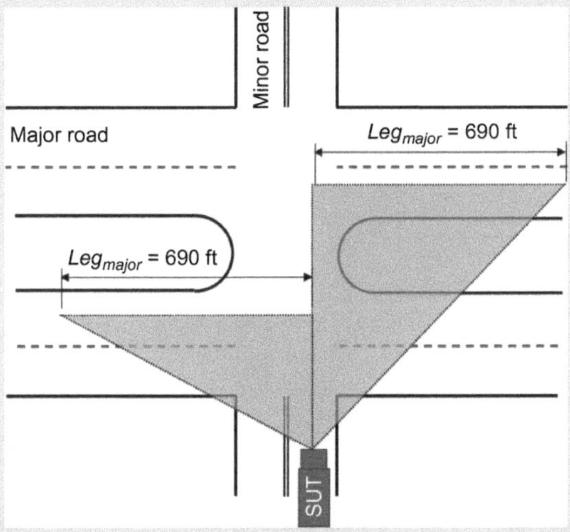

Case C2—Right from Minor Road

The time gap and design speed are needed to calculate the length of the leg of the sight triangle along the major road (for the right turn from the minor road, this distance will be needed to the left of the vehicle along the major road). The base time gap for a single-unit truck is 10.0 s, and no adjustments for lanes are needed because the nearest lane on the major roadway will be used for the right-turn maneuver.

$$Leg_{major} = 1.47\,V_{major}t_g$$

$$Leg_{major} = 1.47(55\text{ mph})(10.0\text{ s})$$

$$Leg_{major} = 809\text{ ft} \approx 810\text{ ft}$$

(*Continued*)

EXAMPLE 4.5 Sight Distance for Yield-Controlled Minor Street Approach—(Continued)

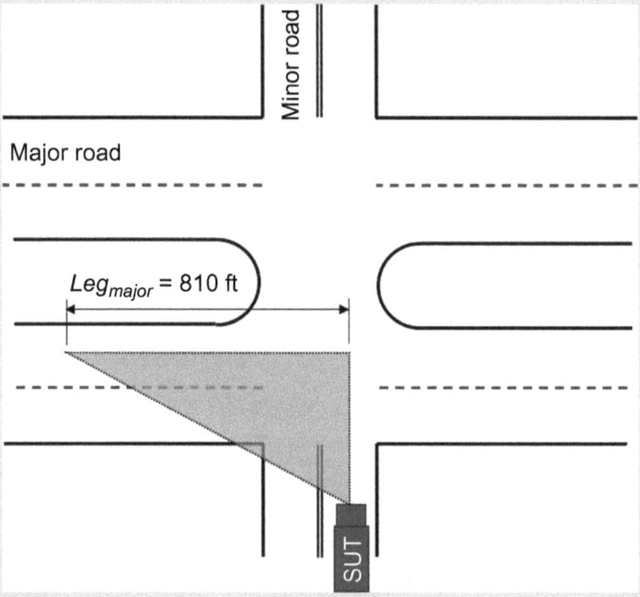

Case C2—Left from Minor Road

The time gap and design speed are needed to calculate the length of the leg of the sight triangle along the major road (for a left turn onto the major road from the minor road, this distance will be needed in both directions along the major road). The base time gap for a single-unit truck is 10.0 s, and 0.7 s is needed for each additional lane or equivalent median width (if the median is too narrow to store the design vehicle). For the left-turning movement, the vehicle will have to cross an extra lane of traffic and the median (with a width equivalent to two lanes of traffic). Therefore, the additional time gap is 2.1 s (3 × 0.7 s).

$$Leg_{major} = 1.47 V_{major} t_g$$

$$Leg_{major} = 1.47(55 \text{ mph})(10.0 \text{ s} + 2.1 \text{ s})$$

$$Leg_{major} = 978 \text{ ft} \approx 980 \text{ ft}$$

(*Continued*)

EXAMPLE 4.5 Sight Distance for Yield-Controlled Minor Street Approach—(Continued)

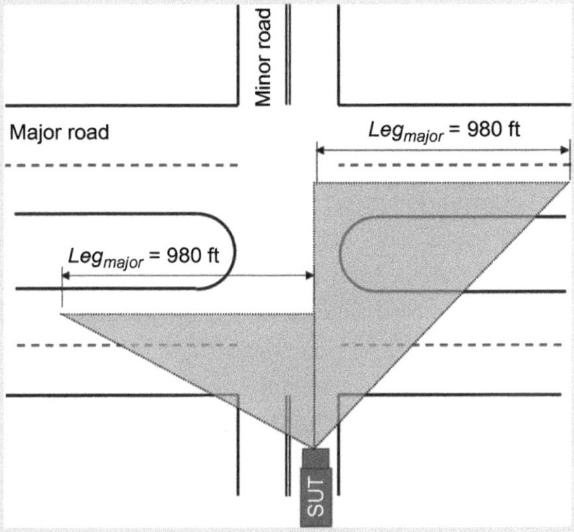

Summary of Case C—Minor Road Yield-Controlled Intersection

The maximum length of the leg of the sight triangle along the major road should be used for each sight triangle (to the left and to the right of the vehicle). In this example, Case C2 (left from the minor road) resulted in the largest values (980 ft) in each direction, which define the length of the sight triangle along the major road.

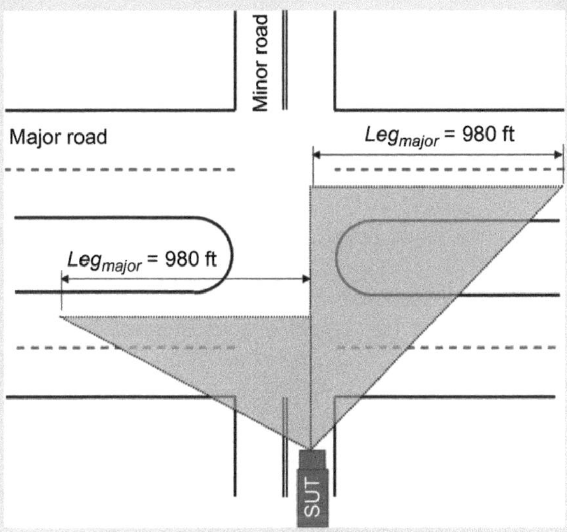

Case D—Traffic Signal Control

The general sight distance requirement at signalized intersections is that the first vehicle on each approach should be visible to the first vehicle at all other positions. Beyond this broad guidance, three specific operational conditions may warrant additional sight distance evaluations:

- Left turns: If left turns are permissive at the signal (left-turning vehicles may turn while conflicting through traffic is proceeding through the intersection), sufficient sight distance should be available for drivers to perceive and react to gaps in traffic.
- Flashing operation: If the signal will operate in two-way flashing mode (major road has flashing yellow and minor road has flashing red to operate as a stop-controlled approach), adequate sight distance as defined by Case B should be provided.
- Right turns: If right-turning movements during the red indication are permitted, the appropriate sight distance determined by the process for Case B2 should be available for drivers.

Case E—All-Way Stop Control

The sight distance requirement at all-way stop-controlled (stop signs at each approach) intersections is that the first vehicle on each approach should be visible to the first vehicle at all other positions. Only very unusual intersection alignments create sight distance concerns at all-way stop-controlled intersections.

Case F—Left Turns from the Major Road

For left-turning movements from the major road, the following equation and associated table can be used to determine the recommended minimum sight distance along the major road. Table 4.19 contains the base

Table 4.19 Sight triangle approach leg time gap for stop-controlled minor road left movement

Design vehicle	Base time gap (s)	Supplement to base time gap: extra lanes to cross on major road (in excess of one lane to cross)
Passenger car	5.5	0.5 s for each additional lane
Single-unit truck	6.5	0.7 s for each additional lane
Combination truck	7.5	0.7 s for each additional lane

Adapted from AASHTO (2011a). Table 9-5.

time gap by design vehicle that can be used in the analysis for a two-way, two-lane road with approach grades of less than 3%. Adjustments should be made to increase the base time gap when additional lanes must be crossed (in excess of one). An evaluation of the provision of adequate sight distance for Case F is particularly relevant near horizontal curves, crest vertical curves, and along divided highways with potential median obstructions.

$$Leg_{major} = 1.47\, V_{major} t_g \quad \text{(Adapted from AASHTO, 2011a, Equation 9-1)}$$

where

Leg_{major} = length of the leg of the sight triangle along the major road in feet

V_{major} = design speed of the major road in miles per hour

t_g = time gap necessary for the vehicle stopped on the minor road to turn left onto the major road in seconds (see Table 4.19 for details to determine time gap for specific site conditions)

4.6 INTERCHANGE DESIGN

Interchanges are grade-separated intersections of roads that use structures to separate conflicting streams of traffic. Connections between the roads are made possible with the use of ramps or loops. Interchanges eliminate the need for at-grade intersections on freeways, which improve safety performance and increase capacity. Interchanges can also be implemented on roadways without full-access control to similarly improve operations. Although interchanges offer significant advantages, they are also expensive and can degrade overall system performance if not designed and implemented properly because they are the source and destination of traffic.

4.6.1 Interchange Type

As with intersections, left-turning traffic movements are the most challenging to accommodate at interchanges. Right-turning movements typically allow a direct, free-flowing maneuver. Through movements are handled with structures to separate the conflicting traffic streams. At interchanges between freeways, or other full-access control facilities, a directional interchange offers the highest level of service by directly serving all movements with minimal or no reductions in speed. Figure 4.41 shows a three-leg directional interchange that allows for high-speed

Figure 4.41 Directional three-leg interchange.

Figure 4.42 Interchange with directional ramps from one roadway.

connections in each direction. Figure 4.42 shows a four-leg interchange that has two directional ramps (serving left movements from one roadway, while the other roadway is served by loop ramps). Figure 4.43 shows a four-leg interchange with one directional ramp and three loop ramps.

The majority of interchanges connect a freeway with a roadway of a lower functional classification, which means that directional ramps are not necessary. The most common interchange type is the diamond interchange, named for its diamond shape when viewed from above

Figure 4.43 Interchange with directional ramps for one movement and loop ramps.

Figure 4.44 Diamond interchange.

(Figure 4.44). With the diamond interchange configuration, exiting traffic diverges from the freeway mainline prior to crossroad and intersects with the crossroad on the same side of the diverge point. Entering traffic accesses the freeway at the same location and merges with the freeway mainline after the crossroad.

Figure 4.45 Single point interchange.

The diamond interchange is common because of its economical design and construction, but is limited in capacity. The single point interchange, Figure 4.45, condenses the interchange into a single point by controlling both ramps with a single set of traffic signals. The single point requires a large and more complex bridge structure, which increases the construction cost relative to a diamond interchange. However, the single point is more efficient than a diamond interchange, particularly with heavy left-turn volumes. The single point interchange is most commonly used in urban areas with restricted right of way and is also known as a single point urban interchange (SPUI). For similar heavy left-turning vehicle movements, the double crossover diamond (DCD) interchange can perform better than the diamond interchange. At the signal on each end of the bridge, traffic crosses over to the opposite side of the roadway to allow for free-flowing or reduced conflict left turns and then crosses back to the right side of the roadway at the second traffic signal. Figure 4.46 shows the crossover point at a DCD. The DCD is also known as a diverging diamond interchange (DDI). DCD can be installed as a new construction project or as a retrofit, which significantly reduces the project costs. Loop ramps can also efficiently serve left-turning traffic without conflicting with both directions of traffic on the crossroad. An interchange with four loop ramps is known as a cloverleaf, while interchanges

Figure 4.46 Crossover point at double crossover diamond interchange.

with fewer than four are known as a partial cloverleaf or parclo. Cloverleaf interchanges can create weaving segments because of the intertwining entering and exiting traffic streams, which can disrupt overall traffic operations and safety.

4.6.2 Interchange Spacing

Entering and exiting traffic can significantly disrupt traffic on the roadway; to improve the flow of through traffic, entering traffic can be restricted with operational features, such as ramp metering. The decision of where to install interchanges and how many to install should consider the balance of mobility and access. Additional access points provide drivers in the local area with additional and multiple options to access the roadway, but through traffic may be negatively affected by the disruption caused by merging and diverging traffic. Therefore, the spacing between interchanges should be as large as practical, while still serving traffic needs. The Green Book (pp 10–68) recommends a minimum interchange spacing of 1 mile in urban areas and 2 miles in rural areas. Grade-separated ramps or collector-distributor roadways can allow less than ideal spacing by removing the weaving movement from the through traffic.

The general advantages of additional interchanges include:

- Ability to service higher traffic volumes than a single interchange, particularly if large traffic generators can utilize multiple interchanges.
- Less travel time spent on other roadways, which leads to lower traffic volumes and higher level of service on the adjacent roadway network.
- General increase in convenience for drivers.
- Additional commercial development opportunities for land in the vicinity of the interchange.
- Increased system redundancy, which can benefit the network if the rerouting of traffic is necessary during an emergency, construction/reconstruction, collision, or other disruptive event.

The general disadvantages of additional interchanges include:

- Merging and diverging traffic degrades the operation of the roadway, including the capacity and safety.
- Additional construction and use of land can negatively impact the environment.
- The funds used for the additional interchange may have a larger benefit for other uses.

4.6.3 Vertical Separation: Over or Under

The type of vertical separation, whether the major road goes under or over the crossroad, is a major design element of an interchange. The topography is generally the driving factor, but the roadway alignment may also influence the type of vertical separation. An overpass, the major roadway over the crossroad, offers the advantages of shorter structural spans on the bridge, the elimination of any height restrictions for the major roadway, improved drainage for the major roadway, and fewer construction interruptions to the crossroad. An underpass, the major roadway under the crossroad as shown in Figure 4.47, offers the advantages of

Figure 4.47 Interchange with underpass (major road under).

having a single bridge (typically), gravitational aids for entering and exiting traffic (entering traffic is aided by gravity for acceleration and exiting traffic is aided by decelerating), a visual cue for exiting drivers that the crossroad is approaching, and a lower noise impact on adjacent land uses because the major roadway is depressed.

4.6.4 Ramp Design

Ramps for right-turning vehicles generally provide a direct link between the crossing roadways at an interchange with a minimal travel distance. Larger radii allow for higher speeds, which are necessary for directional interchanges. Left-turning vehicles can be accommodated by a direct ramp with the use of traffic control devices at the conflicting roadways or with the construction of flyovers or loop ramps. Flyovers and loop ramps are more expensive to construct, but offer additional capacity to the interchange. Loop ramps typically have relatively low design speeds with longer travel distances and larger right-of-way requirements.

The operation of an interchange and adjacent roadways are optimized when drivers' general expectations about ramps are satisfied, including:

- A consistent pattern and directional signage for ramps is provided at successive interchanges.
- Exits diverge and entrances merge from the right side of the roadway.
- Each interchange is served by a single exit ramp prior to the crossroad.
- Each interchange is served by a single entrance ramp beyond the crossroad.

The minimum distance between consecutive ramps depends on the type of ramp and facility. On a standard freeway segment, the minimum recommended distance between an exit ramp and an entrance ramp is 500 ft (AASHTO, 2011a, Figure 10–68). For two consecutive entrance or exit ramps, the minimum recommended spacing is 1000 ft (AASHTO, 2011a, Figure 10–68). An entrance ramp that is followed within a short distance by an exit ramp and connected by an auxiliary lane creates a weaving segment, which is common at cloverleaf interchanges. Recommended values from the width of ramps depends on the radius of the turn, design vehicle, ability to pass a stalled vehicle, type of shoulder, and type of curb (AASHTO, 2011a, Table 3–29). Table 4.4 presents summary information about lane widths for ramps. Adequate acceleration and deceleration lengths are essential for safe maneuvers at interchanges. The ramp speed, roadway speed, and grade influence the minimum acceleration length (AASHTO, 2011a, Table 10-3) and deceleration length (AASHTO, 2011a, Table 10-5). The terminal of the acceleration/deceleration lane can be tapered or parallel.

4.7 PRACTICE PROBLEMS

Problem 4.1

Along an urban arterial, utility poles are near the edge of the roadway and have been struck by errant drivers repeatedly. What options should be considered to address these potential hazards?

Solution

Using the prioritized hierarchy of options, the following options are possible. The higher-rated options should be considered first with a consideration of effectiveness and cost-efficiency.

1. Remove the obstacle: utility poles can be removed if the ability to place the utility lines underground is feasible.
2. Redesign the obstacle to include traversable features: this option is not possible for utility poles.
3. Relocate the obstacle to a further offset location from the edge of the roadway: utility poles could be moved further from their current location if available right of way for the utilities exist beyond their current location.
4. Reduce the severity of impact with inclusion of an appropriate breakaway device: utility poles can be designed with a breakaway base to lessen the forces on the vehicle.
5. Shield the obstacle with a longitudinal traffic barrier designed for vehicle redirection or use a crash cushion: barriers can be installed to shield the utility poles from errant vehicles, however, the barriers also pose a risk to errant vehicles.
6. Delineate the obstacle if the preceding alternatives are not appropriate: if no other alternatives are feasible, the utility poles can be painted or marked to increase their visibility to drivers.

Problem 4.2

What radius should be used to accommodate a three-axle single-unit truck in the right-turning maneuver shown at the intersection in the figure?

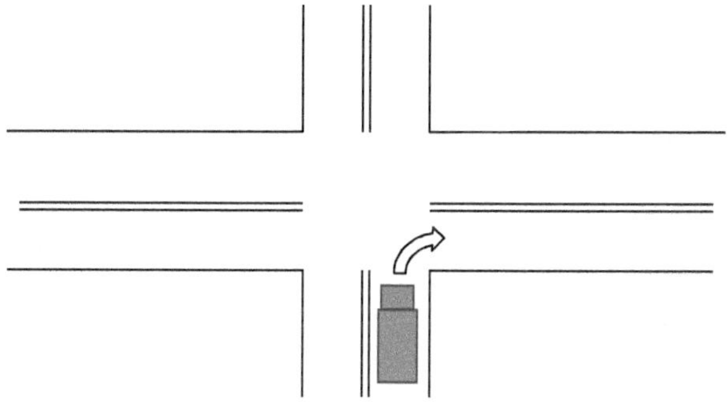

Solution

Three likely options exist for accommodating a three-axle single-unit truck (SU-40) at the right turn. Depending on the needs of the local environment, including constructability, local agency requirements, and site conditions, one of these options may be favored over another:

1. Simple curve: 80 ft
2. Simple curve radius with a taper: radius of 45 ft with an offset of 4 ft on a 10:1 taper
3. Three-center compound curve: large radii of 200 ft and small radius of 30 ft (200−30−200) with a symmetric offset of 7 ft

Problem 4.3

Determine if the following roadside design is consistent with the clear zone concept. The roadway has a 50-mph design speed and a design year AADT of 15,000 vehicles per day. If the design is insufficient, recommend a solution to improve the design.

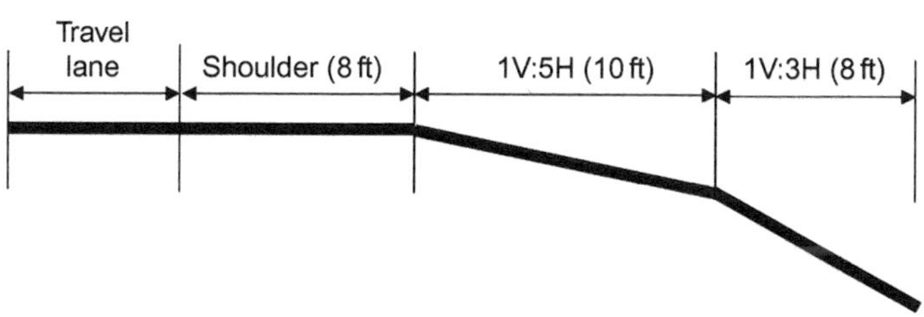

Solution

From Table 4.8 (50 mph design speed, 15,000 vehicles per day, and 1V:5H slope), the recommended clear zone distance is 24–28 ft. The shoulder counts in the determination of the clear zone width, so the foreslope should be 16–20 ft wide. The 1V:3H slope (nonrecoverable slope) interrupts the clear zone distance. The proposed design does not satisfy the recommended clear zone distance. A clear runout area can be added beyond the nonrecoverable slope to improve the roadside design. The clear runout area should include the remainder of the recommended clear zone distance with a minimum value of 10 ft. The remainder of the clear zone is 6–10 ft, and because the minimum value is 10 ft, the recommended clear runout area should have a width of 10 ft.

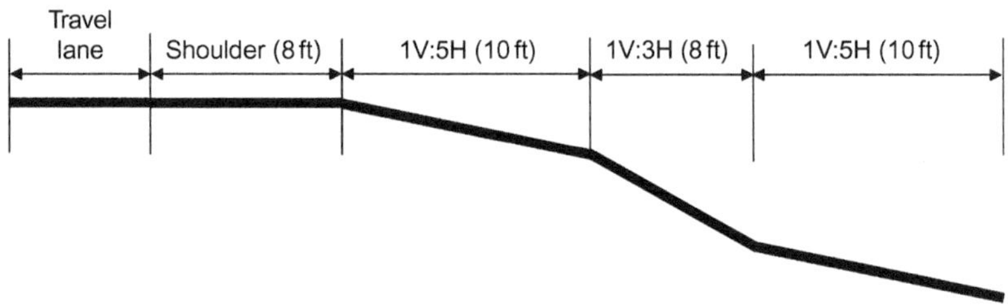

Problem 4.4

Determine the maximum backslope steepness for the rounded-bottom drainage channel configuration shown in the following image.

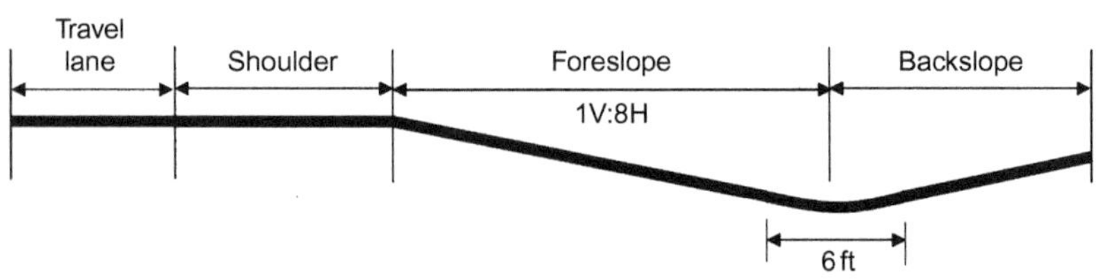

Solution

This drainage channel is defined as an abrupt slope change because it has a rounded-bottom shape with a width of less than 8 ft. From Table 4.9, the recommended maximum steepness of the backslope is 1V:3.5H.

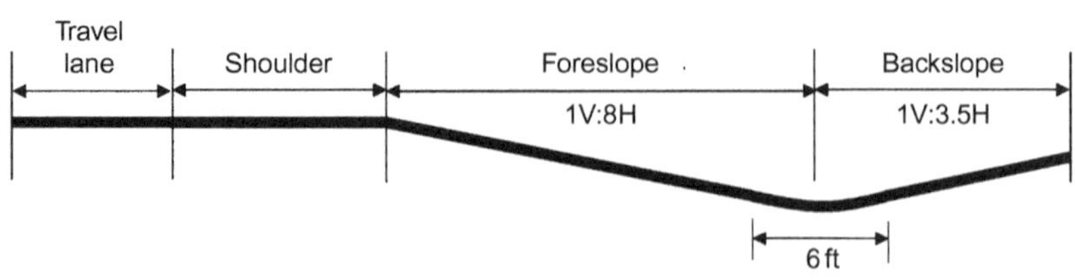

Problem 4.5

Determine the minimum sight distance for a minor street approach to a stop-controlled intersection on a two-way, four-lane divided roadway with a design speed of 45 mph and 11-ft-wide lanes. The roadway has a median that is 30-ft wide. The minor street approach grade is 5% and the design vehicle is a passenger car.

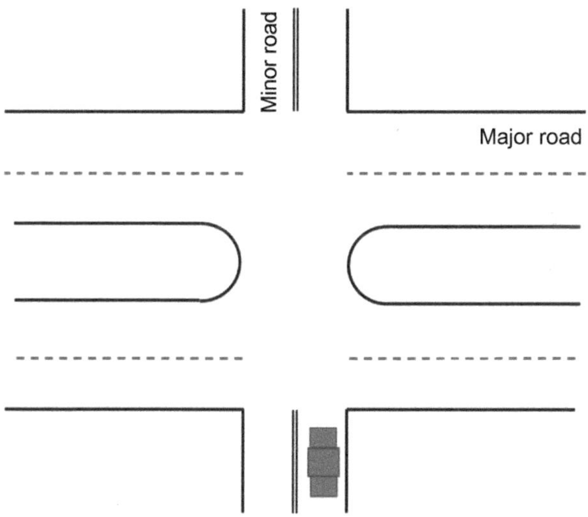

Solution

Case B1—Left from Minor Road

The time gap and design speed are needed to calculate the length of the leg of the sight triangle along the major road (for the left turn from the minor road, this distance will be needed in both directions along the major road). The base time gap for a passenger car is 7.5 s, and 0.5 s is needed for each additional lane or equivalent median width (if the median is too narrow to store the design vehicle). For the left-turn movement, the vehicle will have to cross one extra lane of traffic, but not the median in one movement because a two-stage crossing is possible (the median with a width of 30 ft can storage a passenger vehicle with a width of 19 ft; at least 3 ft should be allowed on each end of the stored vehicle). An additional time gap is needed for the grade of the approach because it is greater than 3% (0.2 s is needed for each additional percent beyond 3%, for a total of 0.4 s for this scenario). Therefore, the additional time gap is 0.9 s (0.5 + 0.4 s).

$$Leg_{major} = 1.47\,V_{major}t_g$$

$$Leg_{major} = 1.47(45\text{ mph})(7.5\text{ s} + 0.9\text{ s})$$

$$Leg_{major} = 556\text{ ft} \approx 560\text{ ft}$$

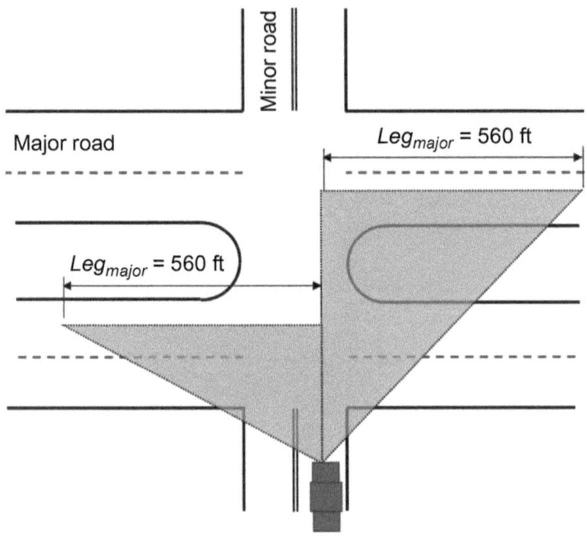

Case B2—Right from Minor Road

The time gap and design speed are needed to calculate the length of the leg of the sight triangle along the major road (for the right turn from the minor road, this distance will be needed to the left of the vehicle along the major road). The base time gap for a passenger car is 6.5 s, and no adjustments for lanes are needed because the nearest lane on the major roadway will be used for the right-turn maneuver. An additional time gap is needed for the grade of the approach because it is greater than 3% (0.2 s is needed for each additional percent beyond 3%, for a total of 0.4 s for this scenario).

$$Leg_{major} = 1.47\,V_{major}t_g$$

$$Leg_{major} = 1.47(45\text{ mph})(6.5\text{ s} + 0.4\text{ s})$$

$$Leg_{major} = 456\text{ ft} \approx 460\text{ ft}$$

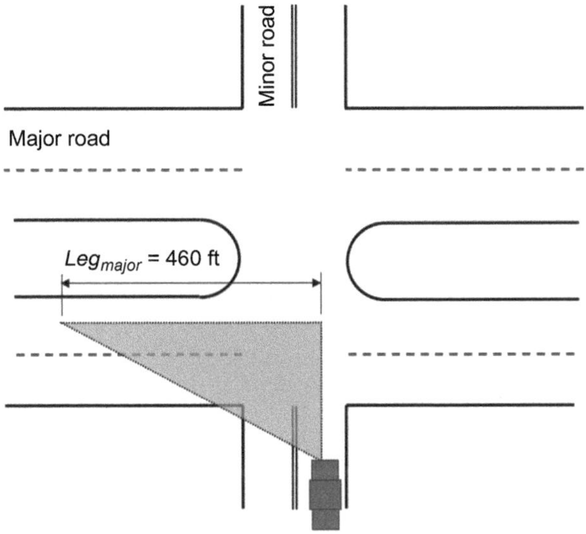

Case B3—Minor Road Crossing

The time gap and design speed are needed to calculate the length of the leg of the sight triangle along the major road (for the crossing of the major road from the minor road, this distance will be needed in both directions along the major road). The base time gap for a single-unit truck is 6.5 s, and 0.5 s is needed for each additional lane or equivalent median width (if the median is too narrow to store the design vehicle). For the crossing movement, the vehicle will have to cross one extra lane of traffic, but not the median in one movement because a two-stage crossing is possible (the median with a width of 30 ft can storage a passenger vehicle with a width of 19 ft; at least 3 ft should be allowed on each end of the stored vehicle). An additional time gap is needed for the grade of the approach because it is greater than 3% (0.2 s is needed for each additional percent beyond 3%, for a total of 0.4 s for this scenario). Therefore, the additional time gap is 0.9 seconds (0.5 + 0.4 s).

$$Leg_{major} = 1.47\,V_{major}t_g$$

$$Leg_{major} = 1.47(55 \text{ mph})(6.5 \text{ s} + 0.9 \text{ s})$$

$$Leg_{major} = 490 \text{ ft}$$

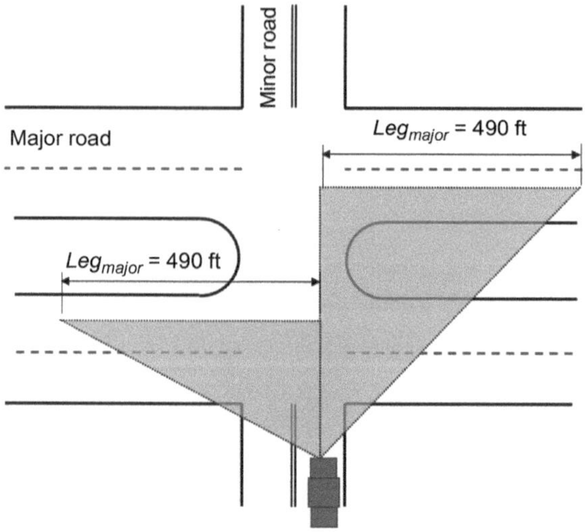

Summary of Case B—Minor Road Stop-Controlled Intersection
The maximum length of the leg of the sight triangle along the major road should be used for each sight triangle (to the left and to the right of the vehicle). In this example, Case B1 (left from the minor road) resulted in the largest values (560 ft) in each direction that define the length of the sight triangle along the major road.

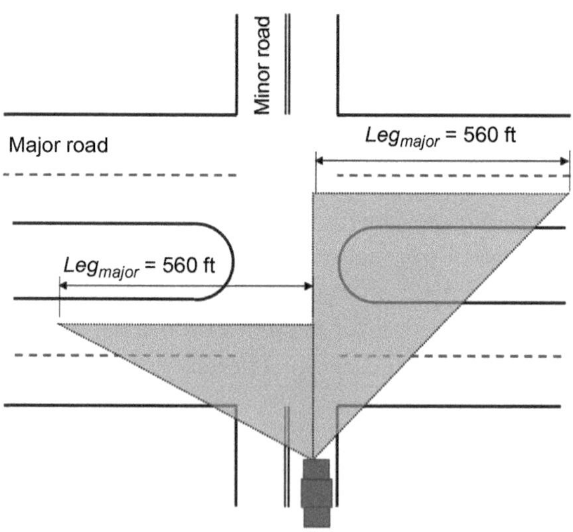

REFERENCES

American Association of State Highway and Transportation Officials (AASHTO), 2009. Manual for Assessing Safety Hardware, 1st ed.

American Association of State Highway and Transportation Officials (AASHTO), 2011a. A Policy on Geometric Design of Highways and Streets, sixth ed.

American Association of State Highway and Transportation Officials (AASHTO), 2011b. Roadside Design Guide, fourth ed.

North Carolina Department of Transportation, 2014. Typical Highway Cross Sections. May 5, Raleigh, NC.

PART 5

Traffic Operations

Bastian J. Schroeder, PhD, PE
Director of Highway Systems, Institute for Transportation Research and Education (ITRE)

Photo by Bastian Schroeder

Contents

5.1 INTRODUCTION

Traffic operations is the analysis of volume and capacity relationships of transportation facilities, with the goal of predicting operational performance in the form of travel time, speeds, delay, queues, and various other measures. As traffic volumes approach, and in some cases exceed, capacity, these performance measures deteriorate (higher delay, lower speed), up to potentially reaching a failure criterion, such as a queue spilling back beyond the physical storage space available for the movement at an intersection. So, as traffic volume, or the load on a transportation segment or approach to an intersection, increases, it is expected that the performance at some point starts to degrade, and eventually break down. Analysts also refer to the performance of a transportation facility as its *level of service* (LOS) and sometimes its *quality of service* (QOS), which are both important concepts discussed in this part of the book.

The field of traffic operations relies on a combination of field data and established analysis methods to provide analysts with methodologies for predicting the performance of an intersection or segment. The methodologies used to conduct operational analyses are typically derived from either broad field observations at representative sites (e.g., through regression), or from traffic flow theory relationships. Those traffic flow relationships in turn can often be calibrated to better fit local operating conditions. The methods then employ a series of analytical steps, each with equations, tables, or charts, to allow for the estimation of performance from available inputs. The analyst needs to have a good understanding of these methodologies and their empirical or theoretical basis to make sure that they are applied correctly and within their intended scope.

This part provides an overview of the basics of transportation operations, including fundamental relationships of traffic flow, and established analysis methods for various transportation facility types. However, the reader is encouraged to also refer to more detailed resources in the literature that exclusively focus on this large field (Roess et al., 2004; TRB, 2015).

5.1.1 Purpose of Traffic Operations

Transportation operational analysis methods are applied to estimate the current or, more often, a future state of the transportation system. The methods provide a structured process for estimating performance from a set of traffic inputs. These inputs most fundamentally include the traffic volume level and the roadway geometry, with the first expressing the *demand side* of how many cars, trucks, pedestrians, or bicycles aim to travel through a point, and the latter the *supply side*, which describes how many cars, trucks, pedestrians, or bicycles can travel through that point. Between the supply and demand side lie empirically derived relationships in the form of equations, tables, and charts that estimate *performance* based on those inputs. At the most disaggregated level, these methods exist for points (e.g., how much traffic can pass through a gate or a bridge), but are more practically useful once aggregated to the intersection, segment, or facility level.

The *demand side* of a traffic operation problem is quantified through the number of objects (cars, trucks, pedestrians, bicycles, etc.), the mix of the traffic stream (e.g., percent trucks), and the temporal distribution of traffic over say the course of a day. The *supply side* is expressed through the type of intersection, the number of lanes, or the radius of curves for turning movements. Researchers and transportation analysts have then developed relationships to estimate the *capacity* of the network element under study, as well as methods to predict *performance* of those elements through measures that typically include the average speed, travel time, density, queue length, or number of stops.

Traffic operational analyses are performed by transportation professionals with training on these operational methods and an understanding of the relationships between volume and capacity (demand and supply) for various traffic network elements. Commonly, the actual analysis is facilitated by software, as the equations can be complex, and because their application can be repetitive, as an analyst evaluates multiple approaches, to multiple time intersections, over multiple time periods, and for multiple scenarios.

One final key element in traffic operations is *data*. Data are needed to quantify both demand and supply levels, before being able to estimate the operational performance. In their most basic form, data take the form of traffic counts—demand—and a count of the number of lanes—supply. But as we'll see in this part, the nuances of demand and how traffic is measured, what the vehicle mix is, how traffic is distributed temporally, and so on, tend to be more complicated. Similarly, the supply characteristics are much more complex to quantify than just counting the number of lanes, as the capacity of each lane is impacted by physical characteristics (lane width, shoulder clearance, radius, etc.) and, more importantly, attributes of traffic control devices that include traffic signals, yield lines, and stop lines. The capacity of a transportation system element is also different for different road users, as one can, for example, fit more cars per hour than trucks per hour, or more pedestrians than bicycles through the same point. Put differently, the capacity or supply, in turn, can be a function of the demand on the system.

5.1.2 Highway Capacity Manual

The U.S. *Highway Capacity Manual*, or *HCM* (TRB, 2015), is the primary reference for traffic operational analysis, methodologies, and level of service (LOS) concepts in the United States, as well as many other countries. The *HCM* is a collection of concepts and methods that guide analysts on how to evaluate a particular type of intersection or roadway segment, based on what can be extensive national or international datasets of operational performance. The *HCM* is also the primary source for defining the capacity of different roadway elements that are used in many applications beyond traffic operations, including transportation planning and even safety analyses. The objectives of the *HCM* are to:

- Define performance measures and describe survey methods for key traffic characteristics.
- Provide methodologies for estimating and predicting performance measures.
- Explain methodologies at a level of detail such that readers can understand the factors that have an effect on multimodal operation.

While the *HCM* is the primary traffic operations resource developed for the United States, many other countries have adopted the *HCM* for their own use, often with some country-specific modification and calibration of methods to better suit local conditions. Some countries have developed their own traffic operations manuals, such as, for example, the German HBS (manual for measuring street systems, FGSV, 2001).

The *HCM*, first produced in 1950, is updated regularly based on new research supported by the Transportation Research Board (TRB). The manual is updated through the Highway Capacity and Quality of Service Committee of TRB. *HCM 2010*, the most recent edition, was distributed in early 2011; a major update of the 2010 *HCM* is expected for publication in late 2015.

The *HCM* is principally organized into four volumes, covering (1) general concepts, (2) methods for uninterrupted flow (freeways), (3) methods for interrupted flow (arterial streets), and (4) supplemental information to further document the methods in the second and third volumes. Each volume is organized into chapters that describe a particular element of the transportation system, ranging from basic freeway segments to signalized intersections, to modern roundabouts, to shared-used pedestrian and bicycle paths. Table 5.1 shows the high-level organization of the 2010 *Highway Capacity Manual* in the three primary printed volumes. Volume 4 is an online-only volume with additional reference materials in support of the printed chapters.

Table 5.1 Overview of *HCM* 2010 organization

Volume 1: Concepts	**Volume 2: Uninterrupted Flow**	**Volume 3: Interrupted Flow**	**Volume 4: Supplemental Material and Applications**

Ch. 1—*HCM* User's Guide Ch. 2—Applications Ch. 3—Modal Characteristics Ch. 4—Traffic Flow and Capacity Concepts Ch. 5—Quality and Level of Service Concepts Ch. 6—*HCM* and Alternative Analysis Tools Ch. 7—Interpreting *HCM* and Alternative Tool Results Ch. 8—*HCM* Primer Ch. 9—Glossary and Symbols	Ch. 10—Freeway Facilities Ch. 11—Basic Freeway Segments Ch. 12—Freeway Weaving Segments Ch. 13—Freeway Merge and Diverge Segments Ch. 14—Multilane Highways Ch. 15—Two-Lane Highways	Ch. 16—Urban Street Facilities Ch. 17—Urban Street Segments Ch. 18—Signalized Intersections Ch. 19—Two-Way Stop Controlled Intersections Ch. 20—All-Way Stop Controlled Intersections Ch. 21—Roundabouts Ch. 22—Interchange Ramp Terminals Ch. 23—Off-Street Pedestrian and Bicycle Facilities	Ch. 24—Concepts: Supplemental Ch. 25—Freeway Facilities: Supplemental Ch. 26—Freeway and Highway Segments: Supplemental Ch. 27—Freeway Weaving: Supplemental Ch. 28—Freeway Merges and Diverges: Supplemental Ch. 29—Urban Street Facilities: Supplemental Ch. 30—Urban Street Segments: Supplemental Ch. 31—Signalized Intersections: Supplemental Ch. 32—Stop-Controlled Intersections: Supplemental Ch. 33—Roundabouts: Supplemental Ch. 34—Interchange Ramp Terminals: Supplemental Ch. 35—Active Traffic Management Technical Reference Library HCMAG—6 case studies

5.1.3 Operational Analysis

Operational analysis is one aspect of the overall transportation system's management process. This process starts with designing the facilities. The design process lays out the physical transportation network and describes the types of intersections or interchanges used to control the interaction of multiple links in the system. The process continues with the forecasting of travel demand in a transportation planning context, where analysts estimate how much traffic is expected on any given link and intersection. This demand is then influenced by a host of factors that can promote or depress demand, or shift it over time and/or space. Traffic operations involves measuring or predicting the performance of this demand for the system, individual links and nodes, and ultimately the user of the system in the form of quantifiable measures of effectiveness (MOEs). This general process is illustrated in Figure 5.1.

Capacity

Capacity is a reflection of the ability of a transportation facility to accommodate a moving stream of people or vehicles. Capacity of any highway system element is defined as the maximum number of vehicles that have a

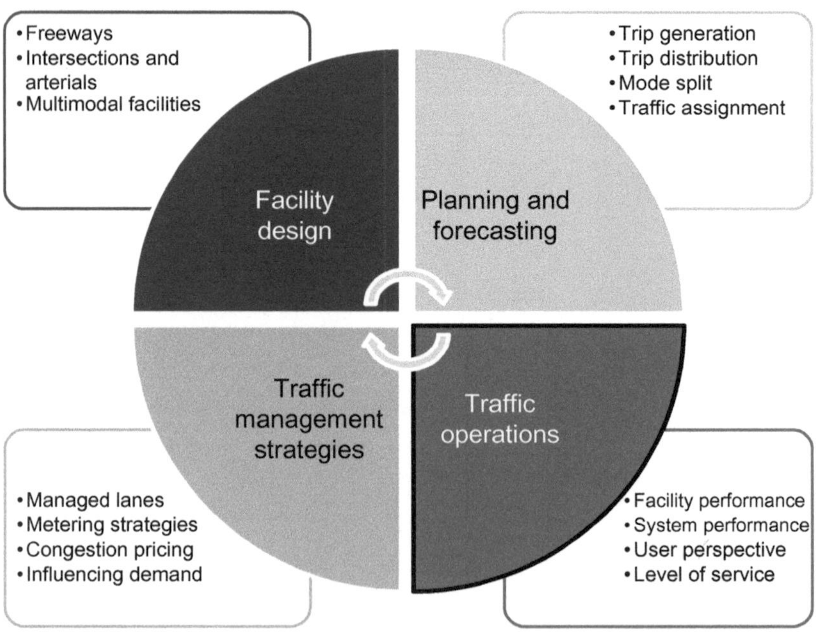

Figure 5.1 Traffic operations in the transportation systems management process.

Table 5.2 Illustration of flow regimes

Regime	Undersaturated	Saturated	Oversaturated
Illustration	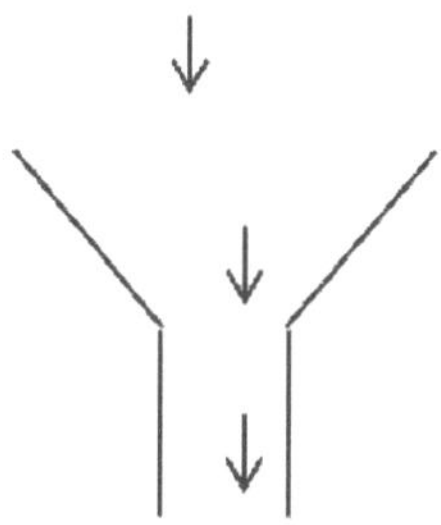	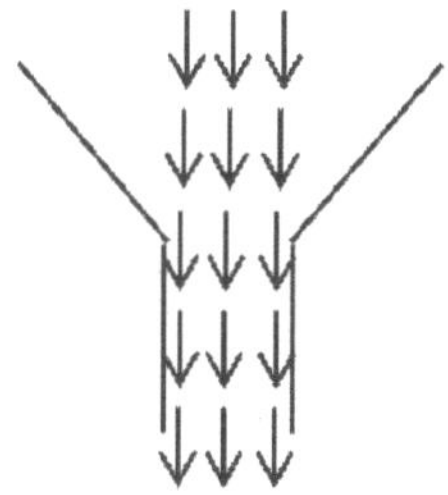	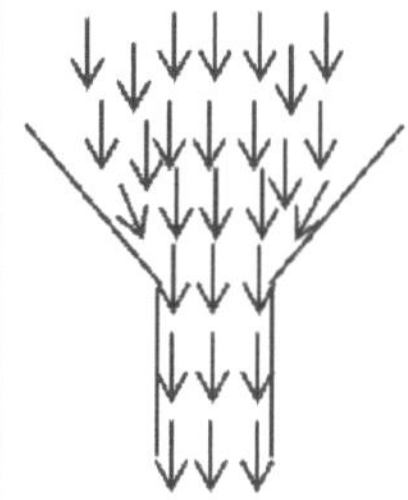
Flow rate	$v_{\text{arriving}} < v_{\text{served}}$	$v_{\text{arriving}} = v_{\text{served}}$	$v_{\text{arriving}} > v_{\text{served}}$

reasonable expectation of passing over a section (in either one or both directions) during a given time period under prevailing roadway and traffic conditions. In other words, capacity is the "supply" measure of transportation facilities. Capacity analysis provides tools for the analysis of existing facilities and for the planning and design of improved or future facilities.

In the *Highway Capacity Manual* (TRB, 2015), capacity is defined as follows:

> *Capacity is the maximum sustainable flow rate at which vehicles or persons reasonably can be expected to traverse a point or uniform segment of a lane or roadway during a specified time period under given roadway, geometric, traffic, environmental, and control conditions; usually expressed as vehicles per hour, passenger cars per hour, or persons per hour.*

The capacity determines a key operating threshold for two traffic flow regimes, which are concepts used throughout traffic operational analysis. They are described as follows and illustrated in Table 5.2:

Undersaturation or *uncongested flow* is a traffic condition in which the arrival flow rate is lower than the capacity or the service flow rate at a point or uniform segment of a lane or roadway.

Oversaturation or *congested flow* is a traffic flow condition in which the arrival flow rate is greater than the capacity. Congested flow is often caused by a downstream bottleneck, which limits throughput, and results in queuing upstream of the bottleneck or choke point.

Typical capacities for highway systems are given in Table 5.3.

Table 5.3 Typical capacities of highway systems

Facility	Definition	Capacity in passenger cars
Freeways and expressways away from ramps and weaving sections	Per lane of freeway per hour	2400
Two-lane highway	Total in both directions, per hour	2800
Urban signalized intersection	Total per lane for through movement per hour of continuous green (urban areas with population over 250,000)	1900
Small town or rural signalized intersection	Total per lane for through movement per hour of continuous green	1750
Modern roundabout	Total per approach lane without any conflicting traffic in circle, depending on roundabout geometry and configuration	1400–1600

The capacities in Table 5.3 should be treated with care, as these represent the base or ideal conditions of the particular transportation system element. As such, the listed capacity values are rarely observed, as they are typically reduced by a range of factors that affect the capacity of roadways and intersections. These factors are broken down into three main categories:

1. *Roadway conditions:* These refer to the geometric characteristics of the street or highway including the type of facility, the surrounding development, the number of lanes, lane and shoulder widths, lateral clearances, design speed, and horizontal and vertical alignment.
2. *Traffic conditions:* These refer to the characteristics of the vehicles using the facility. This includes the distribution of vehicle types, the amount of traffic in the available lanes, and the directional distribution of the traffic.
3. *Control conditions:* These refer to the types and specific design of control devices and traffic regulations in use on a roadway or intersection. The location, type, and timing of traffic signals have a significant impact on capacity. Other important controls include stop and yield signs, lane restrictions, and turning restrictions.

Traffic engineers calculate the capacities of roadways and intersections by using the procedures presented in the *Highway Capacity Manual* (TRB, 2015), which are often implemented in software to facilitate what can be rather complex methodologies.

The volume-to-capacity (v/c) ratio identifies how close traffic volumes are to the calculated capacity. This unitless ratio is calculated by dividing the existing traffic volume by the calculated capacity. The value of the v/c ratio will never exceed 1. As it approaches 1, traffic congestion and delays are expected.

Level of Service

Level of service (LOS) is a measure of the quality of flow along a highway. In the *HCM*, there are six defined levels of service, designated LOS A (best operating conditions) through LOS F (worst operating conditions). Each of these levels of service represents a range of operational conditions within a traffic stream. These operational conditions are characterized by such factors as speed and travel time, freedom to maneuver, traffic interruptions, and comfort and convenience. For example, during the evening rush hour period for a particular stretch of highway, the users of that highway may experience LOS D; but a user traveling along that same exact roadway at 3 am would probably experience LOS A.

In applying and interpreting LOS, it is important to consider that it is a step function with thresholds determining the boundaries between different letter grades. These boundaries and thresholds have been set by the Transportation Research Board's (TRB) Committee on Highway Capacity and Quality of Service, based on expert judgment and longstanding expertise in the application of the *HCM* methods. The thresholds, however, are fixed limits to describe a phenomenon that is highly variable due to day-to-day traffic fluctuations and estimation errors. Analysts should take great care when applying LOS, as its step-function nature may push the transportation element under analysis into the next LOS category, despite an only marginal change in the underlying service measure. To illustrate this point, Figure 5.2 shows the LOS thresholds for signalized intersections, which are based on the average control delay per vehicle in seconds.

The figure illustrates that a 12-s change in the average control delay can lead to an LOS changing by two letter grades from A to C, by one letter grade from B to C, or can leave the LOS entirely

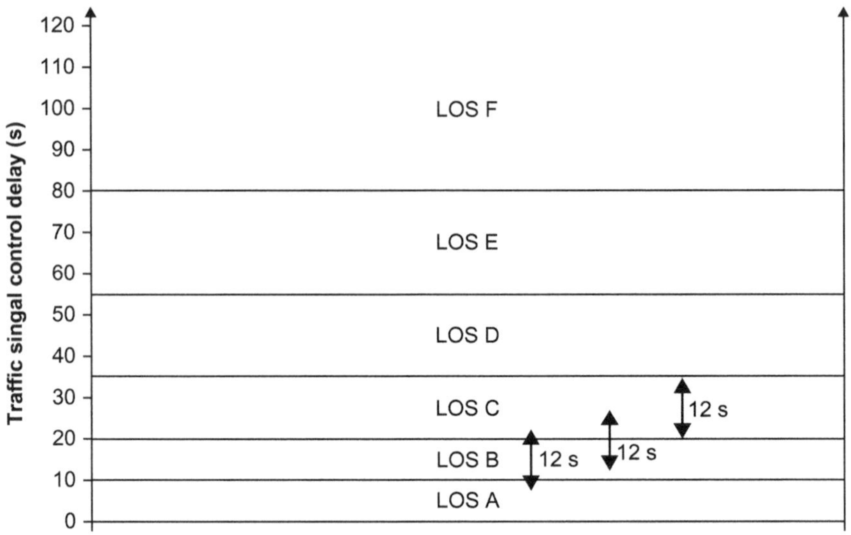

Figure 5.2 Illustration of LOS as a step function for traffic signals. *Source: TRB, 2015.*

unchanged for LOS C and above. As such, an analyst should always consider the net effect in the service measure (and other performance measures, as discussed in the next section) and be cautious whenever the numerical value of a service measure is very close to an LOS boundary.

Measure of Effectiveness and Service Measure

A *measure of effectiveness* (MOE), sometimes referred to as *performance measure* (PM), is a metric designed to describe and quantify the operations of a traffic systems element. Common MOEs include delay, travel time, speed, number of stops, queue lengths, and density. Different MOEs are useful to quantify different aspects of the traffic operations of a system element, and often analysts look at more than one to get the full picture of how something operates. For example, the delay for vehicles at a signalized intersection is interesting and important, but a full picture of the intersection's operations is not possible without also looking at the number of stops (do drivers make it through with one or fewer stops, or are multiple stops—also known as *cycle failures*, as described later—observed in an approach?) or the queue lengths (are queues contained in the provided storage space, like a turn pocket, or do they spill onto other travel lanes?). Similarly, density is a useful measure for analysts to quantify a freeway system, but one could argue that other measures such as the average travel time or queue lengths are more directly tied to the user perception and "bottom line" for drivers traveling along a freeway facility.

The *service measure* is the property one looks at to determine what LOS a user is experiencing. The service measure varies depending on the type of facility that is being analyzed. For example, for freeway segments, the service measure is density. For signalized intersections and roundabouts, the service measure for vehicles is control delay, but for pedestrians it is an index that combines not only delay, but also aspects of comfort, safety, and convenience. Once again, the *Highway Capacity Manual* is a valuable reference that explains the methods for estimating capacity and determining LOS. The service measures for the 2010 *HCM* are summarized in Table 5.4.

Spatial Analysis Scope

The *HCM* contains methods for varying spatial scopes. At the most disaggregated level, analyses are performed for individual *point* or *nodes*, which represent a single intersection. In fact, the *HCM* typically evaluates each approach to these intersection points separately, as traffic demands and geometric characteristics can easily differ from one leg of an intersection to the next. Points also include interchanges, which are the junctions between the surface streets and freeway networks.

The next aggregation level is the *HCM segment*, which for arterial streets represents the combination of an intersection and the link immediately upstream of it (in the direction of travel). Segments are also the lowest aggregation interval for freeways, where freeway, merge, diverge, and weaving segments are the basic building blocks. For both arterials and freeways, multiple segments can be aggregated to the level of facilities, which represent extended urban streets or sections of freeway that are of interest for analysis.

The *facility* is the highest level of analysis that can be readily performed in the *HCM*. However, other, more aggregated levels of spatial scope exist. For example, a system of two parallel facilities, an arterial and a freeway, are referred to as a *corridor*. A corridor analysis scope assumes that there is some interaction between the two facilities, for example, in the form of traffic diversion when an incident or general congestion impacts the operations on one or the other. From the corridor level, an *area* analysis may refer to the level of an entire neighborhood or suburb that

Table 5.4 Service measures by system element in the *HCM*

Transportation element	Service measure by mode				System measure
	Auto	Pedestrian	Bicycle	Transit	
Freeway facility	Density	—	—	—	Speed
Basic freeway segment	Density	—	—	—	Speed
Multilane highway	Density	—	LOS score	—	Speed
Freeway weaving segment	Density	—	—	—	Speed
Merges and diverges	Density	—	—	—	Speed
Two-lane highway	PTSF[1], speed	—	LOS score	—	Speed
Urban street facility	Speed	LOS score	LOS score	LOS score	Speed
Urban street segment	Speed	LOS score	LOS score	LOS score	Speed
Signalized intersection	Delay	LOS score	LOS score	—	Delay
Two-way stop intersection	Delay	Delay	—	—	Delay
All-way stop intersection	Delay	—	—	—	Delay
Roundabout	Delay	—	—	—	Delay
Interchange ramp terminal	ETT[2]	—	—	—	Travel time
Alternative intersection	ETT[2]	—	—	—	Travel time
Off-street pedestrian/bike facility	—	Space	LOS score	—	Speed

PTSF = percent time spent following; ETT = experienced travel time (sum of delay and extra distance travel time)

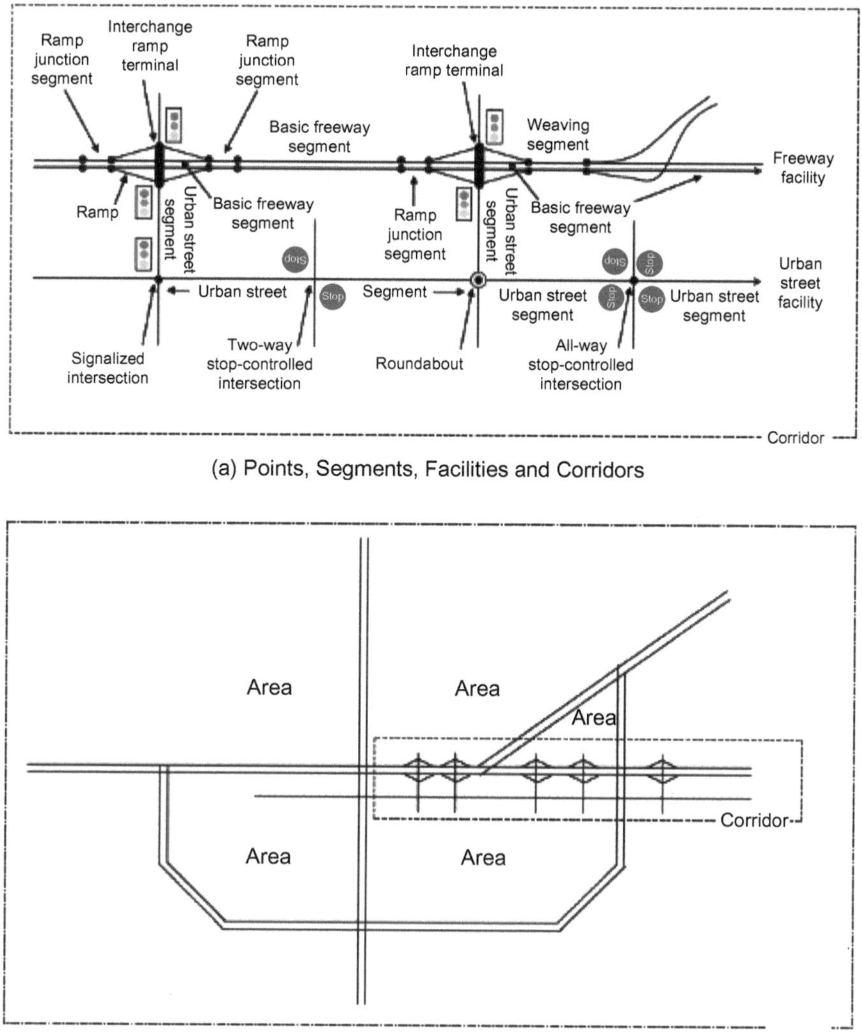

(a) Points, Segments, Facilities and Corridors

(b) Corridors, Areas and Systems

Figure 5.3 Illustration of spatial analysis scope and system elements. *Source: TRB, 2015.*

combines multiple links, nodes, and segments. Finally, multiple areas make up the overall transportation *system* of, say, a metropolitan area. The analysis of corridors, areas, and system typically requires the use of modern software tools that either integrate the methods from the *HCM* or use traffic simulation principles to estimate network performance, as discussed later in this part of the book. The various system elements are illustrated in Figure 5.3.

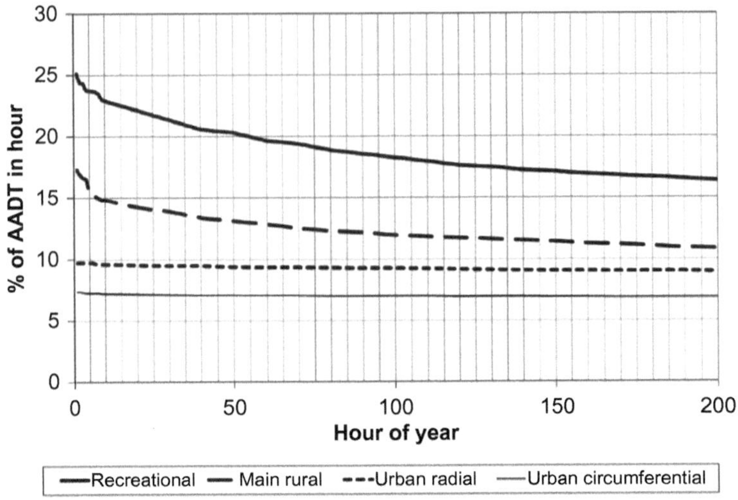

Notes: Recreational, US-2 near Stevens Pass (AADT = 3,862); main rural, I-90 near Moses Lake (AADT = 10,533); urban radial, I-90 in Seattle (AADT = 120,173); urban circumferential, I-405 in Bellevue (AADT = 141,550).

Figure 5.4 Illustration of traffic volume by hour of year. *Source: TRB, 2015.*

Temporal Analysis Scope

Equally important to the spatial scope is the temporal analysis scope. Traditional *HCM* analyses have focused on the analysis of the peak 15 min of the 30th highest hour of the year in terms of traffic demands. This 30th highest hour is also referred to as the *design hour.* The design hour approach acknowledges that it is likely inefficient and cost-prohibitive to base transportation analyses (and the resulting investment decisions) on the single highest hour of the year. In other words, the maximum hour of the year is by definition a very rare event, and has been found to not be reflective of "typical" congestion levels (Figure 5.4).

In practice, analysts typically perform a traffic count on a typical weekday, which often avoids Monday and Friday, as well as any holidays. It is often cost-prohibitive to count traffic for multiple days or weeks at the same intersection, resulting in the general assumption that the measured peak period volume is a reasonable approximation of the *design hourly volume* (DHV). However, the analyst should take great caution to select the observation period carefully, especially when evaluating a site with seasonal variability.

For the purpose of analysis, the peak 15-min flow rate of the DHV is used traditionally as the volume input for operational analyses. The peak 15-min flow rate can be measured directly in the field (preferred), or can be estimated from the DHV using a *peak hour factor* (PHF). The PHF was described in detail in *Part 2: Transportation Planning*, and is defined as the peak hour volume divided by the product of four times the peak 15-min volume. The resulting factor is less than or equal to 1.0, and is used to convert the DHV to a peak 15-min flow rate, by diving the hourly volume by the PHF. This yields a flow rate that is likely greater than the hourly volume, and represents the 15-min peak.

Reliability Analysis

The traditional operational analysis scope in the *HCM* focuses on the operations of the peak 15-min flow rate of the 30th highest hour of the year, as described. This convention is often needed to manage limited resources for traffic counts in support of projects, and to obtain analysis data in a timely fashion. But by definition, this (simplified) approach of single-day counting ignores day-to-day variability of traffic. More importantly, it does not consider effects of nonrecurring sources of congestion that are known to impact the reliability of the system. The Federaly Highway Administration[1] (FHWA) defines seven common sources of unreliable travel as:

- Physical bottlenecks or capacity limitations
- Weather impacts
- Work zones
- Traffic incidents
- Traffic control devices, including signal timing
- Special events
- Day-to-day variability and fluctuations in normal traffic

A sample of peak travel times for a route in Seattle, Washington is shown in Figure 5.5. The figure illustrates that using any single estimate for the purpose of analysis fails to capture the true day-to-day variability in travel.

Significant research in recent years has focused on defining performance measures and developing methodologies to quantify travel time reliability for traffic operational analysis. Travel time reliability is also integrated in the *HCM* as a key new concept in the latest edition, with the *HCM* now offering methodologies to estimate reliability for both surface streets and freeways. At the heart of travel time reliability research and the *HCM* methodologies is the *travel time distribution*, which is the basis of all reliability performance measures. An example travel time distribution is shown in Figure 5.6, which illustrates several key reliability performance measures,

[1] www.ops.fhwa.dot.gov/congestion_report/executive_summary.htm

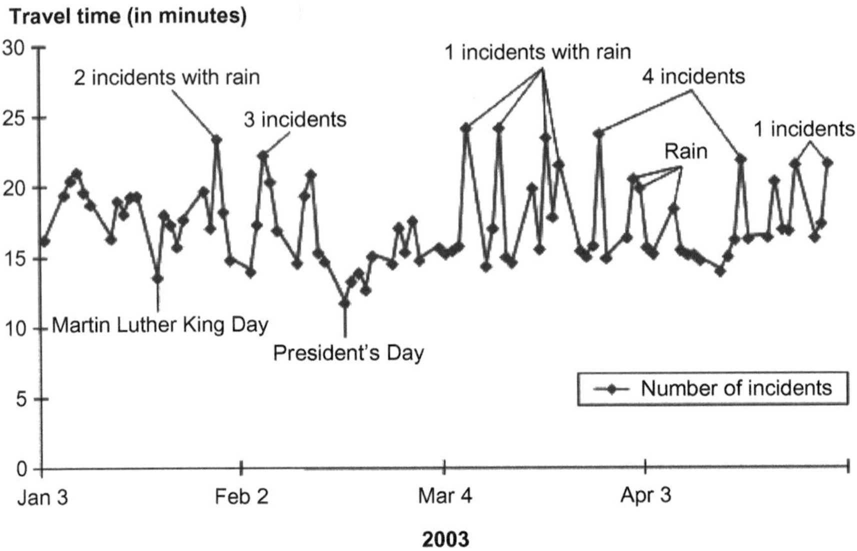

Figure 5.5 Example of day-to-day fluctuation in peak travel time. *Source: FHWA, 2005a; http://www.ops.fhwa.dot.gov/congestion_report/executive_summary.htm.*

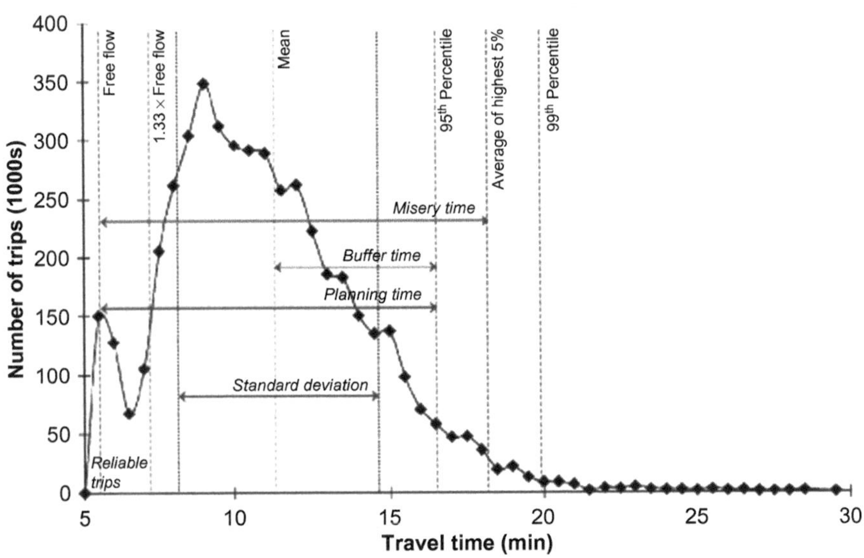

Figure 5.6 Example travel time reliability distribution. *Source: Zegeer et al., 2014; http://onlinepubs.trb.org/onlinepubs/shrp2/SHRP2_S2-L08-RW-1.pdf.*

including the standard deviation of travel time, the 95th percentile of travel time, planning time (comparing 95th percentile to free-flow travel time), and buffer time (comparing 95th percentile to average travel time).

5.2 TRAFFIC FLOW FUNDAMENTALS

Before presenting the analysis methodologies for evaluating the operational performance of transportation facilities, we need to cover some fundamental principles and relationships of traffic flow.

5.2.1 Fundamental Relationship

Speed-Flow-Density

Traffic flow theory relies on three entities to describe the operations of a transportation system element: speed, flow, and density.

The *average speed* of the traffic stream is a fairly intuitive measure that describes how fast or slow, on average, traffic moves on a segment of roadway, and is typically measured in miles per hour (mph) or kilometers per hour (km/h). Traffic flow theory further distinguishes between time mean speed (TMS) and space mean speed (SMS) as described in the following, with the latter being the preferred and correct term to describe operations on a roadway segment.

The *traffic flow* is a measure of throughput through the system element, and is typically measured in vehicles per hour (veh/h), or sometimes vehicles per hour per lane of travel (veh/h per lane). The traffic flow is closely tied to the demand on the segment, as well as its capacity, which may impact and limit the throughput if demand exceeds that capacity.

Finally, *density* is a measure of how many vehicles are found on a given length of roadway segment, and is typically expressed in vehicles per mile per lane (veh/mi per lane) or vehicles per kilometer per lane (veh/km per lane).

Any of these measures can be calculated from the *fundamental relationship of traffic flow*, as follows:

Flow (F) = speed (S) × density (D)

$$F = S \times D$$

$$S = F/D$$

$$D = F/S$$

Each can also be observed and quantified in the field, with the measurement of speeds and flow rates often being much simpler than direct measurements of density. Therefore, density is often calculated from speed and flow, rather than being measured directly.

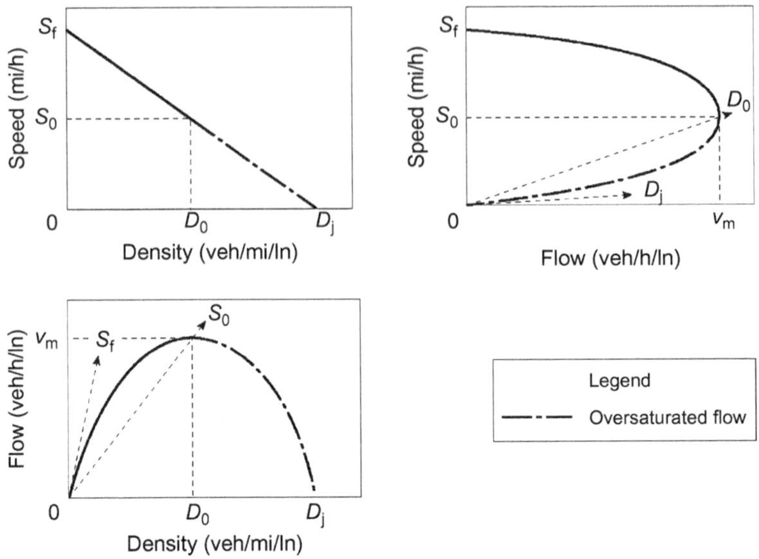

Figure 5.7 Simplified representation of speed-flow-density relationship. *Source: TRB 2015, Highway Capacity Manual.*

In a simplified or generalized traffic flow relationship, it is often assumed that speed varies linearly with density and vice versa, where an increase in density leads to an inversely proportional decrease in speed. This in turn leads to a parabolic relationship between speed and flow, as well as between flow and density, as illustrated in Figure 5.7.

The figure introduces additional concepts, including the free-flow speed (S_f), the flow rate at capacity (F_c), the speed and density at capacity (S_c and D_c), and the jam density (D_j). By definition, the density at S_f is zero and the speed at D_j is zero. In the speed-flow relationship, density is described by radial lines from the origin to a point on the curve, where the slope of that line describes the density. Similarly, speed can be found in the flow-density relationship through the slope of radial lines from the origin. Capacity is shown to separate undersaturated and congested flow regimes, with the flow rate being maximized at the capacity.

The simplified representation of the speed-flow-density relationship is useful for illustration, but it is acknowledged here that the strict linear relationship between speed and density does not entirely hold true in field observations, as illustrated in Figure 5.8.

It is noted here, that the letters used to describe speed, flow, and density can vary across the literature. This book uses the abbreviations "*S*" for

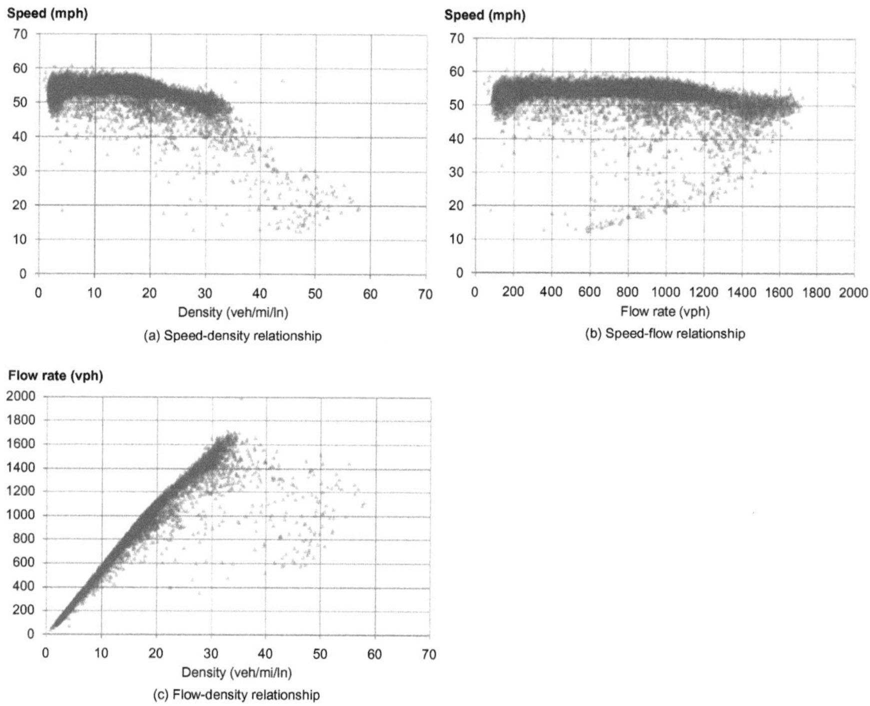

Figure 5.8 Field data illustrating speed-flow-density relationship.

speed, "*V*" for volume, and "*D*" for density, which is consistent with the *HCM*. However, it is also common for textbooks and research papers to refer to speed as "*v*" (for velocity), to volume as "*F*" (for flow rate) or sometimes as "*q*," and density as "*k*." The reader is therefore urged to carefully review variable definitions to assure that variables are used correctly.

Time Mean Speed versus Space Mean Speed

In traffic operations, two different definitions exist for the measurement of speed: time mean speed (TMS) and space mean speed (SMS). In most *HCM* methodologies, the SMS is the correct metric to, for example, describe the average speed over a freeway segment. However, field studies often measure speeds at a point (through sensors, radar speed measurements, etc.), and the commonly used arithmetic average of those speed observations yields the TMS, not SMS. In other words, field studies often average speeds over *time*, while *HCM* methods require the speeds to be averaged over *space*. The TMS is generally slightly larger than the SMS, although the difference is typically on the order of 2% or less.

The time mean speed is defined as the arithmetic average speed of vehicles passing through a point, and is calculated using Eq. (5.1):

$$S_{TMS} = \frac{\sum_{i=1}^{N} S_i}{N} \tag{5.1}$$

where
S_{TMS} = time mean speed (mph)
S_i = individual vehicle speed measured at a point (mph)
N = sample size of speed measurements

The space mean speed (SMS) can be calculated using a straight average of speeds, *only if speeds were measured over a segment*. For example, if the analyst has access to a record of individual vehicle travel times and converts those to individual speeds by dividing the segment length by these travel times, then an average of the resulting speeds yields the space mean speed. This can also be achieved by first summing the individual travel times t_i, and factoring out the segment length, L, as is shown in Eq. (5.2):

$$S_{SMS} = \frac{N \times L}{\sum_{i=1}^{N} t_i} \tag{5.2}$$

where
S_{TMS} = space mean speed (mph)
t_i = individual vehicle travel time measured over a segment (h)
L = segment length (mi)
N = sample size of speed measurements

Alternatively, the space mean speed can be calculated directly from point measurements of vehicles passing through a point by using the harmonic mean (as opposed to the arithmetic mean) as shown in Eq. (5.3):

$$S_{SMS} = \frac{N}{\sum_{i=1}^{N} \frac{1}{S_i}} \tag{5.3}$$

where
S_{SMS} = space mean speed (mph)
S_i = individual vehicle speed measured at a point (mph)
N = sample size of speed measurements

In traffic operation applications, analysts generally use the space mean speed. Depending on whether speed measurements are available as point speeds, or in the form of segment speeds (from travel times), the analyst should therefore use Eq. (5.3) and Eq. (5.2), respectively.

Example 5.1

A speed study collected speeds from nine vehicles at a fixed point as follows: 35 mph, 40 mph, 32 mph, 41 mph, 38 mph, 37 mph, 40 mph, 36 mph, and 39 mph.

Calculate the time mean speed and space mean speed. Also, assuming these vehicle speeds are fixed over a ½-mile segment, calculate the corresponding travel times and show that the space mean speed calculated from these travel times matches the point estimate.

Solution

$$\text{TMS} = 1/9 \times (35 + 40 + 32 + 41 + 38 + 37 + 40 + 36 + 39) = 37.56 \text{ mph}$$

$$\text{SMS} = 9/(1/35 + 1/40 + 1/32 + 1/41 + 1/38 + 1/37 + 1/40 + 1/36 + 1/39) = 37.35 \text{ mph}$$

Travel time (TT) over ½ mile for a 35-mph vehicle is

$$\text{TT} = \tfrac{1}{2} \text{ mi}/35 \text{ mph} = 0.0143 \text{ h} = 51.4 \text{ s}$$

The travel times for all nine vehicles are 0.0143 h, 0.0125 h, 0.0156 h, 0.0122 h, 0.0132 h, 0.0135 h, 0.0125 h, 0.0139 h, and 0.0128 h.

$$\text{SMS} = (9 \times 0.5)/(0.0143 + 0.0125 + 0.0156 + 0.0122 + 0.0132 + 0.0135 + 0.0125 + 0.0139 + 0.0128) = 37.35 \text{ mph}$$

Macro versus Micro Characteristics of Traffic

The concepts of speed, flow, and density are key *macroscopic* characteristics of the traffic stream. They are immensely useful when describing the aggregated performance of a number of vehicles (say over a 15-min period), and follow useful relationships that allow for making predictions of future and changing traffic states. However, it is important to consider the *microscopic* characteristics of individual vehicles or better drivers, as well as the relationship between the individual driver and the aggregated speed-flow-density characteristics of the traffic stream.

A principal microscopic characteristic of traffic is the time *headway*, which is defined as the time between successive vehicles traveling through a fixed point. It is measured as the time between when the front bumper (or other fixed element) of vehicle 1 passes through a point until the front bumper of vehicle 2 (same element) passes through that same point.

The minimum headway between vehicles on an arterial street has been shown through research to be approximately 2 s. In other words, it takes a minimum of 2 s per vehicle to pass through a fixed point on an arterial street. Summing 2 s per vehicle to a full hour of observation, yields the relationship:

$$3600\frac{\text{s}}{\text{h}} \times \frac{1}{2.0\frac{\text{s}}{\text{veh}}} = 1800\frac{\text{veh}}{\text{h}}$$

The resulting 1800 veh/h is the maximum number of vehicles that can pass through a point in one hour, provided there are no interruptions such as traffic signals. For traffic signals, the resulting flow rate is also referred to as the *saturation flow rate*, which is equivalent to the maximum flow rate per hour of continuous green, and will emerge as a key parameter in describing and analyzing the capacity of a signalized intersection. In other words, the maximum (saturation) flow rate is equal to the inverse of the minimum (saturation) headway between vehicles.

For arterial streets, the saturation headway is around 2 s (varying between about 1.9 s and 2.1 s depending on driver aggressiveness), resulting in a saturation flow rate of 1800 veh/h (or a range from 1900 to around 1700 veh/h). For basic freeway segments, the minimum headway can be as low as 1.5 s, which yields the freeway capacity of (3600 s/h)/(1.5 s/veh) = 2400 veh/h that was described in Table 5.3.

Any characteristics that may increase the minimum headway including curvature (e.g., left and right turn at signals or travel through a roundabout), grade, lane widths, or frictional effects from weather, work zones, or incidents will thus reduce the saturation flow rate or capacity of the transportation system element. To measure the capacity of a system element, one could thus measure the aggregated throughput over 15 min or 1 h, or one could measure the saturation headways, and then convert to the hourly flow rates.

Another critical microscopic traffic parameter is the *spacing* between vehicles, which is the distance between two vehicles when traveling at some headway and prevailing speed. If the headway is the time between successive vehicles passing a fixed point, the spacing is the distance between those two vehicles, obtained conceptually from taking a photograph of the two vehicles and measuring the distance between them (or technically from front bumper to front bumper). The spacing can be

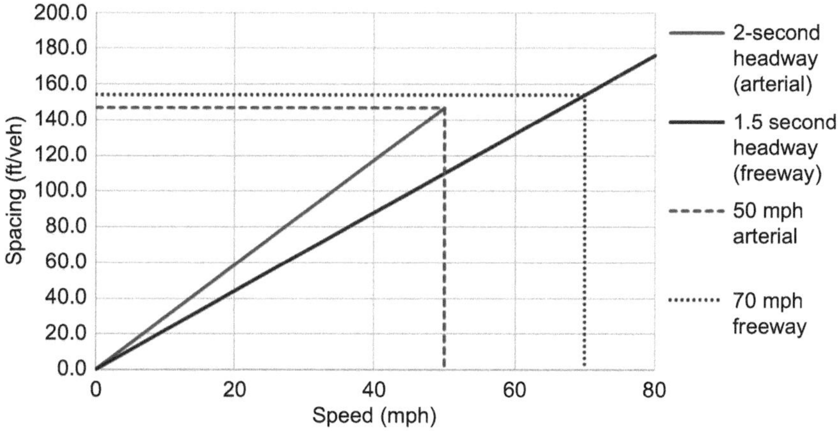

Figure 5.9 Illustration of spacing versus speed for arterials and freeways.

calculated by multiplying the headway by the vehicle speed. For example, a 50-mph arterial with a minimum headway of 2 s/veh results in a spacing at that speed of:

$$\text{mph} \times 1.467 \frac{\text{ft/s}}{\text{mph}} \times 2.0\text{s/veh} = 146.7\text{ft/veh}$$

With increasing vehicle speed, the spacing also increases, provided the headway stays constant. One simple way to rationalize why the capacity of a freeway is higher than that of an arterial street (i.e., why the minimum headway for a freeway is lower than the minimum headway for an arterial) is to look at spacing. Figure 5.9 illustrates that the aforementioned spacing for a 50-mph arterial and a headway of 2 s is essentially the same as the spacing for a 70-mph freeway with a headway of 1.5 s (146.7 ft/veh compared to 154.0 ft/veh).

While this oversimplifies the explanation (e.g., driver reaction times and visual abilities are impacted by speed), it nonetheless illustrates how important it is to understand the relationship between micro and macro characteristics of the traffic stream. Flow rate (F) is the inverse of headway (h); spacing (x) is the product of headway (h) and speed (S); and density (D) can be estimated from the inverse of the spacing (s) of vehicles. A summary of these relationships, corresponding equations, and associated units is shown in Table 5.5.

Table 5.5 Summary of macro and micro relationships

Relationship	Equation	Units
Flow = speed × density	$F = S \times D$	[veh/h] = [mph] × [veh/mi]
Flow = inverse of headway	$F = 3600/h$	[veh/h] = [h/s]/[s/veh]
Spacing = headway × speed	$X = h \times 1.467 \times S$	[ft/veh] = [s/veh] × ([ft/s]/[mph]) × mph
Density = inverse of spacing	$D = 5280/h$	[veh/mi] = [ft/mi]/[ft/veh]

5.2.2 Vehicle Kinematics

Many traffic operational relationships rely on basic vehicle kinematics and physics principles to estimate position, speed, acceleration, and their interrelation. A more detailed discussion of these concepts can be found in various physics textbooks or online resources, but the basic principles are repeated here for quick reference. The discussion uses concepts and variables of position (x), time (t), speed (S), and acceleration (a), as well as subscripts "i" for initial and "f" for final.

In the most basic kinematics relationship, speed is defined as position over time:

$$S = \frac{x}{t} \tag{5.4}$$

Similarly, (uniform) acceleration is defined as the change in speed over a time interval as:

$$a = \frac{\Delta S}{t} = \frac{S_f - S_i}{t} \tag{5.5}$$

Final speed can be estimated from an initial speed and uniform acceleration over time as:

$$S_f = S_i + at \tag{5.6}$$

Similarly, final position or displacement can be estimated from an initial position, initial speed, and uniform acceleration over time as:

$$x_f = x_i + S_i t + \frac{1}{2}at^2 \tag{5.7}$$

These relationships are useful for applications in traffic operations. For example, Eq. (5.4) can be solved for x_{PRT} to calculate the reaction distance of a vehicle traveling at a given initial speed, S_i, and a driver perception reaction time, t_{PR}, as follows:

$$x_{PRT} = 1.47 \times t_{PR} \times S_i \tag{5.8}$$

where

x_{PRT} = distance traveled during perception reaction time (ft)
t_{PR} = perception reaction time (s)
S_i = initial speed of vehicle (mph)

This equation is very useful to estimate the reaction distance before drivers even begin to decelerate for an obstacle. The concept is also used for sight distance calculations, to, for example, determine how far a driver needs to be able to see to accept a certain size gap in traffic. For example, if trying to accept a 5-s gap in a traffic stream traveling at 35 mph, the resulting minimum sight distance is:

$$1.47 \frac{\text{ft/s}}{\text{mi/h}} \times 5s \times 35\text{mi/h} = 257.3\text{ft}$$

Another very useful derivative of the kinematics relationships is the concept of breaking distance, which is given by:

$$x_f = x_i + \frac{S_i^2 - S_f^2}{2a} \tag{5.9}$$

Combining Eq. (5.8) and Eq. (5.9), and setting the final speed, S_f, and initial distance, x_i, equal to zero, yields an equation to estimate the total distance traveled to come to a full stop from some initial speed, as a combination of perception reaction distance and breaking distance.

$$x_{total} = 1.47 \times t_{PR} \times S_i + \frac{(1.47S_i)^2}{2a}$$

where

x_{total} = distance traveled during perception reaction and breaking (ft)
t_{PR} = perception reaction time (s)
S_i = initial speed of vehicle (mph)
a = uniform deceleration rate (ft/sec^2)

Example 5.2

Estimate the total distance required for a vehicle to come to a full stop if traveling 25 mph, assuming a perception reaction distance of 1 s and a deceleration rate of 11.2 ft/sec^2.

Repeat the calculation for speeds up to 45 mph in 5-mph increments, as well as assuming a reaction time of 2.5 s.

Solution

$$X = 1.47 \times 1\ \text{s} \times 35\ \text{mph} + (1.47 \times 35)^2/(2 \times 11.2\ \text{ft}/s^2) = 97.0\ \text{ft}$$

Repeating the calculation for varying initial speeds and changing perception reaction times, yields the following table and chart.

Reaction time (s)	Initial speed (mph)	Stopping distance (ft)
1	25	97.0
1	30	130.9
1	35	169.6
1	40	213.2
1	45	261.5
2.5	25	152.2
2.5	30	197.1
2.5	35	246.8
2.5	40	301.4
2.5	45	360.7

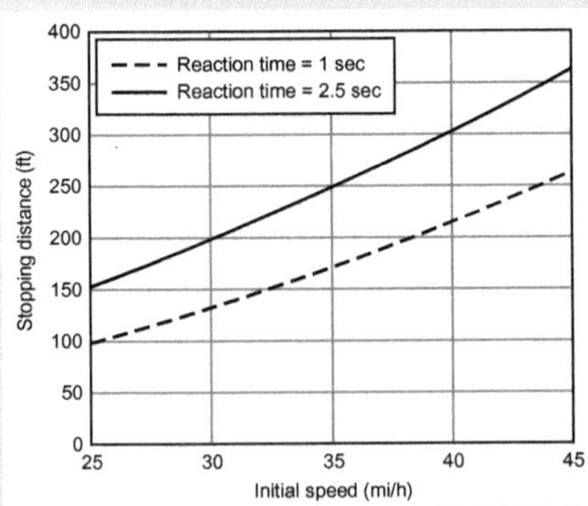

This sensitivity analysis illustrates that a 20-mph increase in speed from 25–45 mph (80% increase) results in a 170% increase in stopping distance from 97.0–261.5 ft.

Similarly, an added perception reaction distance of 1.5 s (a common result of distracted driving, cell phone use, or unexpected incidents), increases the 25-mph stopping distance by 57% from 97.0–152.2 ft, and

(Continued)

Example 5.2—(Continued)
the 45 mph stopping distance by 38% from 261.5–360.7 ft—an increase of 61.5 and 99.2 ft, respectively. For a deceleration event that requires stopping at a traffic signal or a pedestrian crosswalk, these additional travel distances (during the increased perception reaction time) can have dramatic safety implications.

5.2.3 Shockwaves

Shockwaves are an important concept in traffic operational analysis. They can be used to describe the propagation of queues on a freeway, as well as their dissipation. Shockwaves are also used in the theory describing operations at signalized intersections, and thus find application across both uninterrupted and interrupted flow domains in traffic operations.

Shockwaves occur at the transition from one traffic state to another, say, free-flow to stopped, or from stopped back to flowing traffic. This application is common for freeway bottlenecks and incidents, which cause vehicles to drastically slow down or come to a full stop. On arterial streets, the same pattern is found at signalized intersection, where a red signal forces a transition from one traffic state (moving) to another (stopped). But shockwaves also occur between two moving regimes. For example, a slow-moving tractor-trailer on a freeway may force a slower flow regime in its immediate proximity, distinguishing it from faster-speed, freely moving traffic in front or behind it. Another example would be a freeway section, in which vehicles tend to slow down (e.g., due to an upgrade, a bridge, a tunnel, sunglare, etc.), resulting in a changed traffic state and (depending on the demand level) congestion upstream of this section.

Figure 5.10 illustrates accumulating and dissipating waves at a signalized intersection, drawn on a time-space diagram of individual vehicle trajectories. As vehicles approach the (red) signal, they have to slow down until the signal turns green, and can then reaccelerate. The result of multiple stopping vehicles in sequence is a shockwave that describes the propagation of the back of the queue upstream of the signal. Similarly, once the light turns green, another wave describes the rate at which vehicles discharge from congestion at the signal. It is very evident that the queue clears from the front (signal stop bar) toward the upstream end of the queue. This is also referred to as a *front-clearing* queue.

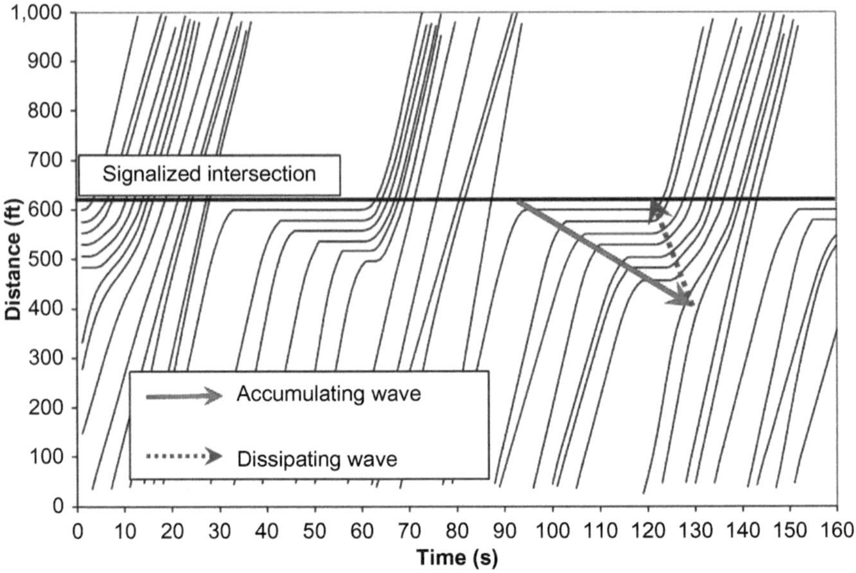

Figure 5.10 Shockwaves at a signalized intersection.

From the simple shockwave diagram in Figure 5.10, the maximum back of queue is found mathematically by the intersection of the accumulating and dissipating waves. Further, the total delay of all vehicles in one cycle is described by the area of the triangle, with the average delay per vehicle being defined as the total area divided by the number of vehicles per cycle. This example is simplified, showing an undersaturated signal (the queue clears each cycle) and a random arrival profile. In reality, arrivals are expected to be at least partially platooned, resulting in multiple flow regimes that yield multiple accumulating waves, and an area of the delay region that is more complicated than a simple triangle. These concepts are discussed further in a later section.

For freeway operations, congestion often forms through the interaction of different drivers and vehicles in near-capacity situations. So rather than having a clear bottleneck (and as a result, clear congestion patterns), the turbulence in a heavy traffic stream on freeways is often more random. Sometimes, this results in congestion without apparent cause, a phenomenon that many freeway commuters are all too familiar with. The random variation of vehicle headways and random turbulence can result in breakdown, even without any apparent bottleneck, incident, or other

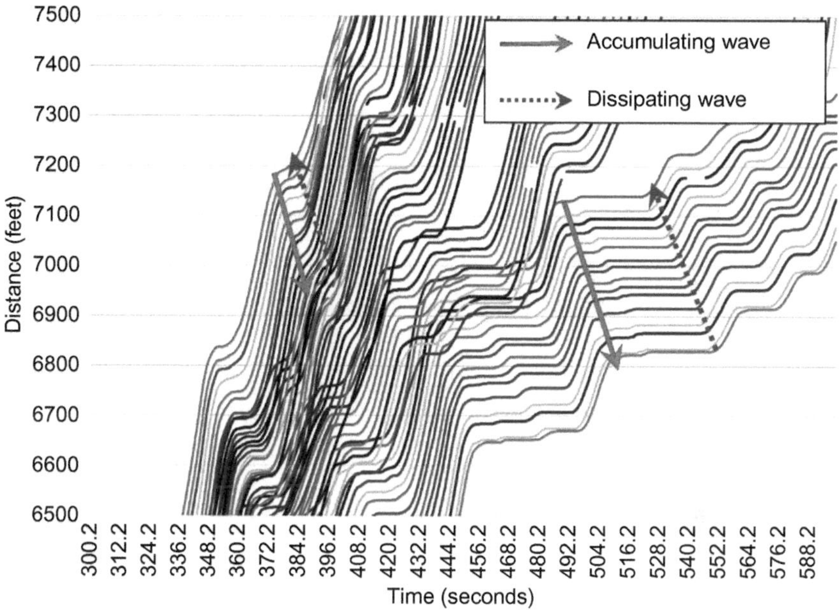

Figure 5.11 Freeway trajectories and shockwaves.

"reason" for the congestion. Figure 5.11 illustrates this with a plot of simulated vehicle trajectories (space vs. time) on a congested freeway. Rather than having a clear bottleneck location, and clean resulting shockwaves as was the case in the signalized intersection example, the congested freeway shows multiple shockwaves resulting from turbulence and localized breakdown events. From an aerial view, this could be visualized as multiple waves of brake lights propagating upstream through the traffic stream.

The speed of the congestion shockwave then describes the speed at which congestion, or more precisely the back of queue, travels backward from the bottleneck location. Shockwaves are often readily observed visually through a wave of brake lights, an effect that is even more dramatic when observed from a bridge, tall building, or a helicopter. In fact, the title page of this part of the book shows a freeway section with congestion on the far hill crest, which, over time, spills back to the position from where the picture is taken. Some isolated brake lights are already visible.

Mathematically, the speed of a shockwave can be calculated from the flow rates and densities of the upstream and downstream traffic states.

Recall the fundamental relationship of traffic flow, as flow (F) = speed (S) × density (D). The speed of the wave between regimes 1 and 2 then is calculated as:

$$v_w = \frac{F_2 - F_1}{D_2 - D_1} \quad (5.10)$$

where

v_w = wave speed (mph)

F_1, F_2 = flow rates in regimes 1 and 2 (veh/h)

D_1, D_2 = density in regimes 1 and 2 (veh/mi)

The number of vehicles crossing from regime 1 into regime 2 in a time interval, t, (or vice versa from regime 2 into regime 1) can be calculated as:

$$\begin{aligned} n_1 &= (S_1 - v_w) \times D_1 \times t \\ n_2 &= (S_2 - v_w) \times D_2 \times t \end{aligned} \quad (5.11)$$

where

n_1, n_2 = number of vehicles crossing wave boundary from regimes 1 and 2 (veh)

v_w = wave speed (mph)

t = time interval (h)

S_1, S_2 = space mean speed in regimes 1 and 2 (veh/h)

D_1, D_2 = density in regimes 1 and 2 (veh/mi)

From Eq. (5.10), it is evident that the shockwave speed can be either positive or negative. The sign of the shockwave speed goes with the direction of traffic, to where a positive shockwave speed travels downstream, while a negative shockwave speed means the wave is traveling upstream. Most congestion wave speeds (from a bottleneck) are thus negative, as traffic accumulates and forms a queue upstream of the bottleneck, and the back of queue travels further upstream over time at the wave speed.

Dissipating shockwaves when congestion clears can either be positive or negative. A positive shockwave is one that occurs when demand at a fixed bottleneck location begins to drop after the peak congestion, and the queue therefore clears from the back. These "back-clearing" queues have a positive wave speed. On the other hand, "front-clearing" queues occur when the bottleneck itself is lifted, while the demand stays fixed. In this case, the queue clears from the front with a negative wavespeed, meaning that the front of the queue travels upstream. Front-clearing queues are common on freeways when the bottleneck is due to an incident that at

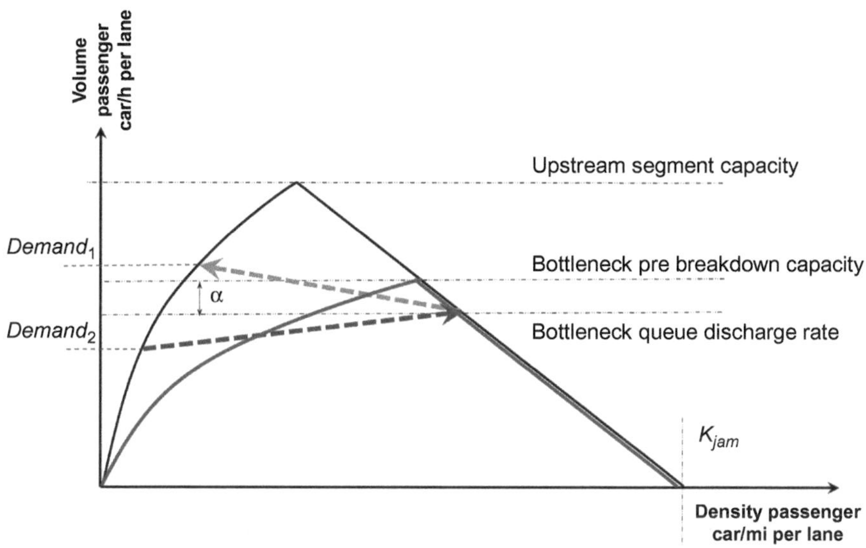

Figure 5.12 Illustration of shockwaves for freeway bottleneck.

some point clears. Front-clearing queues are also common at signalized intersections, where the bottleneck is lifted once the signal turns green. This is apparent in Figure 5.10, where both the forming and dissipating waves travel upstream from the bottleneck with a negative wave speed. In this case, the dissipating wave (lower dashed line) is faster than the forming wave (upper dashed line), which means the queue fully clears when the faster wave catches up with the slower one.

An illustration of shockwave theory applied to a freeway bottleneck is shown in Figure 5.12. It shows a hypothetical case of an upstream segment (thin black line) and a downstream bottleneck with lower capacity (thick gray line). On the *y*-axis, the figure shows two demands: D_1, which is a high demand above the bottleneck capacity that causes congestion, and D_2, which is a lower demand below the bottleneck capacity that causes congestion to dissipate. The figure further distinguishes between the prebreakdown capacity and the queue discharge rate, recognizing a drop in throughput after a bottleneck is activated.

In the figure, two shockwaves are illustrated. First the forming or accumulating wave that is shown as a light gray dashed arrow between the bottleneck queue discharge rate and the upstream segment (high) demand, D_1. This shockwave moves against the direction of traffic and thus has a negative sign. The second wave is the dissipating wave, which is drawn as a dark gray dashed arrow between the (lower) demand D_2 in the upstream segment, and the queue discharge rate. This wave travels in the direction of traffic and therefore has a positive sign.

Example 5.3

A freeway segment has a flow rate of 1500 veh/h (per lane) traveling at a speed of 60 mph. This traffic stream arrives at a crash location that results in a full closure of the facility, resulting in a queue at the jam density (maximum density) of 180 veh/mi (per lane). Estimate the speed at which the shockwave grows upstream of the crash location.

Solution

First, estimate the density of the upstream regime from the fundamental equation, as:

$$D_1 = F_1/S_1 = 1500/60 = 25 \text{ veh/mi}$$

Second, calculate the speed of the wave from Eq. (5.10) using the flow and density in regimes 1 and 2.

$$v_w = \frac{F_2 - F_1}{D_2 - D_1} = \frac{0 - 1500}{190 - 25} = -9.1\text{mi/h}$$

The shockwave is traveling upstream (negative wave speed) at a speed of 9.1 mph

5.2.4 Gap Acceptance

Another fundamental theoretical concept in traffic operations is gap acceptance. Gap acceptance applies whenever the interaction of two traffic streams is not fully controlled through grade separation or signalization. Common examples of where gap acceptance theory is applied are in evaluating the capacity and operations of a modern roundabout, a stop or yield-controlled intersection, or a permissive right- or left-turning movement at a signalized intersection. Gap acceptance theory equally applies to pedestrians trying to get across the street, or bicycles trying to cross or merge into a traffic stream from a side street.

The parameters describing gap acceptance are critically linked to driver behavior and Human Factors Principles. Similar to the concept of headway and saturation headway introduced in Section 5.2.1, gap acceptance is linked to the level of comfort and safety perception of drivers. In other words, a more aggressive driver is likely to accept shorter gaps than a more conservative driver. Vehicle dynamics also play into gap acceptance, to where a slower vehicle (e.g., a truck or another heavy vehicle) is likely to require a longer gap than a small sports car. Similarly, a runner is likely to require a shorter gap than a pedestrian walking more slowly. When trying to describe gap acceptance for a traffic stream, this diversity of characteristics and dynamics attributes needs to be taken into account.

The two fundamental parameters in gap acceptance theory are (1) critical gap, and (2) follow-up headway. They are defined as follows:

1. *Critical gap* is the gap time in seconds at which drivers are equally likely to accept or reject the gap and enter the conflicting traffic stream. It is mathematically used to estimate the time needed for the first vehicle to enter the conflicting traffic stream.
2. *Follow-up headway* is the additional time beyond the critical gap needed for each additional vehicle to enter the conflicting traffic stream.

For example, if the critical gap (for the first vehicle) is 5 s, and the follow-up headway is 2 s, it is expected that two queued vehicles could enter in a gap of 7 s, three vehicles in a gap of 9 s, and so on.

The critical gap is different from the minimum gap needed for *any* vehicle to enter the traffic stream, although the two are commonly confused. The critical gap concept recognizes that the traffic stream is nonhomogeneous, and that different drivers have different risk thresholds. As such, a study measuring all accepted gaps would not find one value, but a distribution of values. Similarly, that study would find a distribution of rejected gaps. If plotting these two cumulative distributions, the critical gap is defined as the intercept of the two curves, or the gap size where drivers are equally likely to accept or reject the gap. This is illustrated graphically in Figure 5.13 based on a pedestrian gap acceptance study

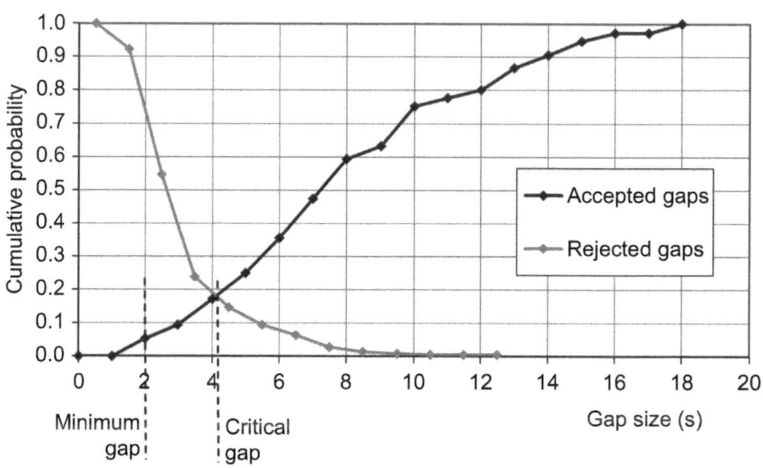

Figure 5.13 Sample critical gap study using graphical method.

(Schroeder, 2008). The figure further shows the minimum gap (i.e., the smallest observed accepted gap), which is mathematically (and practically) very different from the critical gap.

The illustration in Figure 5.13 is also referred to as the *graphical method* for estimating critical gap as described in (Schroeder et al., 2010). Other methods for estimating critical gap are a method based on *maximum likelihood estimation* (MLE), which estimates both a mean and standard deviation of the critical gap assuming a normal distribution, and the *Ramsey-Routledge method*, which can be used to estimate any distribution of critical gaps, including bimodal distributions (e.g., a pedestrian stream with half walkers and half runners). All three methods are described in detail in (Schroeder et al., 2010).

5.2.5 Simple Queuing Theory

Queuing theory finds frequent application in transportation operations to estimate queue lengths at intersections. Queue lengths are an important performance measure, as they directly reflect the experience of drivers, and can further be linked to important design considerations. For example, the estimated queue length of a left turn at a signalized intersection is a key piece of information to design the length of the left-turn pocket. Similarly, queue lengths are critically important for two closely spaced intersections to estimate the potential for the queue from one signal to spillback into the upstream signal, thereby potentially impacting its operations significantly. Queue spillback is also a key consideration for freeway off-ramps, where the queue from the interchange ramp terminal should be contained within the off-ramp length if at all possible, and should not spill back onto the freeway, which would have big safety and operational implications.

Queuing theory has many different applications beyond transportation, and different queueing models exist to describe queue as a function of the number of channels approaching the queuing location, the number of servers able to process the queue, the arrival demands, the time needed to process each entity, and the priority order of processing entities. But while queuing theory can get quite complex, luckily most queues in transportation are simple *first in/first out* (FIFO) queues, and are often assumed to have fixed service times (on average).

In traffic operations, queues (and most other performance measures) can either be obtained deterministically (through the use of analytical equations), or can be simulated. The focus in the following sections is on the former, although simulation concepts are introduced in a later section.

5.2.6 Heavy Vehicle Effects

The traffic stream consists of both cars and various types of heavy vehicles, including single-unit trucks, buses, recreational vehicles, and tractor trailer trucks. In performing an operational analysis of a transportation system element, the proportion of heavy vehicles is an important variable for some of the following reasons:

- Heavy vehicles are larger than cars, and take up more physical space in a queue.
- Heavy vehicles are slower to accelerate, and therefore are slower in accelerating from a stop at a signal, or require larger gaps when, for example, entering a roundabout.
- Heavy vehicles have wider turn radii, causing them to typically conduct turning maneuvers more slowly and carefully.
- Heavy vehicles may have physical constraint on long and/or steep grades, causing them to reduce their speed relative to cars, even at low traffic volumes.

For the reasons just mentioned, it is important to (1) know the percentage of heavy vehicles in the traffic stream, and (2) account for them in operational analysis. The Federal Highway Administration categorizes vehicles into 13 classes, which vary by length, number of axles, and weight as shown in Table 5.6. In the table, motorcycles, cars, and small pick-up trucks are represented by classes 1 through 3, buses are class 4, and single-unit trucks are classes 5 through 7 depending on whether they have 2, 3, or 4 axles. Classes 8 through 13 then represent various forms of heavy trucks with anywhere from 3 to 7 or more axles.

For operational analysis, the *HCM* uses the concept of *passenger car equivalents* (PCEs) to account for heavy vehicles. The PCE value equates the impact of each heavy vehicle to a certain number of passenger cars. For low grades and otherwise ideal conditions, the PCE value may be as low as 1.5 or 2 (the latter substituting two cars for each heavy vehicle), while the PCE for steep grades can be at a value of 5 or more. The PCE values vary by facility type (freeways vs. arterial streets, etc.); the specific way to account for trucks will be discussed in the corresponding discussion in Sections 5.3 and 5.4 of this chapter. The methods also differ by the grouping of heavy vehicles, where uninterrupted flow methods distinguish between single-unit trucks (including buses) and tractor trailers, while interrupted flow methods tend to group all trucks into one group but treat buses and recreational vehicles as a separate group.

Table 5.6 Heavy vehicle classification scheme

	CLASS GROUP		Description	NO. OF AXLES
	1		Motorcycles	2
	2		All cars cars	2
			Cars W/1-Axle Trajler	3
			Cars W/2-Axle Trajler	4
	3		Pick-ups & vans 1 & 2 Axle Trajlers	2, 3, & 4
	4		Buses	2 & 3
	5	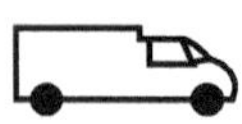	2-Axle, Single Unit	2
	6	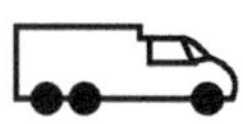	3-Axle, Single Unit	3
	7	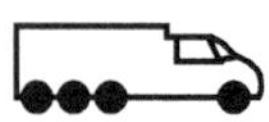	4- Axle, Single Unit	4
HEAVY TRUCKS	8	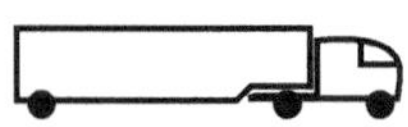	2-Axle, Tractor, 1-Axle, Trailer (2&1)	3
			2-Axle, Tractor, 2-Axle Trailer (2&2)	4
			3-Axle, Tractor,1-Axle Trajler (3&1)	4
	9	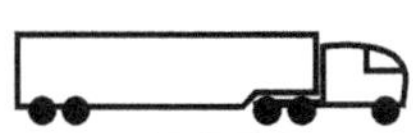	3-Axle, Tractor, 2-Axle Trajler (3&2)	5
			3-Axle, Truck W/2-Axle Trailer	5
	10	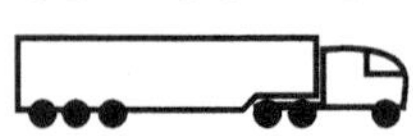	Tractor W/Single Trailer	6&7
	11	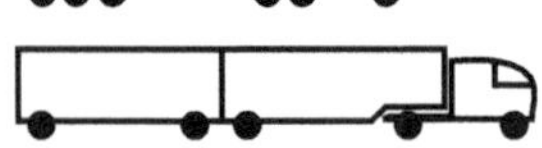	5-Axle Multi-Trailer	5
	12	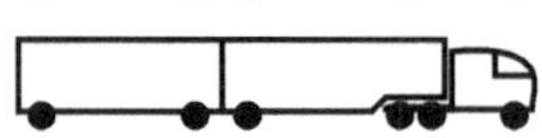	6- Axle Multi-Trailer	6
	13	ANY 7 OR MORE AXLE		7 or more

Source: FHWA; http://www.fhwa.dot.gov/environment/air_quality/conformity/research/improving_data/taqs03.cfm.

5.3 UNINTERRUPTED FLOW

This section presents methods for uninterrupted flow. Uninterrupted flow represents travel on facilities without at-grade intersections and traffic control devices that stop traffic to control movement priority. The interstate system consists of uninterrupted flow facilities and segments, with freeways being generally access controlled and entering or exiting traffic processed through grade-separated interchanges. Uninterrupted flow conditions also apply to two-lane or multilane highway segments, if they are outside of the influence of traffic signals or other traffic control devices, typically at distances greater than 2 mi or 3.2 km.

5.3.1 Concepts

The operations and performance of uninterrupted flow is generally governed by the interaction of vehicles or, more specifically, their drivers. Uninterrupted flow facilities typically do not have traffic control devices to control traffic flow at junctions, but rather process entering and exiting traffic at grade-separated ramps at interchanges. These interchanges can represent bottlenecks or choke points on the freeway system, as turbulence due to lane-changing, merging, and weaving maneuvers reduces the capacity relative to a "basic" freeway segment.

Access Control and Interchange Operations

Uninterrupted flow facilities are access controlled, meaning that any traffic entering or exiting the freeway does so through grade-separated ramps at interchanges. The operational maneuver performed at these on-ramp or off-ramp locations are described as either merging or diverging, respectively. Often, these merge and diverge maneuvers occur at high speeds, and therefore function very differently from signalized intersections.

There are a variety of types of interchanges that all differ in their operational patterns, and can range widely in their capacities and ability to handle specific traffic maneuvers. Interchanges represent junctions between interrupted and uninterrupted flow, but are generally treated as interrupted flow, except for the merge and diverge points. In other words, the bottom ends of the on-ramps and off-ramps where traffic merges with or diverges from freeway traffic are treated as uninterrupted flow, while the top of the ramps (where traffic may, for example, turn left at a signalized intersection or roundabout to get to the on-ramp) are treated as interrupted flow.

Flow Regimes on Uninterrupted Flow Facilities

At low traffic volumes, the flow on uninterrupted facilities is governed by the prevailing speed limit, but has also been shown to be influenced by geometric conditions on the facility, including lane widths and shoulder clearances. The desired speed by drivers at these low flow conditions, also referred to as the *free-flow speed*, is a function of these geometric attributes and the level of comfort of drivers to travel at such speed. Traffic flow under low-volume conditions is thus less impacted by vehicle-to-vehicle interaction (as there aren't many cars), and more by the geometric attributes and "feel" of the roadway.

As traffic flow increases, the interaction of vehicles begins to govern the operations on the facility. Eventually, the density on the facility increases to a point where speeds begin to drop, as drivers are no longer comfortable maintaining high speeds with the now limited maneuver space (recall the discussion of spacing and Figure 5.9). Eventually, traffic flow conditions reach the maximum sustainable flow rate before reaching breakdown conditions, which is referred to as the *capacity*. The flow regime between free flow and capacity is also referred to as undersaturated or uncongested flow.

If traffic demand continues to increase and exceeds the available capacity, breakdown occurs, where the flow becomes unstable and congested. The results are continually increasing densities and decreasing speeds, which are associated with a decrease in throughput as vehicles are queuing upstream of the bottleneck point. Theoretically, traffic densities can increase up to a point where vehicles are literally spaced "bumper to bumper" and traffic flow and speeds approach zero. The flow regime between capacity and this *jam density* is also referred to as the oversaturated or congested flow.

The various flow regimes on an uninterrupted flow facility are illustrated in Figure 5.14. The figure shows free-flow conditions at low flow rates, and prebreakdown (but still stable) conditions as traffic volumes increase. The free-flow regime shows a fixed speed as volumes increase, while the prebreakdown regime exhibits a steady decrease in speed as the flow approaches capacity.

The queue discharge flow regime typically occurs downstream of a bottleneck, where traffic begins to discharge from a queue. It is noted that the queue discharge flow rates are lower than the capacity value. This is commonly observed, and research has shown that the average

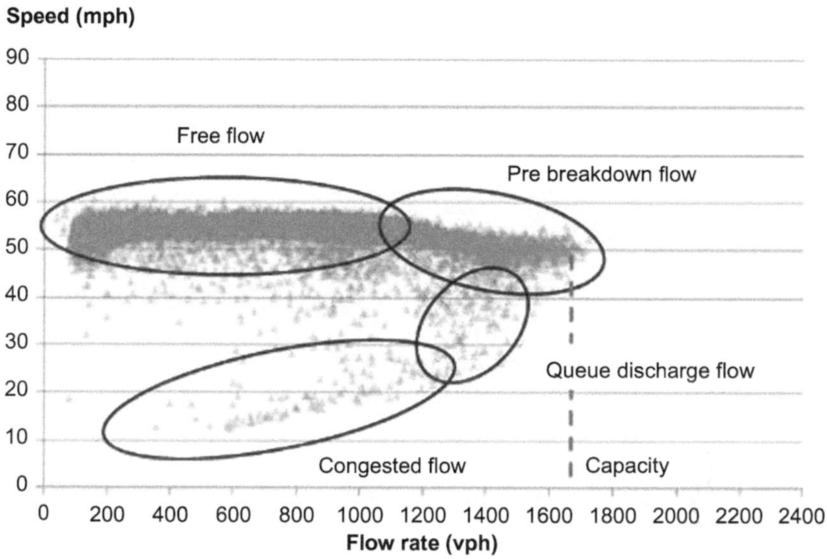

Figure 5.14 Freeway speed-flow data example.

queue discharge flow rate is approximately 7% lower than the prebreakdown capacity (Hu et al., 2012), but with some facilities showing drops in throughput on the order of 20% when transitioning from prebreakdown flow to queue discharge.

The final flow regime in Figure 5.14, congested flow, represents flow in a queue upstream of a bottleneck. As is evident in the figure, the speeds and flow rates in this regime are much deteriorated relative to the other regimes.

Terminology

The following list introduces several key terms that will be used throughout the section on uninterrupted flow.

- *Prebreakdown capacity (veh/h)*: The maximum sustainable flow rate on a freeway that can be achieved without the facility breaking down and transitioning to congested flow.
- *Queue discharge rate (veh/h)*: The flow rate immediately following breakdown as traffic discharges from a queued freeway segment. The queue discharge rate is typically less than the prebreakdown capacity, by an average of approximately 7% (Hu et al., 2012)

- *Jam density (veh/mi)*: The maximum achievable density on a freeway, corresponding to essentially stand,still conditions. The jam density is the inverse of the minimum spacing between vehicles as discussed in Section 5.2.1.
- *Free-flow speed (mph)*: The speed of vehicles at low-volume conditions, impacted only by the geometry of the facility (and the speed limit), but without any speed-reducing effects due to traffic interaction.
- *Speed-flow curve*: A mathematical formula (or series of formulas) describing how speed changes as a function of increasing flow rates.

5.3.2 Basic Freeway and Multilane Highway Segments

The methodologies to analyze basic freeway segments and multilane highways are conceptually very similar, and so are presented together here (as well as in the *HCM*). A basic freeway segment is defined as a divided highway with full control of access and two or more lanes for the exclusive use of traffic in each direction that is outside of the influence of on-ramps or off-ramps (defined in the *HCM* as greater than 1500 ft). A multilane highway segment similarly provides uninterrupted flow on an access-controlled facility with no signalized or stop-controlled at-grade intersections intersecting the mainline. However, multilane highways sometimes allow for isolated driveway access and are generally not held to the same high design standards as freeways. Accordingly, their capacities are expected to be lower than a basic freeway segment. In the *HCM*, a multilane highway segment is considered as an uninterrupted flow segment if it is more than 2 miles from a signalized intersection or other at-grade junction point.

Capacity and Level of Service

Freeway capacity is defined as the maximum sustained 15-min flow rate, expressed in passenger cars per hour per lane, that can be accommodated by a uniform freeway segment under prevailing traffic and roadway conditions in one direction of flow. The calculation of this 15-min flow rate, as well as the conversion from vehicles to passenger cars, was discussed previously. The service measure for freeway LOS is the average segment density in passenger cars per mile per lane. Table 5.7 lists the thresholds for each LOS range for basic freeway and multilane highway segments in the *HCM*.

Therefore, a freeway with a density of 22 passenger cars/mi per lane, for example, would be experiencing an LOS of C. The upper limit of 45

Table 5.7 LOS thresholds for basic freeway and multilane highway segments

Level of service	Description	Density range (passenger cars/mi per lane)
A	Completely free-flowing condition with efficient operating speeds	0–11
B	Stable flow for a freeway or major highway	>11–18
C	Reasonable and uniform flow but with lower operating speeds	>18–26
D	Approaching unstable flow with low operating speeds	>26–35
E	Unstable flow	>35–45
F*	Forced flow or the stop-and-go movement	>45

*LOS = F also applies any time demand exceeds capacity.

passenger cars/mi per lane for LOS E is the maximum density at which sustained flows at capacity are expected to occur. Figure 5.15 shows example images for LOS for a basic freeway segment, adopted from the *HCM*.

Common questions for a freeway operations analysis include:

- How good or bad is the facility performing now?
- How will the facility perform in future?
- Under what conditions will the facility break down?
- How do we design a new facility?
- What is the capacity?
- What is the average travel speed?

Methodology

The basic freeway segment methodology in the *HCM* follows a series of six computational steps as shown in Figure 5.16. Multilane highway segments generally follow the same procedure, and any differences between the two are discussed in more detail in the respective steps.

Step 1: Gather Input Data

To use the method described in the *HCM* to determine the LOS of a basic freeway or multilane highway segment, three main categories of data are needed:

- Geometric data (e.g., lane width, shoulder width, number of lanes, etc.)
- Field-measured free-flow speed (FFS) or a base free-flow speed (BFFS)
- Volume/flow rate data

Figure 5.15 LOS examples for freeways.

These input data are typically collected in traffic studies or, in some cases, are estimated from a transportation planning model. Geometric data are typically obtained from maps, aerial photography, or design drawings for a new facility. Free-flow speed can be field measured for an existing facility or can be estimated using the equation described in Step 2. Traffic volumes are typically field measured and include the percentage of heavy vehicles to convert the traffic stream to passenger car equivalents for the purpose of analysis in Step 3.

For a roadway to perform at its optimal efficiency, it must be designed using *ideal conditions*. In principle, an ideal condition is one for which

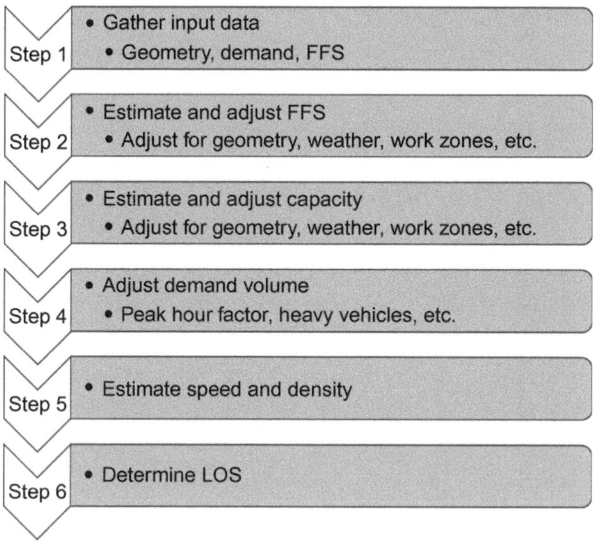

Figure 5.16 Basic freeway segment methodological steps.

Table 5.8 Base conditions for basic freeway segments

Factor	**Base condition**
Lane width	12 ft
Lateral clearance	6 ft on the right shoulder 2 ft in the median
Ramp density	No on-ramps or off-ramps within 3 miles upstream and downstream of segment
Terrain	Grade = 2% or less
Driver population	Passenger cars only, with predominately commuter traffic
Weather	Clear weather, good visibility, dry pavement, no sun glare
Incidents	No broken-down vehicles, police activity, or crashes
Work zones	No construction activity

further improvement will not achieve any increase in capacity. These conditions are also referred to as base conditions. Table 5.8 lists many factors that can affect traffic flow and the base condition for each.

Often, due to costs, terrain, development around the roadway, and many other factors, it is difficult to design a roadway using these ideal conditions. Because of this, it is important in LOS analysis to gather geometric data for the roadway to determine how close or far it is from these ideal conditions.

Step 2: Estimate and Adjust Free-Flow Speed

The free-flow speed (FFS) is considered to be the average passenger car travel speed that most drivers will choose under low flow conditions. For existing roadways, FFS can be measured directly in the field when flows are less than 1000 passenger cars/h per lane. Otherwise, FFS can be estimated by adjusting a base free-flow speed (BFFS) downward to reflect the influence of four geometric factors. In the *HCM*, these factors are lateral clearance, number of lanes, and total ramp density. The base condition for each of these factors was given in Table 5.8. Deviating from these base conditions causes drivers to slow down and thus affects capacity. Reducing lane width or lateral clearance forces drivers closer together than they would prefer. To accommodate for the lack of space, they decrease speed. Likewise, increasing interchange density causes increased weaving and merging, which disrupts flow and causes drivers to decrease speed.

The equation for estimating FFS using BFFS and adjustment factors for basic freeway segments is given in Eq. (5.12):

$$FFS = BFFS - f_{LW} - f_{RLC} - 3.22 \times TRD^{0.84} \tag{5.12}$$

where

FFS = free-flow speed of basic freeway segment (mph)

$BFFS$ = base FFS for basic freeway segment (mph)

f_{LW} = adjustment for lane width (mph) Table 5.9

f_{RLC} = adjustment for right-side lateral clearance (mph) Table 5.10

TRD = total ramp density (ramps/mi)

The equation for estimating FFS for multilane highway segments is given in Eq. (5.13):

$$FFS = BFFS - f_{LW} - f_{TLC} - f_M - f_A \tag{5.13}$$

where

FFS = free-flow speed of multilane highway segment (mph)

$BFFS$ = base FFS for multilane highway segment (mph)

f_{LW} = adjustment for lane width (mph) Table 5.9

f_{TLC} = adjustment for total lateral clearance (mph) Table 5.11

f_M = adjustment for median type (mph) Table 5.12

f_A = adjustment for access-point density (mph) Table 5.13

Table 5.9 Adjustments for lane width (f_{lw}) for freeways and multilane highways

Lane width (ft)	Reduction in free-flow speed (mph)
12	0.0
11	1.9
10	6.6

Source: TRB, 2015.

Table 5.10 Adjustments for right-shoulder lateral clearance on freeways (f_{lc})

Right-side lateral clearance (ft)	Reduction in FFS (mph) for number of lanes in one direction			
	2	3	4	≥5
≥6	0.0	0.0	0.0	0.0
5	0.6	0.4	0.2	0.1
4	1.2	0.8	0.4	0.2
3	1.8	1.2	0.6	0.3
2	2.4	1.6	0.8	0.4
1	3.0	2.0	1.0	0.5
0	3.6	2.4	1.2	0.6

Source: TRB, 2015.

Table 5.11 Adjustment for total lateral clearance (f_{TLC}, left plus right side) for multilane highways

Four-lane highways		Six-lane highways	
TLC (ft)	Reduction in FFS (mph)	TLC (ft)	Reduction in FFS (mph)
12	0.0	12	0.0
10	0.4	10	0.4
8	0.9	8	0.9
6	1.3	6	1.3
4	1.8	4	1.7
2	3.6	2	2.8
0	5.4	0	3.9

Source: TRB, 2015.

The base free-flow speed (BFFS) in Eq. (5.10) is 75.4 mph, which allows for estimation of FFS up to facilities signed at 70 mph, which often have FFS around 75 mph under ideal conditions. For multilane highways, limited research exists on estimating the BFFS. It is recommended that

Table 5.12 Adjustment for median type (f_M) for multilane highways

Median type	Reduction in FFS (mph)
Undivided	1.6
Two-Way Left Turn Lane (TWLTL)	0.0
Divided	0.0

Source: TRB, 2015.

Table 5.13 Adjustment for access point density (f_{AP}) for multilane highways

Access-point density (access points/mi)	Reduction in FFS (mph)
0	0.0
10	2.5
20	5.0
30	7.5
≥ 40	10.0

Source: TRB, 2015.

the roadway design speed (not necessarily the speed limit) is used for those facilities as an initial estimate of BFFS, and that local data be considered whenever possible.

The other adjustment factors used in Eq. (5.12) and Eq. (5.13) are given in the Tables 5.9 through 5.13. The effect for total ramp density used in Eq. (5.12) for basic freeway segments is estimated by directly plugging the total ramp density for the segment under study into the equation. The total ramp density is measured over a distance of 6 miles (3 miles upstream and 3 miles downstream of the segment), and calculated by the number of ramps divided by distance (in units of ramps per mile). This adjustment term implies that a full cloverleaf interchange (four ramps per direction) has a proportionally higher impact on the FFS than a diamond interchange with only two ramps per direction.

The various adjustments for geometric conditions of the roadway in the *HCM* are limited to reductions in FFS, rather than having a direct on capacity. Research on the effects of roadway geometry on freeway capacity is limited (other than in specialized applications, such as work zones), but it is intuitive that reduced geometric standards would also impact the capacity, as discussed further in the following.

Example 5.4

A four-lane freeway is located in an urban area and due to space limitations has lanes only 11-ft wide with lateral clearance of only feet. Being in a dense urban area, a 6-mile section surrounding the segment has a total of seven interchanges with 16 ramps total.

Using this information, come up with an estimate of the FFS for this freeway.

Solution

$$FFS = BFFS - f_{LW} - f_{RLC} - 3.22 \times TRD^{0.84}$$

$BFFS = 75.4$ mph
$TRD = 16/6 = 2.67$ ramps/mi

From Table 5.6 and Figure 5.14:
$f_{lw} = 1.9$ mph
$f_{lc} = 2.4$ mph

So
$FFS = 75.4 - 1.9 - 2.4 - 3.22 \times 2.67^{0.84} = 63.8$ mph

Step 3: Estimate and Adjust Capacity

With all geometric and volume data obtained, an estimate of the roadway capacity is needed to calculate volume-to-capacity (v/c) ratio, performance measures, and ultimately LOS for a basic freeway or multilane highway segment. The ideal or base capacity of the segment can be calculated from Eq. (5.14) and Eq. (5.15) as a function of the free-flow speed (FFS) as follows:

$$c\,(basic\ freeway\ segment) = 2200 + 10 \times (FFS - 50) \qquad (5.14)$$

$$c\,(multilane\ highway\ segment) = 1900 + 20 \times (FFS - 45) \qquad (5.15)$$

The capacities obtained from these equations are for segments under ideal conditions, and do not consider a variety of potential capacity-reducing effects. Some of these effects are implicitly accounted for in the FFS estimation, including lane widths, lateral clearance, and total ramp density. Because these attributes (if not in ideal or base condition) lead to a reduction in FFS, they also implicitly impact the resulting capacity. But various other factors can lead to a reduction in capacity, including:

- Lane width and lateral clearance effects in addition to those included in the FFS estimation
- Lane drops that create turbulence as drivers have to merge into the lanes that continue past the lane-drop point
- Poor visibility due to horizontal and vertical curvature, or due to weather conditions (fog, sun glare, etc.)
- Poor pavement conditions, especially if rutting or potholes are frequent on the facility
- Travel across bridges, through tunnels, or adjacent to landmarks and other points of interest that divert driver attention
- Presence of a significant portion of unfamiliar drivers in the traffic stream
- Lane changes and turbulence resulting from downstream on- or off-ramps
- Presence of incidents or work zones that either reduce lanes or result in onlooker effects
- Poor weather conditions in the form of rain, snow, or ice precipitation

Given the variety of potential capacity-reducing factors, it is critical to properly calibrate the capacity for prevailing conditions for the segment under study. This calibration can occur through local observations and sensor data on freeways, which is the preferred approach. Alternatively, the analyst may refer to various sources in the literature to estimate the magnitude of the capacity-reducing effects. Some of these sources are referenced in the next section. These factors are then incorporated in the estimation of freeway performance through a capacity-reduction factor (CAF) that is multiplied with the ideal capacity to obtain the locally calibrated prevailing capacity, as shown in Eq. 5.16.

$$c_{adj} = c \times CAF \tag{5.16}$$

where

c_{adj} = adjusted capacity of segment (passenger cars/h)
c = base capacity of segment (passenger cars/h)
CAF = capacity adjustment factor

The importance of calibrating the capacity, especially for a known freeway bottleneck, cannot be underestimated. In fact, the ideal capacities (e.g., 2400 passenger cars/h per lane for a facility with FFS of 70 mph) may be rarely observed in reality. Take, for example, the speed-flow data shown in Figure 5.11, which had a capacity closer to 1650 passenger cars/h per lane at an FFS of 55 mph. The ideal capacity at 55 mph FFS according to Eq. (5.13) is $2200 + 10 \times (55 - 50) = 2250$ passenger cars/h per lane, making the field-observed capacity 26.7% lower than the ideal or base values.

Step 4: Adjust Demand Volume

Once the geometric and FFS data are gathered, the only data remaining to be collected are the volume and flow rate data. The flow rate (v_p) is based on the volume and several other factors and can be calculated using Eq. 5.17:

$$v_p = \frac{V}{(PHF \times N \times f_{HV})} \tag{5.17}$$

where

v_p = flow rate, in passenger cars/h per lane
V = directional analysis volume, in vph
PHF = peak hour factor
N = number of lanes in the direction of travel
f_{HV} = heavy vehicle adjustment factor

Heavy vehicles, including trucks, buses, and recreational vehicles create less-than-ideal flows. Longer and more frequent gaps of excessive length form in front of and behind heavy vehicles, vehicles in adjacent lanes are often disrupted, and heavy vehicles tend to be about two to three times the length of a passenger car. Accordingly, the heavy vehicle factor, f_{HV}, is introduced into the flow rate equation to convert the vehicle mix using the facility into equivalent passenger cars.

The heavy vehicle factor is estimated from an equivalency factor that equates each heavy vehicle to a number of passenger car equivalents, or PCEs. PCEs are defined for general terrain (level or rolling). Level terrain is defined as a segment that allows trucks to generally obtain and maintain the same speeds as passenger cars, which is expected for grades less than 2%. Rolling terrain is defined as a segment that causes trucks to slightly reduce their speed over passenger cars, but without steep grades

Table 5.14 General terrain PCE equivalents for heavy vehicles

Passenger car equivalent (PCE) factor	Type of terrain	
	Level	**Rolling**
E_T	2.0	3.0

(above 4%) that may cause some trucks to operate at crawl speeds. The PCE equivalency for trucks factors (E_T) for level and rolling terrain are shown in Table 5.14.

From these equivalency factors and from the percentage of trucks in the traffic stream, the heavy vehicle adjustment factor for volume is calculated from Eq. (5.18):

$$f_{HV} = \frac{1}{1 + P_T(E_T - 1)} \tag{5.18}$$

where

f_{HV} = heavy-vehicle adjustment factor

P_T = proportion of heavy vehicles in traffic stream

E_T = passenger-car equivalent (PCE) of one heavy vehicle

For mountainous terrain and generally steeper grades, no general terrain factors for PCEs exist, and so the analyst has to refer to the *specific grade adjustments* found in the *HCM* and repeated here as Table 5.15, Table 5.16, and Table 5.17. The tables refer to a mix of single unit trucks (SUT) and tractor trailer trucks (TT) of 30%/70%, 50%/50%, and 70%/30%, respectively. The SUT category encompasses heavy vehicles classes 4 and 5 as defined by FHWA (see Figure 5.10), as well as large recreational vehicles (RVs). The TT category includes any vehicles of FHWA class 6 and higher. The resulting E_T values from these tables can then be plugged into Eq. (5.12) to estimate the heavy vehicle adjustment factor. Values not contained in the tables directly can be obtained through linear interpolation. The *HCM* also offers a specific mixed-flow model methodology for estimating the speed of trucks on long steep grades that is beyond the scope of this text.

Table 5.15 Passenger car equivalency factors for 30% SUT and 70% TT fleet mix

% Grade	Length (mi)	Percentage of trucks and buses (%)								
		2%	4%	5%	6%	8%	10%	15%	20%	>25%
−2	0.125	2.62	2.37	2.30	2.24	2.17	2.12	2.04	1.99	1.97
	0.375	2.62	2.37	2.30	2.24	2.17	2.12	2.04	1.99	1.97
	0.625	2.62	2.37	2.30	2.24	2.17	2.12	2.04	1.99	1.97
	0.875	2.62	2.37	2.30	2.24	2.17	2.12	2.04	1.99	1.97
	1.25	2.62	2.37	2.30	2.24	2.17	2.12	2.04	1.99	1.97
	1.5	2.62	2.37	2.30	2.24	2.17	2.12	2.04	1.99	1.97
0	0.125	2.62	2.37	2.30	2.24	2.17	2.12	2.04	1.99	1.97
	0.375	2.62	2.37	2.30	2.24	2.17	2.12	2.04	1.99	1.97
	0.625	2.62	2.37	2.30	2.24	2.17	2.12	2.04	1.99	1.97
	0.875	2.62	2.37	2.30	2.24	2.17	2.12	2.04	1.99	1.97
	1.25	2.62	2.37	2.30	2.24	2.17	2.12	2.04	1.99	1.97
	1.5	2.62	2.37	2.30	2.24	2.17	2.12	2.04	1.99	1.97
2	0.125	2.62	2.37	2.30	2.24	2.17	2.12	2.04	1.99	1.97
	0.375	3.76	2.96	2.78	2.65	2.48	2.38	2.22	2.14	2.09
	0.625	4.47	3.33	3.08	2.91	2.68	2.54	2.34	2.23	2.17
	0.875	4.80	3.50	3.22	3.03	2.77	2.61	2.39	2.28	2.21
	1.25	5.00	3.60	3.30	3.09	2.83	2.66	2.42	2.30	2.23
	1.5	5.04	3.62	3.32	3.11	2.84	2.67	2.43	2.31	2.23
2.5	0.125	2.62	2.37	2.30	2.24	2.17	2.12	2.04	1.99	1.97
	0.375	4.11	3.14	2.93	2.78	2.58	2.46	2.28	2.19	2.13
	0.625	5.04	3.62	3.32	3.11	2.84	2.67	2.43	2.31	2.23
	0.875	5.48	3.85	3.51	3.27	2.96	2.77	2.50	2.36	2.28
	1.25	5.73	3.98	3.61	3.36	3.03	2.83	2.54	2.40	2.31
	1.5	5.80	4.02	3.64	3.38	3.05	2.84	2.55	2.41	2.32

(Continued)

Table 5.15 (Continued)

% Grade	Length (mi)	Percentage of trucks and buses (%)								
		2%	4%	5%	6%	8%	10%	15%	20%	>25%
3.5	0.125	2.62	2.37	2.30	2.24	2.17	2.12	2.04	1.99	1.97
	0.375	4.88	3.54	3.25	3.05	2.80	2.63	2.41	2.29	2.22
	0.625	6.34	4.30	3.87	3.58	3.20	2.97	2.64	2.48	2.38
	0.875	7.03	4.66	4.16	3.83	3.39	3.12	2.76	2.57	2.46
	1.25	7.44	4.87	4.33	3.97	3.50	3.22	2.82	2.62	2.50
	1.5	7.53	4.92	4.38	4.01	3.53	3.24	2.84	2.63	2.51
4.5	0.125	2.62	2.37	2.30	2.24	2.17	2.12	2.04	1.99	1.97
	0.375	5.80	4.02	3.64	3.38	3.05	2.84	2.55	2.41	2.32
	0.625	7.90	5.11	4.53	4.14	3.63	3.32	2.90	2.68	2.55
	0.875	8.91	5.64	4.96	4.50	3.92	3.56	3.07	2.82	2.67
	1	9.19	5.78	5.08	4.60	3.99	3.62	3.11	2.85	2.70
5.5	0.125	2.62	2.37	2.30	2.24	2.17	2.12	2.04	1.99	1.97
	0.375	6.87	4.58	4.10	3.77	3.35	3.09	2.73	2.55	2.44
	0.625	9.78	6.09	5.33	4.82	4.16	3.76	3.21	2.93	2.77
	0.875	11.20	6.83	5.94	5.33	4.56	4.09	3.45	3.12	2.93
	1	11.60	7.04	6.11	5.47	4.67	4.18	3.51	3.17	2.97
6	0.125	2.62	2.37	2.30	2.24	2.17	2.12	2.04	1.99	1.97
	0.375	7.48	4.90	4.36	3.99	3.52	3.23	2.83	2.63	2.51
	0.625	10.87	6.66	5.79	5.21	4.46	4.01	3.39	3.08	2.89
	0.875	12.54	7.54	6.51	5.81	4.94	4.40	3.67	3.30	3.08
	1	13.02	7.78	6.71	5.99	5.07	4.51	3.75	3.37	3.14

Table 5.16 Passenger car equivalency factors for 50% SUT and 50% TT fleet mix

% Grade	Length (mi)	Percentage of trucks and buses (%)								
		2%	4%	5%	6%	8%	10%	15%	20%	>25%
−2	0.125	2.67	2.38	2.31	2.25	2.16	2.11	2.02	1.97	1.93
	0.375	2.67	2.38	2.31	2.25	2.16	2.11	2.02	1.97	1.93
	0.625	2.67	2.38	2.31	2.25	2.16	2.11	2.02	1.97	1.93
	0.875	2.67	2.38	2.31	2.25	2.16	2.11	2.02	1.97	1.93
	1.25	2.67	2.38	2.31	2.25	2.16	2.11	2.02	1.97	1.93
	1.5	2.67	2.38	2.31	2.25	2.16	2.11	2.02	1.97	1.93
0	0.125	2.67	2.38	2.31	2.25	2.16	2.11	2.02	1.97	1.93
	0.375	2.67	2.38	2.31	2.25	2.16	2.11	2.02	1.97	1.93
	0.625	2.67	2.38	2.31	2.25	2.16	2.11	2.02	1.97	1.93
	0.875	2.67	2.38	2.31	2.25	2.16	2.11	2.02	1.97	1.93
	1.25	2.67	2.38	2.31	2.25	2.16	2.11	2.02	1.97	1.93
	1.5	2.67	2.38	2.31	2.25	2.16	2.11	2.02	1.97	1.93
2	0.125	2.67	2.38	2.31	2.25	2.16	2.11	2.02	1.97	1.93
	0.375	3.76	2.95	2.77	2.64	2.47	2.36	2.20	2.11	2.06
	0.625	4.32	3.24	3.01	2.84	2.63	2.49	2.29	2.19	2.12
	0.875	4.57	3.37	3.11	2.93	2.70	2.55	2.33	2.22	2.15
	1.25	4.71	3.45	3.17	2.99	2.74	2.58	2.36	2.24	2.17
	1.5	4.74	3.47	3.19	3.00	2.75	2.59	2.36	2.24	2.17
2.5	0.125	2.67	2.38	2.31	2.25	2.16	2.11	2.02	1.97	1.93
	0.375	4.10	3.13	2.92	2.77	2.57	2.44	2.26	2.16	2.10
	0.625	4.84	3.52	3.23	3.03	2.77	2.61	2.38	2.26	2.18
	0.875	5.17	3.69	3.37	3.15	2.87	2.69	2.43	2.30	2.22
	1.25	5.36	3.79	3.45	3.22	2.92	2.73	2.47	2.33	2.24
	1.5	5.40	3.81	3.47	3.24	2.93	2.74	2.47	2.33	2.25

(Continued)

Table 5.16 (Continued)

% Grade	Length (mi)	Percentage of trucks and buses (%)								
		2%	4%	5%	6%	8%	10%	15%	20%	>25%
3.5	0.125	2.67	2.38	2.31	2.25	2.16	2.11	2.02	1.97	1.93
	0.375	4.89	3.54	3.25	3.05	2.79	2.62	2.39	2.26	2.19
	0.625	6.05	4.15	3.75	3.47	3.11	2.89	2.58	2.42	2.32
	0.875	6.58	4.43	3.97	3.66	3.26	3.01	2.67	2.49	2.39
	1.25	6.88	4.58	4.10	3.77	3.35	3.09	2.72	2.53	2.42
	1.5	6.95	4.62	4.13	3.80	3.37	3.10	2.73	2.54	2.43
4.5	0.125	2.67	2.38	2.31	2.25	2.16	2.11	2.02	1.97	1.93
	0.375	5.83	4.03	3.65	3.39	3.05	2.84	2.55	2.39	2.30
	0.625	7.53	4.92	4.38	4.01	3.53	3.24	2.83	2.62	2.50
	0.875	8.32	5.34	4.72	4.29	3.75	3.42	2.97	2.73	2.59
	1	8.53	5.45	4.81	4.37	3.81	3.47	3.00	2.76	2.62
5.5	0.125	2.67	2.38	2.31	2.25	2.16	2.11	2.02	1.97	1.93
	0.375	6.97	4.63	4.14	3.81	3.38	3.11	2.74	2.55	2.43
	0.625	9.37	5.89	5.16	4.68	4.05	3.67	3.14	2.88	2.72
	0.875	10.49	6.48	5.65	5.09	4.37	3.93	3.34	3.03	2.85
	1	10.80	6.64	5.78	5.20	4.46	4.01	3.39	3.08	2.89
6	0.125	2.67	2.38	2.31	2.25	2.16	2.11	2.02	1.97	1.93
	0.375	7.64	4.98	4.43	4.05	3.56	3.26	2.85	2.64	2.51
	0.625	10.45	6.45	5.63	5.07	4.36	3.92	3.33	3.03	2.85
	0.875	11.78	7.16	6.20	5.56	4.74	4.24	3.56	3.22	3.01
	1	12.15	7.35	6.36	5.69	4.85	4.33	3.62	3.27	3.05

Table 5.17 Passenger car equivalency factors for 70% SUT and 30% TT fleet mix

% Grade	Length (mi)	Percentage of trucks and buses (%)								
		2%	4%	5%	6%	8%	10%	15%	20%	>25%
-2	0.125	2.39	2.18	2.12	2.07	2.01	1.96	1.89	1.85	1.83
	0.375	2.39	2.18	2.12	2.07	2.01	1.96	1.89	1.85	1.83
	0.625	2.39	2.18	2.12	2.07	2.01	1.96	1.89	1.85	1.83
	0.875	2.39	2.18	2.12	2.07	2.01	1.96	1.89	1.85	1.83
	1.25	2.39	2.18	2.12	2.07	2.01	1.96	1.89	1.85	1.83
	1.5	2.39	2.18	2.12	2.07	2.01	1.96	1.89	1.85	1.83
0	0.125	2.39	2.18	2.12	2.07	2.01	1.96	1.89	1.85	1.83
	0.375	2.39	2.18	2.12	2.07	2.01	1.96	1.89	1.85	1.83
	0.625	2.39	2.18	2.12	2.07	2.01	1.96	1.89	1.85	1.83
	0.875	2.39	2.18	2.12	2.07	2.01	1.96	1.89	1.85	1.83
	1.25	2.39	2.18	2.12	2.07	2.01	1.96	1.89	1.85	1.83
	1.5	2.39	2.18	2.12	2.07	2.01	1.96	1.89	1.85	1.83
2	0.125	2.67	2.32	2.23	2.17	2.08	2.03	1.94	1.89	1.86
	0.375	3.63	2.82	2.64	2.52	2.35	2.25	2.10	2.02	1.97
	0.625	4.12	3.08	2.85	2.69	2.49	2.36	2.18	2.08	2.02
	0.875	4.37	3.21	2.96	2.78	2.56	2.42	2.22	2.11	2.05
	1.25	4.53	3.29	3.02	2.84	2.60	2.45	2.24	2.13	2.07
	1.5	4.58	3.31	3.04	2.86	2.61	2.46	2.25	2.14	2.07
2.5	0.125	2.75	2.36	2.27	2.20	2.11	2.04	1.95	1.90	1.87
	0.375	4.01	3.02	2.80	2.65	2.46	2.33	2.16	2.06	2.01
	0.625	4.66	3.35	3.08	2.88	2.64	2.48	2.26	2.15	2.08
	0.875	4.99	3.52	3.21	3.00	2.73	2.56	2.32	2.19	2.12
	1.25	5.20	3.64	3.30	3.08	2.79	2.60	2.35	2.22	2.14
	1.5	5.26	3.67	3.33	3.10	2.80	2.62	2.36	2.23	2.15

(*Continued*)

Table 5.17 (Continued)

% Grade	Length (mi)	Percentage of trucks and buses (%)								
		2%	4%	5%	6%	8%	10%	15%	20%	>25%
3.5	0.125	2.93	2.45	2.34	2.26	2.16	2.09	1.98	1.92	1.89
	0.375	4.86	3.46	3.16	2.96	2.69	2.53	2.30	2.18	2.10
	0.625	5.88	3.99	3.59	3.32	2.98	2.76	2.46	2.31	2.22
	0.875	6.40	4.26	3.81	3.51	3.12	2.88	2.55	2.38	2.28
	1.25	6.74	4.43	3.96	3.63	3.21	2.96	2.60	2.42	2.32
	1.5	6.83	4.48	3.99	3.66	3.24	2.98	2.62	2.44	2.33
4.5	0.125	3.13	2.56	2.43	2.34	2.21	2.13	2.01	1.95	1.91
	0.375	5.88	3.99	3.59	3.32	2.98	2.76	2.46	2.31	2.22
	0.625	7.35	4.75	4.22	3.85	3.39	3.10	2.71	2.51	2.39
	0.875	8.11	5.15	4.54	4.13	3.60	3.27	2.83	2.61	2.47
	1	8.33	5.27	4.63	4.21	3.66	3.33	2.87	2.64	2.50
5.5	0.125	3.37	2.69	2.53	2.42	2.28	2.19	2.05	1.98	1.94
	0.375	7.09	4.62	4.11	3.76	3.31	3.04	2.66	2.47	2.36
	0.625	9.13	5.68	4.97	4.49	3.88	3.51	3.00	2.74	2.59
	0.875	10.21	6.24	5.43	4.88	4.18	3.76	3.18	2.89	2.71
	1	10.52	6.41	5.57	5.00	4.27	3.83	3.24	2.93	2.75
6	0.125	3.51	2.76	2.59	2.47	2.32	2.22	2.08	2.00	1.95
	0.375	7.78	4.98	4.40	4.01	3.51	3.20	2.78	2.56	2.44
	0.625	10.17	6.23	5.42	4.87	4.17	3.75	3.18	2.88	2.71
	0.875	11.43	6.88	5.95	5.32	4.53	4.04	3.39	3.06	2.86
	1	11.81	7.08	6.11	5.46	4.64	4.13	3.45	3.11	2.90

Example 5.5

For a directional volume of 3100 vehicles per hour, determine the flow rate in passenger car per hour per lane for a 6-lane freeway if the heavy truck adjustment factor is 0.83 and the PHF is 0.92.

Solution

$$v_p = \frac{V}{(PHF \times N \times f_{HV})}$$

$$v_p = \frac{3100}{(0.92 \times 3 \times 0.83)} = 1353 \text{ passenger cars/h per lane}$$

Steps 5 and 6: Estimate Speed, Density, and LOS

With all necessary input parameters and adjustments obtained, the prevailing speed on the basic freeway segment can be estimated for undersaturated flow using a two-regime model shown in Eq. (5.19). The model predicts an initial portion of the speed-flow regime where the speed is fixed at the adjusted FFS. After a breakpoint (estimated from Eq. (5.20)) the speed then begins to deteriorate as volumes approach capacity. The two equations are fundamentally a function of FFS, capacity, and input volume, as well as SAF and CAF adjustments.

$$S = \begin{cases} FFS \times SAF & V_p \leq BP_{adj} \\ FFS \times SAF - \dfrac{\left(FFS \times SAF - \dfrac{c \times CAF}{45}\right)}{(c \times CAF - BP_{adj})^2} \times (V_p - BP_{adj})^2 & V_p > BP_{adj} \end{cases} \quad (5.19)$$

$$BP_{adj}(FFS \times SAF, CAF) = [1000 + 40 \times (75 - FFS \times SAF)] \times CAF^2 \quad (5.20)$$

where

S = segment space mean speed, mph
V_p = segment flow rate, passenger cars/h per lane
SAF = free-flow speed adjustment factor
CAF = capacity adjustment factor
BP_{adj} = adjusted breakpoint flow rate, passenger cars/h per lane

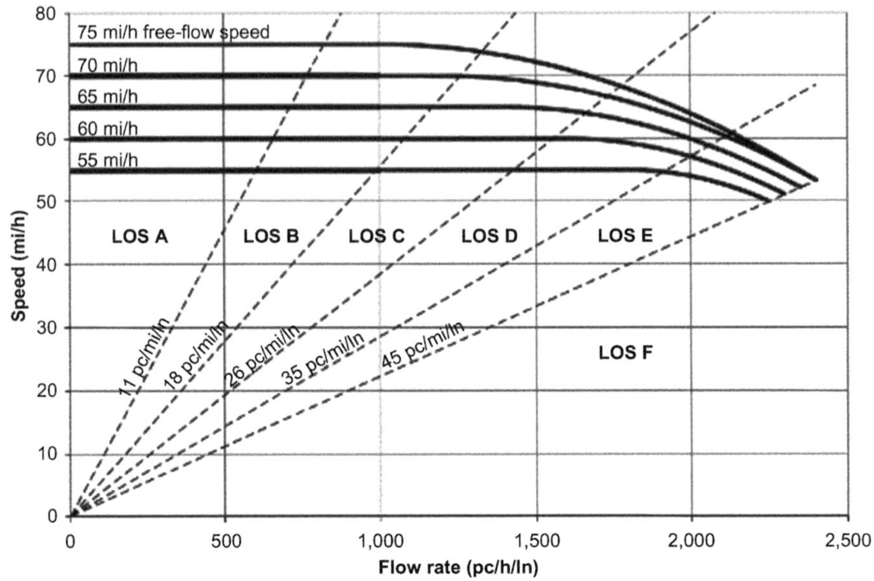

Figure 5.17 Speed-flow curves for basic freeway segments.

From this speed, density on the segment can be estimated from Eq. (5.21):

$$D = \frac{v_P}{S} \tag{5.21}$$

where

D = density (passenger cars/mi per lane)

v_p = demand flow rate (passenger cars/h per lane)

S = mean speed of traffic stream under base conditions (mph)

Figure 5.17 depicts the speed-flow curves for a basic segment obtained from Eq. (5.19). These curves are used to find the LOS for a freeway. Curves are drawn in for 75 mph, 70 mph, 65 mph, 60 mph, and 55 mph average passenger-car travel speed. The speed can be estimated either from the table directly, or more accurately calculated from Eq. (5.19). From the speed and volume, Eq. (5.21) can be used to calculate the segment density. The LOS can be found graphically by locating the intersection of the FFS and flow rate of the facility on Figure 5.17 and reading off the LOS of that region. The LOS can also be obtained directly by comparing the results of Eq. (5.21) with the LOS thresholds in Table 5.7.

Table 5.18 Free flow speeds and corresponding capacities

Free flow speed (mph)	Capacity (passenger cars/h per lane)	Average passenger-car speed (mph) at capacity
70	2400	53
65	2350	52
60	2300	51
55	2250	50

Figure 5.17 graphically shows level of service ranges A through E as defined by the diagonal dashed lines. These lines are not arbitrary. The slope of the line is the density in passenger cars per mile per lane. The density values for the LOS thresholds between each LOS are shown on the figure. The capacity points for the FFS curves drawn on the figure are described in Table 5.18.

Note the shape of the FFS curves in Figure 5.17. They are straight for a significant portion and then begin to curve down from the breakpoint to their ultimate capacity point at the end of the curve. The straight portion of each curve is the free-flow speed portion. That is, drivers will travel at the free-flow speed when the flow rate is within the range shown. For example, at an FFS of 75 mph, drivers will travel at this speed until the flow rate exceeds 1000 passenger cars/h per lane. Beyond this flow, speed will drop off with an increase in flow rate until capacity is reached, at 2400 passenger cars/h per lane and a speed of 53 mph. Thus, the curved portion of the FFS curve shows the average passenger car travel speed at those flow rates beyond the breakpoint, and this speed will be lower than the FFS.

Example 5.6

An existing four-lane freeway is located in an urban area with very restricted geometry. General inputs are:

FFS = 70 mph

2000 vph peak hour volume (one direction, pm peak)

6% trucks with 30/70 SUT/TT fleet mix

2.5% grade over 1.25-mile segment

PHF = 0.92

What is the LOS during the peak hour?

(*Continued*)

Example 5.6—(Continued)

Solution

Estimate the heavy vehicle equivalency factor from Table 5.10 for 2.5% grade, segment length of 1.25 miles, and 6% trucks.

$E_T = 3.36$

Calculate heavy vehicle adjustment factor from Eq. (5.13):

$$f_{HV} = \frac{1}{1 + 0.03(3.36 - 1)} = 0.937$$

Calculate the 15-min peak flow rate from Eq. (5.12)

$$v_p = \frac{V}{(PHF \times N \times f_{HV})} = \frac{2000}{(0.92 \times 2 \times 0.937)} = 1160 \text{ passenger cars/h per lane}$$

Using Table 5.16, estimate the LOS for a flow rate of 1160 passenger cars/h per lane and a FFS of 70 mph as LOS = B.

Adjustments for Weather and Incidents, and Work Zones

Both the free-flow speed and the capacity on freeways have been shown to be impacted by nonrecurring sources of congestion, including weather, incidents, and work zones. Especially in the context of a freeway reliability analysis as discussed in the following, it is important that these factors be considered in the operational analysis methodology. Specifically, these effects can be incorporated using speed adjustment factors and capacity adjustment factors that are multiplied by the prevailing FFS and capacity values to obtain the adjusted, calibrated FFS and capacity estimates. Table 5.19 presents SAF and CAF factors for inclement weather conditions for a free-flow speed of 70 mph. Table 5.20 presents CAF factors for incidents on a freeway with varying levels of severity (SAF for all incidents is 1.0).

For work zones, the CAF and SAF are estimated through equations developed in national research (NCHRP, 2015). The research explored a variety of factors believed to impact work zone performance, including barrier type, area type, lane closure patterns, shoulder widths, work intensity, police presence, speed limit, and lighting conditions. The resulting model to estimate the queue discharge capacity of a freeway work zone is shown in Eq. (5.22).

Table 5.19 Capacity and speed adjustment factors for inclement weather for 70 mph FFS

Weather event	Weather event definition	Speed adjustment factor (SAF)	Capacity adjustment factor (CAF)
Medium rain	>0.10–0.25 in/h	0.96	0.9276
Heavy rain	>0.25 in/h	0.94	0.8587
Light snow	>0–0.05 in/h	0.94	0.9571
Low–medium snow	>0.05–0.10 in/h	0.94	0.9134
Medium–high snow	>0.10–0.50 in/h	0.90	0.8896
Heavy snow	>0.50 in/h	0.88	0.7757
Severe cold	< − 4°F	0.95	0.9155
Low visibility	0.50–0.99 mi	0.96	0.9033
Very low visibility	0.25–0.49 mi	0.95	0.8833
Minimum visibility	<0.25 mi	0.95	0.8591
Nonsevere weather	All conditions not listed above	1.00	1.0000

Source: TRB, 2015.

Table 5.20 Capacity adjustment factors for remaining open lanes during incidents

Number of lanes (one direction)	Shoulder closure	One-lane closure	Two-lane closure	Three-lane closure	Four-lane closure
2	0.81	0.70	NA	NA	NA
3	0.83	0.74	0.51	NA	NA
4	0.85	0.77	0.50	0.52	NA
5	0.87	0.81	0.67	0.50	0.50
6	0.89	0.85	0.75	0.52	0.52
7	0.91	0.88	0.80	0.63	0.63
8	0.93	0.89	0.84	0.66	0.66

NA = not applicable.
Source: TRB, 2015.

$$QDR_{wz} = 2093 - 154 \times f_{LCSI} - 194 \times f_{Br} - 179 \times f_A + 9 \times f_{LAT} - 59 \times f_{DN} \quad (5.22)$$

where

QDR_{wz} = average 5-min queue discharge rate

$f_{\mathrm{LCSI}} = \frac{1}{OR \times N_o}$

OR = open ratio (number of open lanes divided by original lanes)

N_o = number of open lanes

$f_{Br} = 0$: concrete, 1: soft barrier
$f_A = 0$: urban, 1: rural
f_{LAT} = lateral distance of 12 ft
$f_{DN} = 0$: day, 1: night

One of the most critical parameters in this equation was found to be the lane closure severity index (LCSI), which takes into account both the number of closed lanes and the number of original lanes. As such, a work zone with a lane closure from 5 original lanes to 2 final lanes (5-to-2) is estimated to have a lower per-lane capacity than a 3-to-2 work zone. These two configurations are illustrated in Figure 5.18. The LCSI values for the 5-to-2 and 3-to-2 scenarios are 1.25 and 0.75, respectively.

Equation (5.22) estimates the queue discharge flow rate of a freeway work zone. The corresponding prebreakdown capacity of a freeway work was found to be on average 13.4% higher than the queue discharge flow

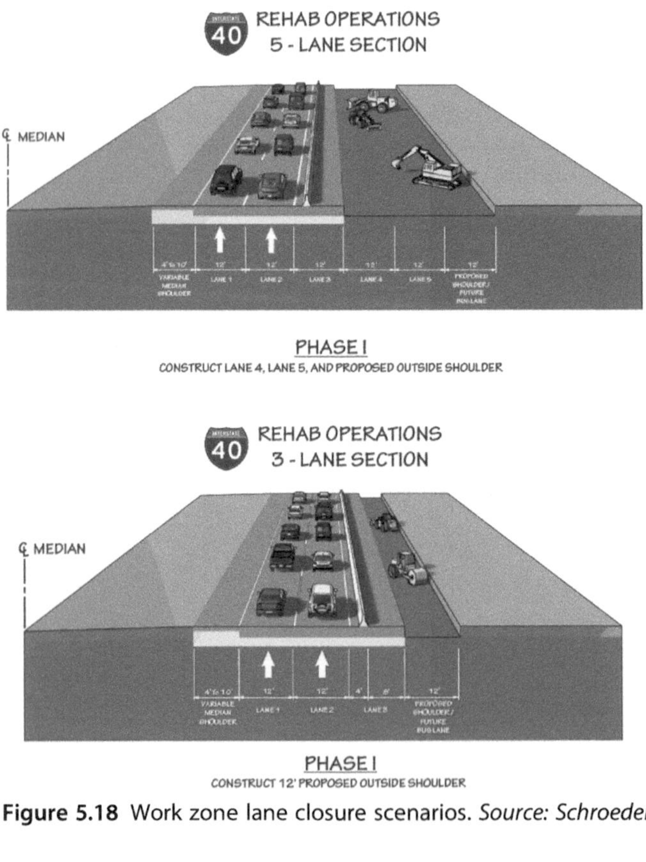

Figure 5.18 Work zone lane closure scenarios. *Source: Schroeder et al., 2014.*

rate. In other words, a work zone loses 13.4% of throughput from its prebreakdown flow rate after breakdown. As such, the capacity adjustment factor (CAF) for a work zone (WZ) is calculated by Eq. (5.23):

$$\mathrm{CAF}_{\mathrm{WZ}} = QDR_{WZ/}(0.864 \times c_{base}) \tag{5.23}$$

where

CAF_{wz} = capacity adjustment factor for a freeway work zone

c_{Base} = base capacity of freeway segment

QDR_{wz} = average 5-min queue discharge rate

Similarly, the work zone FFS can be calculated directly from Eq. (5.24).

$$FFS_{wz} = 9.95 + 33.49 \times f_{Sr} + 0.53 \times f_S - 5.60 \times f_{LCSI} - 3.84 \times f_{Br} - 1.71 \times f_{DN} - 1.45 \times f_{Nr} \tag{5.24}$$

where

f_{sr} = work zone to non-work zone speed ratio

f_s = work zone posted speed limit (mph)

f_{Nr} = number of ramps

(other parameters are as defined previously)

Designing with LOS in Mind

In design problems, the number of lanes is determined to provide a certain target LOS and a given design volume of traffic. The LOS chosen is generally LOS C in rural areas and LOS D in urban areas. LOS E is not used for design purposes. The same figures are used for the design process. The solution is trial and error, where you select a number of lanes and see if that provides the desired LOS. Start with $N = 2$, as this is the minimum number of lanes per direction for a freeway.

Example 5.7

A new suburban freeway is being designed in a level terrain area. The heavy vehicle factor, f_{HV}, is = 0.93, and the PHF = 0.88. FFS is estimated to be 65 mph. The freeway needs to accommodate 4000 vph during the peak hour in one direction.

How many lanes are needed to provide LOS D during the peak hour? LOS C?

(Continued)

Example 5.7—(Continued)
Solution
Select 65 mph FFS curve in Figure 5.17.
Calculate the flow rate trying out various N values:

$$v_p = \frac{V}{(PHF \times N \times f_{HV})}$$

Try $N = 2$:

$$v_p = \frac{4000}{(0.93 \times 2 \times 0.88)} = 2444$$

Flow rate is greater than capacity → two lanes are insufficient.
Try $N = 3$:

$$v_p = \frac{4000}{(0.93 \times 3 \times 0.88)} = 1629$$

Read LOS = D with $N = 3$ → need 6-lane freeway for LOS D.
Try $N = 4$:

$$v_p = \frac{4000}{(0.93 \times 4 \times 0.88)} = 1222$$

Read LOS = C with $N = 4$ → need 8-lane freeway for LOS C.

5.3.3 Merge and Diverge Segments

Merge and diverge segments are found where an on-ramp or off-ramp results in vehicles entering or exiting the freeway. The lane-change maneuvers needed for vehicles to enter or exit the freeway often cause turbulence that reduce the capacity of these segments, relative to a uniform basic freeway segment. Further, it is important to note that merge and diverge segments impact the demand on the freeway, as either more traffic enters or some traffic leaves the facility.

Methodology

In the *HCM*, the merge and diverge segment methodology contains five computational steps as shown in Figure 5.19. Each step is discussed in more details in the following.

Step 1: Gather Input Data and Adjust Demand Volumes

In the *HCM*, merge and diverge segments are defined by a maximum *ramp influence area (RIA)* length of 1500 ft, measured downstream of the

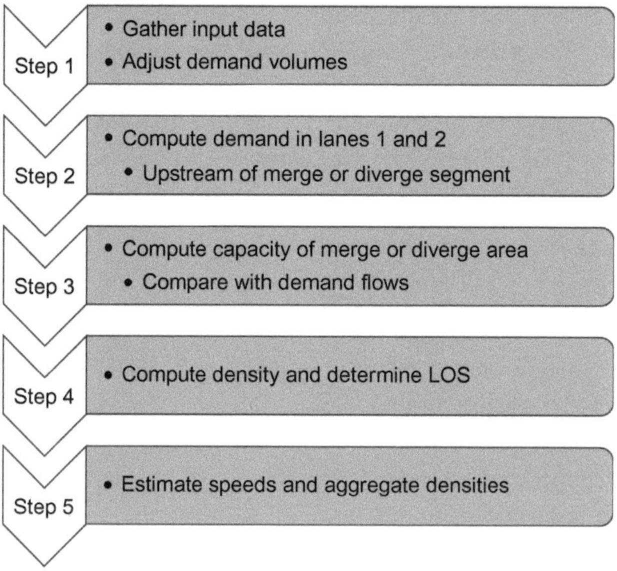

Figure 5.19 Methodology steps for merge/diverge analysis.

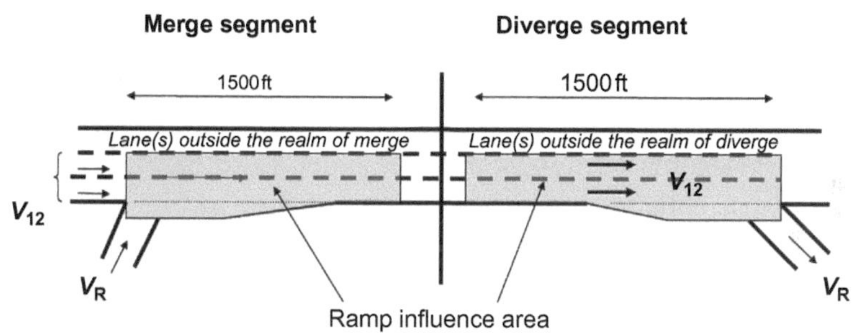

Figure 5.20 Schematic of merge and diverge segments.

gore point at merges, or upstream of the gorge point for diverges. Outside of this 1500-ft RIA, research suggests that traffic has stabilized to where most of the merge/diverge turbulence has subsided, and the operations of the freeway again are categorized as a basic segment. The analysis focus in the *HCM* is further on the two rightmost lanes of the segment, plus the acceleration or deceleration lanes. Research has shown that typically any additional freeway lanes (i.e., the third or third and fourth lanes on six-lane or eight-lane freeways, etc.) show more stable flow conditions without the influence of on-ramp and off-ramp turbulence. Figure 5.20 illustrates the geometry of both merge and diverge segments, as well as the key traffic volume parameters in the form of the volume in the two rightmost lanes in the ramp influence area, V_{12}, and the volume on the on-ramp or off-ramp, V_R.

In the first step, the entering hourly demand volume in vehicles per hour is converted to the peak 15-min low rate expressed in passenger cars per hour, as shown in Eq. (5.25).

$$v_f = \frac{V}{(PHF \times f_{HV})} \tag{5.25}$$

where

v_f = peak 15-min flow rate on the freeway, in passenger cars/h

V = directional analysis volume, in vph

PHF = peak hour factor

f_{HV} = heavy vehicle adjustment factor

The heavy vehicle effects for merge and diverge segments mirror those described previously for basic segments.

Step 2: Compute Demand in Lanes 1 and 2

The second step in the procedure is to estimate the amount of traffic volume in lanes 1 and 2 of the merge and diverge segment. For a four-lane freeway (two lanes per direction) the total demand by definition has to be the same as the demand in lanes 1 and 2. But for six-lane and eight-lane freeways, some portion of traffic is expected to be in the third and/or fourth lane, both presumed in the methodology to be unaffected by the merge/diverge turbulence.

The demand in lanes 1 and 2 is a function of the total freeway demand and the total flow on the ramp. It is also affected by the geometry of the ramp itself, namely by the length of the acceleration or deceleration lane, and the ramp's free-flow speed. Finally, the distribution of traffic across the lanes is impacted by the presence of upstream and downstream ramp. For example, a downstream on-ramp is expected to result in more traffic in lanes 1 and 2 (especially if closely spaced), as the prior on-ramp vehicles have not yet have had a chance to switch lanes into the outside lanes. Similarly, downstream (closely spaced) off-ramps are expected to increase the flow in lanes 1 and 2 of the subject segment, as vehicles are likely to preposition in preparation for the downstream off-ramp.

Mathematically, the flow rate in lanes 1 and 2 is calculated by multiplying the total entering flow rate (from step 1) by a factor describing the proportion of vehicles expected to be in lanes 1 and 2.

$$v_{12,merge} = v_F \times P_{FM} \tag{5.26}$$

$$v_{12,diverge} = v_R + (v_F - v_R)P_{FD} \tag{5.27}$$

where

v_{12} = estimated flow rate in lanes 1 and 2 of ramp influence area (passenger cars/h)

v_F = flow rate in all lanes of freeway just upstream of the merge or diverge (passenger cars/h)

P_{FM} = proportion of freeway vehicles expected in lanes 1 and 2 of merge area (passenger cars/h)

P_{FD} = proportion of freeway vehicles expected in lanes 1 and 2 of diverge area (passenger cars/h)

The equations used for estimating P_{FM} for merge areas are given in Table 5.21. In the case of six-lane freeways, multiple equations exist depending on the separation of the subject ramp and adjacent ramps. Guidance for which equations should be used is summarized in Table 5.22. Equation (5.28) refers to the base condition, or a merge area that is isolated from other ramps. Equations (5.29) and (5.30) then describe the cases where the subject ramp is impacted by an upstream off-ramp and a downstream off-ramp, respectively. The condition for whether the base equation or the adjusted equation is applied depends on the spacing between the two ramps. Specifically, if the distance between the two ramps is larger than or equal to a calculated *equilibrium distance,* L_{EQ}, then the isolated equation is used. For distances less than L_{EQ}, the adjusted equations are used. No impacts for adjacent on-ramps have been found for adjacent on-ramps, and thus the base equation is always used in those cases.

Table 5.21 Estimating proportion of traffic in lanes 1 and 2 for merge area

Number of freeway lanes[a]	**Equations for determining P_{FM}**	**Equation number**
4	$P_{FM} = 1.000$	
6	$P_{FM} = 0.5775 + 0.000028L_A$	Eq. (5.28)
	$P_{FM} = 0.7289 - 0.0000135(v_F + v_R) - 0.003296S_{FR} + 0.000063L_{UP}$	Eq. (5.29)
	$P_{FM} = 0.5487 + 0.2628\left(v_D/L_{DOWN}\right)$	Eq. (5.30)
8	For $v_F/S_{FR} \leq 72$: $P_{FM} = 0.2178 - 0.000125v_R + 0.01115(L_A/S_{FR})$	Eq. (5.31)
	For $v_F/S_{FR} > 72$: $P_{FM} = 0.2178 - 0.000125v_R$	Eq. (5.32)

Table 5.22 Selecting equations for P_{FM} for six-lane freeways

Adjacent upstream ramp	Subject ramp	Adjacent downstream ramp	Equation(s) used as a function of separation distance between ramps, *D*
None	On	None	Eq. (5.28)
None	On	On	Eq. (5.28)
None	On	Off	Eq. (5.28) for $D \geq L_{EQ}$ Eq. (5.30) for $D < L_{EQ}$
On	On	None	Eq. (5.28)
Off	On	None	Eq. (5.28) for $D \geq L_{EQ}$ Eq. (5.29) for $D < L_{EQ}$
On	On	On	Eq. (5.28)
On	On	Off	Eq. (5.28) for $D \geq L_{EQ}$ Eq. (5.30) for $D < L_{EQ}$
Off	On	On	Eq. (5.28) for $D \geq L_{EQ}$ Eq. (5.29) for $D < L_{EQ}$
Off	On	Off	Eq. (5.28) or Eq. (5.29) or Eq. (5.30)

In the special case where both an adjacent upstream off-ramp and downstream off-ramp are present and are located within the equilibrium distance, the P_{FM} is calculated twice, and the larger of the two values used in subsequent computations.

The equilibrium distances for merge areas are calculated from Eq. (5.33) for adjacent upstream off-ramps and Eq. (5.34) for adjacent downstream off-ramps. In general, if the distance between the subject ramp and the adjacent off-ramps is greater than or equal to the equilibrium distance, the subject ramp will be treated as isolated, and the base Eq. (5.28) is used to estimate the flow in lanes 1 and 2. If the distance is less than L_{EQ}, the respective adjusted equation is used.

$$L_{EQ,Merge,\ Upstream\ OFR} = 0.214(v_F + v_R) + 0.444L_A + 52.32S_{FR} - 2,403 \tag{5.33}$$

$$L_{EQ,\ Merge,\ Downstream\ OFR} = \frac{v_D}{0.1096 + 0.000107L_A} \tag{5.34}$$

where

v_F = flow rate in all lanes of freeway just upstream of the merge (passenger cars/h)

v_R = flow rate on subject ramp (passenger cars/h)

v_D = flow rate on downstream ramp (passenger cars/h)
L_A = length of acceleration lanes (ft)
L_{UP} = distance to the adjacent upstream ramp (ft)
L_{DOWN} = distance to the adjacent downstream ramp (ft)
S_{FR} = ramp free-flow speed (mph)

The equations used for estimating P_{FD} for diverge areas are given in Table 5.23. Similar as the case for merge areas, six-lane freeways have different equations describing the proportion of flow in lanes 1 and 2. Guidance for which equations should be used is summarized in

Table 5.23 Estimating proportion of traffic in lanes 1 and 2 for diverge area

Number of freeway lanes	Equations for determining P_{FM}	Equation number
4	$P_{FD} = 1.000$	n/a
6	$P_{FD} = 0.760 - 0.000025v_F - 0.000046v_R$	Eq. (5.35)
	$P_{FD} = 0.717 - 0.000039v_F + 0.604(v_U/L_{UP})$	Eq. (5.36)
	$P_{FD} = 0.616 - 0.000021v_F + 0124(v_D/L_{DOWN})$	Eq. (5.37)
8	$P_{FD} = 0.436$	n/a

n/a = not applicable.

Table 5.24 Selecting equations for P_{FD} for six-lane freeways

Adjacent upstream ramp	Subject ramp	Adjacent downstream ramp	Equation(s) used as a function of separation distance between ramps, *D*
None	Off	None	Eq. (5.35)
None	Off	On	Eq. (5.35), Eq. (5.28)
None	Off	Off	Eq. (5.35) for $D \geq L_{EQ}$ Eq. (5.37) for $D < L_{EQ}$
On	Off	None	Eq. (5.35) for $D \geq L_{EQ}$ Eq. (5.36) for $D < L_{EQ}$ → *Note:* when $v_u/L_{up} <= 0.2$ always use Eq. (5.35)
Off	Off	None	Eq. (5.35)
On	Off	On	Eq. (5.35) for $D \geq L_{EQ}$ Eq. (5.36) for $D < L_{EQ}$ → *Note:* when $v_u/L_{up} <= 0.2$ always use Eq. (5.35)
On	Off	Off	Eq. (5.35) or Eq. (5.36) or Eq. (5.37)
Off	Off	On	Eq. (5.35)
Off	Off	Off	Eq. (5.35) for $D \geq L_{EQ}$ Eq. (5.37) for $D < L_{EQ}$

Table 5.24. Equation (5.36) refers to the base condition, or a diverge area that is isolated from other ramps. Equation (5.36) is used when there is an adjacent on-ramp upstream of the subject diverge segment, while Eq. (5.37) is used when a downstream off-ramp is present. Similar to merge areas, the condition for whether the base equation or the adjusted equation is applied, depends on the spacing between the two ramps. Specifically, if the distance between the two ramps is larger than or equal to a calculated *equilibrium distance,* L_{EQ}, then the isolated equation is used. For distances less than L_{EQ}, the adjusted equations are used.

In the special case where both an adjacent upstream on-ramp and downstream off-ramp are present and are located within the equilibrium distance, the P_{FD} is calculated twice, and the larger of the two values used in subsequent computations.

The equilibrium distances for diverge areas are calculated from Eq. (5.38) for adjacent upstream on-ramps and Eq. (5.39) for adjacent downstream off-ramps. In general, if the distance between the subject ramp and the adjacent ramp is greater than or equal to the equilibrium distance, the subject ramp will be treated as isolated, and the base Eq. (5.35) is used to estimate the flow in lanes 1 and 2. If the distance is less than L_{EQ}, the respective adjusted equation is used.

$$L_{EQ,\ Diverge,\ Upstream\ ONR} = \frac{v_U}{0.071 + 0.000023v_F - 0.000076v_R} \tag{5.38}$$

$$L_{EQ,\ Diverge,\ Downstream\ OFR} = \frac{v_D}{1.15 - 0.000032v_F - 0.000369v_R} \tag{5.39}$$

where

v_F = flow rate in all lanes of freeway just upstream of the merge (passenger cars/h)

v_R = flow rate on subject ramp (passenger cars/h)

v_U = flow rate on adjacent upstream ramp (passenger cars/h)

v_D = flow rate on adjacent downstream ramp (passenger cars/h)

L_{UP} = distance to the adjacent upstream ramp (ft)

L_{DOWN} = distance to the adjacent downstream ramp (ft)

As a final check in step 2, the analyst should verify the reasonableness in the lane distribution arrived at from Table 5.21 and Table 5.23, as the underlying regression equations may sometimes predict values outside the calibrated and reasonable range. Specifically, the following two limitations are applied to all predictions of flow rates in lanes 1 and 2:

1. The average flow in the outer lanes (lanes 3 and 4) has to be less than or equal to 2700 passenger cars/h per lane.
2. The average flow per lane in the outer lanes has to be less than or equal to 1.5 times the average flow per lane in lanes 1 and 2.

If these conditions do not hold, special adjustments are needed. For six-lane and eight-lane freeways, if condition 1 is violated, an adjusted $v_{12,\text{adjusted}}$ is calculated as shown in the following:

$$v_{12,adj,six-lane} = v_F - 2700$$

$$v_{12,adj,eight-lane} = v_F - 5400$$

Similarly, if condition 2 is violated, the adjusted flow in lanes 1 and 2 is calculated as:

$$v_{12,adj,six-lane} = \left(\frac{v_F}{1.75}\right)$$

$$v_{12,adj,eight-lane} = \left(\frac{v_F}{2.50}\right)$$

Step 3: Compute Capacity of Merge or Diverge Area

With the flow in lanes 1 and 2 determined, the capacity of the merge or diverge area is estimated, and that capacity is compared to the predicted demand flows. In particular, capacity is estimated (and checked against demands) for three different components:

1. The capacity of the ramp roadway itself
2. The capacity of the freeway entering or exiting the merge or diverge area
3. The maximum flow rate of the freeway entering the merge or diverge area, which in the case of a merge area constitutes the sum of mainline and on-ramp demands

The capacities of ramp roadways as a function of the ramp free-flow speed are shown in Table 5.25. The capacity of the freeway segment upstream or downstream of the merge/diverge area, as well as the maximum desirable entering flow rates, are shown in Table 5.26 for freeways, and in Table 5.27 for multilane highways and ramps on collector-distributor (C-D)

Table 5.25 Capacity of ramp roadways

Ramp FFS S_{FR} (mph)	Single-lane ramps
>50	2200
>40–50	2100
>30–40	2000
≥20–30	1900
<20	1800

Table 5.26 Capacity and maximum flow rates for freeway merges and diverges

FFS (mph)	Capacity of upstream or downstream freeway segment (per lane)	Maximum desirable flow rate (v_{R12}) entering merge influence area	Maximum desirable flow rate (v_{12}) entering diverge influence area
≥70	2400/ln	4600	4400
65	2350/ln	4600	4400
60	2300/ln	4600	4400
55	2250/ln	4600	4400

Table 5.27 Capacity and maximum flow rate for multilane highways and C-D roads

FFS (mph)	Capacity of upstream or downstream highway or C-D segment (per lane)	Maximum desirable flow rate (v_{R12}) entering merge influence area	Maximum desirable flow rate (v_{12}) entering diverge influence area
≥60	2200/ln	4600	4400
55	2100/ln	4600	4400
50	2000/ln	4600	4400
45	1900/ln	4600	4400

roadways. In these two tables, it is emphasized that demands exceeding the shown capacities result in LOS = *F*, while demands exceeding the shown maximum desirable flow rates merely indicate that the conditions in the merge or diverge area are likely to be undesirable and worse than predicted by the *HCM* method.

Similar to basic freeway segments, the capacity of merge or diverge areas can be adjusted for the impacts of weather, incidents, and work zones, as well as to calibrate the segment capacity to local conditions. The capacity adjustment factor (CAF) approach and adjustment factors for this are the same as described in Section 5.3.2 for basic segments.

Table 5.28 Example of merge segment capacities in various U.S. cities

Location	**No. of lanes**	**Average (standard deviation)**	
		Max prebreakdown flow or capacity (veh/h/lane)	**Average queue discharge flow (veh/h/lane)**
Minneapolis, MN	2	2181 (163)	1644 (96)
Portland, OR	2	2238 (161)	1741 (146)
Toronto, Canada	3	2330 (162)	1865 (124)
Sacramento, CA	3	2174 (107)	1563 (142)
Sacramento, CA	4	2018 (108)	1567(115)
San Diego, CA	4	2075 (113)	1665 (85)
San Diego, CA	5	1928 (70)	1635 (66)

It is emphasized here that the capacities shown in Tables 5.26 and 5.27 are ideal capacities, and that especially merge segments often have much lower capacities due to the high degree of turbulence in these segments. Research is limited on predicting the capacity of these ramp segments, but frequent observations have been documented in the literature that suggest that the capacity of especially merge segments are much below the values shown. As such, local calibration is absolutely necessary to adequately reflect the operations of merge segments. As one example, a national research study summarized some observed capacities and queue discharge flows in merge segments, as shown in Table 5.28 (Elefteriadou, 2015).

The capacity values determined in step 3 of the merge/diverge procedure are used to check the demand volumes against. If any demands (ramp, upstream segment, or downstream segment) exceed the capacity, the LOS for the merge or diverge segment is F and no further speed and density estimation is possible. Instead, a freeway facility analysis should be conducted to estimate queuing upstream of the merge/diverge bottleneck. If the demand exceeds the maximum desirable flow rate entering the ramp influence area, the LOS is not necessarily *F*, but performance may be worse than predicted by the methodology. In that case, a freeway facility analysis is again recommended to check for queueing impacts.

Step 4: Compute Density and Determine LOS

Assuming the various segment demands pass all capacity checks, the density of the merge or diverge influence area can be calculated using Eq. (5.40) and Eq. (5.41) for merge and diverge segments, respectively.

Table 5.29 LOS for merge or diverge segment

LOS	Density (passenger cars/mi per lane)
A	≤10
B	>10–20
C	>20–28
D	>28–35
E	>35
F	Demand exceeds capacity

$$D_{Merge} = 5.5475 + 0.00734 v_R + 0.0078 v_{12} - 0.00627 L_A \quad (5.40)$$

$$D_{Diverge} = 4.252 + 0.0086 v_{12} - 0.009 L_D \quad (5.41)$$

From these densities, the LOS of the merge or diverge segment can be determined from Table 5.29. LOS F is defined as demand exceeding capacity for either the ramp itself, the segment upstream of the merge/diverge, or the downstream segment.

Step 5: Estimate Speeds and Aggregate Densities

The final step in the methodology is to estimate the speeds. Speeds can be estimated in lanes 1 and 2 of the ramp influence area, as well as in the outer lanes, separately. While the primary analysis focus in this methodology is on the ramp influence area, speeds in the outer lanes may be of interest, especially if evaluating a merge/diverge segment in the context of a freeway facility. The two speeds are estimated separately, and can further be aggregated to a total average segment speed. The latter can be used to obtain an aggregated density across the entire segment (note that the step 4 density applies only for lanes 1 and 2 and the acceleration or deceleration lanes). The equations for the various speeds in merge and diverge segments are given in Table 5.30. In the table, SAF refers to a speed adjustment factor to account for impacts of weather, incidents, work zones, or local calibration. The default values for SAF are the same as were shown for basic segments in Section 5.3.2.

With the speeds in the ramp influence area and the outside lanes estimated, the combined average segment speed can be estimated as the weighted average speed across all lanes. To do this, we need to first calculate the volume in the outside two lanes, which is estimated from Eq. (5.42).

Table 5.30 Speed estimation for merge and diverge areas

Segment type	Average speed in	Equation	Condition
Merge	Ramp influence area	$S_R = FFS \times SAF - (FFS \times SAF - 42)M_S$, where $M_S = 0.321 + 0.0039e^{(v_{R12}/1,000)} - 0.002(L_A \times S_{FR} \times SAF/1000)$	
	Outer lanes of freeway	$S_O = FFS \times SAF$	$v_{OA} < 500$passenger care/h
		$S_O = FFS \times SAF - 0.0036(v_{OA} - 500)$	$500 \leq v_{OA} \leq 2300$passenger car/h
		$S_O = FFS \times SAF - 6.53 - 0.006(v_{OA} - 2300)$	$v_{OA} > 2300$passenger car/h
Diverge	Ramp influence area	$S_R = FFS \times SAF - (FFS \times SAF - 42)D_S$, where $D_S = 0.883 + 0.00009v_R - 0.013S_{FR} \times SAF$	
	Outer lanes of freeway	$S_O = 1.097 \times FFS \times SAF$	$v_{OA} < 1000$ passenger car/h
		$S_O = 1.097 \times FFS \times SAF - 0.0039(v_{OA} - 1000)$	$v_{OA} \geq 1000$ passenger car/h

$$v_{OA} = \frac{v_F - v_{12}}{N_O} \quad (5.42)$$

where

v_{12} = demand flow rate in lanes 1 and 2 of the freeway immediately upstream of the ramp influence area (passenger cars/h)

v_{OA} = average demand flow per lane in outer lanes adjacent to the ramp influence area (not including flow in lanes 1 and 2) (passenger cars/h per lane)

v_F = demand flow rate on freeway immediately upstream of the ramp influence area (passenger cars/h)

N_O = number of outer lanes on the freeway (1 for a six-lane freeway; 2 for an eight-lane freeway)

From this, the aggregated speed in a merge or diverge area is estimated from Eq. (5.43) and Eq. (5.44), respectively.

$$S = \frac{v_{R12} + v_{OA}N_O}{\left(\frac{v_{R12}}{S_R}\right) + \left(\frac{V_{OA}N_O}{S_O}\right)} \quad (5.43)$$

$$S = \frac{v_{12} + v_{OA}N_O}{\left(\frac{v_{12}}{S_R}\right) + \left(\frac{V_{OA}N_O}{S_O}\right)} \quad (5.44)$$

where

S_R = average speed of vehicles within the ramp influence area (mph); for merge areas, this includes all ramp and freeway vehicles in lanes 1 and 2; for diverge areas, this includes all vehicles in lanes 1 and 2

S_O = average speed of vehicles in outer lanes of the freeway, adjacent to the 1500-ft ramp influence area (mph)

S = average speed of all vehicles in all lanes within the 1500-ft length covered by the ramp influence area (mph)

v_{12} = demand flow rate in lanes 1 and 2 of the freeway immediately upstream of the ramp influence area (passenger cars/h)

v_{R12} = total demand flow rate entering the on-ramp influence area, including v_{12} and v_R (passenger cars/h)

v_{OA} = average demand flow per lane in outer lanes adjacent to the ramp influence area (not including flow in lanes 1 and 2) (passenger cars/h per lane)

N_O = number of outer lanes on the freeway (1 for a six-lane freeway; 2 for an eight-lane freeway)

Table 5.31 LOS for merge and diverge segments

Level of service	Density range (passenger cars/mi per lane)
A	≤ 10
B	$>10-20$
C	$>20-28$
D	$>28-35$
E	>35
F	Demand exceeds capacity

If the merge or diverge segment is to be used in the context of a freeway facility analysis, one last step is the aggregation of densities across all lanes on the freeway. With the total volume and average speed known, the best way to estimate density is to use the fundamental relationship of traffic flow as shown in Eq. (5.45).

$$D = \frac{v}{S} \tag{5.45}$$

where

D = density including all lanes of the ramp influence area (passenger cars/mi per lane)

V = total flow rate through the merge or diverge area, all lanes (passenger cars/h per lane)

S = average speed of all vehicles through the merge/diverge area, all lanes (mph)

The level of service for a merge or diverge segment is then obtained from Table 5.31.

In addition to this basic procedure, special cases exist for two-lane merge or diverge areas, left-hand on-ramps and off-ramps, merge or diverge areas on a 10-lane freeway, and for merge and diverges with a continuous lane add or lane drop. These special cases are beyond the scope of this part, but are described in detail in the *HCM*.

5.3.4 Weaving Segments

A typical freeway weaving segment is formed when an on-ramp is followed by a downstream off-ramp, and the two ramps are connected by an auxiliary lane. In this configuration, turbulence is created from on-ramp traffic merging into the freeway mainline, while off-ramp traffic diverges from the mainline to the off-ramp. The two ramp flows (cross-) weave, which can result in significant reductions in speed. Additionally, some on-ramp traffic may be destined for the downstream off-ramp, and thus may not change lanes at all.

Many other variations of weaving segments exists, including those with multilane on-ramps and/or multilane off-ramps. There are also two-sided weaving segments, where the on-ramp and off-ramp are located on opposite sides of the freeway mainline. But in all cases, the operations of the weaving segment are impacted by a few key variables:

- The number of lane changes taking place, which is primarily a function of the varying movement demands and the lane geometry
- The number of lanes available for those lane changes
- The length of the weaving segment, or the distance over which these lane changes have to occur

The *HCM* methodology for weaving segments estimates the parameters just mentioned and translates the various inputs into a number of lane changes per mile. This measure is then translated to varying performance measures, including speed, density, and LOS.

Methodology

The *HCM* weaving methodology consists of eight basic steps that are illustrated in Figure 5.21. Each step is discussed in more details in the following.

Step 1: Gather Input Data and Adjust Demand Volumes

The first step in every methodology is gathering input data. For the weaving method, the input data are fundamentally divided between geometric data describing the weaving segment, and demand data. The following geometric data are needed:

- *Number of lanes* in the weaving segment, counting the auxiliary lane.
- *Short length* of the weaving segment (in feet), defined as the distance from ramp gore point at the on-ramp to the gore point at the off-ramp.
- *Location of ramps* on the segment, which could be either on the left or right side of the segment.
- *Number of required lane changes* from ramp to freeway, from freeway to ramp, and from ramp to ramp, which are used to eventually determine

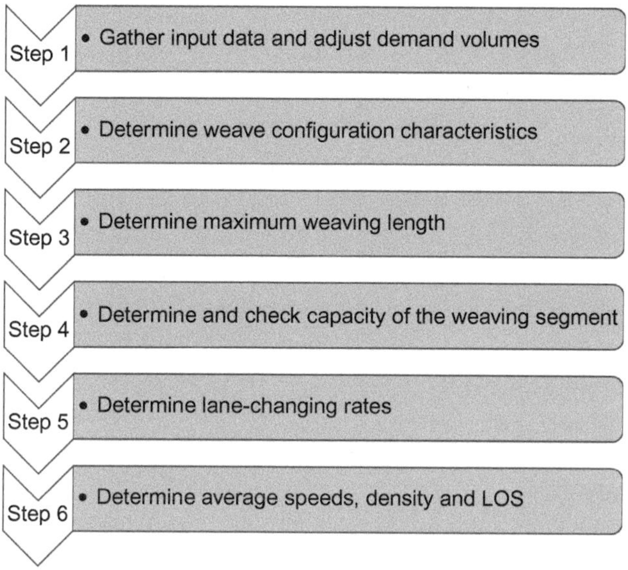

Figure 5.21 Methodology steps for weaving methodology.

total lane change rates. With an increasing number of required lane changes, the turbulence in the weave segment is expected to increase and operations to deteriorate.

- *Number of weaving lanes* defined as the number of lanes from which a weaving movement can be completed with zero or one lane change. The number of weaving lanes, N_{WL}, essentially describes how many lanes are available for the weaving maneuver, with a higher N_{WL} resulting in improved performance.
- *Interchange density* in units of interchanges per mile.
- *Terrain type*, classified as either level, rolling, or a specific grade to account for truck effects.
- *Equivalent capacity of a basic segment*, which is used as an anchor point in the methodology to estimate the weave segment capacity.

In addition to these geometric inputs, several demand inputs are needed, including hourly demands on the mainline, on-ramp, off-ramp, and the specific demand from on-ramp to off-ramp. As with other operational methods, peak hour factor and heavy vehicle adjustments are needed to convert volumes in vehicles per hour to hourly flow rates in passenger car units. The volume conversion is identical to the process used for merge/diverge segments, as shown in Eq. (5.25).

Finally calibration factors in the form of speed and capacity adjustment factors (SAF and CAF) can be used as calibration tools to match the methodology to locally observed conditions.

Step 2: Determine Weave Configuration Characteristics

The weave methodology is fundamentally tied to the configuration of the weaving segments. The two key parameters are the minimum number of lane changes needed in the weaving segment, LC_{min}, and the number of weaving lanes, N_{WL}, which describe in essence how many lanes are available to complete the weaving maneuvers. The number of weaving lanes is defined as the number of lanes from which a weaving segment can be completed with zero or one lane change, resting on the assumption that this is where most weaving vehicles are likely to be positioned to complete their weaving maneuvers.

The minimum number of lane changes is a simple summation of the lane changes needed for each of the weaving movements multiplied by its corresponding flow rate, as given in

$$LC_{MIN} = (LC_{RF} \times v_{RF}) + (LC_{FR} \times v_{FR}) \tag{5.46}$$

where

LC_{RF} = the minimum number of lane changes for a vehicle to move from the on-ramp to the freeway mainline

LC_{FR} = the minimum number of lane changes for a vehicle to move from the freeway mainline to the off-ramp

v_{RF} = the flow rate from ramp to freeway (passenger cars/h)

v_{FR} = the flow rate from freeway to ramp (passenger cars/h)

The estimation of LC_{RF} and LC_{FR} emerges from the geometric configuration of the weaving segment as illustrated in Figure 5.22. Part (a) of the figure shows a very common ramp weave with four lanes in the weaving segment. The number of lane changes from ramp to freeway and from freeway to ramp are each one. The number of weaving lanes is two, as for both of the two outside lanes a weaving movement can be completed with zero or one lane change as indicated by the dashed and dotted arrows.

Figure 5.22(b) illustrates a more complicated weaving segment. Here the ramp-to-freeway movement needs to complete two lane changes, while the freeway-to-ramp movement actually does not need to make any lane changes. Further, the number of weaving lanes is three, as all three outside lanes meet the definition. For this particular weaving segment, it becomes very evident how the configuration alone does not fully

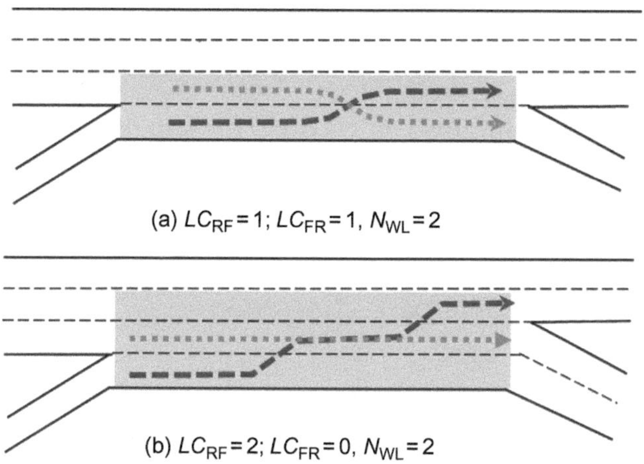

Figure 5.22 Illustrative example of weave configuration.

predict the operations. In fact, the example shown in Figure 5.22(b) may function very well if the freeway-to-ramp volume is heavy and the ramp to freeway volume is low. But if the two flows are reversed, the expected operations are worse, as Eq. (5.46) predicts more lane changes and therefore more turbulence in the segment.

Step 2 is similar for two-sided weaving segments, with two exceptions. First, only the (left-to-right or right-to-left) ramp-to-ramp flow is considered to be a weaving maneuver, simplifying the computation of the minimum lane changes as shown in Eq. (5.47). Second, the number of weaving lanes is always defined to be zero for a two-sided weave, regardless of the number of lanes on the mainline. Presumably, this is because the weaving maneuver always needs to make more than one change when weaving from an on-ramp on one side to an off-ramp on the opposite site, given that freeways typically have two or more mainline lanes.

$$LC_{MIN} = LC_{RR} \times v_{RR} \tag{5.47}$$

where

LC_{RR} = the minimum number of lane changes for a vehicle to move from the on-ramp to the off-ramp on the opposite side of the freeway

v_{RF} = the flow rate from on-ramp to off-ramp (passenger cars/h)

Step 3: Determine Maximum Weaving Length

The next step in the methodology is to determine the maximum length of the weaving segment, which conceptually is the length at which the resulting weaving maneuvers no longer result in any reduction in performance of the segment, as compared to a basic segment. In other words, the weaving turbulence is negligible relative to the overall segment operations. Essentially then, this step checks if the segment in fact operates as a weave, or if it is so long that it should simply be analyzed as a basic segment. The maximum weaving length is calculated from Eq. (5.48).

$$L_{MAX} = [5728(1+VR)^{1.6}] - [1566N_{WL}] \tag{5.48}$$

where

L_{MAX} = maximum weaving segment length (feet)

VR = volume ratio, which is calculated from Eq. (5.49)

N_{WL} = number of weaving lanes as defined previously

The volume ratio is calculated as the volume of weaving vehicles to the total number of vehicles on the segment as shown in Eq. (5.49).

$$VR = \frac{v_W}{v_{total}} = \frac{v_W}{v_W + v_{NW}} = \frac{v_{RF} + v_{FR}}{v_{RF} + v_{FR} + v_{FF} + v_{RR}} \tag{5.49}$$

where

v_W = weaving demand flow rate in the weaving segment (passenger cars/h), $v_W = v_{RF} + v_{FR}$

v_{NW} = nonweaving demand flow rate in the weaving segment (passenger cars/h), $v_{NW} = v_{FF} + v_{RR}$

The maximum weaving length increases with an increase in the volume ratio, which is intuitive as a higher volume ratio means more turbulence and therefore a higher maximum length limit. The maximum weaving length decreases with a greater number of weaving lanes, as more space is available for a given number of lane changes. This effect is illustrated in Figure 5.23. In the procedure, if the short length of the weaving segment, L_s, is greater than the maximum weaving length, the analysis should stop and the segment instead be analyzed as a basic segment.

Step 4: Determine and Check the Capacity of the Weave Segment

Next, the capacity of the weaving segment is checked using two different definitions. The first definition is based on density, and defines capacity as the segment reaching a density of 43 passenger cars/mi per lane, which is

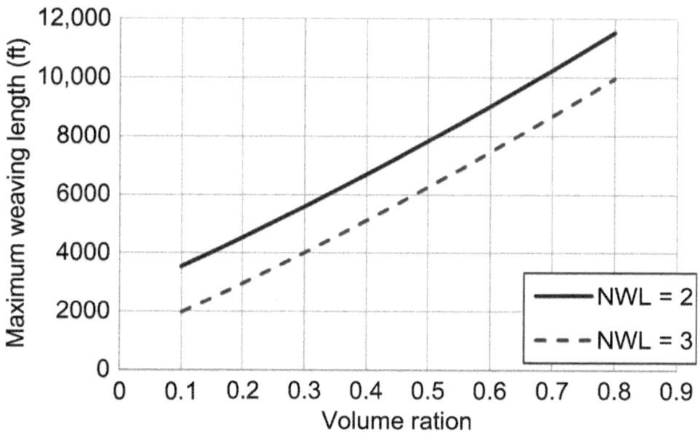

Figure 5.23 Illustration of maximum weaving length.

also the LOS E-F boundary for weaving segments. The second definition is based on weaving demand flows, which recognizes that at some point there are simply too many lane changes to be accommodated in the available number of weaving lanes.

For the first definition, the capacity is estimated from Figure 5.22

$$c_{IWL} = c_{IFL} - [438.2(1+VR)^{1.6}] + [0.0765L_S] + [119.8N_{WL}] \quad (5.50)$$

where

c_{IWL} = capacity of the weaving segment under ideal conditions, per lane (passenger cars/h per lane)

c_{IFL} = equivalent capacity of a basic freeway segment with the same FFS as the weaving segment under ideal conditions, per lane (passenger cars/h per lane)

L_S = the short length of the weaving segment measured from gore to gore (ft)

(all other variables are as previously defined)

The capacity of the weaving segment mentioned decreases with an increasing volume ratio (larger negative number subtracted from base), and increases with more available space for weaving, either in terms of segment length, or in terms of the number of weaving lanes. When comparing this capacity under prevailing conditions (in units of passenger cars/h per lane), back to total volume inputs (in veh/h), it needs to be converted back using Eq. (5.51), which is the inverse of Eq. (5.25).

$$c_W = c_{IWL} \times N \times f_{HV} \tag{5.51}$$

where

c_W = capacity of the weaving segment under prevailing conditions, (veh/h)

N = number of lanes in the weaving segment

f_{HV} = heavy vehicle flow rate adjustment factor

Because Eq. (5.50) represents a regression model, it inevitably runs a risk of being applied at some boundary conditions that are beyond the observed range of data used to build the model in the first place. As such, a second definition of capacity is used in this methodology, which essentially limits Eq. (5.50) by the maximum weaving demand flows that can realistically be accommodated in the segment. This limiting condition is shown in Eq. (5.52).

$$c_{IW} = \begin{cases} \dfrac{2400}{VR} \text{ for } N_{WL} = 2 \text{ lanes} \\ \dfrac{3500}{VR} \text{ for } N_{WL} = 3 \text{ lanes} \end{cases} \tag{5.52}$$

where

C_{IW} = the capacity across *all lanes* in units of passenger cars/h

(all other terms are as defined previously)

Similar to the first condition, this capacity under prevailing conditions (across all lanes, but in units of passenger cars/h) needs to be converted to a capacity under ideal conditions (in veh/h) by multiplying by the heavy vehicle adjustment factor as shown in Eq. (5.53).

$$c_W = c_{IW} \times f_{HV} \tag{5.53}$$

With these two capacities, the analyst compares the total volume to the total capacity in the segment. Both terms should be in the same units (passenger cars/h) to assure a valid comparison. In addition, the capacity could be calibrated for local conditions, or to reflect the impact of rain or incidents, by first multiplying the capacity with a capacity adjustment factor (CAF) as was described in Section 5.3.2 for basic segments.

If the volume exceeds capacity, the analysis stops, and the analyst should instead refer to the freeway facilities methodology to evaluate the (congested) weaving segment and any queue spillback issues resulting from the overcapacity segment.

Step 5: Determine Lane Changing Rates

This step is the key computational step in the methodology, in which lane-changing rates (in units of lane changes per mile) are estimated, with the underlying premise that more lane changes result in more turbulence, and thus generally degraded performance.

Lane changes in a weaving segment fall into three categories:

1. Lane changes *required* by weaving vehicles to continue on their desired path
2. *Optional* lane changes by weaving vehicles, which considers the fact that not every vehicle is correctly prepositioned when entering the segment
3. *Optional* lane changes by nonweaving vehicles to, for example, merge to the left to avoid the weave turbulence area to the right

The number of required plus optional lane changes by weaving vehicles (1 and 2 in the list) is estimated from Eq. (5.54).

$$LC_W = LC_{MIN} + 0.39[(L_S - 300)^{0.5} N^2 (1 + ID)^{0.8}] \qquad (5.54)$$

where

LC_W = hourly rate at which weaving vehicles make required lane changes within the weaving segment (lane change/h)

LC_{MIN} = minimum hourly lane change rate within the weaving segment for required lane changes from step 2 (lane change /h)

L_S = short length of the weaving segment, which is constrained to be greater than or equal to 300 ft in this equation (ft)

N = number of lanes in the segment

ID = interchange density, the number of interchanges within 3 mi upstream and downstream of the center of the subject weaving segment divided by 6, in interchanges per mile (interchanges/mi)

Conceptually, the first term of Eq. (5.54) represents the required lane changes, while the second represents the optional lane changes. Optional lane changes increase with increasing segment length, more lanes, and increasing interchange density. The last term accounts for the fact that lane changes are generally more frequent with more upstream and downstream interchanges, as is often the case in urban areas.

In addition, the lane-changing rate for nonweaving vehicles has to be estimated. A total of three equations exist for this, describing different flow regimes distinguished by an index term shown in Eq. (5.55).

$$I_{NW} = \frac{L_S \times ID \times v_{NW}}{10,000} \qquad (5.55)$$

Table 5.32 Estimating lane change rates for nonweaving vehicles

Equations for determining LC_{NW}	Condition	Equation number
$LC_{NW1} = (0.206v_{NW}) + (0.542L_S) - (192.6N)$	$I_{NW} \leq 1300$	Eq. (5.56)
$LC_{NW2} = 2135 + 0.223(v_{NW} - 2000)$	$I_{NW} \geq 1950$	Eq. (5.57)
$LC_{NW3} = LC_{NW1} + (LC_{NW2} - LC_{NW1})\left(\frac{I_{NW} - 1300}{650}\right)$	$1300 < I_{NW} < 1950$	Eq. (5.58)

This index is calculated first as a function of segment length, interchange density, and the nonweaving vehicle volume as summarized in Table 5.32.

As a final boundary condition, if LC_{NW2} is greater than or equal to LC_{NW1}, then LC_{NW2} is used as the final lane change rate for nonweaving vehicles. The total lane change rate in the weaving segment is the sum of weaving vehicles (from Eq. (5.54) and nonweaving vehicles (from Table 5.32), as shown in Eq. (5.59).

$$LC_{ALL} = LC_W + LC_{NW} \tag{5.59}$$

Step 6: Determine Average Speeds, Density, and LOS

The average speeds in the weaving segment are determined under consideration of the speed of weaving and the speed of nonweaving vehicles. First, the speed of weaving vehicles is estimated from Eq. (5.60) and Eq. (5.61).

$$S_W = 15 + \left(\frac{FFS \times SAF - 15}{1 + W}\right) \tag{5.60}$$

$$W = 0.226\left(\frac{LC_{ALL}}{L_S}\right)^{0.789} \tag{5.61}$$

where

S_W = average speed of weaving vehicles within the weaving segment (mph)

W = weaving intensity factor

(all other terms are as defined previously)

Note that the weaving intensity factor takes the total number of required and optional lane changes, and scales them by the segment length to essentially get an entity in units of lane changes per foot of length in the segment. Intuitively, a higher weaving intensity factor corresponds to more lane changes per distance, which can be thought of as a measure of lane change density across the segment.

Table 5.33 LOS for weaving segments

Level of service	Density range (passenger cars/mi per lane)
A	0–10
B	>10–20
C	>20–28
D	>28–35
E	>35–43
F	>43, or demand exceeds capacity

The speed of nonweaving vehicles is estimated from Eq. (5.62) as a function of free-flow speed, number of lane changes, and the total volume per lane.

$$S_{NW} = FFS \times SAF - (0.0072 LC_{MIN}) - \left(0.0048 \frac{v}{N}\right) \tag{5.62}$$

The total average space mean speed on the segment is obtained from Eq. (5.63), mirroring the procedure for merge/diverge segments in Section 5.3.3. The density of the segment is then estimated from Eq. (5.64). That density is in conjunction with Table 5.33 to determine the LOS for the weaving segment.

$$S = \frac{v_W + v_{NW}}{\left(\frac{v_W}{S_W}\right) + \left(\frac{v_{NW}}{S_{NW}}\right)} \tag{5.63}$$

$$D = \frac{(v/N)}{S} \tag{5.64}$$

5.3.5 Freeway Facility Analysis

The *HCM* segment methodologies are useful for planning purposes and to evaluate the operations and LOS of isolated segments that are below capacity. The segment methods previously mentioned are not suitable for evaluating oversaturated and congested conditions. As such, they are not suitable for most operational analysis of especially urban freeway segments, as most freeways do have some segments and time periods that are above capacity. An example freeway facility with 12 segments is shown in Figure 5.24. It consists of 7 basic segments, and a total of 5 merge, diverge, or weaving segments. Rather than analyzing each one of

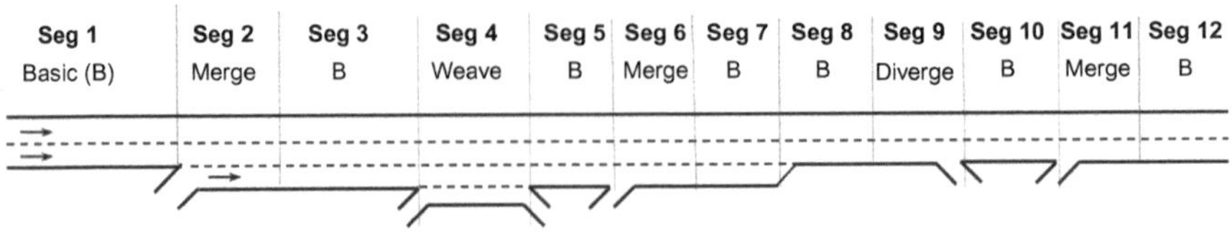

Figure 5.24 Freeway facility example with 12 segments.

these 12 segments in isolation, a freeway facility analysis evaluates all segments jointly, and explores whether there is any interaction (i.e., queue spillback) between segments over space and/or time.

The capacity of a freeway segment can be exceeded due to *recurring* congestion (e.g., known bottlenecks where the demand exceeds the available capacity for some parts of the day), or *nonrecurring* congestion, which refers to incidents, weather impacts, work zones, and seasonal or random fluctuations in the demand levels. One potential recurring bottleneck in Figure 5.24 is the lane drop in basic freeway segment 8. However, whether or not this segment is indeed a bottleneck, and whether other segments are above capacity depends on the demand patterns on the various segments, as well as on any additional capacity restrictions on the facility. In addition, nonrecurring sources of congestion, including incidents, weather, and work zones can occur anywhere on the facility.

A freeway facilities analysis is able to identify and analyze congestion due to both recurring and nonrecurring sources, and evaluate the impacts of that congestion over both time and space. In other words, the *HCM* freeway facilities analysis is a multisegment and multitime period methodology that can track onset, accumulation, and dissipation of congestion over time and space. As such, it is highly computationally complex and intensive, and is therefore typically implemented in a computational engine or software. This section therefore only provides a high-level overview of the method, with computational details found in the *HCM*.

Facility Concepts and Segmentation

The freeway facility methodology works by evaluating each segment and each time period in sequence, starting with the first segment in the first time period, and then moving through all segments, before then going to the second time period. In other words, the facility is evaluated in 15-min chunks, which is the typical *HCM* analysis period.

If in a time period, all segments have a demand-to-capacity (d/c) ratio less than or equal to 1.0, and if there isn't any queuing from prior congestion on the facility, then the operation of the overall facility is a simple summation or aggregation of the individual segments. If the sample facility in Figure 5.24 is generally uncongested, the analyst could have performed 12 individual *HCM* analyses and then summed the results. Use of a computational engine or software in this case brings efficiency, but the analysis could still be done by hand.

But once one or more segments on the facility show a d/c ratio greater than 1.0, the *HCM* methodology switches to a macroscopic simulation that is executed in 15-s time steps, and thus dramatically increases the computational burden. For oversaturated operations, the method employs a modification of the cell transmission model that can model queue propagation and dissipation. The computational details are beyond the scope of this text, but are described in the supplemental freeway facilities chapter in the *HCM* (Eads et al., 2000)

Facility Segmentation Rules

The first analyst challenge in applying the freeway facilities methodology is the segmentation of a facility into *HCM* analysis segments. For this, several rules exist, also practice has shown that sometimes analyst judgment is needed if the real-world facility doesn't exactly fit into the *HCM* segment bins. The basic facility segmentation rules are as follows:

1. Create a new segment whenever *demand* changes, which typically means having traffic enter or exit the facility via a ramp.
2. Create a new segment whenever *capacity* changes, which occurs at merge, diverge, or weaving segments due to turbulence, but can also occur at lane drops, or when other conditions impact the capacity. Examples for the latter include a freeway work zone, a steep grade, or segment with sun glare during the morning commute, which increases vehicle headways and thereby reduces capacity.
3. Classify ramps as either a merge, diverge, or weaving segment (as in the following).
4. Classify an adjacent on-ramp and off-ramp pair that is connected by an auxiliary lane as a weaving segment.
5. Specify the remaining (nonweave) on-ramps and off-ramps as merge and diverge segments, respectively. The influence area is typically defined as 1500 ft measured from gore point, and corresponds to the segment length, although sometimes that distance is restricted by the presence of, say, another ramp at a spacing less than 1500 ft.

6. Classify an adjacent on-ramp and off-ramp pair that is *not* connected by an auxiliary lane as three segments (merge-basic-diverge), provided the gore-to-gore length is greater than 3000 ft. For ramps that are spaced at exactly 3000 ft, there will not be an intermittent basic segment. For ramps that are spaced less than 3000 ft, an overlapping ramp segment (*R*) is defined for the length that the two ramp influence areas overlap. This segment is then analyzed assuming that the more restrictive of the two adjacent ramps governs the operations of the *R* segment.

The segmentation process is illustrated in Figure 5.25. Part (a) of the figure shows a facility with three on-ramps and one off-ramp. In Figure 5.25(b), the facility has been divided into sections corresponding to changes in demand (rule 1). With the four ramps, the facility results in five sections. One noteworthy aspect of these sections, is that they largely respond to modern freeway data sources used by many agencies. The more finely aggregated *HCM* segments are often not directly used in these databases, making the sections of Figure 5.25(b) a very useful basis for calibration and validation of the freeway facility. There are no additional changes in capacity (rule 2) on this example facility, other than those already reflected by rule 1.

Next, rules 3 through 6 are applied to complete the segmentation. Rule 3 tells us to specify each on-ramp and off-ramp as a merge, diverge, or weaving segment. Sections 4 and 5 both show an on-ramp, and are thus specified as merge segments. Rule 4 looks for adjacent on-ramps and

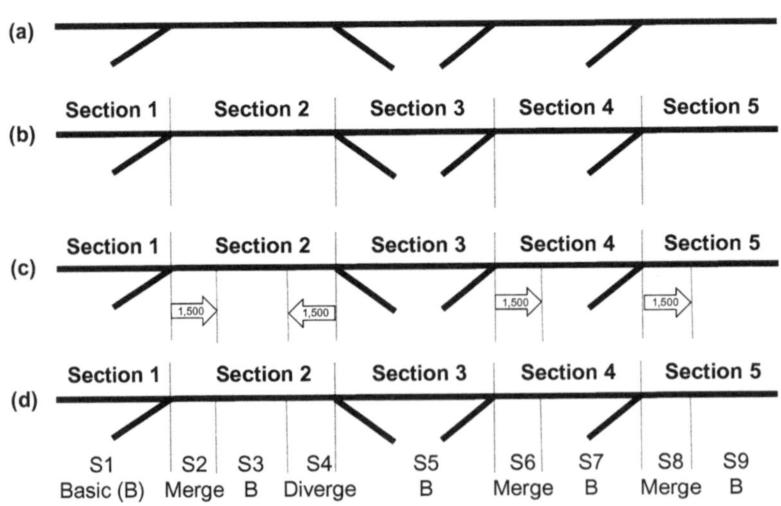

Figure 5.25 Facility segmentation illustrative example.

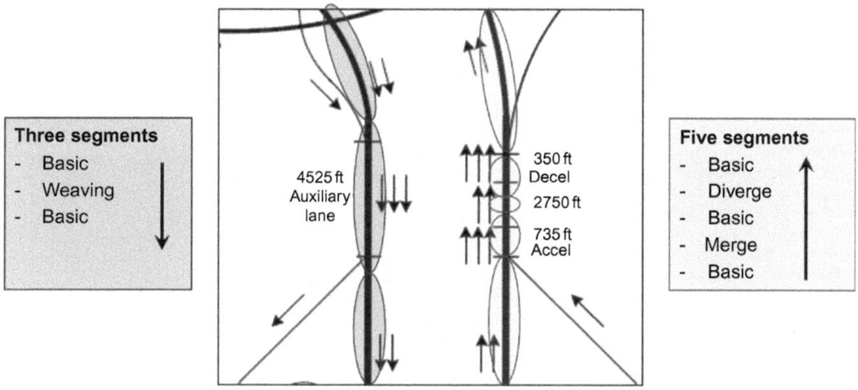

Figure 5.26 Distinguishing weave and merge/diverge segments.

off-ramps connected by an auxiliary lane. Section 2 features an on-ramp followed by an off-ramp, but there is no auxiliary lane connecting the two. As such, there are no weaving segments in this facility. The reader is referred back to Figure 5.24 for a facility with a weaving segment.

Because section 2 is not a weaving segment, we move on to rule number 5, which requires us to measure the distance between the ramps, and specifically check if the two ramp influence areas overlap. Figure 5.25 (c) shows the 1500 ramp influence areas from the on-ramp and off-ramp, and shows that the two do not overlap. As such, section 2 is divided into three segments as merge-basic-diverge using rule 5, while rule 6 does not apply. Figure 5.25(d) shows the completed segmentation into nine *HCM* analysis segments.

Figure 5.26 shows an additional segmentation example using a design drawing. In the example, the southbound direction shows a two-lane freeway with a one-lane on-ramp that transitions into an auxiliary lane to the downstream off-ramp. This on-ramp and off-ramp combination corresponds to a weaving segment. The northbound direction also features an on-ramp followed by an off-ramp, but in this case, no auxiliary lane is used to connect the two ramps. As such, the section between on-ramp and off-ramp is classified as merge-basic-diverge. Because the total length between the two gores is 3835 ft, the segment lengths are 1500 ft, 835 ft, and 1500 ft, respectively.

Illustrating Time-Space Domain

With the completed segmentation of our example facility from Figure 5.25, we will now use this sample example to illustrate the

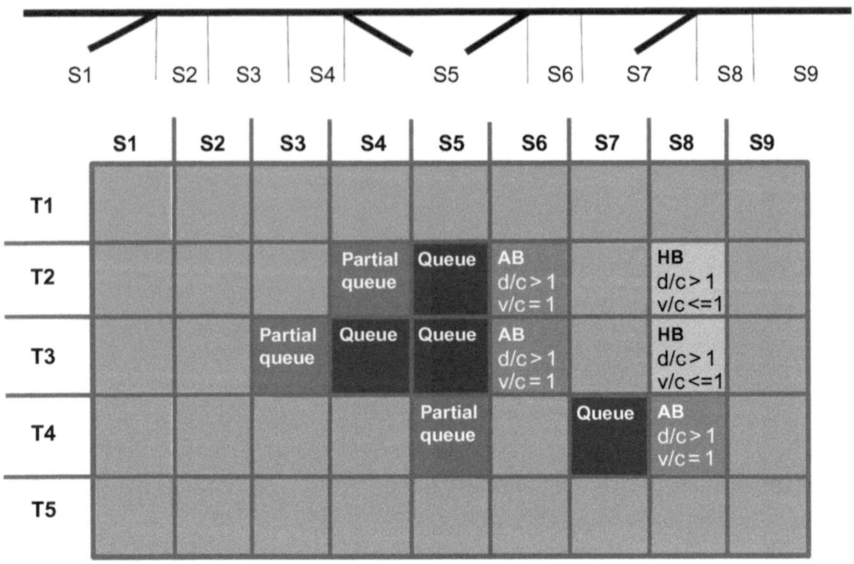

Figure 5.27 Freeway facility time-space domain illustration.

concept of the extended time-space domain, and interaction of (congested) segments over both dimensions. Figure 5.27 shows the nine segments from the sample facility analyzed over five 15-min time periods.

In the first time period (T1) the facility is generally undersaturated, which is illustrated by green shading in all cells. This doesn't necessarily mean that all nine segments are at LOS = A, but rather that no queue interaction occurs between segments. A fully undersaturated initial time period is always good practice in a freeway facility analysis.

In T2, the merge area in segment S6 is shown with a d/c ratio greater than 1.0, making it an *active bottleneck* (AB). As a result, the actual volume processed through this segment is limited to its capacity, making the volume-to-capacity or v/c ratio equal to 1.0. The difference between the demand and the segment capacity is referred to *unserved demand*, which translates into forming a queue upstream of the bottleneck segment. In this example, the unserved demand is such that the queue fully fills segment 5 and partially fills segment 4 at the end of the 15-min period. In the methodology, the actual queue length is estimated through the macroscopic simulation process referenced previously.

Still in T2, it is now noteworthy that segment S8 is also shown as having a d/c ratio greater than 1.0, but a v/c ratio less than 1.0. This means that the bottleneck is less severe than the one on S6, which meters the demand destined for S8. As such, S8 is referred to as a *hidden bottleneck* (HB). A hidden bottleneck has d/c > 1.0, but is metered by a more severe upstream bottleneck. This is one of the key advantages of the facility method, as it flags potential bottlenecks that may not even be evident to operating agencies (the same metering also happens in the field, and so agencies may never see the high demand for segment S8). The implication for practitioners is that if money is spent to improve bottleneck S6, the congestion likely will simply migrate downstream and activate the (hidden) bottleneck in segment S8. A widening or improvement project should therefore encompass both the active and hidden bottleneck.

In T3, the active bottleneck in S6 continues to have d/c greater than 1.0, resulting in more unserved demand and a growing queue, which partially fills segment S3 at the end of the time period. The hidden bottleneck in S8 remains inactive.

In T4, demand in S6 drops to where the segment is no longer above capacity. This results in the queue beginning to dissipate, although it is evident that the queue doesn't fully clear at the end of the time period, still partially filing S5. In this example, the demand on S8 remains above its capacity, resulting in that bottleneck being activated in T3. Segment S8 now has to deal with the discharging queue that was previously stored upstream of bottleneck S6. The result is a queue forming in segment S7.

In the last time period, the example shows all queues as having cleared with all demands having dropped below capacity. This freeway facility example is considered well defined as the entire boundary of the time-space domain (first and last time period, as well as first and last segments) are all under capacity and never contain any queuing. If an analyst finds residual queues at the end of an analysis, or if queues spill over the start of the facility, the analyst is strongly advised to add additional time periods to the end of the analysis and/or additional segments to the beginning of the facility.

Methodology

The freeway facility methodology is summarized in Figure 5.28 in 8 steps. The *HCM* methodology is divided into 17 steps, but several steps are combined here for ease of understanding. As for previous methods, the steps are discussed further in the following sections.

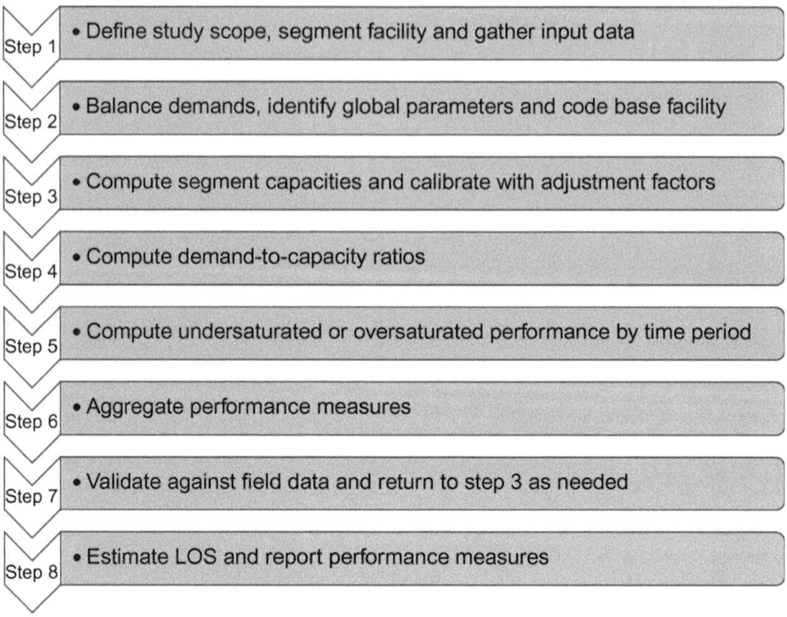

Figure 5.28 Freeway facility methodology steps.

Step 1: Define Study Scope, Segment Facility, and Gather Input Data

The first and arguably most important step in the freeway facility is to adequately scope the study, and then gather input data to support that scope. The previous discussion introduced the concept of the extended time-space domain, and the example associated with Figure 5.27 emphasized that the boundaries of the time-space domain should be uncongested. As such, some local knowledge is needed to decide when and where to start and end the analysis. If the queue ends up spilling beyond the start of the facility, or if it persists after the last time interval, the analysis scope may need to be revised.

Also in this step, the facility is segmented into *HCM* segments as was previously described. This requires some practice and may require assumptions by the analyst for facilities that do not cleanly fit within the *HCM* definition. Finally, input data are gathered for each segment and each time period, which can be time consuming, but is absolutely necessary to accurately model the facility.

Step 2: Balance Demands, Identify Global Parameters, and Code Base Facility

The next step is to balance demands if necessary and to configure the base facility in software or a computational engine tool. Because traffic volumes may be obtained from different sources, demands should be balanced to assure that they are internally consistent. The facility works with entering the mainline entry demand on the first segment, and then adding and subtracting demands from on-ramps and off-ramps as one progresses through the facility. This way, the flow out of one segment and into the next segment are balanced.

The methodology further includes some global parameters used for calibration. The two key global parameters that are also used for calibration are:

1. *Jam density*, which was introduced as a concept in Section 5.2 and determines operations in the congested regime. A higher jam density results in tighter vehicle spacing and, as such, may result in shorter queue lengths. A higher jam density can also result in a reduced shockwave speed, as again vehicles are packed together at higher densities at the same flow rate. A default value of 190 passenger cars/mi per lane is given for this value in the *HCM*.
2. *Queue discharge capacity drop percentage*, also referred to as the "alpha" value. Research has demonstrated that the maximum prebreakdown flow rate or capacity, is higher than the queue discharge flow rate after breakdown and as vehicles are discharging from a queue. The *HCM* uses a default of 7% based on research, although ranges from 0% to 20% have been observed in different studies. A higher alpha value reduces the capacity, and as such is expected to increase queue lengths and extend the duration of congestion.

Both parameters are important for calibration of the facility and apply to the entire facility. With inputs obtained and global parameters set, the analyst then enters the facility data into a computational engine tool for analysis, as the computations are too complex for calculation by hand.

Step 3: Compute Segment Capacities and Calibrate with Adjustment Factors

From the input data, segment capacities are computed based on the segment methodologies for basic, merge, diverge, and weaving segments consistent with their descriptions previously outlined. In addition, the analyst may calibrate capacities using capacity adjustment factors. Research has shown that the capacity of some bottlenecks can be lower than the ideal capacity given in the *HCM*, due to frictional effects or other circumstances. As such, it is not uncommon to have to reduce the capacity of a bottleneck merge segment from say 2400 passenger cars/h per lane to 2000 passenger cars/h per lane, which is a capacity adjustment factor (CAF) of 0.83. The analyst also can specify speed adjustment factors (SAF) and demand adjustment factors (DAF) for each segment or time period.

It is emphasized here that this does not apply to all merge segments (or diverge or weaves or unique basic segments) uniformly, but rather becomes an iterative process in the calibration and validation cycle. In practice, the analyst starts with one set of assumptions, proceeds through the methodology, compares results with available field data, and if necessary returns to this step to further calibrate bottleneck capacities. More research is needed to provide default values on these bottleneck capacities.

Step 4: Compute Demand-to-Capacity Ratios

From the calibrated capacities and given demands, the demand-to-capacity or d/c ratios are computed for each segment and time period. This step determines whether there are any active bottlenecks, which correspond to segments with d/c greater than 1.0 that are not metered by a more severe active bottleneck. In the computations, a time period where all d/c ratios are less than 1.0 is evaluated by batch processing the individual segment methodologies. But once any one segment has d/c greater than 1.0, the computations change to the modified cell transmission model approach, which essentially is a macroscopic simulation process to estimate queue accumulation and dissipation.

Step 5: Compute Undersaturated or Oversaturated Performance by Time Period

This step is executed by the computational engine or software. It computes performance in each segment and time period using either the *HCM* segment methods (if undersaturated) or through the macroscopic simulation for congested periods. The computational details for the latter are described in the *HCM*. A common computational tool for this method is the FREEVAL tool developed at North Carolina State University, but other implementations exist.

Step 6: Aggregate Performance Measures

After computing performance for each segment and time period, the performance measures are aggregated across the entire facility for each time period. The aggregation of speeds and densities is performed by calculating a lane-mile weighted average as given in Eq. (5.65) and Eq. (5.66).

The facility space mean speed (S_{Fac}) in time interval p is calculated for the number of segments (NS) as a function of the segment flow (SF), length (L), and speed (S) for each segment i.

$$\text{Flow(F)} = \text{speed(S)} \times \text{density(D)}:$$

$$S_{Fac}(NS,p) = \frac{\sum_{i=1}^{NS} F(i,p) \times L(i)}{\sum_{i=1}^{NS} F(i,p) \times \dfrac{L(i)}{S(i,p)}} \tag{5.65}$$

The average facility density (D_{Fac}) in time interval p is calculated from the individual segment densities $K(i,p)$ for segment i in interval p, the segment length (L), and the number of lanes (N) of each segment in that time interval:

$$D_{Fac}(NS,p) = \frac{\sum_{i=1}^{NS} D(i,p) \times L(i)}{\sum_{i=1}^{NS} L(i) \times N(i,p)} \tag{5.66}$$

Step 7: Validate against Field Data and Return to Step 3 As Needed

This step corresponds to the validation of facility and segment performance against field data. Depending on whether or not an acceptable match is obtained, the analyst may return to earlier steps to calibrate specific segment capacities (step 3), or even to revise global inputs (step 2) if it becomes evident that an incorrect assumption was made.

The question of what performance measures to emphasize in the calibration/validation process, and what calibration thresholds (percent difference from field data) are acceptable, is governed by agency practice, and not prescribed by the *HCM*. The most common performance measures for calibration and validation are:

- *Facility travel time* across the facility in time period
- *Maximum queue length* in the most congested time interval
- *Start and end of congestion* across the study period
- *Size and extent of the congested regime*

While the first three measures are very quantitative, the last one is a visual comparison of the speed contours across the entire time-space analysis domain. These speed contours can be color-coded to show congested

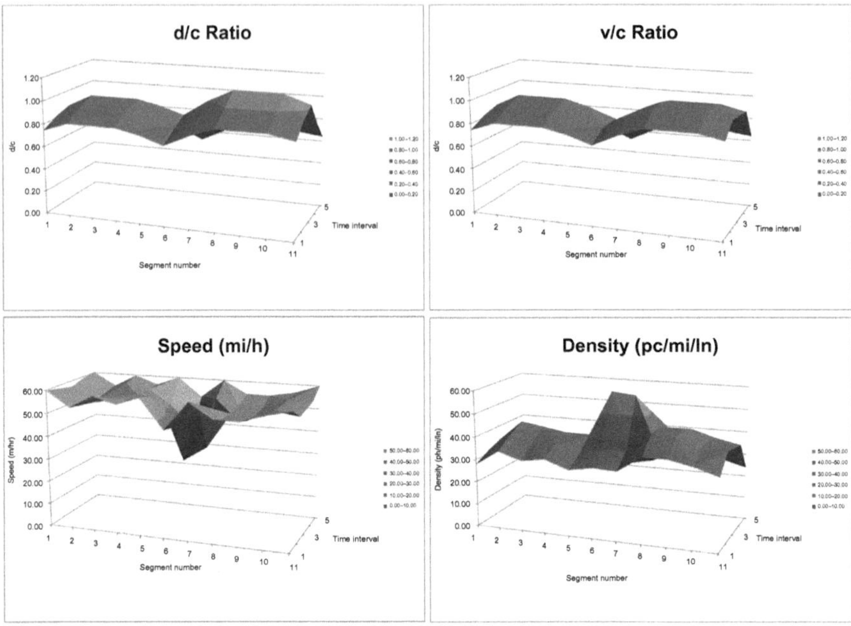

Figure 5.29 Various contour plots from a freeway facility analysis.

periods in red and free-flow speeds in green. If scaled consistently, comparing two sets of speed contours can be a very powerful way to compare to congested freeway facilities, and confirm that the methodology adequately captures the congestion patterns. Contours of other performance measures are equally useful. Figure 5.29 shows four example contours for d/c ratio, v/c ratio, speed, and density across all segments and time periods of a freeway facility.

Step 8: Estimate LOS and Report Performance Measures

As a final step, the analyst estimates the LOS for each time interval using the lane-mile weighted average density and the thresholds in Table 5.34.

Note that in Table 5.34, LOS F includes the condition of "any component d/c ratio $>$ 1.0." In other words, if any segment has LOS = F, then the entire facility will result in LOS = F. This is intentional, as the average density tends to decrease for longer facilities, especially if there are uncongested segments on either end. By flagging any component LOS = F as facility LOS = F, the methodology avoids the potential for "hiding" a bottleneck in a long facility. LOS is not defined across multiple time periods.

Table 5.34 LOS threshold for freeway facilities

LOS	Facility density (passenger cars/mi per lane)
A	≤ 11
B	$>11-18$
C	$>18-26$
D	$>26-35$
E	$>35-45$
F	>45 or any component segment d/c ratio > 1.00

5.3.6 Advanced Freeway Analysis

In addition to the core freeway facility analysis described in Section 5.3.5, several advanced analysis extensions are available that provide additional detail if needed. The three most common and most powerful analysis extensions are managed lane analysis, freeway reliability analysis, and evaluation of active traffic management strategies. These three methodological extensions to the freeway facility analysis are introduced in the following, with full details available in the *HCM*.

Managed Lanes

Managed lanes is a term referring to travel lanes (typically on freeways) that are restricted for use to only a subset of vehicles on the facility. Managed lanes include high-occupancy vehicle (HOV) lanes that are restricted to cars with at least a minimum number of passengers, and tolled facilities that require users to pay a fee to use the facility. Combination facilities exist in the form of HOT lanes (high occupancy or toll), where HOV vehicles are free to use the facility, while single occupancy drivers can purchase access. An example of a managed lane system with a HOT lane and dynamic congestion pricing system used in Minnesota is shown in Figure 5.30.

Much of the science and understanding of managed lanes involves their geometric design and their economic aspects, including questions of setting the optimum price to maximize use of the facility, while preventing breakdown and congested conditions (after all, most drivers are paying for access to escape the congestion on the general purpose lane freeway alternatives). The design and economic characteristics of managed lanes are beyond the scope of this part, but are discussed in national resources (FHWA, 2005b; FHWA, 2006a; Liu et al., 2012)

Figure 5.30 Example of congestion pricing in Minnesota. *Source: FHWA, 2008.*

Managed lanes have noteworthy operational characteristics for consideration, and a new managed lane method is available in the most recent *Highway Capacity Manual* (TRB, 2015) based on U.S. research. The unique characteristics of managed lane traffic operations include:

- Different composition of the traffic stream, which generally does not include any trucks and heavy vehicles, and which may be more homogeneous than the general-purpose lanes, being made up of mostly commuter drivers and those willing to pay for access.
- Different roadway design features, including facilities with only a single lane per direction. As opposed to most standard freeway facilities, this can result in the inability to pass within the managed lane, exacerbating the effect of any slow-moving vehicle on the facility. As a result, for example, it was found that the speed-flow relationship on these facilities "dips" more quickly, with speeds reducing from free-flow speed even at relatively low flow rates.
- Potential for interaction with the adjacent general-purpose lanes, including friction impacts for managed lanes that are separated from the general-purpose lanes only by paint or a small buffer. For these facilities, research found that the operations in the managed lanes can deteriorate significantly if the adjacent general-purpose lanes start to be congested.

- Turbulence at access points to and from the managed lanes, as well as the beginning and end points of the facility. Like most freeway on-ramps and off-ramps, access points to and from the managed lanes can result in lane-change and weaving turbulence that can deteriorate the operating conditions on the managed lanes and/or the adjacent general-purpose lanes.

Methodology Overview

A managed lane analysis in the *HCM* generally mirrors the analysis process for a standard freeway facility analysis as discussed previously. In the facility definition and segmentation, managed lane segments are treated as a parallel *lane group* to the general-purpose lanes, similar to the lane group concept used for analyzing signalized intersections. In the analysis, the two lane groups are generally analyzed separately, but have to match in overall length and number of segments. However, in some cases there will be interaction between the two lane groups, either in the form of adjacent friction impacts or access points at which traffic enters or leaves the managed lanes. But the general speed, flow, density, and LOS concepts for managed lanes are consistent with that of a standard freeway facility. For additional details, the reader is referred to the description of the managed lane methodology in the *HCM* or other sources (Neudorff et al., 2003; Perez et al., 2012; TRB, 2015).

Reliability

Most freeway analysis methods in this part have focused on a single-day analysis of traffic operations. The traditional *HCM* analysis focus is on the peak 15-min conditions on a single segment, which is expanded in the freeway facility analysis to look at the peak period performance over multiple hours and segments. However, as discussed in Section 5.1.3, traffic demand varies from day to day, and transportation facilities are further subject to various nonrecurring sources of congestion, including weather, incidents, and work zones. The *HCM* offers a freeway reliability analysis methodology that takes the base facility and applies it to a whole-year context (or another reliability reporting period, RRP).

The concept underlying a freeway reliability is illustrated in Figure 5.31. The x-axis and y-axis of the figure shows the typical dimensions of space (multiple segments) and time (multiple time periods), respectively. This traditional freeway facility time-space domain is appropriate for a single-day analysis, and for quantifying the *average* recurring congestion on a typical day, under the assumption of generally fixed

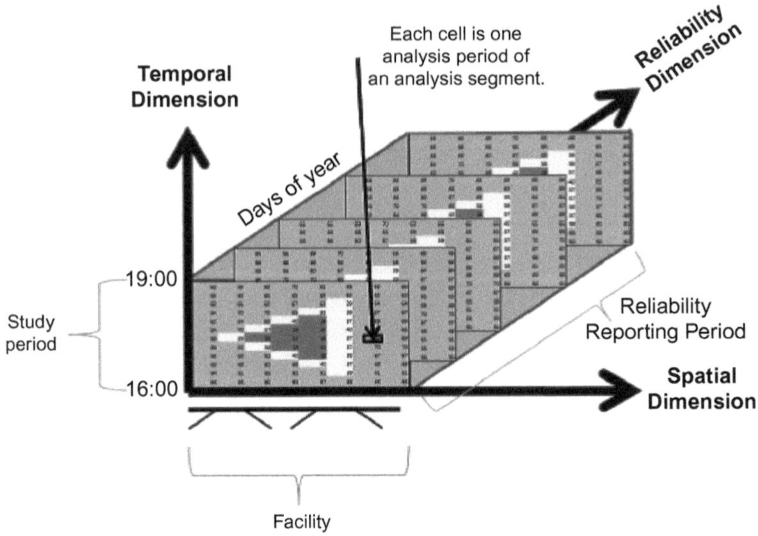

Figure 5.31 Concept of freeway reliability analysis. *Source: Zegeer et al., 2014.*

demand and capacity. But reliability analysis adds a new dimension, pertubating both demand and capacity impacts (from weather, incidents, etc.) over many analysis days. The result of this process is the generation of a reliability distribution as was shown in Figure 5.6.

Methodology Overview

A high-level overview of the *HCM* freeway reliability methodology is shown in Figure 5.32. The method starts with a base dataset that includes segment information (geometry and demands for mainline, ramps, and managed lanes as applicable), and base information about weather, incident, and work zone characteristics of the facility. The latter can be obtained from facility-specific and local data if available, or can be estimated from national defaults provided in the *HCM*.

With these inputs, the method enters a scenario generation process that generates on the order of 250 scenarios to represent the days in the reliability reporting period. The scenario generation process includes both deterministic and stochastic or random components. Demand, for example, is treated deterministically, with fixed (user-defined) variations by day of the week and month of the year. Similarly, work zone effects are user-defined based on local knowledge of the annual construction schedule. But other effects are treated stochastically, with, for example,

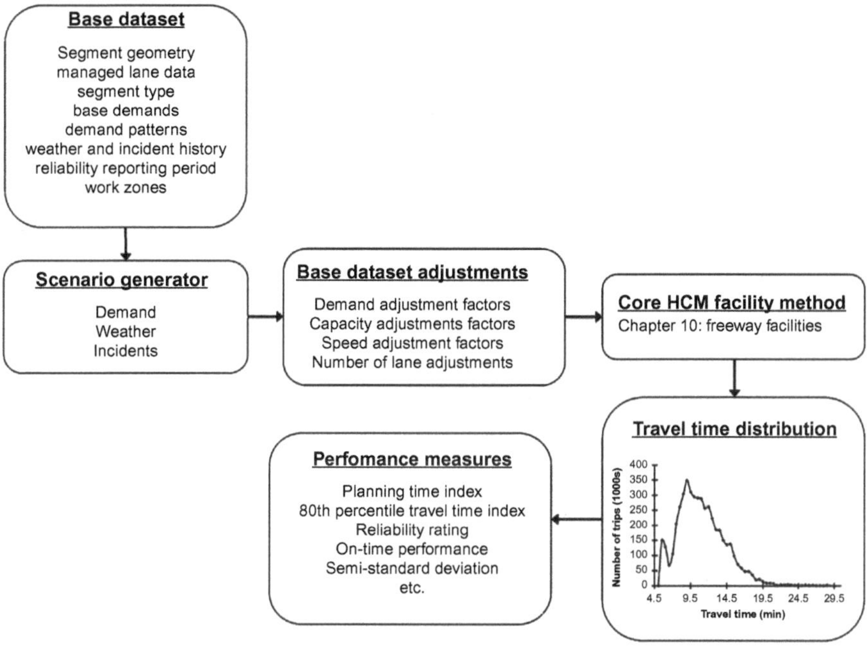

Figure 5.32 Illustration of *HCM* reliability method. *Source: TRB, 2015.*

weather and incidents being assigned randomly to the different scenarios, based on underlying probability distributions for different weather categories. This part of the scenario generation process is computationally complex and is typically implemented in software.

With the scenario listing completed, the scenario-specific demand, and any weather, incident, and work zone effects as applicable, are applied to the base scenario in the form of adjustment factors. The concept of capacity and speed adjustment factors (CAF and SAF) was introduced in Section 5.3.2 and applies equally here. In addition, demand adjustment factors (DAF) are used to inflate or reduce the base model traffic demand levels.

The full scenario set, each with now built-in reliability effects and adjustments, is then batch-processed in the reliability analysis engine, by sequentially processing each scenario in the core facility methodology. Again, a software implementation is needed to do this efficiently. The result is a set of performance measures for each scenario, which can then be aggregated to obtain the reliability travel time distribution, as well as whatever reliability performance measures the analyst is interested in.

Active Traffic Management

Active traffic management (ATM), also referred to as active traffic and demand management (ATDM), essentially refers to the application of strategies that are geared at improving the operations on a transportation facility. ATDM recognizes the reality that most states and countries have a fairly built-out freeway and transportation network, and that the construction of new (freeway) facilities is often challenging due to financial and social implications (the latter referring to impacts on cities and neighborhoods). ATDM encompasses strategies that are intended to improve the operations, throughput, and reliability of an existing facility through other means.

A large list of ATDM strategies exist today, with different ones being more or less popular depending on the state and country. A few of the more common and popular ATDM strategies are listed as follows:

- *Ramp metering:* A strategy that limits access to the freeway mainline at on-ramps through the use of a traffic signal. Ramp meters are geared at maintaining stable flow on the freeway mainline, while sacrificing (some) delay for entering vehicles. Most ramp meters are adaptive to traffic conditions through sensors on the freeway, and many have a queue override feature if on-ramp queues spill back to a point where they may severely impact the surface transportation network. Ramp meters have been shown in research to improve the capacity and reliability of the freeway mainline. Figure 5.33 shows a ramp metering installation used in the United Kingdom on the M6 motorway.

Figure 5.33 Ramp metering installation in the United Kingdom. *Source: http://www.dft.gov.uk/itstoolkit/Tools/T33.php.*

Figure 5.34 Hard running shoulder system on A9 motorway in Germany. *Source: http://www.bmvbs.de.*

- *Hard running shoulder:* A strategy in which the shoulder lane is opened to traffic during certain times of day to increase the capacity of the facility during peak loads. Hard running shoulders are often managed dynamically through overhead gantry signage, and further closely monitored to assure that the shoulder is free of any broken-down vehicles or other obstacles. While the shoulder lane typically does not have the equivalent capacity to a full freeway lane (it may be narrower, and may have reduced lateral clearance), it still represents a potentially significant increase in the facility capacity with minimal construction and right-of-way impacts.

Figure 5.34 shows a dynamic hard running shoulder example used outside of Munich, Germany on motorway A9. The system informs drivers via a side-mounted dynamic sign when the use of the shoulder is allowed. Simultaneously, overhead gantries dynamically control speed, as well as passing restrictions for heavy vehicles.

- *Bus on shoulder:* A variation to the previous strategy, in which the shoulder is opened to transit vehicles when the mainline becomes congested, this strategy has the added potential benefit of encouraging transit use and thereby potentially reducing the total demand in vehicles per hour.
- *Variable speed limit:* A strategy that dynamically changes the speed limit upstream of an incident or other congestion source. Variable speed limits, or VSL, have mostly been linked to safety benefits, in reducing

Figure 5.35 Variable speed limits used in Washington state in the United States. *Source: http://www.wsdot.wa.gov/.*

the number of rear-end incidents as drivers approach a congestion point (at previously high speeds). But other research has linked VSL to enhancing the reliability of the facility, by reducing the likelihood of breakdown. An example variable speed limit installation in the Seattle, Washington area is shown in Figure 5.35.

- *Incident management and response:* Incidents are a documented source of nonrecurring congestion, and a reduction of incident effects is therefore expected to lead to more reliable travel conditions. While the onset or occurrence of incidents can be difficult to avoid (short of safety improvements discussed elsewhere), the response to incidents, their duration, and management of travel conditions surrounding incidents is something within the purview of the operating agency.
- *Congestion pricing and managed lanes:* Managed lanes were introduced previously as a treatment that manages use and access to a transportation facility. Congestion pricing can be used in combination with tolled or HOT managed lane facilities to better manage the demand for and use of the facility. Essentially, this strategy makes travel more expensive (literally) during peak periods, which can have several results including (1) peak spreading to cheaper nonpeak periods, (2) car pooling to qualify for HOV access, (3) modal shift toward transit, (4) use of alternate route, or (5) a net reduction of trips where people either telecommute or decide for another reason that the trip is not

Figure 5.36 Traveler information system used in Frankfurt, Germany. *Photo by Bastian Schroeder.*

necessary. In either case, the net effect is a reduction of the number of vehicles on the road and thus an improvement in performance.

- *Traveler information systems*: This strategy is geared at providing information to travelers about travel conditions on the transportation network, with a particular focus on information about congestion, special events, work zones, and incidents. Traveler information systems include *pretrip* information through websites or news channels, as well as *en-route* information through smart phone apps, radio, and variable message signs. These systems are intended to inform the traveler and allow them to choose alternate routes, or even delay or cancel trips to avoid congestion. Figure 5.36 shows an overhead gantry traveler information system that dynamically provides drivers travel direction, as well as congestion and travel time information.
- *Work zone management*: With the majority of the interstate system in the United States and other countries built in the 1960s and 1970s, construction and rehabilitation activities are common and expected. Most work zones include lane closures to allow workers to complete the rehabilitation or reconstruction tasks, which often occur during off-peak or nighttime conditions, but also may have to be scheduled during daylight hours and peak commuting times. Work zone management refers to a series of ATDM strategies, including traveler

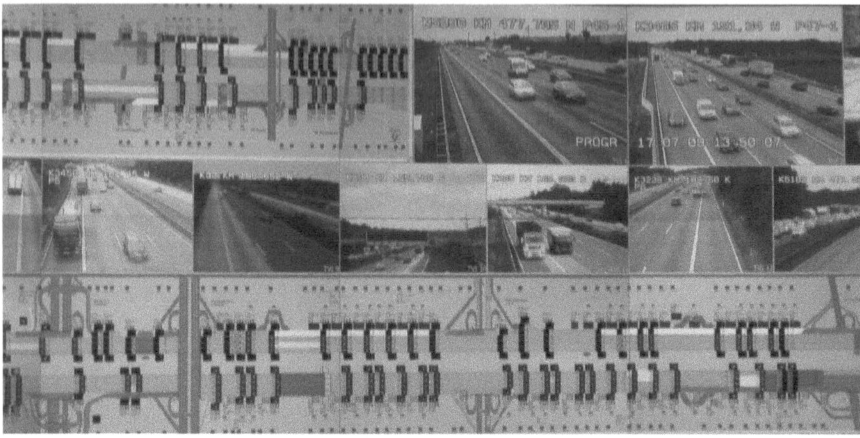

Figure 5.37 TMC video display board used in Hessen, Germany. *Photo by Bastian Schroeder.*

information systems, incident management, and use of shoulder lanes to maximize efficiency and, more importantly, safety of freeway construction activities.

ATDM strategies are typically managed from an advanced traffic management center (TMC), which receives real-time travel and congestion information for the transportation system being monitored. Often, TMCs have access to live camera feeds from pan-tilt-zoom control cameras installed along the interstate system and at interchanges. An example TMC in the state of Hessen in Germany is shown in Figure 5.37.

Methodology Overview

An ATDM analysis can take two different forms. First, it can be applied in a single-day freeway facility context, as the analyst evaluates the impacts of a particular strategy on an average day. Similar to a future scenario analysis, the analyst asks the question of effects of a particular strategy on various congestion measures, and may further choose to compare different strategies in this fashion. However, the challenge with this approach is that most ATDM strategies in their nature are geared at dynamic condition and changing traffic and congestion patterns.

Therefore, the second (and arguably more appropriate) ATDM analysis context is evaluation in a multiday or whole-year reliability analysis. Rather than assessing the ATDM impact on an average day, the reliability framework allows for consideration of variable demand patterns, and random incident and weather impacts as described previously. The reliability context is more appropriate, as the ATDM strategy is evaluated under a set of varying inputs and conditions, ultimately allowing the analyst to look at the impact on the travel time distribution, rather than average conditions.

There are other arguments in favor of evaluating ATDM strategies in a reliability context. For example, Figure 5.5 emphasized the occurrence of untypical or outlier days, and the fact that most travelers tend to remember these "extreme" days more than the average day. Some ATDM strategies, including incident management, are targeted specifically at these extreme days, and so it makes little sense to evaluate them relative to an average day.

Another important consideration is that reliability analysis enables a whole-year cost-benefit analysis of ATDM strategies. As agencies face limited budgets and need to weigh investment decisions carefully, the question of whether to invest in ramp metering (technology investment), hard-running shoulder (infrastructure investment), or incident response teams (staff and equipment investment) is a difficult one to answer. But evaluating these strategies in a whole-year reliability context allows the analyst to directly compare the long-term benefits of these strategies and weigh them against the costs.

5.4 INTERRUPTED FLOW

This section presents concepts and methods for interrupted flow facilities. Interrupted flow refers to travel on elements of the highway system, on which traffic flow is interrupted through traffic control devices at at-grade intersections. Interrupted flow thus encompasses traffic signals, modern roundabouts, stop-controlled intersections, and various variations of these intersection forms. Interrupted flow also occurs along the arterial street of a freeway interchange, which often features one or more intersections. This section presents basic concepts and methods of analysis for interrupted flow, which are in some ways similar to the uninterrupted flow material presented in the last section, but in other ways distinctly different in terms of the methods used to calculate capacity, delay, and other performance measures.

5.4.1 Concepts

Interrupted flow occurs at at-grade intersections, which are locations where two or more roads intersect and vehicles on different approaches can cross paths with other vehicles as they progress through the intersection. The location where vehicle paths cross is known as a conflict point. At a typical intersection of two streets, there are 32 conflict points to consider, while a roundabout can have as little as 8 conflict points for a four-legged intersection. This is illustrated in Figure 5.38.

The purpose of *traffic control devices* is to control the interaction of conflicting vehicle streams at these conflict points, by introducing rules and a hierarchy of movements that govern the interaction of these traffic streams.

Traffic control devices are used to reduce the potential for collisions by assigning the right of way to proceed through the intersection to specific movements and in specific order, and in some cases significantly reducing the number of conflict points. Traffic control devices primarily include traffic signals, yield signs (including those used at entry points to modern roundabouts), and stop signs, but can also include various forms of flashing beacons to alert drivers, pedestrians, or cyclists to the intersection and potential conflicts.

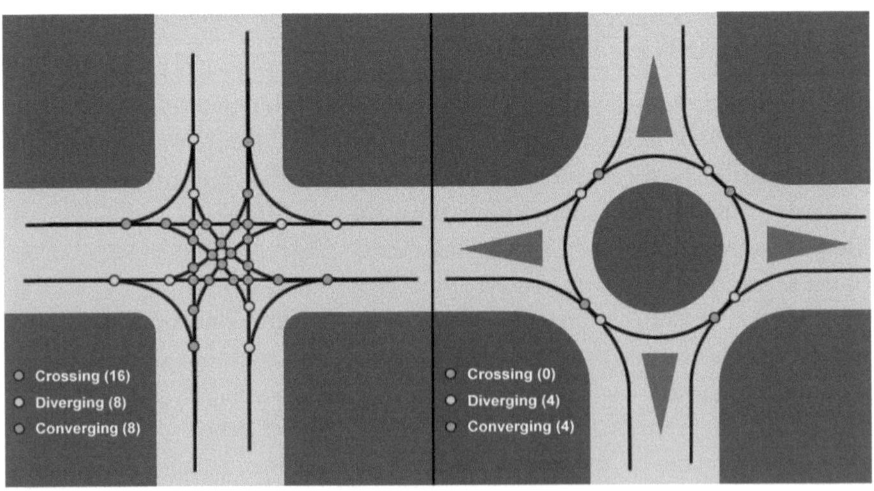

Figure 5.38 Conflict points at signalized intersection and modern roundabout. *Source: http://safety.fhwa.dot.gov/intersection/roundabouts/.*

Some countries further have completely *unsigned* intersections, where the interaction and priority of movements is governed by other rules established in the applicable driving manuals or motor vehicle codes. Examples here include the *right-before-left* rule, which gives priority to the movement to the right. If all approaches of an intersection have vehicles, drivers often communicate through eye contact and hand signals to decide who goes first. This, however, is rare, as many right-before-left intersections are used at very low-volume junctions within residential areas or neighborhoods.

There are even some examples of completely *uncontrolled* intersections, in which case there are no signs, traffic lights, or other traffic control devices, and drivers are left to figure out priority on their own. A prominent example is found in the town of Drachten in the Netherlands, which prominently removed all of the traffic signals in the town of 50,000 or so people, leaving some intersections entirely uncontrolled. Other towns across the world have since copied this approach, first introduced by Dutch traffic engineer Hans Monderman.[2]

But independent of the type of traffic control used, interrupted flow facilities principally differ from uninterrupted flow in that the hourly capacity of each lane is divided up in time, and only a portion of the theoretical hourly capacity remains for the approach under study. So while the capacity of an uninterrupted flow freeway segment is, say, 2000 vehicles/h per lane, the capacity of a traffic signal is the theoretical hourly rate (say, 1800 vehicles/h per lane) multiplied by the fraction of the hour that is actually assigned to that movement. A large focus in interrupted flow analysis is to decide how much time (capacity) to allocate to which movement to safely and efficiently operate the intersection.

Types of Traffic Control at Intersections

Some examples of types of traffic control are shown in the following intersections. The most common forms include two-way stop-controlled intersections (TWSC, Figure 5.39), all-way stop-controlled intersections (AWSC, Figure 5.40), yield-controlled intersections, including modern roundabouts (Figure 5.41), and finally signalized intersections (Figure 5.42).

[2] http://archive.wired.com/wired/archive/12.12/traffic.html

Figure 5.39 Two-way stop-controlled intersection (T-intersection). *Photo by Daniel Findley.*

Figure 5.40 All-way stop-controlled intersection. *Photo by Daniel Findley.*

These intersections are commonly named for the type of traffic control device used to control the interaction, such as a stop sign, a yield sign, or a traffic signal. While the first two are static signs that convey a well-established meaning to the driver, the last is a piece of computer hardware that uses logic to establish, control, and time the sequence of movements. An example of a typical signal controller used in the United States (in this case a TS-2 controller) is shown in Figure 5.43.

Signalized Intersections

Traffic signals are placed at intersections to control the flow of conflicting movements by alternating which of the movements gets the right of way.

Figure 5.41 Yield-controlled intersection (modern roundabout). *Photo by skysiteaerial.com.*

Figure 5.42 Signalized intersection. *Photo by Daniel Findley.*

Figure 5.43 Signalized intersection controller. *Photo by Chris Cunningham*

Signals are most commonly installed either for operational reasons (too high delays without the traffic control device) or for safety reasons (with the goal of eliminating specific crash patterns through signalization). Signals are commonly thought of as controlling the flow of vehicular traffic, but they are also commonly used to accommodate pedestrians and cyclists, or to control the interaction between vehicular traffic and various forms of transit (bus, light rail, heavy rail, etc.).

In the United States, the *Manual on Uniform Traffic Control Devices* (MUTCD) (FHWA, 2009) provides warrants to evaluate the viability of a signal at an intersection, which are evaluated through an engineering study and used to determine if a signal is appropriate at a given location. The MUTCD and its warrants are discussed further in Section 5.5.1. Suppose the decision has already been made to install a traffic signal at an intersection. The question now becomes, "What kind of traffic signal is needed at this intersection?"

Traffic signals (or controllers) are generally categorized as one of three types:

1. *Pretimed*: All operations of the signal remain the same from cycle to cycle and throughout the day or specific times during a day (like am peak or pm peak). This is the way the first signal controllers operated, and can imply some inefficiencies as the traffic signal does not have the ability to react to fluctuations in traffic demand. However, pretimed signals are still quite common and can be very useful in, for example, downtown grid systems, to allow progression of movements in multiple directions.
2. *Actuated-coordinated*: In this scheme, referred to sometimes as *semiactuated*, it is typical that the mainline movement is allocated a specific green time (as well as starting and end points) to facilitate progression of platoons of vehicles along the arterial. In other words, the green "windows" at successive signals are timed such that traffic leaving an upstream intersection in green is likely to be able to proceed through one or multiple downstream intersections in green without stopping. In an actuated-coordinated signal, the side street traffic triggers the signal through various traffic detection technologies to allocate some green time when vehicles are waiting on the side street to proceed through the intersection. In the absence of a "call" on the side-street detectors, that movement may not be given any green time, and unused time reverts to the major movement.
3. *Fully-actuated*: In this scheme, the major street and the side street traffic both trigger the signal to request green time (between a minimum

and a maximum amount) to proceed through the intersection. Fully actuated signals are common in isolated locations, where there are not adjacent signals that would be coordinated with one another. Often agencies also run fully actuated control during nighttime or extremely low-volume conditions to provide fast service to any vehicle wanting to enter the intersection (as opposed to having to wait for a pretimed or coordinated phase to elapse).

In many cases, the signal controller is set to change timing patterns throughout the day as demand dictates. Thus, there may exist one or more "timing plans" such as an am peak hour timing plan, a pm peak hour timing plan, an off-peak weekday timing plan, and a weekend timing plan. There may be timing plans for weekend traffic near a major shopping mall, or special event plans for sporting/concert events for a street network adjacent to an arena. The point here is that signal controllers can respond to a wide variety of traffic patterns on highways and streets. An example of traffic pattern changing over the course of a day, and resulting in alternating timing plans is shown in Figure 5.44. In the example, a total of four timing plans are shown: 1 = free-running, 2 = low-volume off peak, 3 = am peak, and 4 = pm peak plans.

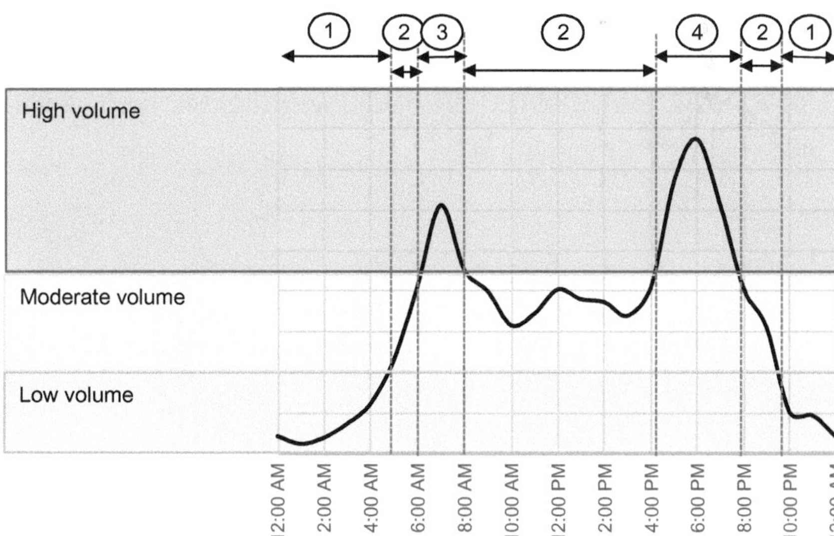

Figure 5.44 Timing plan changes based on traffic patterns.

Typical signalized intersections can provide up to eight protected *movements* that can be shown in various combinations. These phases include four through movements and four left turns, with the four right turns typically being processed concurrently with the adjacent through traffic. In addition, pedestrians are included to be shown in the movement sequence, with bicycles typically being served either as a vehicle (if on street) or as a pedestrian (if on sidewalk).

The 12 vehicular movements (4 through, 4 left, and 4 right) are assigned to different *phases* within the signal control logic. Each phase is assigned a specific time (typically with a minimum and maximum limit for actuated control), and a specific placement in the sequence of phases shown. A full sequence of all phases at an intersection is referred to as a *cycle*. The *cycle length* refers to the total time (in seconds) allocated to complete one sequence of all phases, in which each phase is assigned a *split* time.

While a typical signalized intersections can have eight protected movements, with each allocated to a separate phase, only six movement combinations can take place in a given cycle. Often, an actuated controller provides different variations of this sequence, and these variations are triggered by the demands of traffic on each approach.

Consider the conceptual diagram provided in Figure 5.45. It shows a typical timing sequence separating movements for the east–west direction (left) and north–south direction (right). The sequence first processes (protected) left turns, followed by the through movements. If moving along the center path in the figure, the sequence in a cycle may consist of only four phases: east–west left turns, east–west through, north–south left turns, and north–south through.

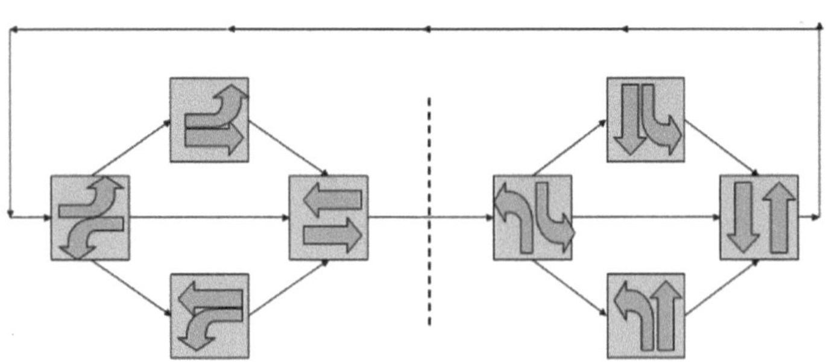

Figure 5.45 Conceptual understanding of an 8-phase intersection.

However, the figure also shows options to, for example, allow the west-to-north left turn (also called westbound left) to continue longer than the eastbound left, presumably due to higher demand levels. If the eastbound left was higher than the westbound left, the bottom sequence may be used, allowing the eastbound through to begin processing traffic after all westbound left vehicles have been processed. The same options can exist for the north–south movements. The example in Figure 5.45 is, of course, just one example for a timing sequence, and many other combinations exist to optimize operations depending on intersection geometry, volume levels, and other considerations.

Signal Terminology

Traffic engineers, signal technicians, and controller designers all need to talk the same language. As such, signalized intersection control uses very specific terminology, as given in the following list and used to describe an intersection movement and timing diagram such as the one shown in Figure 5.46. The example shows a four-legged intersection with protected and *leading left turns* on the east–west mainline. The side street is

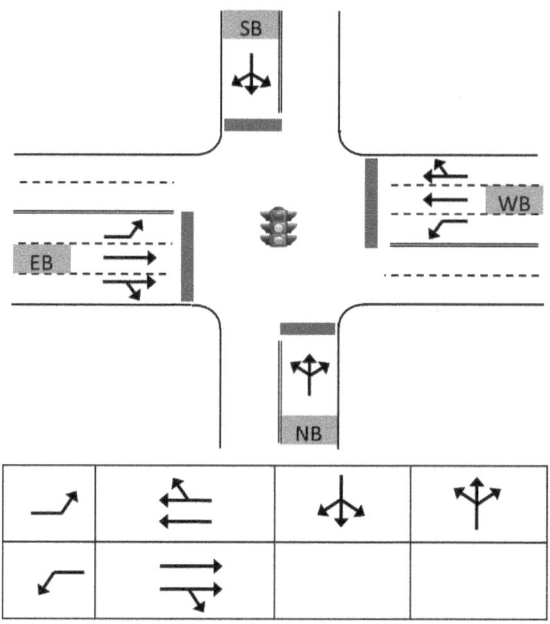

Figure 5.46 Simple intersection and phasing plan.

shown with what is known as *split phasing*, which means that the side street is split between giving green to first the southbound movement and then the northbound movement.

Interval: an amount of time allocated within a phase where signal indications do not change (e.g., green, yellow, red, green arrow, yellow arrow, pedestrian indicators as well).

Phase split (ϕ): an amount of time within the cycle where one or more traffic movements are allowed to move through the intersection. (*Note:* You could have a pedestrian phase where no vehicular movements were allowed.) It is typically the sum of the green, yellow and red intervals associated with a particular movement.

Cycle (C): the total amount of time of all phases that do not overlap whereby one complete sequence of signal indications is achieved.

Lane group: One or more lanes on an approach that operate together under the same phase.

Saturation flow (s): The maximum number of vehicles that can move past the stop bar in one lane group per unit of time, usually per hour (veh/h).

Lane group capacity: The maximum vehicular flow (veh/h) that can be accommodated by a particular lane group under given signal timings.

Volume-to-capacity ratio, v/c (X): The ratio of lane group volume to lane group capacity.

Lost time: Wasted time, when no vehicles are moving through the intersection. Lost time per phase is l; lost time per cycle is L.

Effective green (g): Usable phase time, or total phase time minus lost time per phase.

Critical lane group: Of all the lane groups moving during a phase, this one has the highest volume-to-saturation flow (v/s) ratio.

Overlap: A movement allowed during more than one phase.

Permissive movement: A movement allowed only after opposing movements or pedestrians have cleared, indicated by a green ball signal.

Protected movement: A movement that has the right of way. Protected turns are indicated with green arrow signals.

Protected/permissive movement: A movement made under protected and permissive conditions during different times of the signal cycle.

Leading sequence: When the green arrow appears before the green ball for a particular approach.

Lagging sequence: When the green ball appears before the green arrow for a particular approach.

Roundabouts and Unsignalized Control

Signalized intersections represent just one form of intersection control. While many may think of signals as the "standard" form of intersection, the reality is that there are many more unsignalized intersections across most countries. Just think of the multitude of neighborhood intersections that, in most cases, do not have traffic signals. Similarly, many low-volume intersections can operate just fine under all-way stop-control (AWSC), a common treatment in the United States, or even no formal control in the form of signs or signals, as is common in many European countries (e.g., Germany's "right before left" rule). Similarly, many junctions of major roadways and minor roads can be stop or yield controlled on the minor approaches, a strategy that works up to certain volume levels where the delays on side streets become too large.

A special form of unsignalized intersection is the modern roundabout, which will be discussed in detail in a later section. The roundabout has become a very popular alternative to signalized intersection in many countries (e.g., France, which has more than 20,000 roundabouts). The popularity of roundabouts is largely due to their improved safety performance over signals and two-way stop-controlled intersections, as well as their ability to adapt well to ever-changing traffic patterns and volume levels throughout the day without the need to retime any traffic signals. Roundabouts also have an aesthetic appeal, offering landscaping opportunities, and can provide a safe environment for pedestrians and cyclists largely due to low operating speeds of vehicular traffic. In terms of safety performance, a recent comprehensive study in the United States showed that for a conversion from signalized intersections, roundabouts result in a 48% reduction in total crashes, and, more importantly, a 78% reduction in injury crashes for rural and 60% for urban roundabouts. Similar safety benefits have been documented for roundabouts converted from TWSC intersections. The safety benefits of roundabouts are largely attributed to the slower speeds at which collisions occur (TRB, 2007), as well as the reduced number of conflict points relative to a signalized intersection, as was shown in Figure 5.38.

Figure 5.47 shows typical design features of modern roundabouts, including yield signs at entry, counterclockwise circulation, geometric and physical features that result in slow vehicle speeds, and a lane assignment that eliminates the need for lane changes in the circle and after exit for multilane roundabouts.

Roundabouts can have application even for elevated volume levels, but may require multiple lanes to adequately process the traffic. Figure 5.48

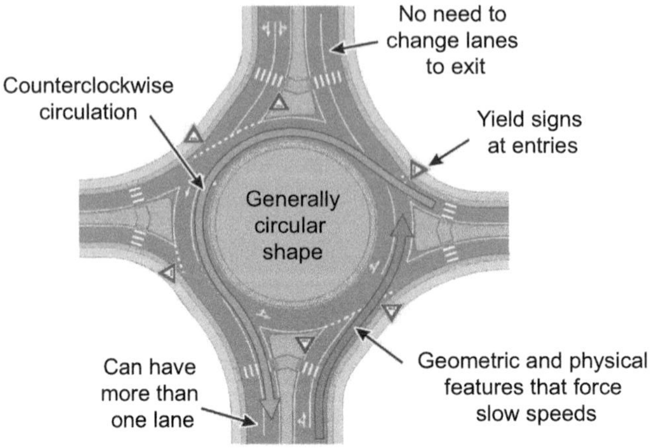

Figure 5.47 Roundabout design features. *Source: NCHRP Report 672.*

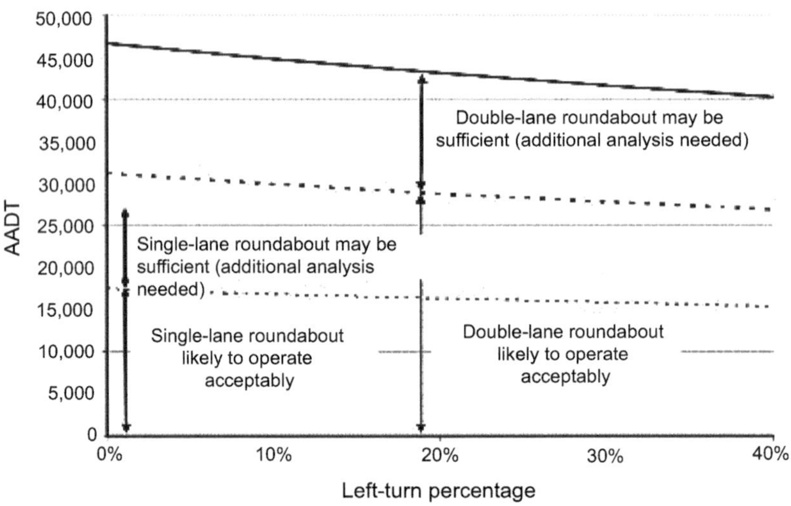

Figure 5.48 Planning-level assessment of roundabouts *Source: Rodegerdts et al., 2010. http://safety.fhwa.dot.gov/intersection/roundabouts/fhwasa10006/ppt/.*

shows the feasibility of a modern roundabout as a function of the average annual daily traffic (AADT) summed across both intersecting roadways, and the percentage of left turns. The various regimes identify where single-lane and double-lane roundabouts are likely to operate acceptably, and where additional analysis is needed to confirm operations of roundabouts are appropriate.

As a result of the safety benefits and operational performance of roundabouts, many states have adopted policies to explicitly consider roundabouts as an alternative for new intersection designs, and even some that have implemented so-called "roundabouts first" policies. These require analysis of roundabouts as the *preferred alternative* over other intersection forms, and only allow construction of, say, a signalized intersection if a roundabout was shown to be infeasible for the location under study.

5.4.2 Critical Movement Analysis

An operational analysis of intersections can be quite complicated, especially in the case of a signalized intersection with actuated timings. In practice, systems of intersections are typically analyzed using software, to automate complex (and at time iterative) computations. Fortunately, a straightforward planning-level methodology exists for the assessment of intersection capacity, which can typically be completed by hand in as little as 15 min as a true "back of the envelope" calculation. The *critical movement analysis* (CMA) was formalized by TRB Circular 212: Interim Materials on Highway Capacity in 1980 (TRB, 1980), but dates back even before that time. Today, CMA is sometimes ignored as a "dated" and "simplistic" approach, which is unfortunate as the straightforward and simple approach is invaluable in developing an understanding of intersection capacity relationships, and in performing quick estimation performance checks of intersection operations.

The basic principle underlying the CMA is that two vehicles cannot be in the same place at the same time, and that within an hour there is thus a limit to the number of vehicles that can be processed through any one point. The discussion in Section 5.2.1 on microscopic characteristics of traffic flow has already introduced the concept of minimum headway between successive vehicles (say, 2 s per vehicle) and how this translates to a maximum flow rate of 1800 vehicles per hour of continuous green in each lane. CMA adapts this concept, though further restraining the maximum hourly flow rate to account for *lost time*, which is time needed to safely transition from one vehicle stream to another. Accounting for lost time, a reasonable hourly "point capacity per lane" is 1400 veh/h, which gives a conservative estimate of intersection operations with a lost time of roughly 20% over the maximum flow rate. In reality, the lost time depends on the number of critical phases, and intersections with fewer phases will also result in lower lost time (and thus higher capacities). But the estimate of 1400 veh/h is a reasonable initial estimate for a planning-level, quick estimation of intersection performance.

In application of the critical movement analysis method, the basic task then involves identifying conflicting movements, estimating the associated traffic streams, and comparing the resulting flow rates across each conflict point to the capacity. The availability of multiple lanes increases the capacity, as well as allows demand to be distributed across multiple lanes.

The only required inputs for applying the CMA are turning movement volumes and approach geometry. Potential applications of the CMA include:

- Planning-level or quick estimation capacity analysis
- Evaluating adequacy of approach geometry (number of lanes)
- Development of initial signal phasing plans and green times
- Conducting reasonableness checks on software calculations
- Performing quick and efficient alternative evaluation

Critical Movement Analysis Step-by Step

The following basic steps are used in a critical movement analysis of an intersection (TRB, 1980)[3].

- *Step 1—Identify movements and hourly volumes per lane:* The analyst reviews the approach geometry (number of lanes) for each approach to the intersection, and needs to obtain the traffic volumes in vehicles per hour (adjusted to peak 15 min as desired). The volumes are then converted to a per-lane basis for analysis for all movements.
- *Step 2—Assign movements to phase sequence:* A traffic signal allows certain phases to run concurrently, while others are run sequentially. The analyst needs to have a sense of what phasing scheme is used or make some assumptions. Phasing scheme options are discussion in more detail in Section 5.5. Consideration is given to the treatments of left turns as either protected or permissive.
- *Step 3—Determine critical volume pair in each interval:* For each conflicting movement pair in each interval, the analyst calculates the sum of volumes on a per-lane basis. The highest movement pair (volume across a point) is identified as the critical volume pair.
- *Step 4—Sum critical volumes for phase sequence:* The critical volumes for all intervals are summed for the overall phase sequence.

[3] http://ops.fhwa.dot.gov/publications/fhwahop08024/chapter3.htm#3.3

Table 5.35 CMA thresholds for v/c ratios

v/c Ratio	Assessment	Implication
<0.85	Under capacity	Expect low-to-medium delays and minimal queuing
0.85–0.95	Near capacity	May expect high delays, but no excessive queuing
0.95–1.0	At capacity	High delays expected and some queues won't clear
>1.0	Over capacity	Very high delays and excessive queueing expected

- *Step 5—Identify maximum critical volume:* This is the theoretical total volume past a point. Earlier, it was recommended to use a conservative volume of 1400 veh/h, which is based on the 1985 *HCM*. This value considers varying effects of lost time, heavy vehicles, grades, unfamiliar drivers, turn geometry, and pedestrians. The 1400 veh/h value gives a conservative estimate of capacity, which is deemed appropriate for planning-level analyses. But higher values can be adopted by agencies as desired. For more detailed analysis, the lost time and maximum volume (capacity) can be estimated using the methodology described in the next section.
- *Step 6—Calculate critical v/c ratio and determine intersection status:* The critical v/c or volume-to-capacity ratio is calculated by dividing the sum of critical volumes from step 4 by the maximum critical volume from step 5. Thresholds for interpreting the resulting v/c ratio are given in Table 5.35.

In application of the CMA, it is noted that the steps just listed are readily combined as needed, but were specifically separated out here to fully describe the method. It is also noted that the *HCM* has adopted a slightly different quick estimation method that modified the CMA approach described here to work more explicitly within the context of the *HCM* operational context. The reader is encouraged to refer to that method as needed, but it is believed that the basic CMA presented here provides a straightforward and more easily understood approach.

Example 5.8

A four-legged intersection has the lane geometry and hourly traffic volumes shown in the following figure and table. It can be assumed that all left turns are protected. Determine if this intersection is likely to operate below or above capacity?

(*Continued*)

Example 5.8—(Continued)

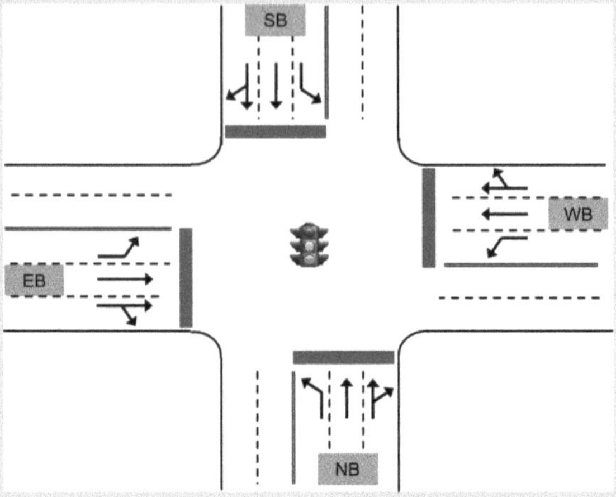

Approach	Left	Through	Right
EB	130	585	275
NB	95	370	64
WB	200	400	96
SB	185	400	130

Solution

Step 1: Lane configurations and volumes are given in the table.

Step 2: It is assumed that the intersection first processes east–west traffic, followed by north–south traffic. Because left turns are protected, each left turn (LT) conflicts with the opposing through (TH) and right-turn (RT) traffic in the CMA.

Step 3: Determine critical volume for east–west street.

EB LT + WB TH/RT = 130 + 0.5 × (400 + 96) = 378 veh/h

WB LT + EB TH/RT = 200 + 0.5 × (585 + 275) = 630 veh/h → critical!

Determine critical volume for north–south street.

NB LT + SB TH/RT = 95 + 0.5 × (370 + 64) = 312 veh/h

SB LT + NB TH/RT = 185 + 0.5 × (400 + 130) = 450 veh/h → critical!

Step 4: Sum critical volumes.

East–west critical + north–south critical = 630 + 450 = 1080 veh/h

Step 5: Assume maximum critical volume of 1400 veh/h.

Step 6: Estimate v/c ratio.

Critical v/c = 1080/1400 = 0.77

According to this, the intersection is expected to be *below capacity.*

5.4.3 Signalized Intersection Operational Analysis

The operational analysis of signalized intersections can be very complex. As was discussed earlier, some intersections are pretimed, while others (more commonly) are running in actuated or actuated-coordination operations. The challenge for the operational analysis of the more common actuated coordination signal system, is that phase durations depend on the arriving traffic volume, which triggers a detector to call or extend different phases. At the same time, the traffic volume, or more specifically the arrival patterns, at the next downstream intersection depend on the phase durations of the upstream intersection. To make this really tricky, consider that arterial streets have traffic moving in two directions, essentially reversing what is upstream and downstream when moving in the opposite direction. In other words, signal phase durations and thereby signal capacity and operations depend on arrivals from the upstream signals, which in turn impacts how much traffic the signal discharges to that same adjacent signal in the opposite direction.

This complication of actuated controllers makes the analysis process iterative for the case of actuated-coordinated controllers, a process that is described in great detail in the *HCM*. This discussion focuses on pretimed signal control as a more straightforward and simplified analysis approach. It is true that almost all signal installations done today rely on actuated control for one or more of the approaches. This achieves the most flexibility in accommodating fluctuating traffic demand throughout the day. However, during the peak hour, traffic demand may be at high-enough volumes for the controller to receive the maximum green time for all approaches. If this happens, the controller is essentially functioning like a pretimed controller. Therefore, we can use timing calculation procedures appropriate for this kind of operation for simplified analysis and in the absence of using software for the more complicated analyses.

Finally, it is emphasized that the signal operations analysis described here is closely interrelated with the signal timing process described in Section 5.5.3, and the reader may be referred to that section for some additional concepts.

Methodology

The *HCM* signalized intersection procedure is illustrated in Figure 5.49, where it is summarized in seven steps. All steps are described in detail in the following.

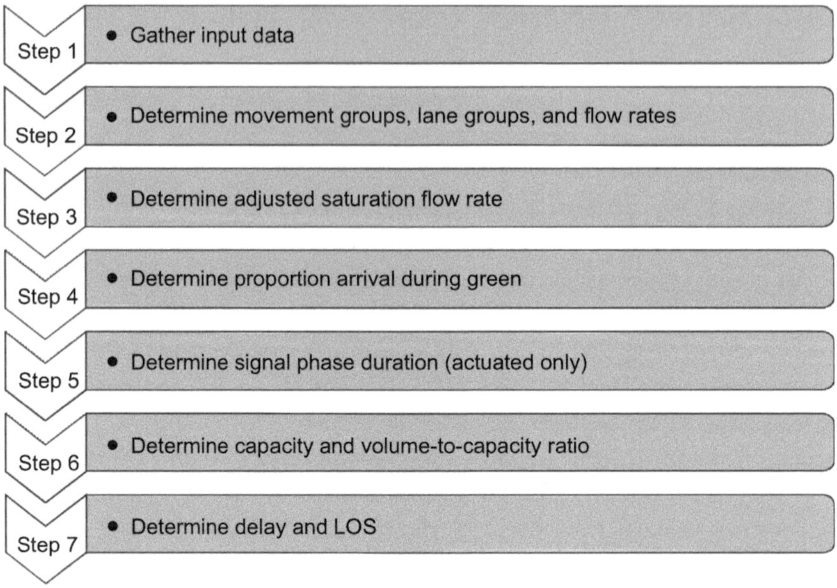

Figure 5.49 Methodology steps for signals analysis.

Step 1: Gather Input Data

The first step in the analysis is to gather input data. This includes the demand or volume data in the form of turning movement counts, including a percentage of heavy vehicles for each movement. Data needs also include geometric data about the intersection, including number of lanes, turn pocket details, lane widths, and intersection widths. Finally, the input data includes signal timing data, with many of the terms and concepts introduced in Section 5.4.1.

Step 2: Determine Movement Groups, Lane Groups, and Flow Rates

As a second step, each intersection approach is divided into movement groups and lane groups. Movement groups generally refer to movements that are progressed together on one approach during a signal phase, while lane groups refer to the smallest capacity analysis unit. Often, movement groups and lane groups are identical, for example, in the case of exclusive left-turn lanes. The difference comes into play for shared lanes.

Take, for example, an intersection approach with one exclusive through lane and one shared through-right lane. Clearly the two lanes move during the same signal through phase. But for capacity analysis purposes, the two lanes are different as the shared through/right lane may experience impedance from slower turning vehicles and pedestrians or

Table 5.36 Illustrating movements, movement groups, and lane groups

Number of lanes	Movements by lanes	Movement groups (MG)	Lane groups (LG)
1	Left, through & right:	MG 1:	LG 1:
2	Exclusive left:	MG 1:	LG 1:
	Thru. & right:	MG 2:	LG 2:
2	Left & thru:	MG 1:	LG 1:
	Thru. & right:		LG 2:
3	Exclusive left:	MG 1:	LG 1:
	Exclusive left:		
	Through:	MG 2:	LG 2:
	Through:		
	Thru. & right:		LG 3 :

Source: TRB, 2015.

cyclists when making a turn. As such, the capacity of the shared lane is likely to be less than the capacity of the exclusive (unimpeded) lane, justifying separate analyses for the two.

Table 5.36 shows some examples of lane configurations, movement groups, and lane groups. As a rule of thumb, any exclusive turn lanes are separated into their own movement groups, and any shared lanes are then separated out into their own lane group.

The total approach volume is assigned to each movement group, which typically emerges naturally from the turning movement patterns. For example, all left-turn demand is assigned to the left-turn movement group, and so forth. Estimating flow rates for lane groups (the smallest analysis unit in the *HCM*) can be a bit more challenging for multilane groups with one or more shared lanes. The analyst may use judgment to decide on the relative distribution of flows, or may turn to a detailed procedure given in the *HCM*.

Step 3: Determine Adjusted Saturation Flow Rate

Next, the analysis estimates the saturation flow rate for each lane group. The saturation flow rate is one of the most critical parameters describing lane group capacity. It was discussed and introduced in Section 5.2.1.3. as the inverse of the saturation headway. Saturation flow rate is defined as the maximum throughput in a lane per hour of continuous green (the second important factor in signal capacity is what proportion of that hour is actually available to the movement, which is estimated later).

The saturation flow rate is a function of a base saturation flow rate, and a whole series of adjustment factors accounting for lane width, heavy vehicles, lane utilization, and several other parameters, as shown in Eq. (5.67).

$$s = s_0 f_w f_{HVg} f_p f_{bb} f_a f_{LU} f_{LT} f_{RT} f_{Lpb} f_{Rpb} f_{wz} f_{ms} f_{sp} \tag{5.67}$$

where

s = adjusted saturation flow rate (veh/h per lane)
s_o = base saturation flow rate (passenger cars/h per lane)
f_w = adjustment factor for lane width
f_{HVg} = adjustment factor for heavy vehicles and grade
f_p = adjustment factor for existence of a parking lane and parking activity adjacent to lane group
f_{bb} = adjustment factor for blocking effect of local buses that stop within intersection area
f_a = adjustment factor for area type
f_{LU} = adjustment factor for lane utilization
f_{LT} = adjustment factor for left-turn vehicle presence in a lane group
f_{RT} = adjustment factor for right-turn vehicle presence in a lane group
f_{Lpb} = pedestrian adjustment factor for left-turn groups
f_{Rpb} = pedestrian–bicycle adjustment factor for right-turn groups
f_{wz} = adjustment factor for work zone presence at the intersection
f_{ms} = adjustment factor for downstream lane blockage
f_{sp} = adjustment factor for sustained spillback

In the *HCM*, the base saturation flow rate, s_O, is equal to 1900 passenger cars/h per lane for large metropolitan areas with a population of more than 250,000, and a default of 1750 passenger cars/h per lane for smaller communities. However, many agencies have performed local studies and developed their own defaults based on local driver characteristics. This practice is strongly encouraged, and details on how to conduct saturation flow (or headway) studies are described in the *ITE Manual of Transportation Engineering Studies* (Schroeder et al., 2010).

For the various adjustments to saturation flow rate, a factor of 1.0 corresponds to standard or ideal conditions, with nonstandard conditions resulting in an adjustment typically below 1.0 to reduce the saturation flow. Some of the more common saturation flow adjustments are summarized in Table 5.37.

Table 5.37 Saturation flow rate adjustments

Factor	Condition	Factor	Comments
Lane width (ft), f_W	<10.0	0.96	n/a
	≥10.0–12.9	1.00	
	>12.9	1.04	
Heavy vehicles and grade, f_{HVg}	Downhill	$\frac{100-0.79P_{HV}-2.07P_g}{100}$	P_{HV} = % heavy veh P_g = % grade
	Level or uphill	$\frac{100-0.78P_{HV}-0.31P_g}{100}$	
Parking, f_p	Parking present	$\frac{N-0.1-\frac{18\times N_m}{3600}}{N}$	N = # of lanes N_m = # parking maneuvers per hour $f_p >= 0.050$
Bus blockage, f_{bb}	Buses and bus parking present	$N-\frac{14.4N_b}{3600}/N$	N_b = Bus stopping rate per hour $f_{bb} \geq 0.050$
Area type, f_a	Central business district (CBD)	0.90	Use judgment before applying
Lane utilization, f_{LU}	Imbalanced utilization	Custom	Special applications in *HCM*
Right turns, f_{RT}	Right turn	$1/E_R$	E_R = right-turn equivalency (=1.18)
Left turns, f_{LT}	Left turn	$1/E_L$	E_L = left-turn equivalency (=1.05)
Pedestrian or bike, f_{pb}	Pedestrians and bikes present	Custom	Special application in *HCM*
Work zone, f_{WZ}	Work zone present	Custom	Special application in *HCM*
Downstream lane blocks, f_{ms}	Downstream lane closure	Custom	Special application in *HCM*
Spillback, f_{sp}	Queue spillback	Custom	Special application in *HCM*

In the table, the more common and straightforward adjustments are given directly. For some more complicated adjustments the reader is referred to the *HCM* or additional details. If these special conditions (e.g., work zones, lane closures, or spillback) are not present, the adjustment factor default for all conditions is 1.0.

Step 4: Determine Proportion Arrival on Green

The procedure next estimates the proportion of arrivals during green. A signal generally runs more efficiently with a greater percentage of arrivals on green, as vehicles can proceed through the signals (and the corridor) without stopping. On the other hand, if arrivals in green is low (and arrivals in red high) the intersection intuitively incurs more delay.

The proportion arrivals in green is estimated from a term called the *platoon ratio* (R_p) and the proportion of the cycle that is green for a movement, which is the g/C ratio (effective green to cycle length ratio).

$$P = R_p \frac{g}{C} \qquad (5.68)$$

The value for the platoon ratio can be estimated using a detailed iterative procedure for arrival flow prediction available in the *HCM*. For the purpose of this discussion, Table 5.38 provides guidance for estimating the platoon ratio as the function of the arrival type at the signal. Arrival type is a value from 1 to 6, which describes how well a signal is coordinated relative to its upstream neighbor. Arrival type 1 in this case is poor, while arrival type 6 is ideal or (near) perfect progression. Arrival type 3 refers to essentially random arrivals.

Step 5: Determine Signal Phase Duration (Actuated Only)

For an actuated controller, this step is where the phase duration would be estimated based on the arrival flow profiles and the signal timing parameters (min green, max green, extension times, etc.). As discussed earlier, the phase duration at one intersection impacts arrivals at the downstream intersection, the arrivals at the downstream intersection impact its phase duration, those phase durations in turn impact the arrivals at the first intersection for traffic in the opposite direction, which in turn impact its phase durations, and so forth. This makes the procedure iterative, and in the case of an arterial street with multiple intersections, quite complicated.

Table 5.38 Arrival types and platoon ratio

Arrival type	Typical signal spacing (ft)	Progression quality	Description of coordination patterns	Platoon ratio
1	≤ 1600	Very poor	Predominant arrivals on red, which can occur on a coordinated two-way street where the nonpeak direction does not receive good progression	0.33
2	>1600–3200	Unfavorable	A less extreme version of arrival type 1	0.67
3	>3200	Random arrivals	Essentially random arrivals at isolated signals or widely spaced coordinated signals	1.00
4	>1600–3200	Favorable	Coordinated operation on a two-way street where the subject direction receives good progression	1.33
5	≤ 1600	Highly favorable	Coordinated operation on a two-way street where the subject direction receives good progression	1.67
6	≤ 800	Exceptionally favorable	Coordinated operation on a one-way street in dense networks and central business districts	2.00

But for the purpose of a pretimed intersection, or a congested actuated signal with most phases extended to their maximum times, this step can be skipped in this discussion. The reader is, however, encouraged to refer to the *HCM* for a more detailed discussion of this method, and to use software to perform the more complicated actuated computations.

Step 6: Determine Capacity and Volume-to-Capacity Ratios

The capacity of each lane group at a signalized intersection is estimated from Eq. (5.69).

$$c = N \times s \times \frac{g}{C} \tag{5.69}$$

where

c = capacity (veh/h)
N = number of lanes in the lane group
s = saturation flow rate (veh/h per lane)
g = effective green time for group (s)
C = cycle length (s)

The equation applies to all protected movement, with the capacity estimation for permitted movements being more complicated due to gap acceptance processes. For details on the capacity of permitted movements, the reader is referred to the *HCM*.

The volume-to-capacity ratio, X, for the lane group is estimated simply by dividing the lane group flow rate by the previously calculated capacity as shown in Eq. (5.70).

$$X = \frac{v}{c} \tag{5.70}$$

where

X = volume-to-capacity ratio
v = demand flow rate (veh/h)
c = capacity (veh/h)

Step 7: Determine Delay and LOS

As the final step, performance measures for each lane group are calculated, including control delay, LOS, and queue storage ratio. The control delay for a signalized intersection is the sum of three delay terms: uniform delay, incremental delay, and initial queue delay, as shown in Eqs. (5.71) through (5.73). Only the uniform delay is discussed in this text (the incremental and initial queue delays are described in detail in the *HCM*).

$$d = d_1 + d_2 + d_3 \tag{5.71}$$

where

d = control delay (s/veh)
$d1$ = uniform delay (s/veh)
$d2$ = incremental delay (s/veh)
$d3$ = initial queue delay (s/veh)

Conceptually, the uniform delay considers the effect of the traffic signal only, without any impacts from adjacent intersections. The incremental delay adds a time-dependent delay component, which considers the effects of other vehicles, signal type, presence of upstream signal, and signal capacity effects. The initial queue delay is added for signals at which a queue is present at the beginning of a signal cycle that impacts the operations. The uniform delay is calculated as shown in Eqs. (5.72) and (5.73).

$$d_1 = PF \frac{0.5C(1-g/C)^2}{1-[min(1,X)g/C]} \tag{5.72}$$

$$PF = \frac{1-P}{1-g/C} \times \frac{1-y}{1-min(1,X)P} \times \left[1 + y\frac{1-PC/g}{1-g/C}\right] \tag{5.73}$$

where

PF = progression adjustment factor

y = flow ratio, $y = \min(1,X)g/C$

P = proportion of vehicles arriving during the green indication (decimal)

g = effective green time (s)

C = cycle length (s)

(all other variables are as previously defined)

The incremental delay and the initial queue delay require advanced computations of nonrandom arrivals, queue spillback, and oversaturated cycles, which are beyond the scope of this text. The reader is referred to the *HCM* for details on those computations.

With the lane group delays calculated, delays can be aggregated to the approach level by calculating the weighted average control delay across all lane groups for that approach. The weighting is done by the lane group flow rates. Similarly, the intersection control delay is obtained through a weighted average of the individual approach delays.

Level of service (LOS) is defined based on the average control delay at the lane group, approach, or intersection level. The thresholds for LOS are shown in Table 5.39.

5.4.4 Modern Roundabouts

The capacity of modern roundabouts is principally a function of the availability of entering vehicles to accept gaps in the conflicting circulating traffic stream. Because roundabouts (typically) do not have traffic signals

Table 5.39 LOS threshold for signalized intersections

LOS	Control delay (s/veh)
A	≤10
B	>10–20
C	>20–35
D	>35–50
E	>50–80
F	>80 or v/c ratio > 1.0

installed, the capacity emerges from the geometric characteristics of the roundabout itself, and the behavioral characteristics of the drivers. In fact, two different schools of thought exist for how roundabout capacity is derived.

One approach to roundabout capacity is to emphasize the effects of geometrics, which is the underlying principle of the British roundabout methodology that has since been adopted by several other countries. In this approach, roundabout entry capacity is described through a regression equation as a function of entry number of lanes and width, circulating lane and width, as well as specific geometric attributes that include flare angles and curvature. This approach is largely independent of driver behavioral characteristics, and as such readily allows the comparison of different geometries. However, it does not take into account the human factors element, and assumes that all drivers behave consistently given a certain roundabout geometry, regardless of their familiarity with the site or level of assertiveness.

The second approach is one anchored in empirical observations of driver behavior and based on gap acceptance theory as introduced in Section 5.2.4. It uses the concepts of critical gap and follow-up headway, which are then used to estimate the capacity. Geometric effects of roundabout geometry (e.g., number of lanes) are not explicitly used in the capacity model, but rather implicitly through their effect on the gap acceptance attributes. The behavioral roundabout capacity was probably first developed in Australia, and has since also been used in the United States and various other countries.

Regardless of the underlying methodology and theory, the roundabout capacity will be sensitive to geometry (at least number of lanes), as well as the conflicting flow rates on the circulating lanes. This last point is critical, as it means that the roundabout entry capacity at a given approach is not fixed over the course of the day, but rather varies as a function of

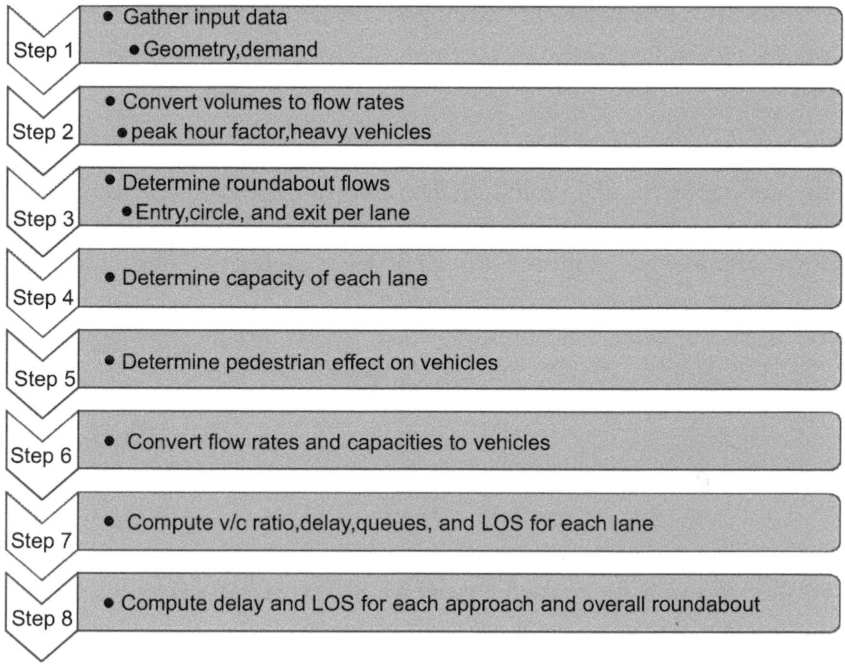

Figure 5.50 Methodology steps for roundabout procedure.

traffic volumes. During off-peak periods with low circulating flows, gaps in that conflicting flow are frequent and large, which results in high entering capacities. During peak periods, however, circulating traffic may increase, which reduces gap availability and thereby capacity. This principle of decreasing entry capacity with increasing circulating flow is common for all roundabout models, both geometry-based and behavioral. One example of the behavior of these models is presented in Figure 5.50 for the *HCM* roundabout procedure.

Methodology

The *HCM* roundabout procedure is illustrated in Figure 5.50, where it is summarized in eight steps. All steps are described in detail in the following sections.

Step 1: Gather Input Data

As with all operations procedures, the first step is to gather input data. For the roundabout procedure this involves geometric data (mostly number of lanes on all approaches and in the circle) and traffic volume data.

The latter are principally the turning movement counts (left, through, right) for all approaches, the heavy vehicle percentage, and global parameters such as the peak hour factor.

Step 2: Convert Volumes to Flow Rates

In the second step, the volumes (in units of vehicles per hour) are converted to 15-min flow rates in units of passenger cars per hour. To do this, the volumes are adjusted for peak hour factor and heavy vehicle percentage, as shown in Eqs. (5.74) and (5.75).

$$v_{i,PCE} = \frac{V_i}{PHF \times f_{HV}} \tag{5.74}$$

$$f_{HV} = \frac{1}{1 + P_T(E_T - 1)} \tag{5.75}$$

where

$v_{\text{i,PCE}}$ = demand flow rate for movement i (passenger cars/h)
V_{i} = demand volume for movement i (veh/h)
PHF = peak hour factor
f_{HV} = heavy-vehicle adjustment factor
P_{T} = proportion of demand volume that consists of heavy vehicles
E_{T} = passenger car equivalent for heavy vehicles, which is 2.0 for roundabouts

Step 3: Determine Roundabout Flows

As a third step, the turning movement flow rates (left, through, right) are used to estimate the circulating and exiting flow rates at the various roundabout approaches. For example, to estimate the capacity of the south (northbound) roundabout entry in Figure 5.51, the conflicting circulating flow, $v_{\text{C,NB}}$ is the sum of several flows (black arrows), including:

- The through, left-turn, and U-turn movement from the west approach
- The left turn and U-turn from the north approach
- The U-turn from the east approach

Similarly, the exiting flows at the south approach is the sum of three flows (gray arrows):

- The right-turn movement from the west approach
- The through movement from the north approach
- The left from the east approach
- The U-turn from the south approach

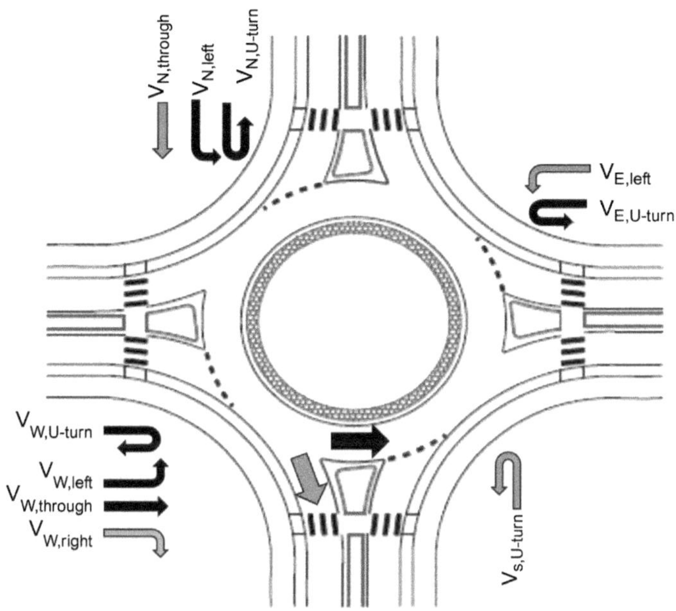

Figure 5.51 Illustration of roundabout flows.

Table 5.40 Lane utilization for multilane roundabouts

Case	Assumed lane assignment	Left lane	Right lane
1	L, TR	$v_U + v_L$	$v_T + v_R$
2	LT, R	$v_U + v_L + v_T$	v_R
3	LT, TR	$(\%LL)v_e$	$(\%RL)v_e$
4	L, LTR	$(\%LL)v_e$	$(\%RL)v_e$
5	LTR, R	$(\%LL)v_e$	$(\%RL)v_e$

For multilane roundabouts, one additional step is needed to distribute the total approach volume across the two lanes. This is a function of the lane assignment (striping of roundabout entry), and the relative turning movement flow rates. The *HCM* distinguishes five cases, summarized in Table 5.40. The volumes are for each respective approach. For approaches with multiple left or right-turn lanes, additional guidance for estimating the percentage of traffic in the left lane (%LL) and right lane (%RL) is found in the *HCM* or can be estimated based on analyst judgment.

Table 5.41 Intercept and slope parameters for roundabout capacity models

	One-lane entry roundabouts		Two-lane entry roundabouts		
	1 × 1	**1 × 2**	**2 × 1**	**2 × 2 (right)**	**2 × 2 (left)**
Number of entering lanes	1	1	2	2	2
Number of circulating lanes	1	2	1	2	2
Intercept (a)	1380	1420	1420	1420	1350
Slope (b)	−0.00102	−0.00085	−0.00091	−0.00085	−0.00092

Step 4: Determine Capacity of Each Lane

The capacity of each roundabout entry is then determined as a function of the conflicting circulating flow using Eq. (5.76). The equation uses an intercept term, *a*, and a slope parameter, *b*, which are shown in Table 5.41 for five different cases:

1. Single entry lane with single circulating lane (1 × 1)
2. Single entry lane with two circulating lanes (1 × 2)
3. Two entry lanes with single circulating lane (2 × 1)
4. Two entry lanes with two circulating lanes - right lane (2 × 2)
5. Two entry lanes with two circulating lanes - left lane (2 × 2)

For the last geometry, the table shows different terms for the right and left entering lanes. For roundabouts with right-turn bypass lanes that are under yield control, the one-lane entry equations are used depending on whether the bypass lane has one or two conflicting lanes (1 × 1 and 1 × 2 models, respectively). Two-lane bypass lanes are uncommon at roundabouts. Bypass lanes with continuous lane adds are assumed to be free-flowing.

$$c_{entering} = a \times e^{-b \times v_{circulating}} \tag{5.76}$$

The five capacity equations are illustrated graphically in Figure 5.52.

Step 5: Determine Pedestrian Effect on Vehicles

The entering capacity is next adjusted for the presence of pedestrians. Research undertaken in Germany found that as drivers yield to pedestrians at roundabouts, the entering capacity is reduced by a reduction factor, f_{ped}. This effect is strongest at low conflicting circulating flow rates, which would otherwise have high entry capacities. As the circulating

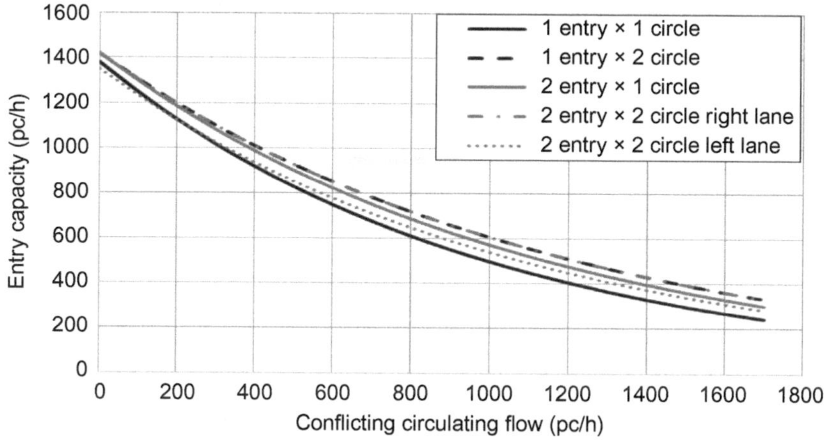

Figure 5.52 Plot of roundabout capacity models.

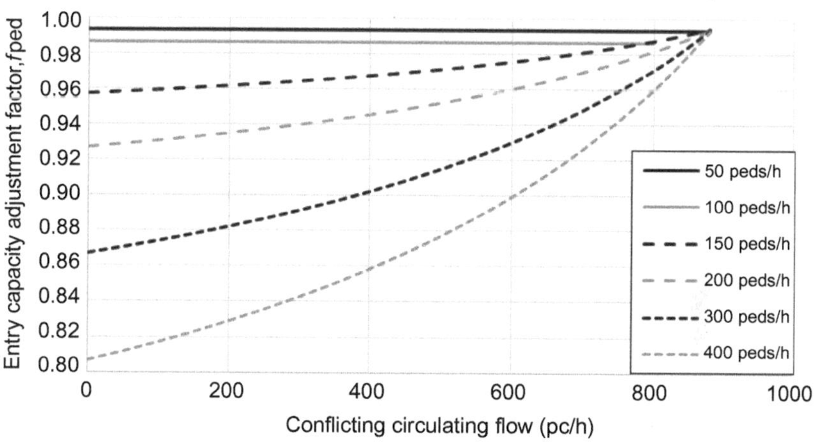

Figure 5.53 Pedestrian effect on roundabout single-lane entry capacity.

vehicle flow increases, entering vehicles are more likely to be delayed or queued at the entry already, thereby offsetting the added effect from pedestrians. The pedestrian effect increases further with increasing pedestrian volumes.

Figures 5.53 and 5.54 show the pedestrian effect on roundabout entry capacity for single-lane and two-lane entries, respectively. One important caveat in the application of these charts is that the research was done in Germany, where the yield compliance is generally very high (close to 100%). As a result, the net effect in a location where yielding is lower (say 50%) is expected to also be less (assume half the base effect).

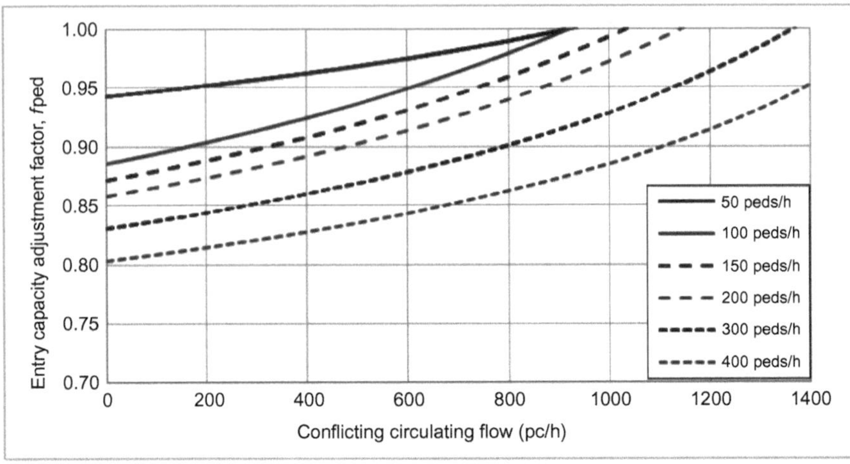

Figure 5.54 Pedestrian effect on roundabout two-lane entry capacity.

Step 6: Convert Flow Rate and Capacities to Vehicles

For step 6, the resulting flow rates and capacities are converted back to units in vehicles per hour before proceeding with the computations. Volumes are obtained by multiplying the flow rates by the heavy vehicle factor, f_{HV}. Capacities are obtained by multiplying the PCE capacities by f_{HV} and by the pedestrian adjustment factor, f_{ped}. In cases where different percentages of trucks enter the roundabout from different approaches, a weighted average heavy vehicle adjustment factor can be calculated, by weighting each factor by its corresponding traffic volume.

Step 7: Compute v/c Ratio, Delay, Queues, and LOS for Each Lane

With volumes, v_i, and capacities, c_i, available for each entry, the volume-to-capacity ratio for each approach, x_i, can be calculated from Eq. (5.77).

$$x_i = \frac{v_i}{c_i} \tag{5.77}$$

The volume-to-capacity ratio is the key input in Eq. (5.78) to estimate control delay, and Eq. (5.79) to calculate the 95th percentile queue length. The level of service for a roundabout is defined based on the average control delay in seconds per vehicle with LOS thresholds given in Table 5.42.

Table 5.42 LOS threshold for roundabouts

LOS	Control delay (s/veh)
A	≤ 10
B	>10–15
C	>15–25
D	>25–35
E	>35–50
F	>50 or v/c ratio > 1.0

$$d = \frac{3600}{c} + 900T\left[x - 1 + \sqrt{(x-1)^2 + \frac{\left(\frac{3600}{c}\right)x}{450T}}\right] + 5 \times \min[x, 1] \tag{5.78}$$

where

d = average control delay (s/veh)
x = volume-to-capacity ratio of the subject lane
c = capacity of the subject lane (veh/h)
T = time period (h) (T = 0.25 h for a 15-min analysis)

$$Q_{95} = 900T\left[x - 1 + \sqrt{(1-x)^2 + \frac{\left(\frac{3600}{c}\right)x}{150T}}\right]\left(\frac{c}{3600}\right) \tag{5.79}$$

where

Q95 = 95th percentile queue (veh)
x = volume-to-capacity ratio of the subject lane
c = capacity of the subject lane (veh/h)
T = time period (h) (T = 1 for a 1-h analysis, T = 0.25 for a 15-min analysis)

Step 8: Compute Delay and LOS for Each Approach and Overall Roundabout
As a final step in the methodology, the control delay can be aggregated to the approach level, by weighting the delay of each lane by its corresponding volume. Similarly, the overall roundabout control delay can be calculated as the weighted average of the approach delays. The LOS for the approach or the intersection is then determined using the same thresholds given in Table 5.42.

5.5 TRAFFIC SIGNALS AND SIGNAL TIMING

Traffic signals are the most commonly analyzed form of intersection control in many countries. They may not be the most common form of control overall, as so many low-volume unsignalized intersections exist in neighborhoods, subdivisions, and industrial parks. But signals are still the most common control form at intersections and along corridors that are likely to experience congestion due to high traffic demands. This may change in the future, as alternative forms of intersection control such as roundabouts are becoming increasingly popular (at the time of this writing, France has more than 20,000 roundabouts). But for now, one might say that traffic signals are still the "bread and butter" of most traffic operations engineers.

The following sections provide an introduction to signal warrants for deciding when to install a signal, the (hardware) components of a signal, a detailed walk through the signal timing process, and basic signal coordination concepts. Those interested in additional information on traffic signals are referred to the *HCM* for analysis procedures, and other resources for information about signal types, timing, and hardware, including the *Signal Timing Manual*[4] (Urbanik et al., 2014) and the *Traffic Detector Handbook* (FHWA, 2006b; Klein et al., 2006).

5.5.1 Signal Warrants and MUTCD

The *Manual on Uniform Traffic Control Devices* (MUTCD) is a set of guidelines and standards published by the Federal Highway Administration (FHWA, 2009) for the installation and maintenance of traffic control devices. The MUTCD is divided into nine parts as follows:

- Part 1: General—Contains introduction and overview of MUTCD
- Part 2: Signs—Contains guidance on a variety of signs in the categories of regulatory signs, warning signs, guide signs on surface streets and expressways, toll road signs, manage lane signs, and other specialized signs such as recreation and cultural interest areas
- Part 3: Markings—Contains guidance for the use of pavement markings for all aspects of the roadway
- Part 4: Highway Traffic Signals—Contains information on the installation and placement of traffic signals
- Part 5: Traffic Control Devices for Low-Volume Roads
- Part 6: Temporary Traffic Control—Contains guidance for signing and marking in work zones on surface streets and interstates

[4] http://ops.fhwa.dot.gov/arterial_mgmt/tstmanual.htm

- Part 7: Traffic Control Devices for School Areas
- Part 8: Traffic Control for Railroad and Light Rail Transit Grade Crossings
- Part 9: Traffic Control Devices for Bicycle Facilities

In the language of the MUTCD, the manual commonly distinguishes between "shall" conditions (actual requirements to implement a traffic control device), "should" conditions (suggestion to implement a traffic control device, but not required), and "may" conditions (option or allowable to install a traffic control device).

One of the more commonly used aspects of the MUTCD for traffic operational analysis are the signal warrants included in Part 4 of the manual. There are a total of nine signal warrants:

- Warrant 1, eight-hour vehicular volume
- Warrant 2, four-hour vehicular volume
- Warrant 3, peak hour
- Warrant 4, pedestrian volume
- Warrant 5, school crossing
- Warrant 6, coordinated signal system
- Warrant 7, crash experience
- Warrant 8, roadway network
- Warrant 9, intersection near a grade crossing

Warrants are intended to be used as guidance to decide when it is appropriate to install a traffic signal. Warrants are *not* requirements, such that even if one or more warrants are met, a traffic signal does not necessarily have to be installed, but is further deemed as an appropriate treatment. If a location or a proposed location does not meet any of the nine warrants, a traffic signal is said to be "unwarranted" and is typically not an appropriate traffic control device for the location under study.

As an example, Figure 5.55 presents the peak-hour volume warrant for vehicular volumes at a signalized intersection. Based on the total major street volume (total of both approaches) and the higher minor street volume, a traffic signal is warranted if the combined volume lands above the lines drawn in the figure. Overall, three lines are shown for different roadway cross sections for major and minor street. This and other warrants can be reduced to 70% for small towns (less than 10,000 population), as well as for high-speed roadways (depending on warrant, greater than 35 or 40 mph).

A second MUTCD example, Figure 5.56, shows both the four-hour and peak-hour pedestrian signal warrant thresholds from Warrant 4. Similarly, a pedestrian signal is warranted when the combination of vehicular volumes (x-axis) and pedestrian volume (y-axis) exceeds the line drawn.

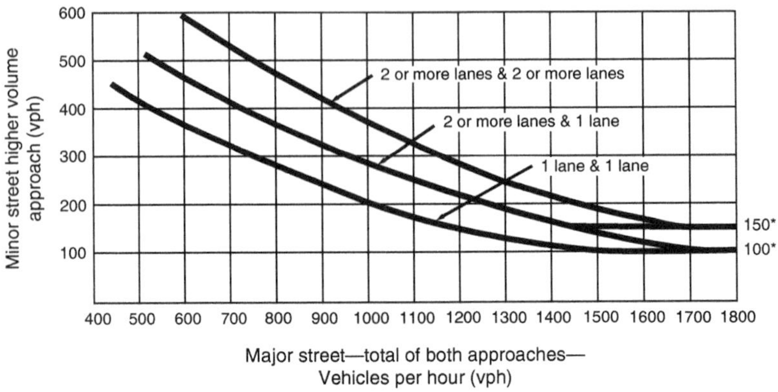

*Note: 150 vph applies as the lower threshold volume for a minor-street approach with two or more lanes and 100 vph applies as the lower threshold volume for a minor-street approach with one lane.

Figure 5.55 MUTCD peak-hour vehicular volume warrant.

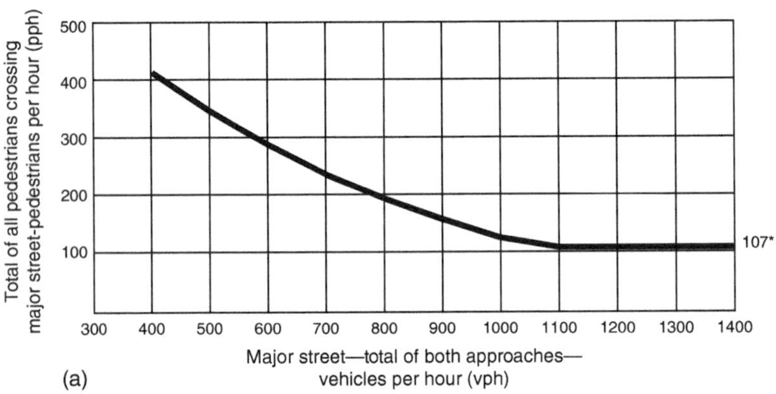

(a)

*Note: 107 pph applies as the lower threshold volume.

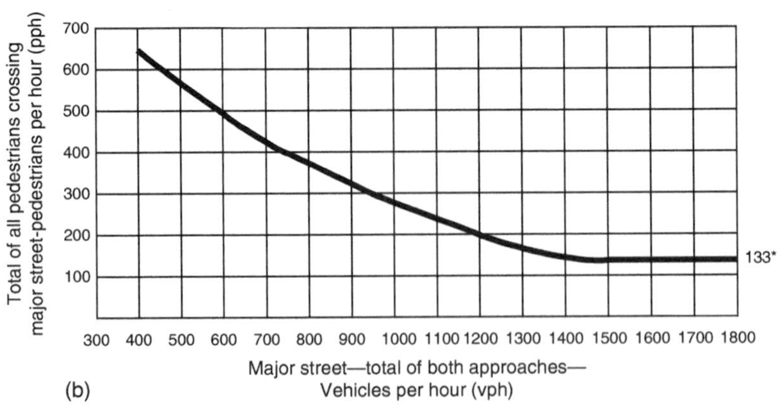

(b)

*Note: 133 pph applies as the lower threshold volume.

Figure 5.56 MUTCD pedestrian volume warrant. (a) Four-hour volume, and (b) peak-hour volume. *Source: http://mutcd.fhwa.dot.gov/pdfs/2009r1r2/pdf_index.htm.*

5.5.2 Components of a Traffic Signal

A traffic signal is essentially a computer system that contains the logic to run and control the operations of the intersection. A typical traffic signal has both hardware and software components. The hardware components are partly contained in the signal control cabinet, and partly installed on, in, or above the roadway in the form of various sensors and detectors. The software is principally contained within the signal controller (the local computer running the intersection), but often also interacts with a traffic management center.

Two examples of the hardware components of a traffic signal cabinet are shown in Figure 5.57, including a TS-2 cabinet (a) and a 2070 cabinet (b)—each referring to different controller standards and specifications.

The principal hardware components of a signal control system are:

- *Signal controller*: The computer containing the logic to run the intersection housed inside the signal cabinet.
- *Malfunction management unit*: Also referred to as a conflict monitor, which physically prevents conflicting phases to show concurrent green phases at the same time. In other words, even in the case of logic error where the software would want mainline and side-street green to be shown at the same time, this hardware system prevents that from happening, and would instead revert the signal into flashing mode.
- *Detector rack*: A space in the cabinet containing the detector cards. These are the hardware components that receive calls from in-road or above-ground detectors and sensors, and translate them into inputs to the signal controller.
- *Field terminal*: The interface between internal and external wiring at the controller.
- *Field transfer relays, load switches, and flashers*: Hardware components that help the software and hardware components of the signal to interact. Relays physically transmit a pulse signal to a hardware device, load switches control the change of the different signal indications (red, yellow, green), and flashers control on/off sequences for beacons or signals in flash mode.
- *Police switches*: A feature that allows law enforcement to access the signal cabinet from a side door and manually transition phases for special events, thereby overriding the preprogrammed phase durations.

The typical flow of information through a signalized intersection, ranging from the moving cars, to detectors, controller, signal indication, and ultimately to moving cars again, is shown in Figure 5.58.

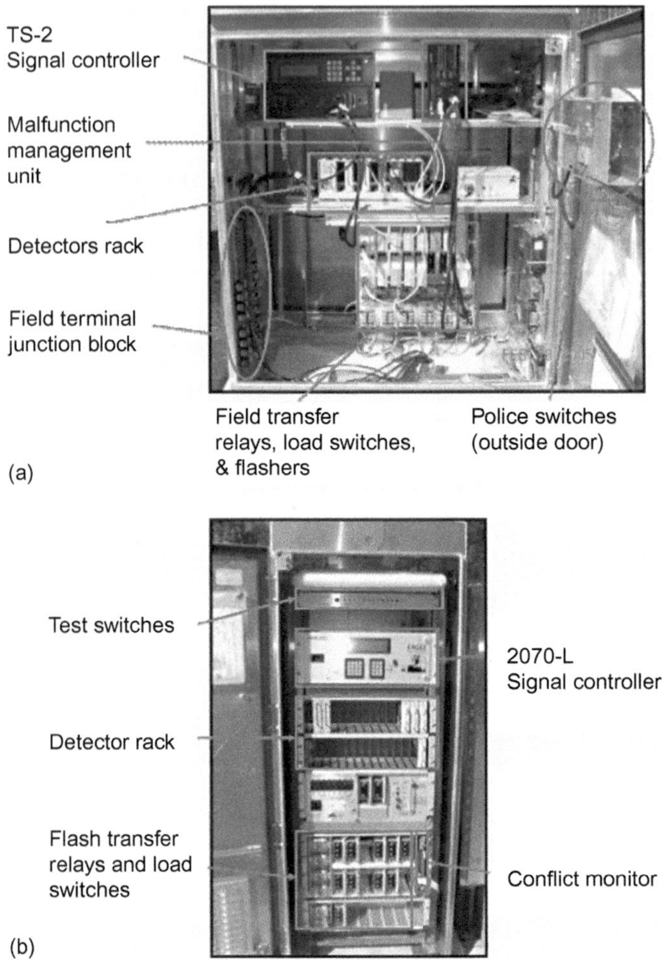

Figure 5.57 Components of a traffic signal cabinet. (a) TS-2 controller, and (b) 2070 controller.

5.5.3 Signal Timing Process

There are numerous ways and approaches to time traffic signals. The process described here is one that reflects preference and standard practice of the author, but it is acknowledged that variations of this process can likely lead to similar performance. The proposed process consists of the following 12 steps.

1. Identify left-turn treatments for all intersection approach legs.
2. Identify lane groups for each approach.

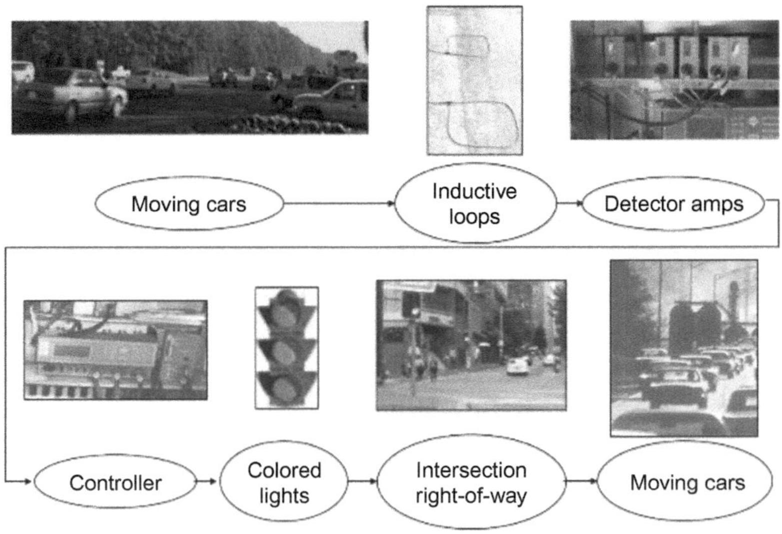

Figure 5.58 Flow of information at a signalized intersection.

3. Determine lane group traffic data, including adjusted volumes, saturation flow rates, and volume-to-saturation flow ratios (v/s).
4. Develop the desired phasing plan for the intersection.
5. Calculate required yellow and all-red times.
6. Determine lost time per phase and lost time per cycle.
7. Select a target volume-to-capacity ratio (X_c) for the intersection.
8. Estimate the cycle length for the intersection.
9. Calculate total effective green time available per cycle.
10. Distribute effective green time across all phases.
11. Calculate actual green time per phase.
12. Check pedestrian clearance time requirements.

Each of the 12 steps is described in more detail in the following.

Identify Left-Turn Treatments for All Intersection Approach Legs

The first question in signal timing is whether left-turn treatments for each approach are (a) permissive, (b) protected, or (c) a combination protected/permissive mode. In general, a protected left turn gives higher capacity for that left-turn movement, as left-turning vehicles don't have to yield to the opposing through traffic as in the case of permissive phasing. However, a protected left turn takes time away from the (often heavier) through movements, and therefore should only be used if absolutely necessary.

As a general rule of thumb, left turns should be protected for high-volume left-turning movements combined with high volumes of opposing through traffic. A useful calculation in this regard is to check the *cross product* of the two traffic streams. From this calculation, left turns should be protected if left-turn demand × opposing through is:

- 50,000 for 1 opposing lane
- 90,000 for 2 opposing lanes
- 110,000 for 3 opposing lanes

There are other reasons to protect left-turning movements, including:

- Sight distance is restricted for left turns to safely judge gaps in opposing traffic.
- Opposing vehicle speed is 45 mph or higher resulting in potentially high-severity collisions in the event of a crash.
- Presence of more than one left-turn lane, making gap selection difficult.
- Having a history of left-turn collisions under permissive control.

Protected/permissive control can be used to further increase the capacity of a protected left-control movement, or to shorten the duration of the protected phases, provided that none of the aforementioned safety considerations apply. For the purpose of the signal timing process, it is easiest to initially assume either protected *or* permissive control.

Identify Lane Groups for Each Approach

Lane groups are defined as one or more lanes on an approach that operate together under the same phase. Multiple through movements, for example, share the same lane group, as they are processed by the same phase. But having an exclusive left-turn lane allows that movement to be processed through its own phase, and it is therefore assigned its own lane group. The same holds for exclusive right-turn lanes. Any shared lanes (left/through, right/through, or left/through/right) are also assigned a separate lane group. The relationship between movements and lane groups is illustrated in Table 5.43. Note that the *HCM* further uses the concept of movement groups introduced in Section 5.4.3., which is omitted here for simplification.

Determine Lane Group Traffic Data

Next, the analyst determines traffic data for each lane group, including adjusted volumes, saturation flow rates, and volume-to-saturation flow ratios (v/s). First, hourly volumes are adjusted to account for peak-hour factor and heavy vehicles, as described in Section 5.4.3. Similarly, saturation flow rates are calculated using equations laid out in Section 5.4.3. or included in the *HCM*.

Table 5.43 Translating movements into lane groups

Number of lanes	Movements by lanes	Lane groups (LG)
1	Left, through, & right:	LG 1:
2	Exclusive left:	LG 1:
	Through & rights:	LG 2:
2	Left & Through:	LG 1:
	Through & right:	LG 2:
3	Exclusive left: Exclusive left:	LG 1:
	Through: Through:	LG 2:
	Through & right:	LG 3:

From the volumes and saturation flow rates, the analyst computes the ratio of volume flow to saturation flow rate for each lane group. This is denoted as the *v/s* ratio. The maximum lane group v/s ratio for phase *i* is denoted as the critical *v/s*. The phase lane group with the highest v/s ratio is referred to as the phase's *critical lane group*.

Develop the Desired Phasing Plan for Intersection

Next, the analyst selects the phasing plan for the intersection, or tests different options for combining movements into phases, as well as sequencing different phases. This step requires an understanding of phasing options and practice. A useful approach to determine the optimum phase sequence is to try and minimize the sum of the critical v/s ratios. Conceptually, each phase duration will be driven by one critical lane group, which is the one with the highest volume-to-saturation flow rate ratio. Summing these v/s ratios provides an estimate of the total congestion level of the intersection (before accounting for lost time). Henceforth, any phasing scheme that can reduce this critical sum can reduce the congestion level and therefore improve performance. As a general rule, it is desirable to combine lane groups with similar v/s ratios in the same phase if at all possible, to make most efficient use of the allocated time (driven by the lane group with the highest v/s ratio).

To facilitate this step, it is helpful to think of phasing in relation to two rings with up to eight phases, as illustrated in the *ring-barrier diagram*

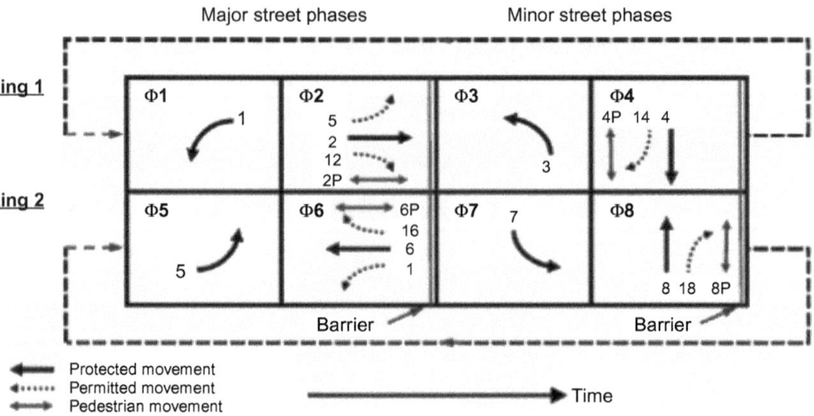

Figure 5.59 Dual-ring structure and sample movements. *Source: TRB, 2015.*

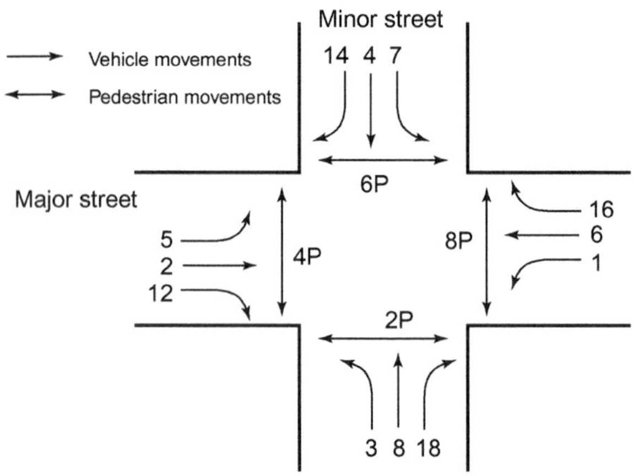

Figure 5.60 Movement numbering convention for traffic signals.

shown in Figure 5.59. In reality, a signal controller can have more than four rings and more than eight phases, but the following example represents the most common base configuration for traffic signals.

The dual-ring concept shows the movements within each phase and how they relate to all other movements. All movements on the major street (in this case east–west movements) must be completed before the movements on the minor street can start (north–south street in the example). This is called crossing the barrier. Figure 5.59 is an example of a standard 8-phase signal. The typical numbering convention for traffic signals is illustrated in Figure 5.60. The major street through movements (east–west in example here, but could be north–south) are assigned numbers 2 and 6, with the major street left turns adjacent to them being numbered 5 and 7, respectively. The side-street through movements are numbered 4 and 8, with the adjacent left turns being assigned numbers 7 and 3, respectively. It emerges that all through movements are even numbered, while left turns are odd numbered. Similarly, adjacent through and left-turn numbers add up to 7 on the main line and 11 on the side street (sometimes referred to as the "7-11 rule"). Right turns are assigned movement numbers 12, 14, 16, and 18, adding 10 to the respective adjacent through movement. Finally, pedestrian movements are assigned numbers concurrent with the through movements they are associated with (e.g., pedestrian phase 4P runs concurrent with vehicle through movement 4).

The ring-barrier diagram in Figure 5.59 is only one example of a traffic signal phase sequence with all left turns protected and leading the through movements. Many variations exist that result from lagging one or more left turns or running left turns in permissive mode. A specialized phase sequence referred to as *split phasing* runs the side-street movements sequentially with a phase for movements 3 and 8 being followed with a phase for movements 7 and 4. Split phasing is common when a heavy left turn exists on a minor approach, but the lane assignment is such that left turns cannot be protected (i.e., approach has a shared through/left lane).

Regardless of the phase sequence, step 4 of the signal timing process sums the critical v/s ratios for each phase to determine the optimum phase sequence. The example in Section 5.5.4 illustrates this further.

At this point in the timing process, the sum of critical v/s ratio should preferably be on the order of 0.8–0.9 or less, as lost time has not yet been accounted for. If the sum of critical v/s ratios approaches 1.0, it is highly unlikely that the timing plan will result in adequate performance, once lost times due to clearance and change intervals have been incorporated into the cycle.

Calculate Required Yellow and All-Red Times

When transitioning from a green indication of one movement to a green indication of a conflicting movement, a change and clearance interval is needed to assure that this transition happens in a safe manner. Generally, this transition includes a yellow time and an all-red time.

The yellow time (Y) is timed to provide *dilemma zone protection* to any approaching driver. The dilemma zone refers to a region in the approach to a traffic signal where the driver is too close and too fast to come to a stop before the signal (when seeing it transition from green to yellow).

The change interval is timed long enough to allow a driver to first react to the changing signal (perception-reaction time), and then to either safely come to a stop before the signal stop bar or to have sufficient time to proceed through the length of the intersection before transitioning to the phase for the next movement. Similarly, the all-red (*AR*) time is timed to allow a vehicle that didn't have time to break and that enters the intersection at the very moment the signal transitions from yellow to red.

Mathematically, the clearance interval (*CI*) is calculated as $CI = Y + AR$, or more precisely as shown in Eq. (5.80):

$$CI = t_{PR} + \frac{S_0}{2 \times (a + Gg)} + \frac{W + L}{S_0} \tag{5.80}$$

where

CI = change interval (s)
t_{PR} = perception-reaction time (s)
S_0 = initial vehicle speed (ft/s)
a = vehicle deceleration rate (default is 11.2 ft/sec^2)
G = approach grade (%)
g = gravitational acceleration (32.3 ft/sec^2)
W = width of the intersection in direction of travel (ft)
L = length of vehicle (default is 20 ft)

The duration of the yellow interval should be greater than or equal to the sum of the first two terms in Eq. (5.80), with the all-red time greater than or equal to the third term.

Determine Lost Time Per Phase and Lost Time per Cycle

The lost time per phase (ℓ) is typically 3–4 s as observed in the field. It is made up of start-up lost time (2.0 s) plus clearance lost time (1–2 s). A common lost time of 4 s per phase is a typical default used by many agencies. The lost-time concept and relationship to the green (*G*), yellow (*Y*), and red (*R*) indication for each phase split is illustrated in Figure 5.61.

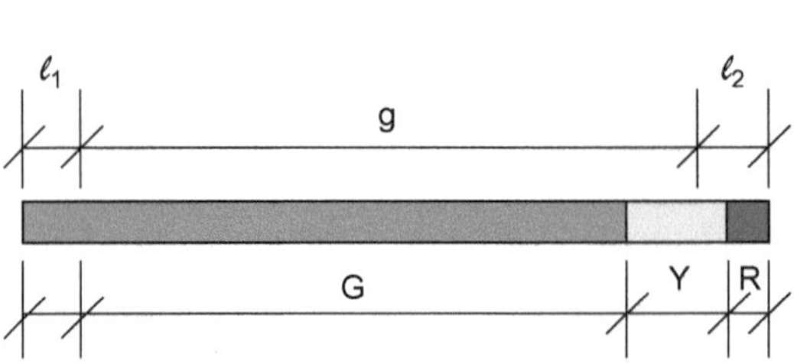

Figure 5.61 Illustration of lost time.

A phase split is made up of the sum of $G + Y + R$. As shown in the figure, after subtracting the start-up lost time (l_1) and end lost time (l_2), the remaining time is referred to as the effective green time, g. The effective green time is the time during which vehicles actually move for the given phase, and it is principally linked to the movement capacity. Notice also, that the effective green extends into the yellow indication, recognizing that some vehicles will be proceeding into the intersection during that time.

For the purpose of signal timing calculations, the total lost time per phase (l) is the sum of the individual lost times for each critical phase as follows:

$$\ell = \ell_1 + \ell_2$$

The total lost time per cycle (L) is the sum of the individual lost time for all critical phases. This sum should consider any overlaps, which are movements that are active in more than one phase. If a critical movement is shown in two successive phases, the lost time only applies once at the end of the (combined) phase indication.

Select a Target Volume-to-Capacity Ratio (X_c)

With lost times calculated, the analyst needs to select a target volume-to-capacity ratio, X_c. Conceptually, an intersection designed to $X_c = 1.0$ would be likely to have the perfect amount of time for each movement based on the predicted average traffic volume levels. However, as traffic volumes are not constant from cycle to cycle, the actual arrivals are expected to be lower and, more importantly, higher in some cycles. As such, it is good practice to not design a signal for operating at (average) capacity, but rather to incorporate a 10–15% safety margin. As such, a target X_c in the range of 0.85–0.90 is typically recommended for pretimed control, although values as high as 0.95 can work for some actuated intersections.

Estimate the Cycle Length

With the target X_c ratio and the sum of critical v/s ratios, the following equation can be used to estimate the cycle length for the intersection.

$$C = \frac{L(X_C)}{X_C - \sum \left(\frac{v}{s}\right)_i} \tag{5.81}$$

where

C = cycle length (s)

L = total lost time per cycle (s)

X_c = target v/c ratio

$\sum(v/s)_i$ = sum of critical v/s ratios

For this method, a desired intersection critical v/c ratio, X_C, must be specified, and entered into the Equation (5.81) to solve for cycle length, C. This becomes the trial cycle length because it will have to be checked against pedestrian requirements if they exist (see step 12). The typical acceptable range for C is 40–180 s, with a general preference for shorter cycle lengths. The resulting cycle length should be rounded to the next highest 5 s for cycle lengths less than 80 s or to the next highest 10 s for cycle lengths greater than 80 s.

Note that X_C is a great measure of the spare capacity at an intersection. An X_C of 0.80 means there is about 20% spare capacity with adjustments in signal timing and cycle length to accommodate additional traffic volumes. X_C is usually 0.80–0.90 for pretimed or semiactuated signals and 0.90–0.95 for actuated signals. If your choice of X_C results in an unacceptable cycle length, increase it and try again.

Alternatively, Webster's optimum cycle length equation can be used as follows:

$$C_0 = \frac{1.5L + 5}{1 - \sum}\left(\frac{v}{s}\right)_i \tag{5.82}$$

where

C_0 = Webster's optimum cycle length (s)

(all other variables are as defined previously)

Calculate Total Effective Green Time Available per Cycle

From the cycle length and total lost time, the total available effective green time per cycle can readily be estimated as $C - L$.

This is the total amount of time in each cycle that is available for traffic to move through the intersection. For example, if a signal has a cycle length of 120 s with four critical phases, each with a lost time of 4 s, the total available green time is $120 - 16 = 104$ s. Expressed as a proportion of time, due to lost time the signal only provides $104/120 = 0.87$, or 87% of the hour for vehicles to actually move through the intersection. Accordingly, any sum of critical v/s ratios from step 4 that are greater than 0.87 in this case, would not be able to be processed through this signal under the selected timing plan and cycle length. This is a useful check before proceeding further with the computations.

Distribute Effective Green Time across All Phases

From the total green time, the analyst can now allocate the effective green time (g_i) to each phase based on its critical v/s. If the available movement time for all phases is $C-L$, the effective green time per phase is estimated as:

$$g_i = (C-L)\frac{\left(\frac{v}{s}\right)_i}{\sum\left(\frac{v}{s}\right)_i} \tag{5.83}$$

where

g_i = effective green time for phase i (s)
(all other times are as defined previously)

Calculate Actual Green Time per Phase

Next, the actual green time per phase (G) is calculated. This is the actual duration of the green indication that is visible to the driver, and also programmed into the signal controller. It is calculated under consideration of the clearance interval ($CI = Y + AR$) and the start and end lost times ($l = l_1 + l_2$). As illustrated in Figure 5.61, the following relationship holds:

$$G + Y + AR = g + l_1 + l_2$$

This yields the following relationship for the actual green time G_i for phase i:

$$G_i = g_i - CI_i + l_i = g_i - (Y + AR) + (l_1 + l_2)$$

This equation is sometimes also expressed in terms of the extension of effective green time, e, which is defined as $e = Y + AR - l_2$. In this case, the actual green time is calculated as:

$$G_i = g_i + l_i - e$$

Note that it is common for major street movements to have a green time of at least 15 s, and minor street movements of at least 7 s. So while it is generally desirable to keep cycle lengths as low as possible, a slight increase in cycle length may be needed to assure that each G_i exceeds these minimum values.

Check Pedestrian Clearance Time Requirements

As the final step in signal timing, a check for pedestrian clearance times needs to be performed. Pedestrian movements are accommodated at many intersections, and are often needed to assure that pedestrians can get across the street safely and efficiently. Pedestrian phases consists of two intervals: First, the *walk* interval is intended to get pedestrians off the sidewalk and into the street. It typically has a minimum time of 5–7 s. Second, the *flashing don't walk* interval is timed long-enough to assure that a pedestrian that steps into the roadway at the end of green can safely make it across the intersection.

The MUTCD requires that the flashing "don't walk" phase be calculated as a function of the pedestrian walking speed and the crossing width. The default walking speed is set at 3.5 ft/s, which corresponds to the 15th percentile speed of pedestrians observed in a national study of pedestrian crossings (TRB, 2006). However, in cases where a significant portion of elderly or disabled pedestrians are expected to cross, that walking speed should be lowered to 3 ft/s or even less based on local data. As such, the minimum pedestrian phase duration is estimated as:

$$t_{ped,min} = t_W + t_{FDW} = t_W + \frac{W}{S_{ped}} \tag{5.84}$$

where

$t_{\mathrm{ped,min}}$ = minimum pedestrian phase duration (s)
t_{W} = duration of walk interval (default = 5–7 s)
t_{FDW} = duration of flashing "don't walk" indication (s)
W = crossing width (ft)
S_{ped} = pedestrian walking speed (default = 3.5 ft/s)

If the total split for a vehicular phase ($G + Y + AR$) is less than the corresponding minimum pedestrian phase duration, the vehicle phase needs to be increased to allow for the minimum. As needed, cycle lengths may be increased to accommodate all phases. The need to make adjustments due to pedestrian clearance time requirements is most frequent for (low volume) side-street phases, which may otherwise only have a short split assigned to process vehicular volumes.

5.5.4 Signal Timing Example

The following example illustrates the signal timing process in detail. Figure 5.62 shows a four-legged intersection of Main Street (running east–west) and 5th Street (running north–south). The figure shows lane assignment and basic assumption. Traffic volumes are given in Figure 5.63.

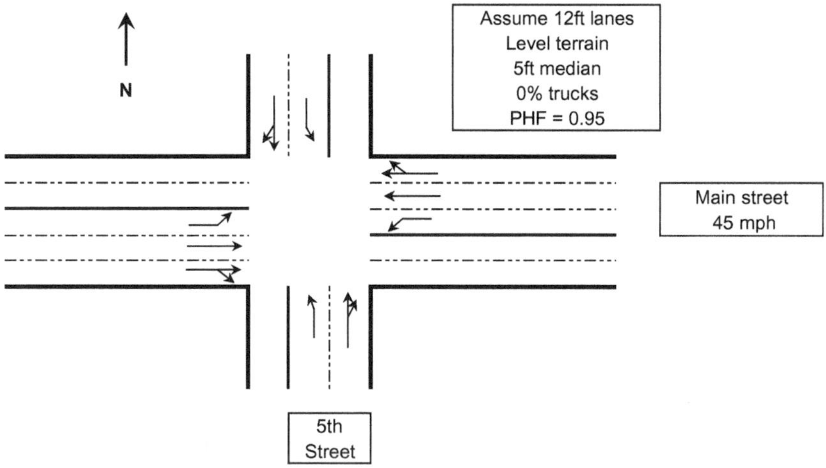

Figure 5.62 Intersection geometry for signal timing example.

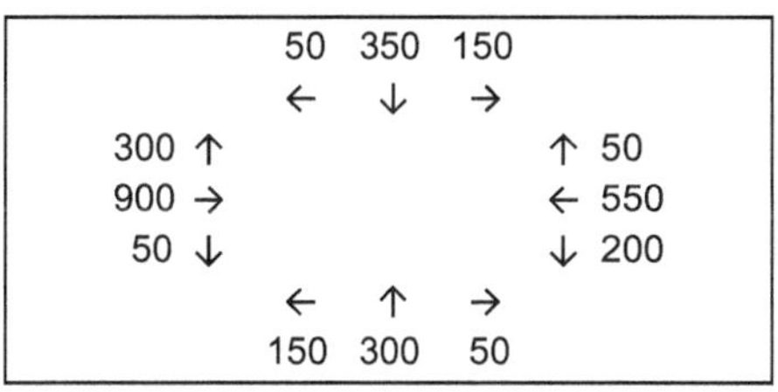

Figure 5.63 Traffic volumes for signal timing example.

Step 1: Identify Left-Turn Treatments

For this example, we will assume all protected left-turn phasing. In reality, some of the left turns may run as permissive (if volumes are low enough), or as protected-permissive, provided that adequate sight distances are provided and the permissive left turn can be completed safely. Permissive left turns would decrease the capacity of the left turn over a protected movement (at the same effective green time), while a protected-permissive movement would add some capacity over the protected-only movement. So in some ways, the assumption of a protected-only movement is conservative, and additional capacity may be gained if permissive turns are eventually allowed at the intersection.

Step 2: Identify Lane Groups for Each Approach

Per *HCM* guidance presented in Section 5.5.3, each exclusive turn lane and each shared lane has to be assigned a separate lane group. As such,

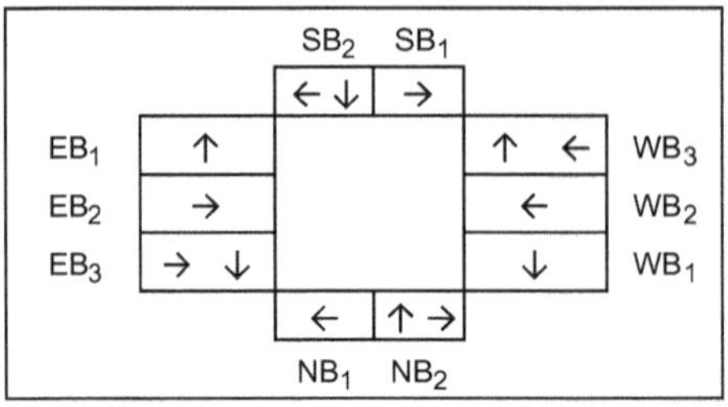

Figure 5.64 Lane groups for signal timing example.

three lane groups emerge each for the east and west approaches, with two lane groups for the north and south approaches. This is illustrated in Figure 5.64.

Step 3: Determine Lane Group Traffic Data

The traffic volumes shown in Figure 5.63 first need to converted into peak 15-min flow rates by dividing them by the peak hour factor (and typically a heavy vehicle adjustment factor and driver population adjustment factor, which do not apply in this example). These flow rates are then assigned to the different lane groups. For the through traffic on the eastbound and westbound approaches, it is assumed that 55% of through traffic chooses the exclusive lane, with 45% in the shared lane. In the absence of field data, this type of assumption has to be made by the analyst. A good reasonableness check in this volume split is to assure similar v/s ratios for both lane groups, which does hold true in this example.

The lane group flow rates are then divided by the saturation flow rate of each lane group to give the v/s (volume to saturation flow) ratio for the lane group. Conceptually, the v/s ratio describes the fraction of an hour worth of green time needed to serve the demand of the particular lane group given its saturation flow rate (capacity). In this example, exclusive through lanes have an assumed saturation flow rate of 1900 passenger cars/h per lane; shared lanes a rate of 1800 passenger cars/h per lane; and exclusive left-turn lanes a rate of 1700 passenger cars/h per lane. The saturation flow rates can also be field measured or be calculated using the appropriate *HCM* equation. This entire process is illustrated in Tables 5.44 and 5.45.

Step 4: Develop the Desired Phasing Plan

Figure 5.65 shows all v/s ratios applied to the lane group diagram. If one were to sum up all v/s ratios, the resulting sum equals 1.79. In other words, 179% of the cycle (or an hour) are needed to practice all these movements.

Table 5.44 Lane group traffic data for signal timing example—traffic volumes

Traffic volumes

	NB			SB			EB			WB		
	L	T	R	L	T	R	L	T	R	L	T	R
Demand	150	300	50	150	350	50	300	900	50	200	550	50
Demand/PHF	158	316	53	158	389	53	310	947	53	211	579	53

Assume 55/45 split for through across exclusive/shared lanes

Table 5.45 Lane group traffic data for signal timing example—lane group volumes

Lane group volumes

	NB_1	NB_2	SB_1	SB_2	EB_1	EB_2	EB_3	WB_1	WB_2	WB_3
Demand/PHF	158	369	158	442	316	521	479	211	318	314
SAT./Flow	1700	1900	1700	1,900	1700	1900	1800	1700	1900	1800
V/S	0.09	0.19	0.09	0.23	0.19	0.27	0.27	0.12	0.17	0.17

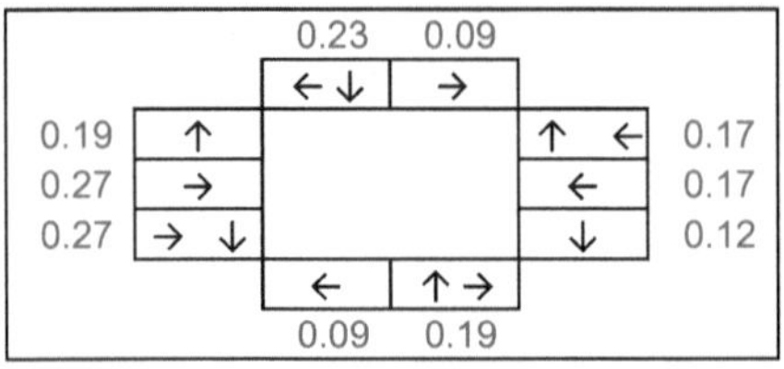

Figure 5.65 v/s ratios applied to movements.

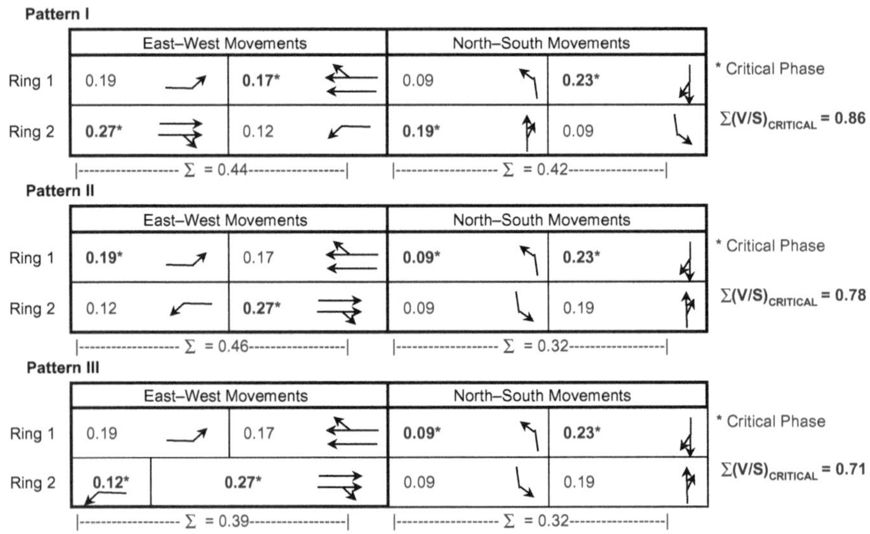

Figure 5.66 Illustration of three phasing patterns for signal timing example.

Luckily, different lane groups and movements can be combined into the same phase. For example, northbound and southbound through movements can move together, thereby combining the 0.23 and 0.19 v/s ratio. The duration of this combined phase is governed by the phase with the higher v/s ratio, which is also referred to as the *critical v/s ratio.* Each phase accordingly will have one critical v/s ratio that determines the required phase duration, which can be summed for the intersection. The objective of step 4 is to come up with the best possible signal timing plan that *minimizes the sum of critical v/s ratios* for all phases.

Figure 5.66 illustrates three potential phasing schemes for this intersection drawn in a ring-barrier diagram representation. The signal timing process needs to find the sum of critical v/s ratios, which follows the *critical path* through the ring-barrier diagram. For each side of the barrier (east–west vs. north–south), one phase sequence is going to have the higher sum of critical v/s ratios.

In the example, pattern I runs each approach sequentially: eastbound, westbound, northbound, and finally southbound. The resulting sum of critical v/s ratios is 0.86. The inefficiency in this scheme becomes apparent when scrutinizing the v/s ratios for left turns and through movements, where the latter tend to be higher.

Pattern II then combines left turns and through movements, as each pair has similar v/s ratios. The result is a more efficient overall phasing scheme and a sum of critical v/s ratios that is reduced to 0.78. Further improvements are possible when allowing one left turn to proceed longer than the other, if v/s ratios are uneven.

This is illustrated in pattern III, where the higher v/s ratio for the eastbound left turn (0.19) is given more time than the westbound left turn (0.12). This in turn allows the higher v/s eastbound through movement to start before the westbound through. The result is that both the heavy eastbound left and through phases are given more time than the lower-volume westbound movement. The resulting sum of critical v/s ratio is thereby reduced to 0.71, making this the preferred phasing scheme for the intersection. Note that for the north–south phases, both left turns have identical v/s ratios, and no additional efficiency gains are therefore possible here.

Step 5: Calculate Required Yellow and All-Red Times

The clearance interval (yellow and all-red) is calculated from Eq. (5.80), which is repeated as follows. First, the calculations are shown for the east–west street. The width of the intersection is calculated by summing 12-ft travel lanes and a 5-ft median.

I - East–west street

$$CI = \underbrace{t_{PR} + \frac{S_o}{2(a + G_g)}}_{\text{yellow}} + \underbrace{\frac{W + L}{S_o}}_{\text{red}}$$

$$W = 3(12\text{ ft}) + 5\text{ ft} = 41\text{ ft}$$

Lanes Median

$$S_o = 45\text{ mph} \times 1.47\text{ fps/mph} = 66.2\text{ fps}$$

$$CI = 1 + \frac{66.2}{2(11.2 + 0)} + \frac{41 + 20}{66.2}$$

$$CI = 1 + 2.96 + 0.92$$

Result: Use yellow > 4 s and AR > 1 s
I – North–south street

$$W = 5(12\text{ ft}) + 5\text{ ft} = 65\text{ ft}$$
$$S_o = 30\text{ mph} \times 1.47\text{ fps/mph} = 44.1\text{ fps}$$
$$CI = 1 + \frac{44.1}{2(11.2 + 0)} + \frac{65 + 20}{44.1}$$
$$CI = 1 + 1.96 + 1.93$$

Result: Use yellow > 3 s and AR > 2 s

Step 6: Determine Lost Time per Phase and Per Cycle

To calculate the lost time per phase, we can assume a start-up lost time of 2 s, as well as an end lost time of 2 s. As a result, the total lost time per (critical) phase is 4 s for both the north–south and the east–west movements.

The total lost time per cycle is then estimated by summing the lost times for all critical phases. Lost time applies every time a critical movement starts or stops along the critical path through the intersection phase sequence. In other words, Figure 5.66 that was used to calculate the sum of critical v/s ratios, can also be used to estimate the total lost time. In the figure, it is evident that each of the three phasing options features a sequence of four critical phases. As such, the total lost time for this eight-movement, four-critical-phase intersection is $L = 4 \times 4\text{ s} = 16\text{ s}$ of total lost time, regardless of the phasing scheme applied.

Step 7: Select a Target Volume-to-Capacity Ratio

Next, we select a target volume-to-capacity ratio. In this example, we will select $X_c = 0.90$, meaning that our intersection will be utilized at 90% of its theoretical capacity, thereby allowing a 10% safety margin for fluctuations in traffic volumes.

Step 8: Estimate the Cycle Length

With most inputs completed, we can now calculate the cycle length for the intersection. We show both options for doing so, starting with the minimum cycle length formula from Eq. (5.81).

$$C = \frac{LX_C}{\left[X_C - \sum\left(\frac{V}{S}\right)_{Critical}\right]} = \frac{16(0.90)}{[0.9 - 0.71]} = 75.8\text{ s}$$

This value is typically rounded up to the nearest 5 s, so we will use a cycle length of 80 s.

Second, we can calculate the optimum cycle length using Webster's equation from Eq. (5.82).

$$C = \frac{1.5L + 5}{\left[1 - \sum\left(\frac{V}{S}\right)_{Critical}\right]} = \frac{1.5(16) + 5}{[1 - 0.71]} = 100 \text{ s}$$

Step 9: Total Effective Green Time Available per Cycle

The total effective green time is estimate by subtracting the 16-s total lost time from the cycle length, giving a time of 64 s and 84 s for cycle lengths of 80 and 100 s, respectively.

Step 10: Distribute Effective Green Times across All Phases

We can now distribute the effective green times across all phases using Eq. (5.83). The resulting phase times for both cycle lengths are shown in Table 5.46. The phases along the critical path for pattern from Figure 5.66 are shown in bold and with "*" symbols. Note that the phases along the critical path by definition have to add up to the total available green time, which in this case is 64 and 84 s for the two cycle length options. The noncritical phases require less green time, and therefore do not add up to the total available effective green. It is common practice to extend the mainline movements for those noncritical movements, to use up any unallocated green time. For example, for the east–west movements, the effective green time for the critical movements (ring 2) in the 80 s cycle add up to 10.8 + 24.3 = 35.2 s. But the noncritical movements in ring 1 add up to only 17.1 + 15.3 = 32.5 s. The difference of 35.2 − 32.5 = 2.7 s is applied

Table 5.46 Phase times for signal timing example

Ring	Movement	v/s	C = 80 s	C = 100 s
1	EBL	0.19	17.1	22.5
	WBT	0.17	15.3 (18.0)	20.1 (23.7)
	NBL*	**0.09**	**8.1**	**10.6**
	SBT*	**0.23**	**20.7**	**27.2**
2	**WBL***	**0.12**	**10.8**	**14.2**
	EBT*	**0.27**	**24.3**	**31.9**
	SBL	0.09	8.1	10.6
	NBT	0.19	17.1 (20.7)	22.5 (27.2)

Table 5.47 Actual green times for signal timing example

Ring	Movement	C = 80 s	C = 100 s
1	EBL	16.1	21.5
	WBT	17.0	22.7
	NBL	7.1	9.6
	SBT	19.7	26.2
2	WBL	9.8	13.2
	EBT	23.3	30.9
	SBL	7.1	9.6
	NBT	19.7	26.2

to the westbound through movement, increasing its effective green time to $15.3 + 2.7 = 18$ s. These modified effective green times are shown in parentheses in the table.

Step 11: Calculate Actual Green Time per Phase

The effective green times from Step 10 are not converted to actual green times programmed into the signal controller, and are seen by the drivers as a green indication. The actual green times are calculated as:

$$G_i = g_i - CI_i + l_i$$

In our example, all clearance intervals were 5 s (4 + 1 for east–west and 3 + 2 for north–south), and all lost times were assumed to be 4 s. The resulting green times are shown in Table 5.47.

These green times should be checked to assure that the minimum green time for all phases are provided, which are typically around 7 s for left turns and side-street phases and 15 s for through movements on the mainline. In this example, these minimum times are provided or both cycle lengths. Similarly, some signal controllers require green times to be rounded to the nearest full second, and rounding is good practice even for more flexible controllers. But before finalizing the phase times, there is one more check that needs to be performed.

Step 12: Check Pedestrian Clearance Time Requirements

The final step in the signal timing process is to check the pedestrian clearance times. Pedestrian movements are typically processed with the adjacent mainline through. Pedestrian clearance time requirements are calculated from Eq. (5.84). We can assume a walk interval of 5 s for this intersection and a walking speed of 3.5 ft/s, per MUTCD.

Table 5.48 Total splits per movement for signal timing example

Ring	Movement	C = 80 s	C = 100 s	Pedestrian clearance	Check
1	EBL	21.1	26.5		
	WBT	22.0	27.7	16.7	OK
	NBL	12.1	14.6		
	SBT	24.7	31.2	21.6	OK
2	WBL	14.8	18.2		
	EBT	28.3	35.9	16.7	OK
	SBL	12.1	14.6		
	NBT	24.7	31.2	21.6	OK

$$t_{ped,min} = t_W + t_{FDW} = t_W + \frac{W}{S_{ped}}$$

Crossing minor street (east—west movements):

$$t_{ped,E-W} = 5 + \frac{41}{3.5} = 5 + 11.7 = 16.7 \text{ s minimum}$$

Crossing major street (north—south movements):

$$t_{ped,N-S} = 5 + \frac{65}{3.5} = 5 + 18.6 = 21.6 \text{ s minimum}$$

The currently assigned splits (green plus yellow plus all-red) for the east—west movements are shown in Table 5.48. The comparison with the required pedestrian clearance times shows that the assigned splits for the through movement are long enough to accommodate pedestrians at a 5-s walk interval plus the required flashing "don't walk" time. As a result, both an 80-s or a 100-s cycle would be feasible for this intersection. The decision of the final intersection timing depends on agency practices, as well as potential interaction with adjacent intersections, where a common cycle length is required if coordination is to be provided along a signalized intersection corridor.

5.5.5 Signal Coordination

Signal coordination means that vehicles traveling along the main road tend to progress or move through downstream signalized intersections without having to stop. There are several situations where signalized intersections adjacent to each other should be coordinated. Coordination is critical for signals at interchange ramp junctions because of queuing concerns between the closely spaced signals for both through traffic and left-turning traffic. Signal coordination along an arterial corridor will significantly reduce both overall travel time and delay for through vehicles using that corridor. Signal coordination along a one-way road allows for smooth progression of traffic, usually in a downtown area where one-way roads surround each block and there are often pairs of one-way roads on opposite sides of the blocks.

Time-Space Diagrams

A time-space diagram shows a platoon of vehicles moving along a street. Signal coordination is established when the platoon can arrive at a signalized intersection under a green interval and continue proceeding along the street without having to slow down or stop. Perfect progression using the entire green interval is rarely possible, especially if two-way progression is desired. The following two figures show a one-way progression pattern (Figure 5.67) and a two-way progression pattern (Figure 5.68).

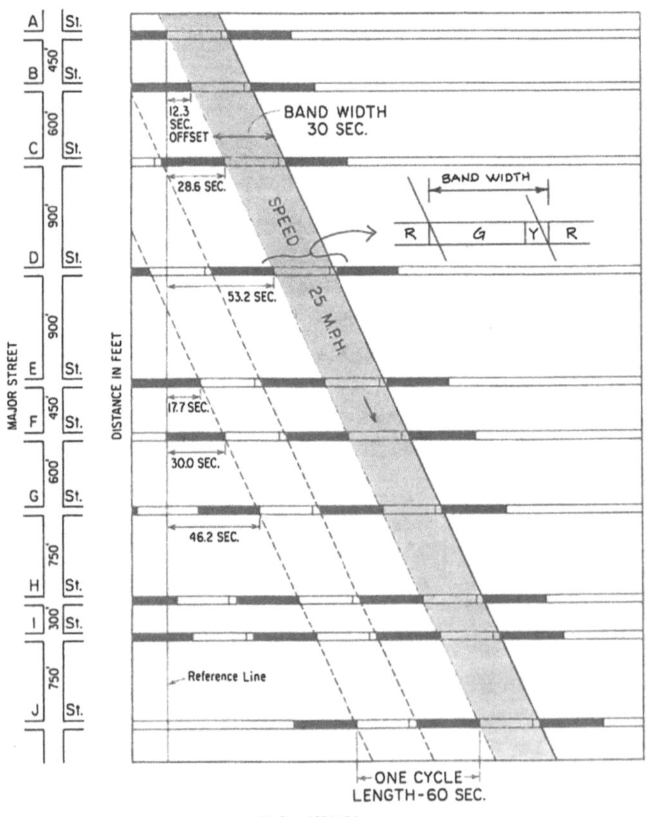

Figure 5.67 Time-space diagram for a one-way street.

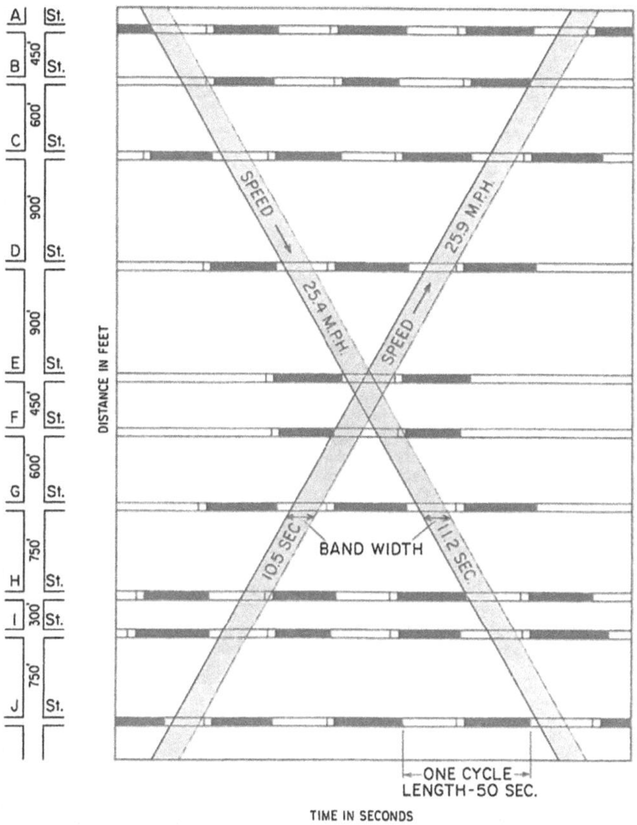

Figure 5.68 Time-space diagram for a two-way street.

Figure 5.67 depicts many new terms, which are defined as follows. These terms are critical in properly describing the progression aspects of the traffic stream.

- *Reference line*: This is a straight line drawn perpendicular to the time scale. The selection of its location is arbitrary, however, a convenient line would be the left edge of the figure. All offset times are measured away from the reference line in the direction of increasing time.
- *Offset*: The time in seconds from the reference line to the start of the green phase for the first full cycle at any intersection, thus, the offset must always be less than one cycle length. Offsets are calculated by taking the total distance (ft) traveled from the starting point to the intersection where the offset is needed and dividing by the travel speed (fps), remembering to subtract as many cycle lengths as necessary to get the offset time to be less than one cycle length.

- *Bandwidth*: Sometimes called progression bandwidth, it is the time between the two parallel progression speed lines. It represents the maximum time in seconds that a platoon of vehicles can progress through each intersection, and because the two lines are parallel, the bandwidth is the same for all intersections. The speed lines must pass through only the green (and/or yellow) interval at every intersection. The maximum bandwidth for an intersection occurs when the first speed line passes through the entering corner of the green interval and the second speed line passes through the exiting corner of the yellow interval resulting in a bandwidth that is slightly shorter than the green plus yellow intervals by the time to travel through the intersection, about 1.5 s. Note that the bandwidth time is measured *along* the time scale and is *not* the perpendicular measure between the two parallel lines.
- *Cycle length*: The time for one complete pass through critical phases in a timing plan. For a time-space diagram to repeat throughout the day, the cycle length must be the same at every intersection.
- *Speed*: The slope of the progression line, usually determined and plotted using fps and converted back and forth to mph as needed. Recall that 60 mph = 88 fps.

For all time-space diagrams, the vertical axis (usually distance) and the horizontal axis (usually time) must be drawn to scale. This includes the width of each intersection. A scaled diagram allows for determining the proper bandwidth and progression speed given a set of input criteria and the desired outcome. One can also calculate the efficiency of the bandwidth for a given direction. The formula is:

$$efficiency = \frac{band\ width}{cycle\ length} \times 100\%$$

An efficiency of 40% to 55% is considered good.

Drawing a Time-Space Diagram

In many cases, time-space diagrams are produced automatically through software. However, it is useful to understand the basic steps involved in drawing a time-space diagram, which are as follows.

1. Establish your distance and time scales to fit the layout on a page if possible.
2. Lay out your street intersections to scale, including the width of the side streets.
3. Select a reference line and draw it in.

4. If laying out offsets, calculate the offset at each intersection using distance (ft) divided by speed (ft/s) to yield seconds. This becomes the start of the green phase at that intersection.
5. Draw in all the green, yellow, and red intervals at each intersection using the offset as the measuring point (both forward and backward).
6. Draw in the speed lines (they must be parallel) to achieve maximum bandwidth and measure the width. If the speed lines are being fitted into an existing situation, calculate the speed by taking distance (ft) divided by travel time (s) giving fps and convert to mph.
7. Label the diagram.

5.6 PRACTICE PROBLEMS

Problem 5.1

An existing four-lane freeway is located in a growing urban area with level terrain. PHF = 0.94, f_{HV} = 0.962, measured FFS = 70 mph, and drivers are everyday commuters. Current peak hour volume is 3650 vph in one direction.

What is the current LOS during the peak period?

Problem 5.2

A six-lane rural freeway has four ramps in 6 miles, 12-ft lanes, 4-ft left shoulder, 10-ft right shoulder, and is on a 3.8% grade for 1.3 miles. The traffic demands are 1500 vehicles in peak 15 min, 10% trucks, and PHF = 0.9.

What is the current LOS?

At 3% per year traffic growth, when will this segment reach capacity?

Problem 5.3

A six-lane freeway has a right-side off-ramp, ideal geometry, level terrain, and 70-mph FFS. The ramp itself is designed at a FFS of 35-mph ramp with a 950-ft deceleration lane. The freeway mainline has 4850 vph and the off-ramp volume is 800 vph. An upstream on-ramp ½ mile away has 900 vph entering volumes. All traffic streams have 9% trucks and PHF = 0.93.

Determine the density, speed, capacity, and LOS at RIA, the outer lane, and for the overall cross section.

Problem 5.4

A standard ramp weave segment with three lanes has an FFS of 65 mph, and a short length of 1850 ft. The mainline volume is

4650 veh/h, entering volume is 875 veh/h, and exiting volume is 730 veh/h. The ramp-to-ramp flow is 175 veh/h. All traffic streams have 10% trucks, the terrain is level, and the interchange density is 0.9 interchanges/mi.

Estimate the capacity and LOS of the weaving segment.

Problem 5.5

Two arterial streets cross at the intersection shown in the diagram. The traffic volumes and saturation flow rates are provided in the table. Determine the cycle length and phase time for an appropriate phasing plan for this signalized intersection. Max $C = 180$ s.

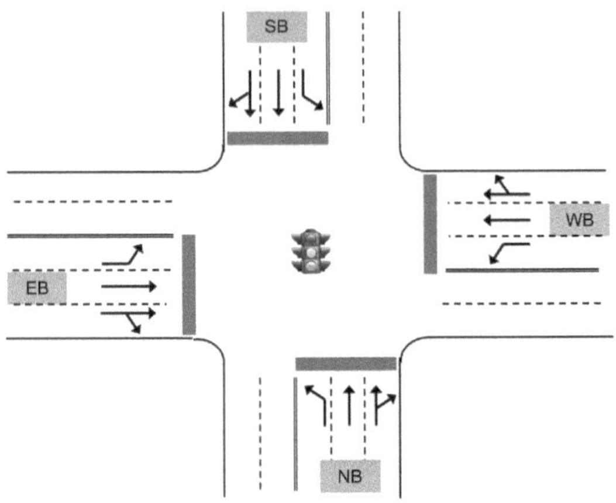

Traffic volumes, passenger cars/h

	SB	
R	T	L
52	695	103

EB					WB
	L	138	81	R	
EB	T	1007	920	T	WB
	R	184	115	L	

124	840	61
L	T	R
	NB	

Saturation flow, passenger cars/h per lane

L	1400
TL	1500
T	1600
TR	1600
R	1400
LTR	1000

Problem 5.6

A crosstown route intersects with a one-way street near the downtown of a city, as shown in the diagram. Adjusted traffic volumes and saturation flow rates are provided in the table. Determine the phasing, cycle length, and green interval and phase times for the signal plan if $X_C = 0.80$.

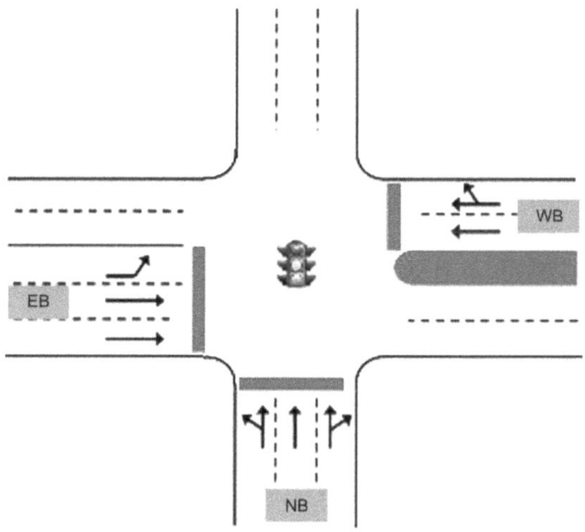

Traffic volumes, passenger cars/h

SB	R	T	L
SB	0	0	0

EB				WB
L	149		66	R
T	887		795	T
R	0		0	L

NB	L	T	R
NB	245	890	206

Saturation flow, passenger cars/h per lane

L	1285
TL	1620
T	1620
TR	1620
R	1390
LTR	1015

Problem 5.7

Caldwell Avenue, a four-lane arterial with a 40-mph speed limit, intersects with Cherry Street, a two-way local street in a small city. Cherry Street currently has stop control at this intersection. Local residents along Cherry Street have complained of difficulty getting out of their neighborhood during rush hours. An investigation of the conditions at the intersection, including a three-year crash analysis, shows that a signal is needed. The lane geometry and adjusted traffic volumes and adjusted saturation flow rates are provided in the table. Using $X_C = 0.85$, determine the phasing plan, cycle length, and green and phase times for the intersection. The minimum cycle length is 90 s.

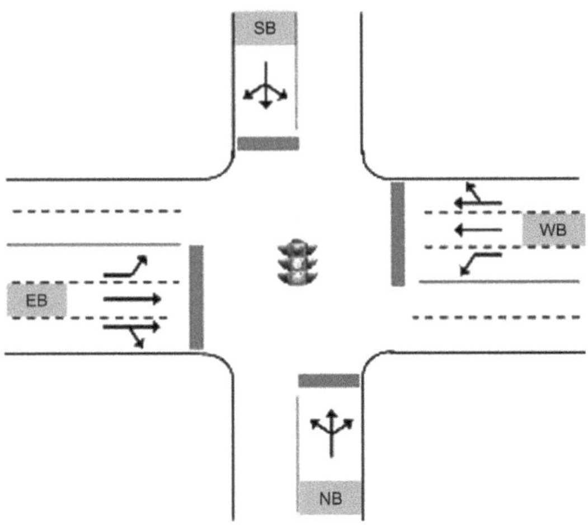

Traffic volumes, passenger cars/h

SB	R	T	L
	57	78	43

EB	
L	53
T	1427
R	63

WB	
R	66
T	1695
L	71

NB	L	T	R
	39	62	22

Saturation flow, passenger cars/h per lane

L	1180
TL	1300
T	1635
TR	1635
R	1500
LTR	975

Problem 5.8

Meadow View Drive is the entrance to a new subdivision off of Holly Ridge Road, a designated thoroughfare route in a city. Holly Ridge Road is a four-lane divided highway with a posted speed limit of 45 mph. Because of the site conditions, the intersection needs to be signalized. Lane geometry along with adjusted traffic volumes and adjusted saturation flow rates are as shown. Using $X_C = 0.80$, determine the phasing plan, cycle length, and green and phase times for the intersection. The minimum cycle length is 60 s.

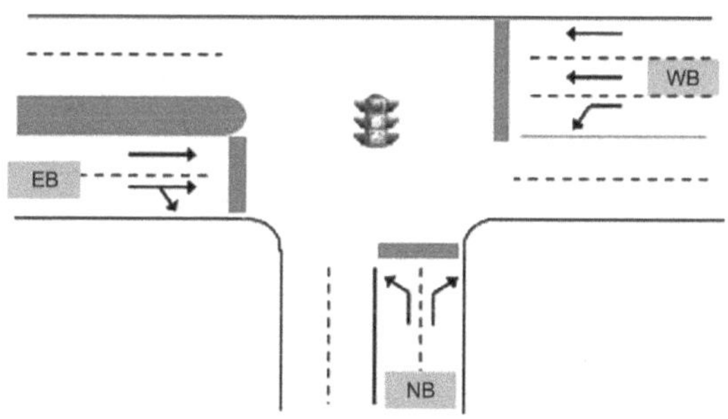

Traffic volumes, passenger cars/h

Approach	L	T	R
SB	0	0	0
EB	0	787	63
WB	108	675	0
NB	245		206

Saturation flow, passenger cars/h per lane

L	1290
TL	1340
T	1545
TR	1545
R	1490
L,R	1415

Problem 5.9

Derby Avenue is a two-way arterial street just outside the downtown of a small city. The figure shows a series of cross streets along Derby Avenue, which are all signalized intersections, and the signals need to be coordinated for efficient progression along Derby Avenue. Northbound (NB) traffic has priority movement in the morning peak hour. The progression speed is 30 mph, with a 60-s cycle length and a 60/40 phase split at each intersection.

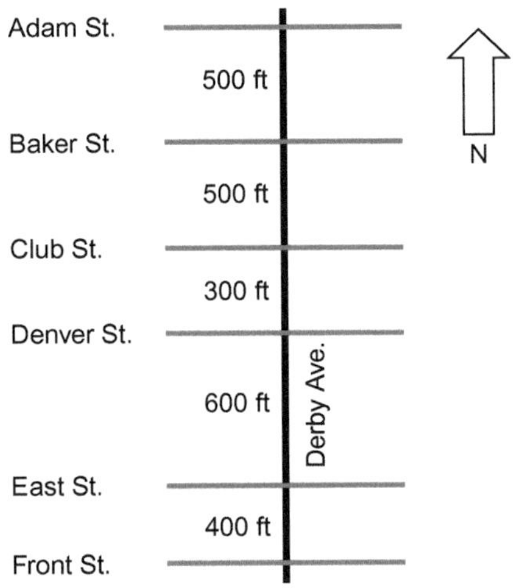

Derby Avenue is a 5-lane highway with a two-way left turn lane (TWLTL) in the middle. All cross streets are 40 ft wide.

1 Using grid paper, construct a time-space diagram for NB traffic including all labels, offset times, bandwidth, and the efficiency.
2 Using the information from (1) as fixed, draw in a maximum bandwidth for SB traffic for the same progression speed as NB traffic, and determine the bandwidth and efficiency. Vehicles may have to stop at one or more intersections, so the bandwidth would not be continuous through all intersections.

Problem 5.10

For the same segment of Derby Avenue as in Problem 5.9, use the offset times specified in the figure to determine the progression speed, bandwidth, and efficiency for NB and SB travel through all intersections without interruption, with about equal priority to each direction. Use a 50-s/30-s split within an 80-s cycle. The posted speed limit is 35 mph and both progression speeds must be at or below the posted speed limit.

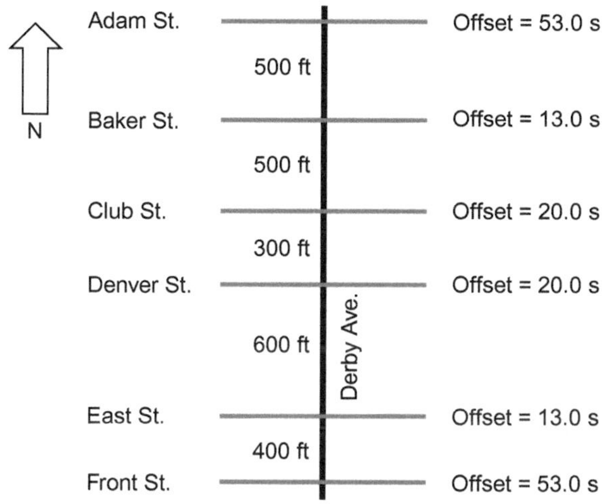

Problem 5.11

A city is considering the installation of a modern roundabout at the intersection of N. Main Street and E. Third Street. Peak-hour traffic volume estimates for the construction year are shown in the following table. Explore the feasibility of a single-lane and a two-lane roundabout at this location in a planning-level analysis and provide your thoughts on design needs for the intersection.

Approach	Volumes (veh/h)		
	Left	Through	Right
South	175	482	215
East	68	172	92
North	124	316	124
West	215	84	86

REFERENCES

Eads, B.S., Rouphail, N.M., May, A.D., Hall, F., 2000. Freeway Facilities Methodology in Highway Capacity Manual 2000. In: Transportation Research Board, National Research Council, Washington, D.C. Transportation Research Record: Journal of the Transportation Research Board. 1710, 171–180.

Elefteriadou, L., 2015. Proactive Ramp Management Under the Threat of Freeway-Flow Breakdown. (NCHRP 03-87) Final Report. National Academies of Science, Washington, D.C.

Federal Highway Administration (FHWA), 2005b. Managed Lanes: A Primer. Report FHWA-HOP-05-031. Federal Highway Administration, Washington, D.C.

Federal Highway Administration (FHWA), 2006a. Ramp Management and Control: A Primer. Report FHWA-HOP-06-080. Federal Highway Administration, Washington, D.C.
Federal Highway Administration (FHWA), 2006b. Traffic Detector Handbook, Vol. 1, third ed. Federal Highway Administration, Washington, D.C, <http://www.fhwa.dot.gov/publications/research/operations/its/06108/06108.pdf>.
Federal Highway Administration (FHWA), 2008. Technologies That Complement Congestion Pricing: A Primer. Report FHWA-HOP-08-043. Federal Highway Administration, Washington, D.C.
Federal Highway Administration (FHWA), 2009. Manual on Uniform Traffic Control Devices (MUTCD). Federal Highway Administration, U.S. Department of Transportation, Washington, D.C.
FHWA, 2005a. Traffic Congestion and Reliability: Trends and Advanced Strategies for Congestion Mitigation, vol. 6. Federal Highway Administration.
Forschungsgesellschaft für Straßen und Verkehrswesen (FGSV), 2001. "Handbuch für die Bemessung von Straßenverkehrsanlagen (HBS)." Köln, Germany.
Hu, J., Schroeder, B., Rouphail, N., 2012. Rationale for incorporating queue discharge flow into highway capacity manual procedure for analysis of freeway facilities. Trans. Res. Rec. J. Trans. Res. Board 2286, 76–83.
Klein, L.A., Mills M.K., Gibson D.R.P., 2006. Traffic Detector Handbook: -Volume II. No. FHWA-HRT-06-139.
Liu, X., et al., 2012. Analysis of Managed Lanes on Freeway Facilities. National Cooperative Highway Research Program, Transportation Research Board of the National Academies.
NCHRP, 2015. Project 03-107. Work Zone Capacity Methods for the Highway Capacity Manual. National Academies of Science, Washington, D.C.
Neudorff, L.G., et al., 2003. Freeway Management and Operations Handbook. No. FHWA-OP-04-003.
Perez, B.G., et al., 2012. Priced Managed Lane Guide. No. FHWA-HOP-13-007.
Rodegerdts, L., Bansen, J., Tiesler, C., Knudsen, J., Myers, E., Johnsonm, M., et al., 2010. Roundabouts: An Informational Guide, NCHRP Report 672. Transportation Research Board of the National Academies, Washington, DC.
Roess, R.P., Prassas E.S., McShane W.R., 2004. Traffic engineering.
Schroeder, B.J., 2008. A behavior-based methodology for evaluating pedestrian-vehicle interaction at crosswalks. ProQuest.
Schroeder, B.J., et al., 2010. Manual of transportation engineering studies.
Schroeder, B.J., et al., 2014. Work Zone Traffic Analysis & Impact Assessment. No. FHWA/NC/2012-36.
Transportation Research Board (TRB), 1980. Interim Materials on Highway Capacity. Transportation Research Circular, 212. Washington, D.C. <http://trid.trb.org/view.aspx?id = 153315>.
Transportation Research Board (TRB), 2006. Improving Pedestrian Safety at Unsignalized Crossings. NCHRP Report 562. <http://onlinepubs.trb.org/onlinepubs/nchrp/nchrp_rpt_562.pdf>.
Transportation Research Board (TRB), 2007. Roundabouts in the United States. NCHRP Report 572. <http://onlinepubs.trb.org/onlinepubs/nchrp/nchrp_rpt_572.pdf>.
Transportation Research Board (TRB), 2015. Highway Capacity Manual. Transportation Research Board of the National Academy of Sciences, Washington, D.C.
Urbanik, T., et al., 2014. Signal Timing Manual. Research Report FHWA-HOP-08-024.
Zegeer, J., et al., 2014. Incorporating Travel Time Reliability into the Highway Capacity Manual. No. SHRP 2 Report S2-L08-RW-1.

PART 6

Traffic Safety

Christopher M. Cunningham, MS, PE

Contents

6.1 INTRODUCTION

Many elements converge to create a good highway facility design and traffic safety is certainly one of the most important. State Departments of Transportation include public safety as the top priority in their mission statements. In 2013 alone, 32,719 people were killed in an estimated 5.7 million police-reported motor vehicle crashes across the United States.[1] That equates to 90 motor vehicle—related deaths each day, or one every 16 minutes. The collection and analysis of data related to collisions or other surrogate measures are fundamental to understanding how to reduce that toll on society. Such data are often utilized to better understand why collisions occur, to help identify crash-prone locations, to aid in installing countermeasures to reduce the likelihood of those crashes in the first place, and even in determining countermeasure effectiveness following treatment.

As transportation engineers, our mission is mobility for all and our mandate is safety. Several topic areas exist in the field of traffic safety, including: (1) roadside clearance analysis, (2) work zone safety, (3) horizontal curve advisory speeds, (4) road safety audits, (5) conflict analysis, (6) identification of hazardous locations, (7) countermeasure deployment,

[1] National Center for Statistics and Analysis. (2015, July) Overview: 2013 data. (Traffic Safety Facts. Report No. DOT HS 812 169). Washington, DC: National Highway Traffic Safety Administration.

and (8) analysis of countermeasures, to name a few. The focus of this part will be on (4) through (8).

6.2 ROAD SAFETY AUDITS

Used extensively in countries such as Australia, New Zealand, and the United Kingdom, a road safety audit (RSA) is a safety analysis tool that provides a formal evaluation of a new or existing roadway by an independent audit team. Recently, the Federal Highway Administration (FHWA) began providing guidelines for conducting general road safety audits[2] and more specific pedestrian road safety audits.[3] Both sets of guidelines provide a formal process for completing the audit, what team member's qualifications should be considered, and any additional tools the team should consider using when conducting the evaluation.

If used correctly, the RSA can be extremely useful to the agency. Because funding is often limited, the most common areas that agencies focus on in roadway design are roadway capacity, right of way, environmental issues, and political interest and constraints. An RSA is useful because it often brings to light the effect these focus areas may have on safety. For this reason, RSAs can be more useful in the planning and design stages of a new project in lieu of existing locations where changes are often hard to make and are most time-consuming and costly as projects move forward. RSAs have other advantages too. For instance, RSAs often uncover potential driver errors that would not readily present themselves during a collision study. In addition, crash reports do not always identify safety problems at sites. The concepts of conflict analysis (discussed in Section 6.3) fall into this category, though at a minimal level as RSAs are typically conducted along long stretches of roads and not single intersections or intersection approaches.

RSAs should not be mistaken for "road safety reviews" conducted by agencies. The primary difference between this and an RSA is that road safety reviews are conducted by internal staff, they almost always look at sites with severe crash problems and/or history, and they are reactive instead of proactive. RSAs, on the other hand, attempt to remove the potential for conflicts or collisions by identifying problem areas of a planned design or recently constructed project that can be addressed immediately. In addition,

[2] Federal Highway Administration. *Road Safety Audit Guidelines*. McLean, VA: Federal Highway Administration, 2006.

[3] Federal Highway Administration. *Pedestrian Road Safety Audit Guidelines and Prompt Lists*. McLean, VA: Federal Highway Administration, 2007.

using an external audit team eliminates potential bias or conflict during site visits, which brings to light problem areas that an agency may ignore.

An RSA consists of eight steps that can be used to formulate a final audit response. These are summarized in Table 6.1, along with responsibilities of the agency/design team and the audit team.

The majority of the work is done during Steps 2 through 4; however, once the process is completed only a couple of times, agencies can quickly replicate the process on other roads, allowing all participating agencies to benefit from the process.

6.2.1 Step 1: Identifying the Study Site(s)

Study sites that can be considered include new or existing sites in pre- or postconstruction. Good candidate sites for preconstruction audits are complex road designs, existing sites with known safety problems, and high-profile sites that are of interest to the public or politicians. Good postconstruction audits include sites with notable collision problems following construction, high-profile projects, or sites where traffic characteristics are expected to change due to construction or new developments in the area.

Table 6.1 Road safety audit process and responsibilities

Eight typical RSA steps		Project owner and/or design team	RSA team
Step 1	Identify project or existing road(s) for RSA	X	
Step 2	Select multidisciplinary RSA team	X	
Step 3	Conduct start-up meeting to exchange information	X	X
Step 4	Perform field reviews under various conditions		X
Step 5	Conduct RSA analysis and prepare report on findings		X
Step 6	Present RSA findings to project owner/design team	X	X
Step 7	Prepare formal response	X	
Step 8	Incorporate findings into project when appropriate	X	

6.2.2 Step 2: Selecting the Audit Team

RSA teams are usually formed through one of two avenues. In many cases, an RSA team is constructed by an outside consulting agency not familiar with the site, while other teams are formed by other local agencies outside the jurisdiction. The latter usually offers more flexibility and is a win–win for both agencies as they can use safety funds toward implementing safety improvements. This arrangement often entails a memorandum of understanding that discusses how each agency can benefit from the process and how often staff are committed to the process. Regardless of the method by which the RSA audit team is formed, each member should be formally trained in the process before embarking on RSA studies.

The audit team usually consist of 3–10 people from an outside group not familiar with the study site(s). The team usually consist of 3–5 engineers having some experience in roadway design, operations, and/or safety. Other auditors should supplement the team, such as local law enforcement, fire and rescue, and pedestrian and bicycle advocates (especially those familiar with ADA issues), to name a few. The local jurisdiction should be provided as much information about the site(s) being audited prior to the kickoff meeting.

6.2.3 Step 3: Kickoff Meeting

During the initial stages of the meeting, a minimum of two auditors are usually assigned to be the secretary and photographer. This allows others to be more active during the field review while others are taking notes to compile following the field review. The audit team should go over the materials provided by the jurisdiction under review, noting the beginning and end points of the project(s), any known problems the agency is having with safety at the site(s), and agreeing on a plan of action prior to conducting the review. The audit team should do its best to remember to consider roadway safety from the viewpoint of any and all road users including motorists, pedestrians, bicycles, transit, emergency responders, maintenance personnel, older drivers, and users with disabilities.

Although the project team is usually provides much of the necessary information related to the site(s) prior to the meeting, the jurisdiction should provide printed copies of any materials for use during the review. Examples of items usually provided by the local jurisdiction include roadway drawings (if in the design process), aerial imagery, condition and collision diagrams, and summarized collision data, to name a few. Any imagery or drawings that may be useful during the field review can be printed on larger paper for note-taking. It is also a good idea to have multiple copies in the event there are a lot of notes or an additional person is asked to take notes at the site(s). It is highly advisable that the audit team review the materials provided for familiarity and understanding prior to the actual site visit(s). This is especially true of any collision data and condition and collision diagrams, which should tell the auditor(s) where many of the problem areas are likely located at the site(s).

There are many equipment items that should be considered prior to visiting the site(s). First and foremost, high-visibility safety vests should be provided, especially where auditors may create a safety hazard or if attention is drawn to the auditing team that may be suspicious. A traffic vest provides a means to be seen and understood in many situations, especially if supplemental nighttime visits are being conducted. Other items that may be useful include: (1) still and video cameras for recording information that can later be used to document problem areas in the report or during any presentations, (2) a measuring wheel to measure distances such as vehicle lane or sidewalk width, (3) a stopwatch for measuring phase or clearance times, or even gather rough speed measurements, and (4) a clipboard and paper to keep notes.

6.2.4 Step 4: Conducting the Field Review

The field review is the primary focus of the audit where team members are asked to consider all roadway user needs at the site(s) of interest to the local jurisdiction. When visiting the site, it can be valuable for the audit team to drive together in a large van to hear each other's input and encourage open discussion. Where multiple modes are present at a site, the audit team should put themselves in the role of each user. For instance, at every site, the audit team should drive through the site as a motorist driving through all possible directions of traffic. Where time permits, other drivers are encouraged to drive the site multiple times during the visit(s) to get a feel for potential problem areas from the drivers viewpoint. Team members should be mindful that motorists are not just the general public, but also include emergency responders. It is advisable that the secretary and photographer sit toward the front to write down all observations and take pictures for the report. Another example would be sites where pedestrian or bicycle activity are expected. In these cases, it is highly advisable that audit team members walk (and bike, where team members are comfortable and have means to do so) the site to get a first-hand view of all facilities along the site. For all audits, it is always best to visit the site during the day and night as the site characteristics and needs vary greatly for all users. In addition, auditing the site during peak and off-peak periods is advisable.

When conducting the field review, a prompt list can be useful to make sure all team members consider all safety aspects of a project for each user type being considered. It is a way to ensure the RSA team gets the most out of a site visit and assist the team so that they do not forget specific features during the review. This is certainly true if RSAs are new to the audit team or if a new team member is added to an existing team already familiar with the process. The prompt list is also useful for reporting issues after the field review. An example of a recommended prompt list for pedestrian road safety audits is provided in Table 6.2.[4]

This prompt list helps remind the audit team to observe pedestrian elements such as marked crossings, audible and accessible push buttons, pedestrian signal heads, cracked or raised crosswalks, and so on. The list also provides guidance on whether the element should be considered during various stages of the project.

6.2.5 Step 5: Conducting Analysis and Preparing Report

The analysis of the site(s) should take place as soon as possible following the field review when all findings are fresh in the minds of audit team members. Typically, the roadway audit team meets in a room, similar to that used for a training environment, where a projector or white board are present, and share thoughts through the use of visuals or notes. The materials provided during the kickoff meeting are a good starting point for looking at the issues at the site(s), especially any aerials or drawings that include points of interest by team members. Pictures and videos can also be viewed during the analysis, with the secretary and photographer providing assistance where necessary.

During the final analysis, the audit team should prioritize the list of safety issues and provide mitigation measures that could be considered to improve the safety at the site(s). One method for prioritizing issues is to consider safety risk based on the likely frequency and severity of crashes at the site(s). Four recommended classifications are provided in Table 6.3:[5]

Using both measures, a matrix can be can be used to determine the relative risk when prioritizing the known safety issues. In the example presented

[4] Federal Highway Administration. *Pedestrian Road Safety Audit Guidelines and Prompt Lists.* McLean, VA: Federal Highway Administration, 2007.

[5] Federal Highway Administration. *Road Safety Audit Guidelines.* McLean, VA: Federal Highway Administration, 2006.

Table 6.2 Example prompt list directed at pedestrian road safety

Master prompt	Detailed prompt		RSA stages			
			Planning	Design	Construction	Postconstruction
Presence, design, and placement	A.1.1	Are sidewalks provided along the street?	✓	✓	✓	✓
	A.1.2	If no sidewalk is present, is there a walkable shoulder (e.g., wide enough to accommodate cyclists/pedestrians) on the road or other pathway/trail nearby?	✓	✓	✓	✓
	A.1.3	Are shoulders/sidewalks provided on both sides of bridges?	✓	✓	✓	✓
	A.1.4	Is the sidewalk width adequate for pedestrian volumes?	✓	✓	✓	✓
	A.1.5	Is there adequate separation distance between vehicular traffic and pedestrians?	✓	✓	✓	✓
	A.1.6	Are sidewalk/street boundaries discernable to people with visual impairments?		✓	✓	✓
	A.1.7	Are ramps provided as an alternative to stairs?	✓	✓	✓	✓
Quality, conditions, and obstructions	A.2.1	Will snow storage disrupt pedestrian access or visibility?	✓	✓	✓	✓
	A.2.2	Is the path clear from both temporary and permanent obstructions?	✓	✓	✓	✓
	A.2.3	Is the walking surface too steep?	✓	✓	✓	✓
	A.2.4	Is the walking surface adequate and well maintained?		✓	✓	✓
Continuity and connectivity	A.3.1	Are sidewalks/walkable shoulders continuous and on both sides of the street?	✓	✓	✓	✓
	A.3.2	Are measures needed to direct pedestrians to safe crossing points and pedestrian access ways?		✓	✓	✓
Lighting	A.4.1	Is the sidewalk adequately lit?	✓	✓	✓	✓
	A.4.2	Does street lighting improve pedestrian visibility at night?	✓	✓	✓	✓
Visibility	A.5.1	Is the visibility of pedestrians walking along the sidewalk/shoulder adequate?	✓	✓	✓	✓
Driveways	A.6.1	Are the conditions at driveways intersecting sidewalks endangering pedestrians?		✓	✓	✓
	A.6.2	Does the number of driveways make the route undesirable for pedestrian travel?	✓	✓	✓	✓

Table 6.3 Relative risk matrix

Likelihood	Severity
Frequent: ≥5 crashes/year	High: fatality likely
Occasional: 1–5 crashes/year	Medium: severe injury likely
Infrequent: <1 crash/year	Low: injury likely
Rare: <1 crash/5 years	Negligible: property damage only

Table 6.4 Relative risk matrix

Risk category		Severity			
		Negligible	Low	Medium	High
Accident frequency category	Frequent	C	D	E	F
	Occasional	B	C	D	E
	Infrequent	A	B	C	D
	Rare	A	A	B	C

in Table 6.4, a risk score of "A" corresponds to a safety issue considered to have low risk while "F" corresponds to a high-risk safety issue.

Once the analysis is finalized, one or more team members prepares a draft report for review by other team members. Following review by all members, the report is finalized and provided to the local agency for review and comment. The report should include all background information related to the site(s) being studied; the audit team members' names, backgrounds, and expertise; site visit dates and times; an explanation of the prioritization method used in the analysis; and a list of the safety issues and suggested mitigation to be considered.

6.2.6 Step 6: Presenting Findings

This step provides a means to discuss with the local agency the findings presented in the RSA and provide supplemental detail to help understand the recommendations being made. The idea is to provide an opportunity for the local agency to ask questions or seek clarification on notable items found in the report in a round-table discussion format. Regardless of the agency's feelings regarding a recommendation, any and all safety concerns should be documented in the report. Similarly, the local agency can provide a formal response to each recommendation, which can be included in the final documentation.

6.2.7 Steps 7 and 8: Preparing Formal Response and Incorporating Findings

The local agency is responsible for responding to the findings presented by the RSA team. This is usually done in the form of a formal response letter, which includes simple responses to each audit issue. The agency should identify what action will be taken for each issue and include a brief explanation for the action being taken. This response letter is provided to the audit team and becomes part of the project record and report.

The purpose of the RSA is to be proactive about implementing safety measures that can reduce the potential for crashes; therefore, agencies should implement changes based on recommendations from the team as soon as possible. The formal response along with the other report documents provided to the local agency should be used as a reference in discussions with other staff and consultants working on the site of interest.

6.3 CONFLICT ANALYSIS

Suppose you are working for a city and complaints are filed concerning an intersection that is considered unsafe for pedestrians. You review the collision data and find no collisions in the last three years between any motor vehicle and pedestrian. What do you do? (1) You would probably schedule a site visit to spot any concerning behavior, and (2) you could also perform an alternative safety study such as a conflict analysis. The *Manual of Transportation Engineering Studies*[6] provides this definition for a conflict:

> *A potentially risky interaction between two or more vehicles or road users when one or more vehicles or road users take evasive action, such as braking or weaving, to avoid a collision. Conflicts are used as surrogate measures for actual crashes, which are rare.*

As noted, conflict analyses are typically used when there is limited collision data available for analysis or collisions of a certain type are not sufficient such that a proper statistical analysis can be accomplished. Traffic conflict studies can also be performed as a supplement to a normal crash investigation to learn more about the extent of a problem and its location in the target area being studied. However, conducting traffic conflict studies is not always simple, and when performed improperly, they may provide misleading information.

[6] Institute of Transportation Engineers. (2010). Manual of Transportation Engineering Studies (2nd ed., pp. 11). Englewood Cliffs, N.J: Prentice Hall.

Traffic conflict studies require a relatively small investment of time and resources and require no special equipment. Trained observers watch traffic and note on a form when a conflict occurs. Observers usually require a week or less of training. At a single intersection approach or small intersection, adequate for a study of 1 to 3 days in length, provided they get a sufficient sample of conflicts.

Conflict studies can be utilized in several ways and these depend on the type of analysis being performed. Typically, conflict studies are conducted to determine the number and type of conflicts at an intersection, the magnitude of a traffic safety problem at a known location, or to evaluate the effectiveness of a safety-related countermeasure that has been implemented and needs studying. This section describes the type of conflicts an analyst should consider studying along with the four methods of conflict analysis used for field studies and their uses in the transportation industry for making sound safety decisions.

6.3.1 Conflict Types

The most common type of conflict study is conducted at intersections, though studies can be conducted at any "spot" or "section" with the right training. Researchers typically define conflicts at intersections using one of 14 possible categorizations. These conflict types are provided in Figure 6.1.[7] These conflicts apply to any type of intersection control: signalized, unsignalized, and even driveway openings, if necessary. Although 14 intersection conflict types are provided, in most conflict studies observers record only the conflict types that are related to the study purpose, rather than all 14 types. Section 6.3.2 describes determining the type of conflicts that should be considered at a site for general study purposes. However, it is entirely plausible that the analyst studying the site is looking at one or two key conflict types based on user input from the general public or another entity.

Traffic conflict types are not well defined for nonintersection locations such as weaving sections, diverges, or merges. Although these are usually harder to capture, they are entirely possible to collect. Conflict studies at sites with unusual geometry or movements that are unfamiliar to analysts can be challenging to collect conflict data. It is advised that analysts

[7] Glauz, W. D., and D. J. Migletz (1980). *Application of Traffic Conflict Analysis at Intersections*, National Cooperative Highway Research Program Report 219, Transportation Research Board, Washington, D.C.

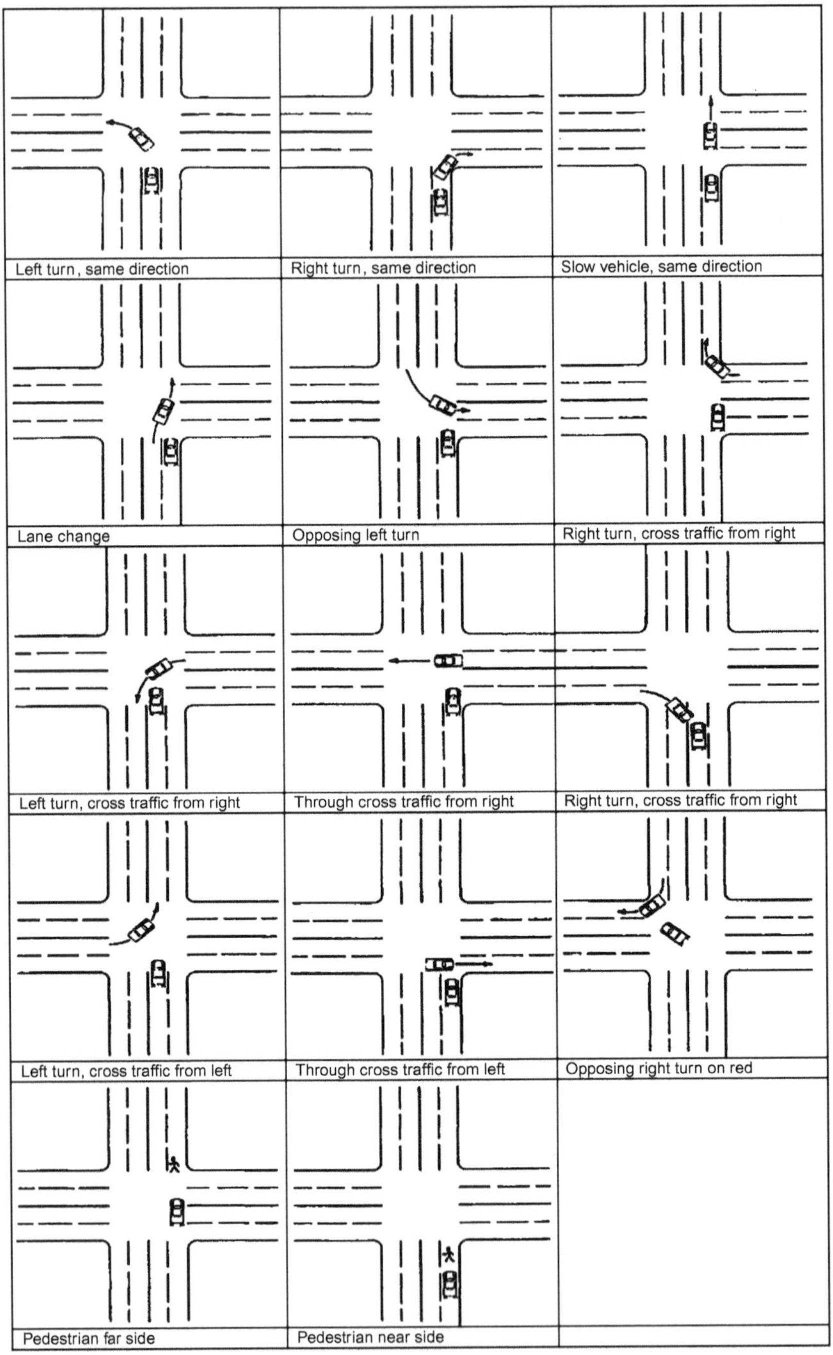

Figure 6.1 Typical conflicts associated with signalized intersections.

conduct preliminary observations or pilot tests to inform observers which conflicts to record at nonintersection locations.

6.3.2 Determining Number and Type

When conducting a conflict study, the analyst will often want to document the number and type of conflicts taking place at an intersection. This exercise is usually done for one of two reasons. First, the engineer or planner may wish to compare the number and type of conflicts between alternative designs during the planning stages of a project to get a sense of the expected safety of multiple designs. More often, an analyst will conduct this type of study prior to an actual field investigation to get a sense of the types of possible conflicts that may need to be studied. Although several will be identified, only a select few may be studied if there is knowledge of the types of problems taking place at the site.

Conflicts are most often classified as either diverging, merging, or crossing. *Diverging* conflicts are typically less problematic and associated with driver confusion; however, these types of conflicts can be extremely dangerous. For instance, a driver in the circulating lanes of a roundabout may accidentally try to exit at the wrong approach and redirect into the circulating lanes. *Merging* conflicts are more prominent and are most often correlated with side-swipe collisions. These conflicts are associated with driver misunderstanding, poor geometric design, or lack of appropriate signage or lane markings. *Crossing* conflicts are extremely problematic. This conflict type is correlated most often with angle and left-turn opposing collisions.

The schematic in Figure 6.2 shows conflict types at two different intersection types: a typical four-leg intersection and a modern roundabout. Notice that the conflict types correlate to the types of collisions you would

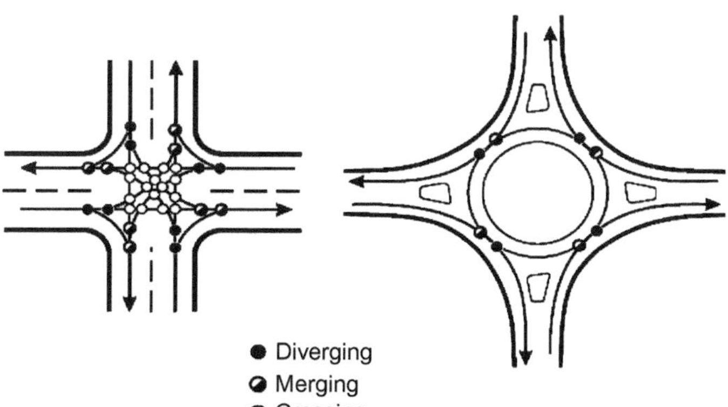

Figure 6.2 Conflict types at a typical four-leg intersection and a modern roundabout.

Table 6.5 Conflict points for a 4-leg intersection and modern roundabout

	Number of conflicts	
Conflict type	**4-leg intersection**	**Modern roundabout**
● Diverging conflict	8	4
◑ Merging conflict	8	4
○ Crossing conflict	16	0
Total conflicts	32	8

expect (Table 6.5). If the analyst were looking at the difference in potential conflict types in addition to other measures (such as operations, life span, maintenance, etc.), they would conclude that the roundabout was the safest option as the dangerous crossing conflicts have been removed and diverging and merging conflicts have been cut in half. These potential crash savings would then need to be weighed along with other tradeoffs.

The conflicts shown in Figure 6.2 only include vehicle-to-vehicle interactions. If pedestrians are allowed to cross at the 4-leg intersection, there are 16 additional conflict points for single-lane approaches on all legs. Conflicts can also occur between bicyclists and vehicles and between bicyclists and pedestrians.

EXAMPLE 6.1 Determining the Number and Type of Potential Conflicts

Determine the number and type of conflicts for a median U-turn intersection that removes all direct left turns at the main intersection and diverts to the downstream U-turn opening. What are the expected safety benefits compared to a standard 4-legged intersection?

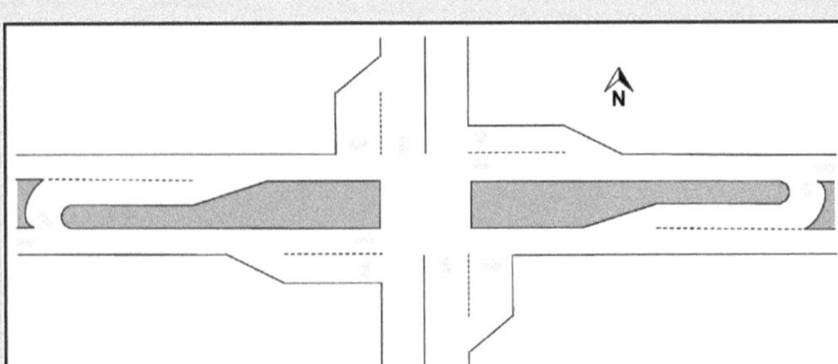

Solution

First, draw the individual movements for the east and westbound directions (solid horizontal line), followed by the movements of the north and

(Continued)

EXAMPLE 6.1 Determining the Number and Type of Potential Conflicts—(Continued)

southbound directions. Then, identify the number and types of conflicts. The primary benefit is the reduced number of crossing conflicts, which should greatly improve safety by removing more severe types of collisions.

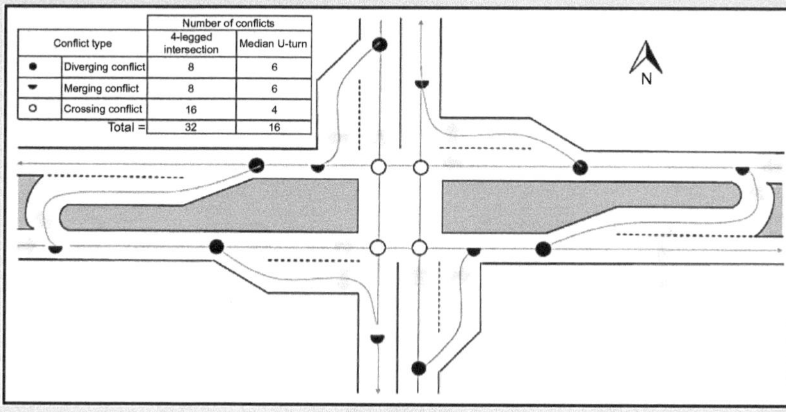

EXAMPLE 6.2 Determining Number of Possible Conflicts

For the T-intersection shown, how many vehicle-to-pedestrian conflict points are present if pedestrians are only allowed to cross the minor street?

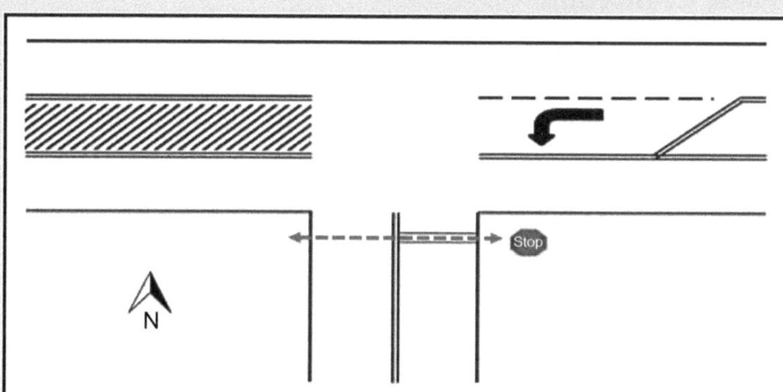

(Continued)

EXAMPLE 6.2 Determining Number of Possible Conflicts—(Continued)

Solution

First, draw the individual movements for the east and westbound directions (solid line with arrow heads), followed by the movements of the north and southbound directions (solid line that passes through encircled 1). There are three pedestrian–vehicle conflicts. Two different vehicular directions approach the exiting lane, while only one direction approaches the entry lane to the intersection.

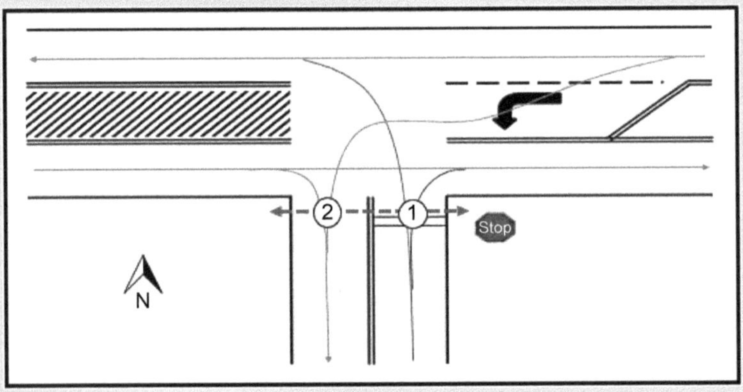

6.3.3 Field Evaluation

An analyst could use conflict analysis to determine the *magnitude* of a potential safety problem at a location of interest. This type of field evaluation is often completed much faster than a collision study, especially at new sites where sufficient collision data will not be readily available for many months. Typically, field evaluations are completed using one of four methods:

1. Before-and-after study: This method is only used at existing sites where a treatment is being employed. The treatment can be a safety or operational countermeasure. Safety countermeasures are intended to reduce conflicts; therefore, a short conflict study could determine if the intended reduction of conflicts is evident prior to conducting a future crash study. On the other hand, a study of an operational

treatment, although it may improve operations, may have the unintended consequence of increasing conflicts. If an operational treatment is suspected to have this issue, it should be studied early before waiting for crashes to take place. The before-and-after analysis method is the least common of the four methods deployed because it requires a staff person to recognize the need to collect the data prior to implementing a treatment.

2. Cross-sectional study: Typically, evaluation of a countermeasure is done using a before-and-after study; however, if the option to conduct a before-and-after study has passed, a cross-sectional study could be implemented provided intersections exist that are similar in nature during the before period. To evaluate countermeasures, the conflict types being studied should be closely related to the countermeasures that have been implemented. The conflict types can be identified using methods described in Section 6.3.1.
3. Reference group: This analysis method is similar in concept to a cross-sectional study; however, the analyst does not have access to a similar site type or has a short time frame to do the analysis using other nearby sites. Instead, this method looks to compare conflict rates from a study site to known rates of similar "spots" or "segments." This method is not used often because there exists little to no information on typical conflict rates, and the information that is available is often deemed unreliable. An example of typical conflict rates for a reference group of intersections with four approaches is provided for reference in Table 6.6.[8]
4. Frequency: This method looks only at conflicts at the site of interest following a treatment installation. The frequency-only method is used when the analyst or engineer has failed to collect before data and/or cannot determine common conflict rates from a reference group. This method is the most common of the four types because conflict studies are almost always conducted at existing sites where one or more problems have already been identified as needing attention, and not at a site where a general safety study is being conducted. The problem areas being studied are usually identified

[8] Migletz, D. J., W. D. Glauz, and K. M. Bauer (1985). *Relationships between Traffic Conflicts and Accidents,* FHWA/RD-84/042, Federal Highway Administration, Washington, D.C.

Table 6.6 Typical conflict rates for intersections with four approaches

Conflict type	Conflicts/hour		Conflicts/day			
	Mean	Variance	Mean	Variance	Percentile	
					90th	95th
Signalized with entry volumes greater than 25,000 vehicles/day						
Left turn, same direction	7.6	22	83	12,000	270	360
Slow vehicle	61	34	670	24,000	870	940
Lane change	1.7	—[b]	18	160	35	43
Right turn, same direction	20	11	220	7.6	470	510
Opposing left turn	2	1.2	22	380	48	60
All same direction[a]	90	74	990	67,000	1300	1500
Signalized with entry volumes 10,000–25,000 vehicles/day						
Left turn, same direction	12	22	130	10,000	270	340
Slow vehicle	34	11	380	4900	470	500
Lane change	0.7	—[b]	8	53	17	22
Right turn, same direction	11	12	120	2400	190	220
Opposing left turn	2.6	1.2	29	210	49	56
All same direction[a]	59	95	640	25,000	860	930
Unsignalized with entry volumes 10,000–25,000 vehicles/day						
Left turn, same direction	12	21	130	12,000	270	350
Slow vehicle	14	5.2	150	5900	260	290
Right turn, same direction	5.6	11	62	1200	100	120
Opposing left turn	0.8	1.2	9	40	17	21
Right turn from right	0.8	1.1	9	99	21	29
All same direction[a]	29	77	320	29,000	540	640
Through cross traffic	0.6	—[b]	7	16	12	14
Unsignalized with entry volumes 2500–10,000 vehicles/day						
Left turn, same direction	6.4	22	71	1000	110	130
Slow vehicle	9.3	5.5	100	9600	220	300
Right turn, same direction	5.3	11	58	2200	120	150
Opposing left turn	0.3	—[b]	4	8	8	9
Right turn from right	0.5	1.1	6	12	10	12
All same direction[a]	21	77	230	18,000	410	490
Through cross traffic	1.1	—[b]	12	75	24	29

Note: Basic intersection conflict types not shown had mean hourly rates less than 0.5. Statistics are based on sample counts conducted in the Kansas City metropolitan area on all four approaches of signalized intersections and on the approaches with the right of way at unsignalized intersections. Counts were taken during the daylight, in dry weather, and do not include secondary conflicts.

[a]"All same direction" includes left turn, same direction; slow vehicle; lane change; and right turn, same direction conflict types. "Through cross traffic" includes cross traffic from left and cross traffic from right conflict types.

[b]Not available.

through direct public input, political involvement, or the media. This method is also used at sites where changes have been implemented but no before data were collected—in this case the usual reason is because there was no expectation that a potential safety problem may surface.

Regardless of the method used, during a field study an analyst should collect two forms of data: (1) the number of observations of driver actions that involve a change in direction and/or speed to avoid a collision and (2) a count of the number of vehicles moving through the intersection (or some other surrogate to determine the rate of conflicts). Hard braking (including skidding) and abrupt lane changing are behaviors that would be counted as they are made to avoid a crash. Normal braking for a vehicle ahead turning right, for example, is not considered a conflict. The study then would tally up all conflicts, comparing the types of conflicts observed and the frequency, where possible.

Although the type and magnitude of conflicts does help define any safety issues present at a site of interest, the rate of conflicts is also helpful during analysis. This is especially true of conducting study types 1, 2, or 3. Ideally, traffic volume data are collected at the site during the analysis period for each of the movements being studied. However, this can be distracting to the analyst. If actual volumes are not gathered onsite, the next best option for calculating a rate of conflicts is to use a recent turning movement count. Another option for determining rate of conflicts is the use of a denominator based on time. For instance, conflict per hour, per minute, per day, or per signal cycle may be used. Regardless of the unit of measure used, the analyst must make sure that all sites use the same method of comparison.

If the study is oriented toward pedestrian and/or bicycle conflicts with vehicles, the study will focus more on risky pedestrian and bicyclist behaviors and determine if corrective action is needed and what might be the best approach to reduce risky behaviors. Regardless of the mode(s) of transportation being studied, it is always good to get a count to determine conflict rate where appropriate.

EXAMPLE 6.3 Field Evaluation of a Diverging Diamond Interchange

An existing diamond interchange is being retrofit with an innovative interchange form, the diverging diamond interchange (DDI). The primary reason for replacing the interchange is to improve operations by reducing queues and delay to the traveling public. Since replacing the interchange, the State Department of Transportation (DOT) has received several phone calls from motorists complaining about unsafe driving conditions around the interchange ramp terminals. (*Note:* DDIs require drivers to "crossover" and drive on the left side of the road between the ramps, allowing free-flow left turns onto the freeway.)

As the safety analyst, you are asked to determine what safety problems may be present at the intersection and recommend potential countermeasures. Because this interchange treatment is new, you decide to conduct a surrogate safety analysis using conflicts. As a starting point, you draw a basic conflict diagram showing that the DDI only has 14 types of conflicts (compared to the 26 known conflict types at a conventional diamond interchange). Using this diagram, you cannot immediately determine what safety problem(s) may be present.

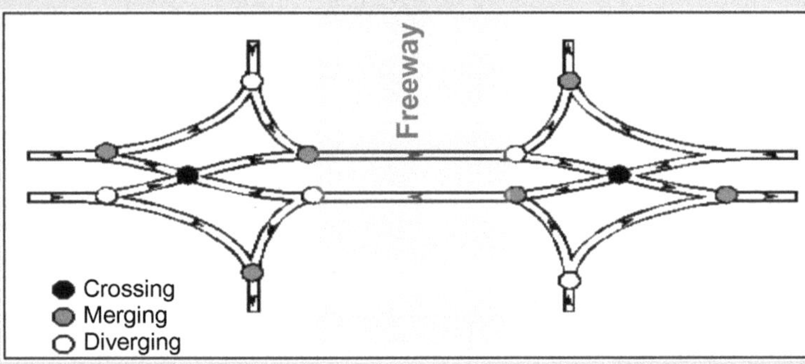

You then take a colleague with you to collect conflict data at each of the ramp terminals. You collect data during the am and pm peak periods on two separate days, each over a two-hour period, switching crossovers on the second day to help remove analyst bias. During the field assessment, you collect conflicts during 86 and 100 cycles at side A and B, respectively (several cycles were not able to be collected on side A). In addition to conflicts, you decide to collect an additional data set called "erratic maneuvers," which are single-vehicle incidents that have the potential to lead to a collision. Your combined conflict/erratic maneuver diagram is shown in the following.

(Continued)

EXAMPLE 6.3 Field Evaluation of a Diverging Diamond Interchange—(Continued)

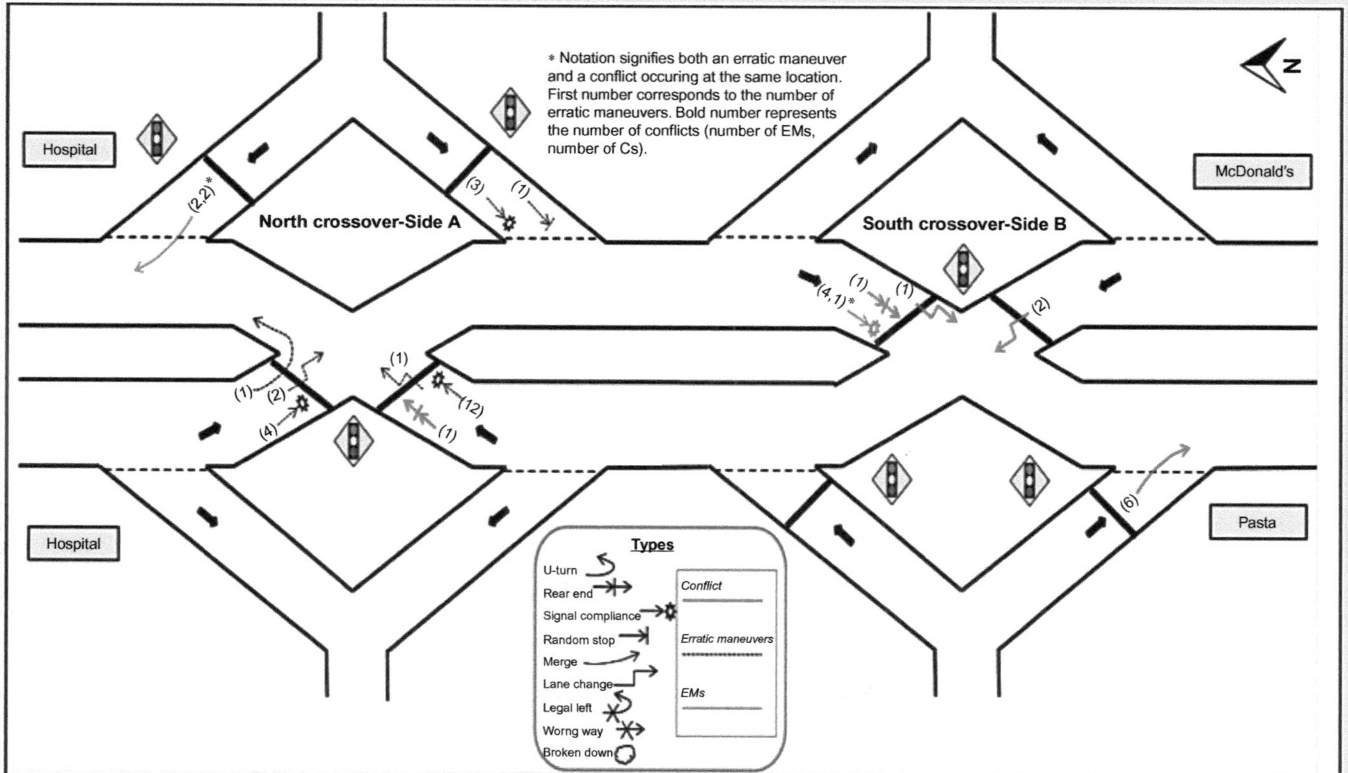

(*Continued*)

EXAMPLE 6.3 Field Evaluation of a Diverging Diamond Interchange—(Continued)

Given that no before data are available and there exist no known sites that can be used as a reference group, what appear to be the problems at this site?

Summary of data		Side A		Side B	
Total cycles		86		100	
Unit		Event	E/cycle	Event	E/cycle
Total		29	0.34	15	0.15
Event	Conflicts	3	0.03	1	0.01
	Erratic maneuvers	26	0.30	14	0.14
Event detail	U-turn	1	0.01	0	0
	Rear-end conflicts	1	0.01	1	0.01
	Signal compliance	19	0.22	5	0.05
	Stop	1	0.01	0	0
	Merge	4	0.05	6	0.06
	Lane change	3	0.03	3	0.03
	Other	0	0.00	0	0

Solution

Looking at the conflict diagram, the majority of conflicts and erratic maneuvers are taking place at the crossover and not the ramps. The data are further summarized using conflicts, erratic maneuvers, and total "events" per cycle and graphed.

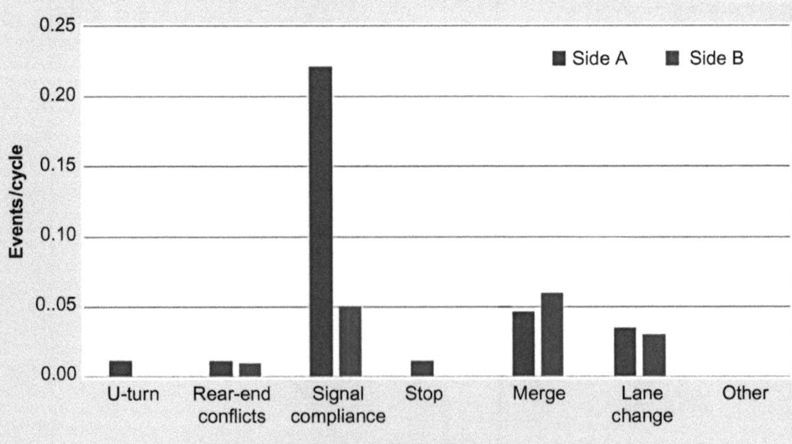

(Continued)

EXAMPLE 6.3 Field Evaluation of a Diverging Diamond Interchange—(Continued)

The data in the graph indicate an obvious signal compliance issue. Based on your notes, this was most likely due to the long cycle lengths present at the two-phase intersections. You recommend to your boss that they consider half cycling the DDI. In addition, there appeared to be a merging problem for right-turning vehicles onto the arterial. Your notes indicate that drivers appeared to be turning into oncoming traffic that came from a direction they were not expecting (the opposite side of the road through the crossover). You recommend looking at ways to improve sight distance and angle. Last, lane changing in the crossover was somewhat problematic at both crossovers. Looking at the aerial diagrams, you tell your boss to consider improving channelization by providing better alignment cues such as curbing or pavement marking symbols at, and through, the crossover.

6.4 COLLISION ANALYSIS

Collision study techniques have improved dramatically in recent years, primarily due to the focus by agencies on robust study methods and the ease of analysis and data storage using computers and spreadsheet tools. New and improved techniques used by traffic engineers in various jurisdictions and DOTs are discussed in this section. Some advanced analysis techniques are not discussed in this part as they require expertise and/or significant effort not typically possible at the local level.

As transportation engineers, our role in designing roadways is to provide the most forgiving transportation system we can within the prescribed budgetary constraints. When studying transportation facilities, analysis of trends in crash types, time of day, and weather conditions should be distinguished from detailed crash causes, which is often a secondary exercise for practitioners who may be involved in an expert witness role. Standard crash reports, at least at present, do not require all necessary data, leaving most of the core causal factors to be debated and resolved on a larger policy level. However, when a significant causal factor is present in a large proportion of crashes (e.g., distracted driving, drowsiness, or elderly crash rates, to name a few), engineers have an obligation to make recommendations to policymakers that could reduce their occurrence.

This section discusses techniques for conducting studies using reported collisions. First, the collision report forms are described along with sources of collision data available to the analyst. Further discussion is provided on how to reduce and merge data sets to prepare for analysis, along with typical problems and limitations that come up when preparing data. Second, collision diagrams are utilized to look for clues on why there may be more collisions at this intersection than other comparable intersections. Last, studies are discussed with particular emphasis on trends, techniques for hazardous site identification, and the possible countermeasures implemented to reduce the potential for problematic collisions at the studied sites.

6.4.1 Data Collection

Crash data used by safety analysts are recorded primarily by the police on hard copy report forms or laptop computers soon after a crash. One police report form is filled out per collision regardless of the number of vehicles or individuals involved. Most states have a standard collision form used by all police forces within the state. The form commonly has as many as 50–200 fields (and more in some states) including such detail as specific information on the drivers and passengers, the vehicles, the roadway, the conditions at the time of the collision, a narrative describing the event, and, in most cases, a sketch of the collision with vehicle paths and objects that are struck.

Collision data are most often available directly from the agency requesting or conducting the analysis. Many municipal and state agencies can provide analysis with copies of, or electronic access to, requested collision data quickly via software designed to poll various parameters noted by the analyst. Agencies providing data will usually erase any personally identifiable data items such as name, driver's license number, or address to protect the privacy of those individuals. In some cases, it may be quicker to access collision data records from an alternate source, especially if the analyst is to analyze a subset of the total population. Several databases are available for reference:

- The Highway Safety Information System (HSIS), developed by the Federal Highway Administration (FHWA), consists of nine relatively high-quality databases from geographically scattered states based on the quality of data available and their ability to merge data from various files. Collision data from police reports are processed to correct errors and to allow easy merges with traffic, roadway, and other files also maintained within HSIS.

- The Fatality Analysis Reporting System (FARS) is maintained by the National Highway Traffic Safety Administration (NHTSA). FARS is an almost complete census since 1975 of all collisions in the United States that have resulted in one or more fatalities.
- The General Estimates System (GES), which is also maintained by NHTSA, contains records from detailed independent investigations of a small, statistically selected sample of collisions across the United States.
- The Model Minimum Inventory of Roadway Elements (MMIRE) is an FHWA undertaking that includes a listing of roadway inventory and traffic elements critical to safety management and proposes standardized coding for each. MMIRE serves as the companion to the Model Minimum Uniform Crash Criteria (MMUCC), which was previously developed to standardize crash data variables used by states and local jurisdictions to improve crash data systems.
- SAFETYNET was designed by the Federal Motor Carrier Safety Administration (FMCSA) to manage and provide appropriate access to crash data, roadside inspection history and data, and motor carrier and shipper identification information. Data records include, but are not limited to, truck/bus driver name, driver social security number, driver license number, driver date of birth, driver and company contact information, and vehicle identification number (VIN).

Once data are collected, they can be further aggregated by collision type and severity to better understand how to address safety at a particular site or sites. *Collision type* is one of the most important fields on the report form for traffic engineers. Jurisdictions code collision type in various ways; however, most states now code collisions by first harmful event and the most harmful event for each individual vehicle involved. Most often, the police officer coding the collision is responsible for coding the collision type. An example of collision codes used by some states is provided in Table 6.7.

Looking at Table 6.7, the police officer filling out the report has 24 choices, including several choices of "other." He or she could code the collision as a noncollision, collision of a motor vehicle with some other object than another vehicle, or a collision between more than one motor vehicle in the traffic stream. For collisions involving more than one motor vehicle, most states employ a typical coding system for collision type, including right-turn, left-turn, rear-end, head-on, side-swipe, angle, and backing collision types.

Table 6.7 Example of collision types used by many states

Collision type	Code number	Specific collision
Noncollision	1	Ran off road right
	2	Ran off road left
	3	Ran off road straight ahead
	4	Jackknife
	5	Overturn/rollover
	6	Other noncollision
Collision of motor vehicle with...	7	Pedestrian
	8	Pedal cyclist
	9	Railway train/engine
	10	Animal
	11	Movable object
	12	Fixed object
Collision of two or more motor vehicles	13	Parked motor vehicle
	14	Rear end, slow, or stop
	15	Rear end, turn
	16	Left turn, same roadway
	17	Left turn, different roadway
	18	Right turn, same roadway
	19	Right turn, different roadway
	20	Head on
	21	Side swipe, same direction
	22	Angle collision
	23	Backing up
	24	Other collision with vehicle

The driver and passenger *injury codes* are also very important for most traffic safety studies. The most common coding scheme used by states is the "FABCO" or "KABCO" scale, which includes five categories:

1. F (fatality) or K (killed): The person died within 30 days of the collision as a direct result of injuries received during a crash.
2. A: The person experienced serious, incapacitating, nonfatal injuries during the collision, for example, broken bones, massive loss of blood, or even more serious injuries.
3. B: The person experienced a visible but not serious or incapacitating injury during the collision.
4. C: The person complained of pain or momentary loss of consciousness due to an injury during the collision, but no visible sign of injury was evident to the investigator.

5. O: No injury, which includes "PDO" or "property damage only" collisions. These are often significantly underreported as they are often handled between the driver(s) of the vehicles.

Given the increasing number of fields present on the collision report form, there are certainly concerns related to the reliability of data. Common errors in collision data are unreported collisions such as property damage only (PDO), incorrect or incomplete location, severity changed after a report is filed (i.e., an occupant dies four days later from injuries sustained from the crash), the collision type is incorrect, or the roadway characteristics may be reported incorrectly.

Last, it is common that a minimum of three years of collision data are utilized for collision analysis. Three years of data usually provide a sufficient sample size to provide statistically sound findings, especially if vehicle-to-vehicle collisions are being analyzed. More data are always better; however, the analyst should make sure that conditions have not changed during any reporting periods. Common examples include changes in reporting thresholds (i.e., PDO crash thresholds increased from \$1000 to \$1500), roadway construction was present that could cause large fluctuations in crashes during certain time periods, or seasonal impacts, which are usually problematic over larger study areas (i.e., statewide hazardous crash study impacted by a random localized weather event on one side of the state but not the other).

6.4.2 Planning/Prescreening Methods

Municipalities and State Department of Transportations (DOTs) are highly focused on making roads safe for the driving public and often have safety programs or groups dedicated to the cause. Collision analyses are almost always the focus of the agencies' safety programs, where dedicated resources are put aside to try and reduce the number of collisions, the severity of collisions, and the liability risk due to collisions. As such, these agencies should regularly identify sites that may need countermeasures to improve safety. Once a hazard has been identified, corrective action can be planned and implemented. There are limited funds available for any transportation program, including safety issues, so analysis is a vital component of any decision-making process. A systematic approach is helpful in identifying these hazardous sites, and no one method is perfect. Several methods are provided in the following subsections.

Many studies focus on one location or a limited set of locations on the highway network. One of the analyst's most important tasks in these studies is to reduce the database to crashes that occurred at locations of interest. Analysts usually summarize collisions into those that occurred at *spots* and those that occurred in roadway *sections*. Spots are short segments of highways that help identify problem "point" locations such as intersections, curves, and short bridges. The highway cross section and other features at a spot should be noticeably different from surrounding spots. Sections are longer, relatively homogeneous segments of highway convenient for studying cross sections, pavement surfaces, and other longitudinal features. Roadway sections typically correspond to tangent sections of roadways. Spot lengths of 0.2–0.3 miles and section lengths of 1–2 miles are recommended for most agencies.

Collision Frequency

The simplest method of identifying hazardous sites is to list locations ranked by the total collisions, the number of collisions by a particular type, or by injury severity. The primary reason agencies like this method is that it is simple to conduct and is intuitive. When specialized collisions, such as pedestrian collisions, are being monitored for frequency, a spot map is often utilized, which provides a method for determining collision-prone locations qualitatively. Minimizing total collisions or high-severity collisions by addressing the highest crash locations in a municipality or other domain is a logical first step in analysis. However, a safety analyst must understand that utilizing this site selection method by itself will almost always lead to the heaviest traveled locations to be chosen because they have the highest vehicle exposure. Using this method alone will ultimately cause agencies to ignore less busy sites that may produce a better effect following countermeasure implementation. In addition, agencies will continue to pour safety funds into the same intersections with very limited results.

Collision Rate

Safety analysts will often identify collision-prone locations by collision rate. This method helps account for vehicle exposure by accounting for traffic volume, which is the primary limitation of the frequency method. Traffic volumes should be utilized during similar periods of time to that of collision data. If turning movements or other similar volume counts are collected during different times of the year, seasonal factors should be applied. These are typically provided by the state Department of Transportation by facility type and/or location within the state. Once seasonal corrections are made, volumes may need to be adjusted for growth or decline if data are used from a different time period. For high-growth areas, traffic volumes outside of five years or more should be used with extreme caution as nearby developments cause fluctuations that are difficult to account for.

The rate method should be carefully calculated based on whether roadway spots or sections are being evaluated. Agencies usually calculate the collision rates for spots in terms of collisions per million entering vehicles (MEV). The following equation can be used for determining collision rates at spots.

$$R_{spot} = \frac{(A)(1,000,000)}{(365)(T)(V)}$$

where

R_{spot} = collision rate for the spot

A = number of reported collisions during the time period

T = number of years considered

V = AADT or annual average daily traffic, vehicles per day

Highway section collision rates should be ranked on the basis of reported collisions per million vehicle miles (MVM).

$$R_{section} = \frac{(A)(1,000,000)}{(365)(T)(V)(L)}$$

where

$R_{section}$ = collision rate for the *section*

A = number of reported collisions during the time period

T = number of years considered

V = AADT or annual average daily traffic, vehicles per day

L = section length, miles

Unlike collision frequency, the major disadvantage of utilizing the rate method is bias toward low exposure sites. For instance, a few unlucky collisions at a spot or section with low exposure will produce a high collision rate.

Agencies that typically use collision rates will usually pair this method with collision frequency to determine the most appropriate allocation of safety funds. For instance, an agency may choose to only consider sites meeting a certain minimum frequency of fatal collisions (a subset of total collisions in this case identifying those most severe crash types). Of this subset, the analyst would then rank the contending sites by their rate (R_{spot} or $R_{section}$). This helps avoid some of the biases present by using both methods.

EXAMPLE 6.4 Collision Rate

The State DOT has set aside $2,300,000 this year for rural spot safety funding. Four divisions submitted proposals for two intersections each, along with the approximate costs of countermeasures and the expected reduction in collisions following treatment installation. The eight rural intersections are shown in the following table, along with total collisions for a three-year period. Which sites would you recommend to be funded?

Intersection	Collisions	AADT		Cost	% reduction
		Road 1	Road 2		
1	25	8200	3500	$400,000	20%
2	21	7000	500	$280,000	15%
3	32	5500	750	$480,000	28%
4	35	6000	3000	$320,000	15%
5	24	8250	1000	$320,000	22%
6	42	9250	500	$800,000	17%
7	28	6700	1250	$400,000	15%
8	22	8300	750	$300,000	25%

Solution

The collision rates are calculated in the following table. Intersection 2, 3, 4, 6, and 7 should be funded at a total cost of $2,280,000.

Intersection	Collisions	AADT		Cost	% reduction	R_{intx}
		Road 1	Road 2			
1	25	8200	3,500	$400,000	20%	1.95
2	21	7000	500	$280,000	15%	2.56
3	32	5500	750	$480,000	28%	4.68
4	35	6000	3000	$320,000	15%	3.55
5	24	8250	1000	$320,000	22%	2.37
6	42	9250	500	$800,000	17%	3.93
7	28	6700	1250	$400,000	15%	3.22
8	22	8300	750	$300,000	25%	2.22

Rate Quality Control

The rate quality control (RQC) method has been used by agencies for many years. This method uses a statistical test to determine if the collision rate for a particular intersection or roadway segment is unusually high when compared to other intersections or roadways with similar characteristics.[9] This method only applies to collision rates and not frequency. It assumes that the number of crashes follows a Poisson distribution, which is appropriate for almost all crash safety studies. This can be verified using a representative sample of sites.

RQC can apply to spots or sections, with crash rates reported as collisions per million vehicles for both. The RQC method flags a location as hazardous if the rate of collisions at a particular site is greater than the critical rate. In short, this method establishes a boundary that is not totally arbitrary, but it is still biased toward low exposure sites just like the rate method presented earlier. The RCQ method calculates a critical rate for similar intersections as follows:

$$R_{critical} = R_a + \frac{0.5}{M} + K\sqrt{\frac{R_a}{M}}$$

where

$R_{critical}$ = critical crash rate for locations with similar characteristics

R_a = mean crash rate for locations with similar characteristics

M = the traffic volume during the analysis period (million entering vehicles, or million vehicle miles)

K = constant corresponding to a level of confidence in the finding

Agencies commonly use 90%, 95%, and 99% confidence intervals for K, which correspond to 1.282, 1.645, and 2.327, respectively. If R_{spot} or $R_{section} > R_{critical}$, analysts should consider the site as hazardous.

EXAMPLE 6.5 Rate Quality Control (RCQ) Method

Analysts are examining a narrow bridge as a spot that may be so hazardous that it needs some safety treatment. A total of 14 collisions have been recorded at the bridge in the past 4 years. The AADT at the bridge has remained steady over the past 4 years at 1000 vehicles per day. A sample of other narrow bridges in the highway district revealed a mean collision rate of 0.60 collisions per million entering vehicles. The analysts wish to use the rate quality control method· to make a decision about whether this bridge is hazardous. Using that method, determine the critical collision rate, above which the site would be declared hazardous at the 95% confidence level.

(*Continued*)

[9] Stokes, R.W. and M.I. Mutabazi (1996). Rate Quality Control Method of Identifying Hazardous Road Locations, TRR 1542. Transportation Research Board. National Research Council. Washington, D.C.

EXAMPLE 6.5 Rate Quality Control (RCQ) Method—(Continued)

Solution

$$R_{critical} = R_a + \frac{0.5}{M} + K\sqrt{\frac{R_a}{M}}$$

where

R_c = critical crash rate for locations with similar characteristics

$$R_a = 0.6\frac{collisions}{MEV}$$

$$M = \frac{(1000\ vpd) * (365\ days) * (4\ years)}{1,000,000}, or\ 1.46\ MEV$$

$$K = 1.645$$

$$R_{critical} = 0.6 + \frac{0.5}{1.46} + 1.645\sqrt{\frac{0.6}{1.46}}$$

$$R_{critical} = 2.0\ collisions\ per\ MEV$$

EXAMPLE 6.6 Rate Quality Control (RCQ) Method

Roadway section C–D had 38 reported collisions in 3 years, and the agency responsible for the section estimated that travel on the section was 20 million entering vehicles (MEV) during that time. The mean collision rate for all sections in the jurisdiction similar to section C–D was 120 per 100 MEV. Should an analyst flag section C–D as hazardous with 90% confidence?

Solution

The RCQ method requires that the same units be used for consistent variables; therefore, the analyst should convert R_a to 1.2 collision per MEV to be consistent with the units given for M. Now the equation for RCQ can be applied directly, as follows.

$$R_{critical} = R_a + \frac{0.5}{M} + K\sqrt{\frac{R_a}{M}}$$

where

R_c = critical crash rate for locations with similar characteristics

$$R_a = 1.2\frac{collisions}{MEV}$$

$$M = 20\ MEV$$

(Continued)

EXAMPLE 6.6 Rate Quality Control (RCQ) Method—(Continued)

$$K = 1.282$$

$$R_{critical} = 1.2 + \frac{0.5}{20} + 1.282\sqrt{\frac{1.2}{20}}$$

$$R_{critical} = 1.539 \ collisions\ per\ MEV$$

and

$$R_{section} = \frac{38\ collisions}{20\ MEV} = 1.9\ collisions\ per\ MEV$$

(problem statement says 20 MEV over the three-year time period).

Because $R_{section} > R_{critical}$, the agency should consider roadway section C–D as hazardous with 90% confidence using the rate quality control method.

Equivalent Property Damage Only

Analysts can adjust collision frequencies or rates to reflect the cost to society. One common method of taking severity into account before ranking locations is to compute the number of equivalent property damage only (EPDO) crashes.[10] This method uses a weighting factor based on the number of "equivalent" PDO crashes that would equal a more severe crash type (K, A, B, or C). Weighting factors are usually based on crash cost estimates provided at the federal or state level if regionally calibrated. For this reason, different agencies may use different equivalency factors. The weighting factor is most often calculated using crash cost estimates as follows:

$$W_y = \frac{CC_y}{CC_{PDO}}$$

where

W_y = weighting factor based on crash severity, y

CC_y = crash cost for crash severity, y

CC_{PDO} = crash cost for PDO crash severity

Using the weighting factors, the EPDO rating is calculated as follows:

$$EPDO\ Rating = W_K(K) + W_A(K) + W_B(B) + W_C(C) + PDO$$

where

$W_{K,A,B,C}$ = weighting factors for each crash type

K, A, B, C, PDO = number of fatal, A, B, C, and PDO collisions, respectively

[10] NCHRP (1986). "Methods for Identifying Hazardous Highway Elements." NCHRP Report 128. Transportation Research Board, National Research Council, Washington, D.C.

Table 6.8 Crash cost estimates for North Carolina

Crash type	Cost per crash (2013 dollars)
Fatal crash (K)	\$10,133,000
Injury crash (A)	\$564,000
Injury crash (B)	\$176,000
Injury crash (C)	\$96,000
Property damage only (PDO)	\$6700
Average crash	\$99,000
Injury crash (K + A + B + C)	\$293,000
Nonfatal injury crash (A + B + C)	\$128,000
Severe injury crash (K + A)	\$4,451,000
Moderate injury crash (B + C)	\$117,000

2013 Crash Cost Estimates, NCDOT, available at https://connect.ncdot.gov/resources/safety/Documents/Crash%20Data%20and%20Information/2013%20Crash%20Costs.pdf

Weighting factors are often based on national crash cost estimates posted by the FHWA.[11] More often, states have regional crash cost estimates that can be applied more directly. An example of crash cost estimates from North Carolina is provided in Table 6.8.

The major problem with EPDO as a screening tool is that it biases toward sites with higher fatal crashes as the weighting factors are significantly higher for this crash type compared to injury and PDO collisions. For this reason, many agencies separate collision severity types differently than others using different categories (e.g., severe (K + A), moderate (B + C), etc.), and sometimes even using engineering judgment or other means that are agreed on by other parties of interest.

One commonly accepted method for calculating EPDO is using the "Kentucky formula." This method significantly reduces the effect of regression to the mean that more severe crashes cause when using this method. The formula uses weighting factors of 9.5 for severe crashes (K + A) and 3.5 for moderate injury crashes (B + C). The equation is provided as follows.

$$\text{EPDO} = 9.5(\text{F} + \text{A}) + 3.5(\text{B} + \text{C}) + \text{PDO}$$

Regardless of the method used, spots or sections of interest are screened by calculating the EPDO and ranking them from highest to lowest.

[11] FHWA (2005). Crash Cost Estimates by Maximum Police-Reported Injury Severity within Selected Crash Geometries, FHWA-HRT-05-051. October.

EXAMPLE 6.7 Equivalent Property Damage Only (EPDO) Method: Kentucky Formula

The Hobsock Police Department recorded the collision data shown in the following table at the intersection of North Creek Ave. and the I-999 eastbound off-ramp during the period 1/1/2013 to 12/31/2015.

Date	Number of People		PDO
	B-injuries	C-injuries	
2/17/2013	0	2	—
3/1/2013	0	1	—
7/2/2013	0	0	1
7/5/2013	0	0	1
12/22/2013	0	0	1
7/14/2014	0	0	1
9/30/2014	0	2	—
1/4/2015	0	0	1
2/21/2015	0	0	1
7/24/2015	1	1	—
$\sum = 10$	$\sum = 1$	$\sum = 3$	$\sum = 6$

The ramp AADT during the period was 8600, while the AADT on North Creek Ave. was 34,100 during that time. There were no fatalities or A-injuries reported. The mean collision rate at similar intersections in Hobsock is 0.15 collisions per million entering vehicles. Using the Kentucky formula, what is the number of EPDO collisions expected at this intersection?

Solution

The key to completing this problem is recognizing that each row represents one collision. To classify each collision by severity, the analyst looks at the worst injury classification for each collision. Using the data compiled on the final row (which represents the total number of collisions and not total injuries), the EPDO number of collisions is calculated as follows:

$$\text{EPDO} = 9.5(\text{F} + \text{A}) + 3.5(\text{B} + \text{C}) + \text{PDO}$$
$$\text{EPDO} = 9.5(0 + 0) + 3.5(1 + 3) + 6$$
$$\text{EPDO} = 0 + 14 + 6 = 20$$

EXAMPLE 6.8 Equivalent Property Damage Only (EPDO) Method

Washington State has provided the crash cost estimates by severity (see following table).

Severity	Comprehensive crash costs
Severe injury (K + A)	$4,650,000
Moderate injury (B + C)	$110,000
PDO (O)	$12,400

Problem 6.8(a)

Four intersections have been provided by four neighboring districts requesting safety funding for improvements. Your job is to determine which district should receive funding. Given crash data from each of the four intersections over the last 2 years, determine the weighting factors and EPDO scores for the four signalized intersections. Based on EPDO alone, which intersection should receive funding for improvements?

Signalized intersection	AADT		Total	K	A	B	C	O
	Major	Minor						
1	14,000	7500	27	2	3	2	7	13
2	22,000	15,000	25	5	4	5	3	8
3	8000	500	20	0	2	4	6	8
4	13,000	4000	34	1	2	2	5	24
Total	57,000	27,000	106	8	11	13	21	53

Solution

The weight factor is calculated by dividing each severity category average cost by the average cost of a PDO collision, as shown in the following.

$$W_{K+A} = \frac{\$4,650,000}{\$12,400} = 375$$

$$W_{B+C} = \frac{\$110,000}{\$12,400} = 9$$

(Continued)

EXAMPLE 6.8 Equivalent Property Damage Only (EPDO) Method—(Continued)

The EPDO value is calculated as follows for the four intersections.

$$\mathrm{EPDO}_n = [\mathrm{W}_{\mathrm{K+A}} \times (\mathrm{K}+\mathrm{A})] + [\mathrm{W}_{\mathrm{B+C}} \times (\mathrm{B}+\mathrm{C})] + \mathrm{PDO}$$

$$\mathrm{EPDO}_1 = [375 \times (2+3)] + [9 \times (2+7)] + 13 = 1969$$

$$\mathrm{EPDO}_2 = [375 \times (5+4)] + [9 \times (5+3)] + 8 = 3455$$

$$\mathrm{EPDO}_3 = [375 \times (0+2)] + [9 \times (4+6)] + 8 = 848$$

$$\mathrm{EPDO}_4 = [375 \times (1+2)] + [9 \times (2+5)] + 24 = 1212$$

Based solely on EPDO, intersection 2 should receive funding with a calculated 3455 EPDO collisions over a two-year period.

Problem 6.8(b)

Calculate the EPDO collision rate per million entering vehicle miles. Based on the EPDO crash rate, which intersection should receive funding for improvements?

Solution

R_{EPDO} is calculated as follows for the four intersections.

$$R_{spot} = \frac{(A \times 1{,}000{,}000)}{(365 \times T \times V)}$$

$$R_{EPDO(1)} = \frac{(1969 \times 1{,}000{,}000)}{[365 \times 2 \times (14{,}000 + 7500)]} = 125$$

$$R_{EPDO(2)} = \frac{(3455 \times 1{,}000{,}000)}{[365 \times 2 \times (22{,}000 + 15{,}000)]} = 128$$

$$R_{EPDO(3)} = \frac{(848 \times 1{,}000{,}000)}{[365 \times 2 \times (8000 + 500)]} = 137$$

$$R_{EPDO(4)} = \frac{(1212 \times 1{,}000{,}000)}{[365 \times 2 \times (13{,}000 + 4000)]} = 98$$

Based solely on EPDO collision rate, intersection 3 should receive funding with a $\mathrm{R}_{\mathrm{EPDO}}$ of 137 collisions per MEV over a two-year period.

Relative Severity Index

The relative severity index (RSI) is a simplistic method of comparing the average crash cost among a list of sites. Although this method can be used as a hazardous site selection tool, it is typically more geared toward determining if a site should be considered for further review. The FHWA and many state and local jurisdictions estimate postcrash costs by crash type, and the reports are fairly easy to come by. The RSI for locations can be calculated using the following equation:

$$RSI_n = \frac{\sum_{j=1}^{n} RSI_j}{N_{0,n}}$$

where

RSI_n = average RSI costs for the location of interest n,

RSI_j = RSI cost for each crash type j, and

$N_{0,n}$ = number of observed crashes at site n

The RSI for each separate location of interest is determined to be hazardous or in need of further review if the RSI is greater than the RSI for the population under study. The RSI for the entire population is calculated as follows:

$$RSI_p = \frac{\sum_{y=1}^{n} RSI_y}{\sum_{y=1}^{n} N_{0,y}}$$

where

RSI_p = average RSI costs for the reference population

RSI_y = RSI cost for site y

$N_{0,y}$ = number of observed crashes at site y

Similar to the EPDO method, which can bias toward sites with higher severity crashes, this method can bias toward higher cost crash types. This makes this method more promising if utilized in conjunction with another method that considered frequency or rate.

EXAMPLE 6.9 Relative Severity Index Method

Crash data on four problematic intersections are provided in the following table. The collision data covers a time period of 3 years.

(Continued)

EXAMPLE 6.9 Relative Severity Index Method—(Continued)

Signalized intersection	Crash type							
	Rear end	Side swipe	Angle	Pedestrian/bicycle	Head on	Fixed object	Other	Total
Cost	$8000	$10,000	$112,000	$540,000	$1,000,000	$11,500	$22,000	
1	14	8	8	0	1	41	37	109
2	28	14	12	0	0	32	21	107
3	19	5	14	0	3	17	22	80
4	34	6	6	2	0	26	35	109
Total	95	33	40	2	4	116	115	405

(Continued)

EXAMPLE 6.9 Relative Severity Index Method—(Continued)

Problem 9(a)

Determine the RSI for the following four intersections of your local jurisdiction.

Solution

First the total cost of each collision type over the 3-year period must be calculated for each of the four intersections. The calculations for each crash type at intersection 1 are shown as follows and input into the following table. The last three intersections follow the same calculations, and the answers alone are provided following the calculation of RSI for intersection 1.

Rear end: 14 × \$8000 = \$112,000
Side swipe = 8 × \$10,000 = \$80,000
Angle: 8 × \$112,000 = \$896,000
Pedestrian/bicycle: 0 × \$540,000 = \$0
Head on = 1 × \$1,000,000 = \$1,000,000
Fixed object: 41 × \$11,500 = \$471,500
Other: 37 × \$22,000 = \$814,000

$$RSI_n = \frac{\sum_{j=1}^{n} RSI_j}{N_{0,n}}$$

$$RSI_1 = \frac{(\$112,000 + \$80,000 + \$896,000 + \$0 + \$1,000,000 + \$471,500 + \$814,000)}{(14 + 8 + 8 + 0 + 1 + 41 + 37)}$$

$$RSI_1 = \frac{(\$3,373,500)}{(109)} = \$30,950$$

(Continued)

EXAMPLE 6.9 Relative Severity Index Method—(Continued)

Signalized intersection	Crash type and costs							Total	RSI
	Rear end	Side swipe	Angle	Pedestrian/bicycle	Head on	Fixed object	Other		
Cost	$8000	$10,000	$112,000	$540,000	$1,000,000	$11,500	$22,000		
1									
2	$224,000	$140,000	$1,344,000	$0	$0	$368,000	$462,000	$2,538,000	$23,720
3	$152,000	$50,000	$1,568,000	$0	$3,000,000	$195,500	$484,000	$5,449,500	$68,119
4	$272,000	$60,000	$672,000	$1,080,000	$0	$299,000	$770,000	$3,153,000	$28,927

(Continued)

EXAMPLE 6.9 Relative Severity Index Method—(Continued)

Problem 9(b)

Determine the RSI for the reference population.

Solution

$$RSI_p = \frac{\sum_{y=1}^{n} RSI_y}{\sum_{y=1}^{n} N_{0,y}}$$

$$RSI_p = \frac{\$3,373,500 + \$2,538,00 + \$5,449,500 + \$3,153,000}{(109 + 107 + 80 + 109)}$$

$$RSI_p = \frac{\$14,514,000}{405} = \$35,837$$

Problem 9(c)

Which intersections should be considered for further evaluation and countermeasure implementation?

Solution

Intersection 3 is the only site whose RSI exceeds the reference population.

Sites with Promise

Each of the hazardous site identification methods presented so far have had inherent flaws. Bias toward higher or lower exposure sites is the most prevalent problem when considering methods that use frequency or rate. Rate-based methods require volume data on top of crash data already collected, which can be cumbersome. EPDO and RSI methods are based on cost by severity and crash type, which can bias decisions toward higher severity or crash types, ignoring sites that may have serious crash problems that don't rise to the top because the crashes are less costly. The method of "sites with promise" (SWP) was developed to overcome several of these flaws and to provide a logical and defensible bias for recommendations of sites that can actually be fixed.[12]

The primary advantage of the SWP method is that it seeks to find fixable sites and not necessarily sites that are hazardous alone, optimizing safety funds set aside by agencies for site improvements. Frequency is the primary building block of this method, which pairs down sites prior to implementation of the five (or less) conditions being considered. The remaining sites are then investigated further based on frequency (total and specific crash type) and rate, along with a look at recent surges in crash frequency. Specific countermeasures are also able to be investigated for the entire pool where appropriate.

The SWP method uses, at most, five conditions in determining promising sites. Although five conditions are present, the analyst should only use conditions applicable to the study at hand. Following application of the method, the analyst uses the basic findings, along with engineering experience and judgment, to choose sites for potential countermeasure implementation. The analyst should ask him or herself the following questions when considered each of the five conditions.

Condition A: "Is there one or more countermeasures we would like to consider installing on a programmatic level?" Often, it is easier to obtain and set aside funds for countermeasures that can be installed at multiple locations. Sites should only be considered if the countermeasure would make sense at that location. Therefore, a crash subset should be analyzed that is representative of the types of crashes the countermeasure would be expected to improve. If funds are not available to support widespread implementation of a countermeasure, this condition should not be investigated.

Condition B: "Are there newly constructed or reconstructed sites that seem to be having a crash problem?" The crash frequency (F_i) at each

[12] Hauer, E. (1996). Identification of Sites of Promise, Transportation Research Record 1542, Transportation Research Board, Washington, D.C., pp. 54–60.

individual location is compared to the mean frequency (F_m) of similar sites using the following equation.

$$\frac{F_i - F_m}{\sigma_F}$$

Usually, the reference population used is the remaining sites that have a long-term pattern for comparison. Typically, very short periods of time are all the analyst has to identify a site. Although whole years are ideal, they are often not possible. For this case, to make a fair comparison, sites with less than one full year should use a multiplier to bring the frequency into a common denominator for comparison purposes. For instance, a newly rebuilt site with only 6 months of data should be doubled. The inherent flaw in using a factor in this way is that it discounts seasonality. In other words, the initial 6 months of data may have been low or high compared to other times of the year, potentially over- or underestimating the problem that one of these sites may be experiencing.

Condition C: "Are there sites that have recently seen a significant increase in collision frequency?" All long-term sites should be checked annually. Where sites appear to have some deficiency, it is advisable to look at the longer-term trend to make sure the site is appropriately flagged.

Condition D: "Are there lower-volume sites that we may be missing because the exposure is not accounted for?" This condition makes sure that low-volume/low-exposure sites are considered and that not all safety funds are poured into high-volume sites that see little to no improvement. Exposure data in the form of traffic volumes will be necessary in this case, but only after the site pool has been trimmed down using collision frequency.

Condition E: "Are there newly constructed or reconstructed sites that seem to be having a crash problem but could be missed because they are low-volume, low-collision sites?" This criterion is similar to Condition B but instead looks at collision rates. The crash rate (R_i) at each individual location is compared to the mean rate (R_m) of similar sites using the following equation.

$$\frac{R_i - R_m}{\sigma_F}$$

For this case, to make a fair comparison, sites with less than one full year should use a multiplier to bring the frequency into a common denominator for comparison purposes prior to calculating collision rates. Also note that the analyst should consider whether "spot" or "sections" rate calculations are appropriate.

The analyst should note that the SWP method does not account for regression to the mean, where sites identified based on high collision frequency will naturally return to their mean rate regardless of the treatment. This method's basic premise is that it uses the combined strengths of the rate and frequency methods to determine where countermeasures have the highest potential.

EXAMPLE 6.10 Sites with Promise (SWP) Method

The Bangerter public works department is conducting its yearly spot safety problem. It has budgeted for $1,000,000 in safety-related countermeasures with a focus on reducing more prominent collisions at signalized intersections, namely angle, left-turn opposing, and rear-end crashes. The department is open to investigating any and all countermeasures to improve sites; however, two particular focus areas this year will be backplates/louvers/12-in LEDs (visibility of signal heads) and proper intersection sight distance to allow drivers to see conflicting traffic and traffic signal heads. Two newly constructed sites include only a single year's worth of data, and the department wants to make sure there is not a safety problem that needs addressing before more data are available. Which of the 11 sites should the city consider for further evaluation?

(*Continued*)

EXAMPLE 6.10 Sites with Promise (SWP) Method—(Continued)

Site	Annual average 2012–2014					Total collisions			Special notes
	Angle collisions	Left-turn collisions	Rear-end collisions	Other collisions	Minor + major ADT	2011	2012	2013	
1	45	21	21	24	42,000	49	36	26	
2	9	4	22	15	38,000	—	—	50	Rebuilt in fall 2012
3	16	22	27	39	19,000	33	37	34	
4	10	18	12	32	28,000	21	27	24	
5	51	12	27	19	36,000	36	29	44	
6	39	26	12	17	31,000	30	29	35	
7	15	9	46	16	32,000	29	24	33	
8	24	13	25	9	26,000	17	20	34	
9	8	8	5	3	25,000	—	—	24	Finished Christmas 2012
10	22	26	22	48	35,000	34	38	46	
11	22	30	16	30	21,000	30	36	32	

(*Continued*)

EXAMPLE 6.10 Sites with Promise (SWP) Method—(Continued)

Solution

The scenario described meets each of the five conditions for SWP. The following table is referenced for several of the conditions.

Site	Angle + left + rear-end collisions	Frequency per year	Rate per million vehicles
1	87	37.0	2.41
2	35	50.0	3.60
3	65	34.7	5.00
4	40	24.0	2.35
5	90	36.3	2.77
6	77	31.3	2.77
7	70	28.7	2.45
8	62	23.7	2.49
9	21	24.0	2.63
10	70	39.3	3.08
11	68	32.7	4.26
Mean (similar sites)		32.0	3.1
Standard deviation (similar sites)		5.6	0.9

The solutions for each condition are given in the following.

Condition A: The department is trying to reduce collisions related to intersection sight distance and traffic control visibility. These safety problems are most directly associated with angle, left-turn opposing, and rear-end collisions (such as those drivers that see the traffic signal late). Looking at the frequency of these collisions in the previous table (excluding newly constructed sites), Sites 1 and 5 look the most promising with 87 and 90 collisions per year from 2011–2013.

Condition B: Compare newly constructed Sites 2 and 9 with comparable sites (1, 3–8, and 10–11) using total collisions. The mean frequency per year and standard deviation of the remaining comparable sites are calculated in the previous table (32.0 ± 5.6 crashes). Site 2 is the only site with an average frequency of crashes greater than zero. A value of 3.2 means that the difference in collisions between Site 2 and the reference population is 3.2 standard deviations, which is very significant!

(Continued)

EXAMPLE 6.10 Sites with Promise (SWP) Method—(Continued)

Site 2

$$\frac{50 - 32.0}{5.6} = 3.2$$

Site 9

$$\frac{24 - 32.0}{5.6} = -1.4$$

Condition C: Looking at the collisions from 2011–2013, crash frequency at Site 8 increased drastically in 2013 by a factor of nearly 2. It would be worthwhile to look at crash trends over a longer period at this site to determine if this increase was isolated.

Condition D: Looking at collision rates for sites over the 3-year period (excluding Sites 2 and 9) in the previous table, Site 3 is much higher than any of the other similar sites. Note that rates here are calculated for "spots" and not "segments," and that total collisions were using in the calculation of "rate per million vehicles."

Condition E: Similar to Condition B, compare newly constructed Sites 2 and 9 with the remaining comparable intersections. Site 2 is the only site greater than zero; however, a value of 0.5 indicates it is less than one standard deviation difference from the reference population. Therefore, this site should not be considered further when looking at collision rates alone.

Site 2

$$\frac{3.6 - 3.1}{0.9} = 0.56$$

Site 9

$$\frac{2.6 - 3.1}{0.9} = -0.56$$

Summary

Sites 1 and 5 appear most promising for implementation of a countermeasure addressing the three identified collision types. At these two sites, angle collisions were most prominent. Site 2, which was recently rebuilt, appears to have a short term-crash problem that may need addressing quickly. The limited crash history seems to indicate a rear-end collision problem. Site 5 appeared to rapidly deteriorate in 2013 and should be investigated further to make sure this was not just a sporadic issue that

(*Continued*)

EXAMPLE 6.10 Sites with Promise (SWP) Method—(Continued)

doesn't need addressing. It would be advisable to look at a larger subset of data going back several more years in time. Although the collision frequency at Site 3 was in the normal range, it had a very high collision rate when accounting for traffic volume. It is not obvious what the issue may be, but further investigation should be conducted. Last, a condition and collision diagram would be helpful in determining the exact safety countermeasure to implement, if any, at each of the sites noted previously.

6.4.3 Countermeasure Development

Following the prescreening analysis of a pool of sites, the next step is consideration of countermeasures to reduce or eliminate collisions at these sites. Options should not only improve the safety at sites, but should also be cost-effective; you are looking for the lowest dollars spent per collision or injury saved. Two primary tools used to summarize factors leading to a collision are collision and condition diagrams. These are described in more detail in the following subsections. These diagrams, along with a good list of potential countermeasures, can be used to improve the safety of roadways with known crash problems.

Collision Diagram

A good collision diagram is very important for selecting countermeasures for projects or programs with multiple sites. Figure 6.3 provides an example collision diagram for reference. This detailed look at the intersection is intended to show were disproportionate numbers of collisions are taking place, or even where or what movements may be prevalent to more severe collisions taking place. The concept of the collision diagram is to provide the analyst with a good understanding of what problem areas may need addressing at this site so that countermeasures can be considered. Collision diagrams contain very few details about the particular location. Street names, outlines of the edges of the pavement, and a direction orientation are all that is necessary besides the symbols for collisions and fixed objects. Sometimes, collision diagrams contain a table or supplementary diagrams for summarizing the collisions by type, severity, light condition, or road condition. This can be particularly helpful when a collision diagram becomes crowded and clusters are hard to identify. Note that the analyst can see the number of collisions, their location and patterns, time of day,

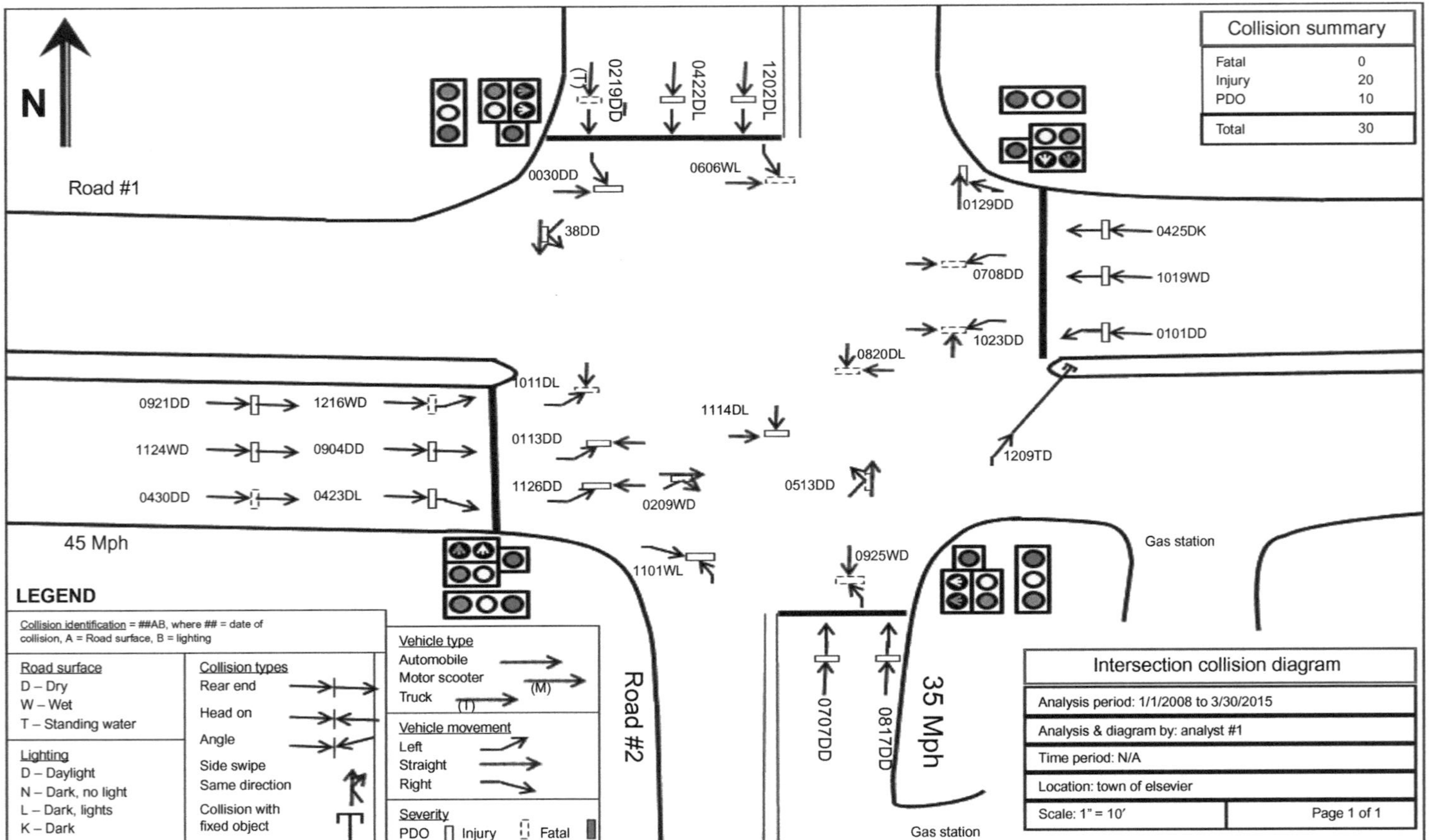

Figure 6.3 Example of a collision diagram.

date, pavement condition, severity, and so on. All of these data are useful in identifying possible causes for the crashes and then developing a list of possible countermeasures, the focus of the next section.

Probably the most important aspect of a collision diagram is that it provides information related to the causal chain, allowing the analyst to look at countermeasures that can address every aspect of the collision from precrash to postcrash. It is estimated that 90% of all collisions take place due to human factors alone, or some combination of human, roadway, and vehicle factors. In many instances, the causal chain may indicate a problem that should be investigated outside the bounds of a project. For instance, run-off-the-road and rear-end collisions may be very high at a particular intersection or roadway segment, but the collision diagram indicates collisions are taking place during raining conditions. In this case, a larger corridor study may reveal a synonymous problem along sections of the entire roadway that needs to be addressed (but the major emphasis is the intersection or link identified during the prescreening evaluation).

In some cases, diagrams may become crowded. If so, the analyst can use symbols to represent collisions of a particular kind; however, one should only consider this when necessary as details are often lost when combining collisions in this manner. For instance, a star next to an arrow or a bolded arrow may indicate 10 or more collisions of a certain type at that location; but, the descriptors of light conditions or roadway surface may be lost if they are not also the same.

Condition Diagram

Although a collision diagram is more prevalent, analysts often use condition diagrams to supplement the analysis before generating countermeasure ideas. Unlike collision diagrams, condition diagrams are drawn to scale providing information on intersection (or segment) layout, roadway and lane widths, sight obstructions, traffic control devices, crosswalks, fixed objects, and other potentially noteworthy features. Condition diagrams can start as rough sketches in the field or be started in the office using aerial imagery, later filling in the details while in the field. An example condition diagram is provided in Figure 6.4. A supplemental checklist of common site descriptors is provided in Figure 6.5.

Unlike collision diagrams, which focus on many prevalent factors leading up to a collision, condition diagrams focus primarily on traffic design and operation components that can easily be missed if the analyst does not leave the office. For instance, a common problem at many

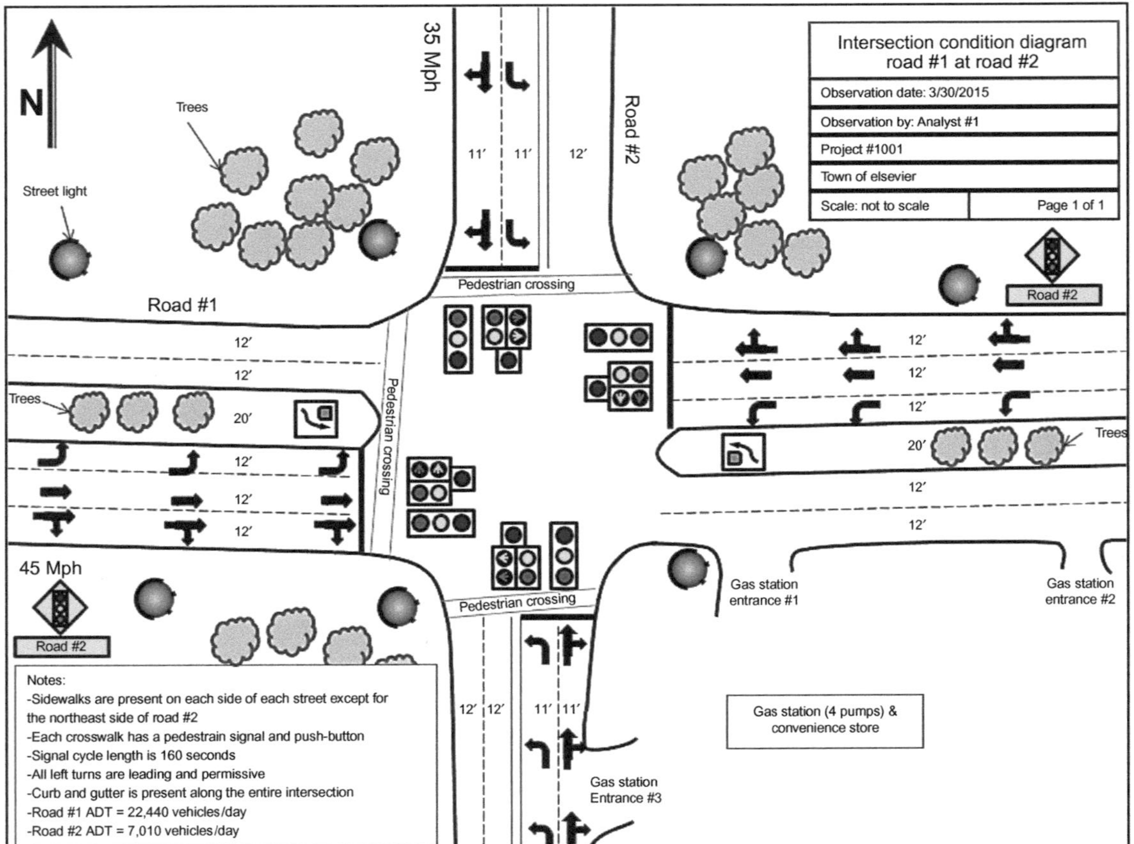

Figure 6.4 Example condition diagram.

Study information	**Area/intersection data**
Day and date of study	Type of intersection (T, 2-way, 4-way, etc.)
Time period of data collection	Street/highway names and/or numbers
General weather conditions	Directional arrow (north arrow)
Investigator's name	Intersection control (signal, stop, roundabout, etc.)
Site location and jurisdiction	Signal phasing
Roadway names/numbers	Cycle lengths
Any special conditions	General area description within 0.1 mile (land use type and/or intensity)
Geometry	
Approach grade	**Signs and markings**
Number of lanes	Traffic signs before intersection
Lane width	Warning signs within 500 feet
Movement for each lane	Speed limit on approach legs
Shoulder width	Lane designation signs
Median type and width	Crosswalks
Sight distance restrictions	Roadway edge markings
Roadway functional classification	Lane separation markings
Potentially distracting conditions	

Figure 6.5 Possible site characteristics to include in condition diagram.

intersections are issues related to intersection and stopping sight distance. Overgrown trees or bushes, especially where unusual horizontal (e.g., intersection skew, curvature, etc.) or vertical (e.g., crest or sag) alignment features are prevalent. Other useful information to the analyst that can be collected in the field is roadway debris from crashes, skid marks, signal light functionality (such as LEDs), and surrogate measures that take place while making observations (such as conflicts or erratic maneuvers).

Countermeasure Selection

Mentioned earlier in the part, traffic collisions occur as a result of one or more contributing circumstances such as human, vehicle, or roadway factors. It is important to determine the causal chain based on these factors when choosing one or more potential countermeasures. Because the traffic analyst does not control the vehicle design or human factor issues such as speeding or drinking and driving, his or her primary focus is on human and roadway factors. Vehicle factors should be addressed by automotive designers or enforcement agencies.

Once a site (or sites) is identified as hazardous and the appropriate collision and/or condition diagrams are drawn, cluster analysis is conducted to see what the predominant issue or issues are at the site of interest.

The issues should align with the earliest known human or roadway factor in the causal chain. Once this is done, one or more countermeasures can be investigated as a potential treatment. The engineer should always keep in mind that collisions are rare events, and countermeasures change constantly. Because of this, generating a list of possible countermeasures for a site of interest is, in many ways, more art than science. Ideas for countermeasures can be drawn from various sources but often come from experience and a thorough understanding of the roadway being investigated.

There are many alternatives that should be considered by the analyst when determining countermeasure implementation. Engineers should always consider the four Es: *engineering*, *enforcement*, *education*, and *emergency response*. It is entirely possible that an engineer may decide to do nothing at all. In some instances, there may be no obvious countermeasure to address the problem, in which case safety funds should be used at other sites where countermeasures are more easily identified. However, the engineer may decide to look at policy and design criteria changes that change how facilities are designed and constructed. These are usually related to larger programmatic issues that arise when conducting safety studies.

The safety analyst should always be aware that certain countermeasures, though chosen for valid reasons, may prove ineffective if installed. For instance, the analyst may be asked to install a traffic signal or guardrail where a known collision problem exists; however, these countermeasures can come with their own inherent risk and should be used as a last resort—only after carefully considering if safety improvements are possible (considering frequency and severity). Similar in concept, it is possible that countermeasures may fix an issue at an identified location, however, at the detriment of the downstream intersection or roadway. For example, updated signal timing at a standalone intersection within a coordinated network may cause problems at downstream intersections that have not also been retimed. Once a countermeasure is chosen, it should be evaluated carefully over an extended period of time to make sure that the expected change in behavior was positive, and if not, the countermeasure should be removed immediately. Countermeasure evaluation is discussed later in this part.

Common crash types and possible countermeasures are provided for a select number of facility types in Figures 6.6–6.12.[13] The Federal Highway Administration administers an online tool, called the Crash Modification

[13] Traffic Engineering Handbook, 5th edition, Washington, D.C.: Institute of Transportation Engineers, 1999.

Potential causal factor	Possible countermeasure
Restricted sight distance	Remove sight obstructions Restrict parking near intersection Provide all-way STOP or signal Install/improve warning signs Install stop line closer to crossroad
Excessive speed	Install/improve warning signs reduce speed limit with enforcement Install rumble strips
Inadequate roadway lighting	Install or improve lighting
Inadequate advance warning signs	Install or improve warning signs
Large traffic volume	Provide traffic signal Reroute traffic
Inadequate traffic control devices	Upgrade traffic control devices Increase enforcement

Figure 6.6 Right-angle crashes at two-way, stop-controlled intersections.

Potential causal factor	Possible countermeasure
Large turn volume	Create one-way street Add left-turn lane prohibit left turns Reroute left-turn traffic provide traffic signal with left-turn phase
Restricted sight distance	Remove sight obstructions Provide left-turn lane Prohibit left turns Provide traffic signal with protected-only left-turn phase
Excessive speed	Reduce speed limit Improve enforcement

Figure 6.7 Left-turn crashes at two-way, stop-controlled intersections.

Factors (CMF) Clearinghouse, which provides an analyst a firsthand look at the effectiveness of various countermeasures on varying facility types and demographics.[14] This tool is the preeminent source for all countermeasure evaluation, providing several sources of documentation to help transportation engineers identify the most appropriate countermeasure for their safety needs.

[14] FHWA Crash Modification Factors Clearinghouse, www.cmfclearinghouse.org.

Potential causal factor	Possible countermeasure
Restricted sight distance	Remove sight obstructions Install/improve warning signs Reduce speed limit with enforcement
Excessive speed	Reduce speed limit with enforcement Adjust phase change interval Install rumble strips
Inadequate roadway lighting	Install or improve lighting
Poor visiblility of traffic signals	Install or improve warning signs Install overhead signal heads Install 12-inch signal lenses Install visors Install back plates Relocate/add signal heads
Inadequate signal timing	Retime signal Adjust phase change interval Provide red clearance interval Provide progression Provide signal actuation with dilemma zone protection
Inadequate advance warning signs	Install/improve warning signs
Large traffic volume	Add lane(s) Retime signal

Figure 6.8 Right-angle crashes at signalized intersections.

Potential causal factor	Possible countermeasure
Large turn volume	Provide left-turn phase Prohibit turns Provide turn lane Increase corner radius for right turn
Slippery pavement	Reduce speed limit with enforcement Overlay pavement provide adequate drainage Groove pavement Provide SLIPPERY WHEN WET signs
Inadequate roadway lighting	Improve lighting
Crossing pedestrians	Install/improve crosswalk devices Provide pedestrian signal devices
Poor visibility of traffic signals	Install or improve warning signs Install overhead signs Install 12-inch signal lenses Install visors Install backplates Relocate/add signal heads
Inadequate signal timing	Adjust phase-change interval Provide red clearance interval Provide progression Provide signal actuation with dilemma zone protection
Unwarranted signal	Remove signal

Figure 6.9 Rear-end crashes at signalized intersections.

Potential causal factor	Possible countermeasure
Excessive speed	Reduce speed limit with enforcement
Slippery pavement	Reduce speed limit with enforcement Overlay pavement Provide adequate drainage Groove pavement Provide SLIPPERY WHEN WET signs
Inadequate roadway lighting	Improve lighting
Poor visibility of curve warning sign	Increase sign size
Inadequate roadway design	Widen lane(s) Realign curve Install guardrail
Inadequate delineation	Install/improve warning signs Install/improve pavement markings Install/improve delineation
Inadequate Shoulder	Upgrade shoulder
Inadequate pavement maintenance	Repair road surface

Figure 6.10 Run-off-the-road crashes on a section of two-lane, rural highway.

Potential causal factor	Possible countermeasure
Excessive speed	Reduce speed limit with enforcement Install median barrier
Inadequate/ improper pavement markings	Install/improve pavement markings
Inadequate roadway design	Widen lane(s)
Inadequate shoulder	Upgrade shoulder
Inadequate pavement maintenance	Repair road surface
Unsafe overtaking	Upgrade signing and marking

Figure 6.11 Head-on crashes on a section of two-lane, rural highway.

Potential causal factor	Possible countermeasure
Restricted sight distance	Remove sight obstruction Install/improve warning signs Provide STOP signs Reduce grade on approach Install gate and/or flashers
Excessive speed	Reduce speed limit with enforcement
Slippery pavement	Improve skid resistance
Poor visibility of traffic control devices	Increase sign size Install roadway lighting
Inadequate pavement markings	Install/improve pavement markings
Inadequate lighting	Install or improve lighting

Figure 6.12 Motor vehicle striking a train at a highway-railroad grade crossing.

When considering countermeasures for crashes with fixed objects, the analyst should first consider removing the object or moving the object. If neither is possible, the engineer should look for options to make the object safer (e.g., breakaway supports for poles). As a last resort, the object could be shielded (e.g., guardrail) and sufficient warning should be provided to the driving public via a sign or other appropriate measure. The engineer should always remember that guardrails and other barrier types should be used as a last resort as they are hazardous in themselves.

EXAMPLE 6.11 Countermeasure Selection

A two-lane, rural highway in your jurisdiction has had two run-off-the-road crashes near the same spot within the past 3 months. The posted speed limit is 55 mph. After investigation, you have determined that excessive speed and slippery pavement are the most likely causal factors. All but which of the following are possible countermeasures to reduce future crashes at this spot?

(A) Overlay with 1.5 in of new asphalt pavement.
(B) Reduce the speed limit to 45 mph.
(C) Install guardrail to redirect vehicles back onto the shoulder.
(D) Provide adequate drainage to ensure water does not contribute to hydroplaning.

Solution

Looking at the tables provided in Figures 6.6 through 6.12, we can determine that all countermeasures should be considered with the exception of (C). Installing guardrail should only be used as a last resort!

EXAMPLE 6.12 Countermeasure Selection—CMF Clearinghouse

Several rural 4-legged, two-way, stop-controlled intersections across the state have been screened as having crash problems. The State Department of Transportation (DOT) is considering installation of roundabouts at several of these crash-prone locations. As the safety analyst for the DOT, you have been asked to make recommendations for treatments based on their likely effectiveness. What information can you provide to your supervisor that would aid in his/her decision to install roundabouts versus another possible treatment?

Solution

The CMF Clearinghouse can be used to evaluate the effectiveness of a potential countermeasure using previous studies in the literature. A quick search of the database, querying "rural roundabouts," yields the following results.

(*Continued*)

EXAMPLE 6.12 Countermeasure Selection—CMF Clearinghouse—(Continued)

▾ Countermeasure: convert high-speed rural intersection (4 leg) to roundabout

CMF	CRF(%)	Quality	Crash type	Crash severity	Area type	Reference
0.32	68	★★★★☆	All	All	Rural	Isebrands, 2012
0.12	88	★★★★☆	All	Serious injury, minor injury	Rural	Isebrands, 2012
0.26	74	★★★★☆	All	All	Rural	Isebrands, 2012
0.11	89	★★★★☆	All	Serious injury, minor injury	Rural	Isebrands, 2012
1.41	−41	★★★☆☆	All	All	Rural	Isebrands, 2012
0.4	60	★★★☆☆	All	Serious injury, minor injury	Rural	Isebrands, 2012

Source: The FHWA CMF Clearinghouse available at http://www.cmfclearinghouse.org/

Six studies are reported, with five of the six studies showing positive collision reduction factors (CRFs), indicating increased safety. Of those five, the minimum expected decrease in collisions was 60%. Looking more carefully at the results, four of the six reports have 4 of out 5 stars, and each of those higher-quality reports show reductions. The lowest reported reduction was 68%. Taking crash severity into account (the focus of this countermeasure evaluation), the second and fourth studies show an 88% and 89% reduction in serious injury collisions. Based on the focus of treatment site locations, the supervisor could expect a reduction of similar magnitude.

6.4.4 Posttreatment Analysis Methods

Highway improvements are often targeted at improving safety; however, these improvements are often left in place with little oversight by the agency that installed them. The most promising safety studies include collision data in lieu of the surrogates discussed in previous sections. However, these studies require that sufficient data after treatment installation be available for study, making surrogate studies a promising first step in evaluating countermeasures on high-profile projects. The focus of this section on postcountermeasure evaluation is intended to give a solid understanding of the *fundamentals* of before-and-after observational studies. Concepts are examined, and some examples of more simplistic methods and their intended use, assumptions, and so on are provided. It is highly recommended that analysts conducting before-and-after safety evaluations of countermeasures consider a more detailed review of all study methods in Hauer's seminal text, *Observational Before-After Studies in Road Safety.*[15]

Fundamentals

In many cases, agencies *do* conduct safety studies of countermeasures; however, these studies are often plagued with serious flaws that provide incorrect, and sometimes very misleading, results. For instance, consider the example provided in Figure 6.13. Here, a countermeasure installed at an intersection in September 2014 is being evaluated to determine if the treatment improved safety at the intersection. Using a simple before-and-after study evaluation comparing average monthly frequency of the before time period to the after time indicates no change in collisions. Although not terribly obvious on first glance, the simple trend analysis in this case is likely providing inaccurate results for one or more reasons.

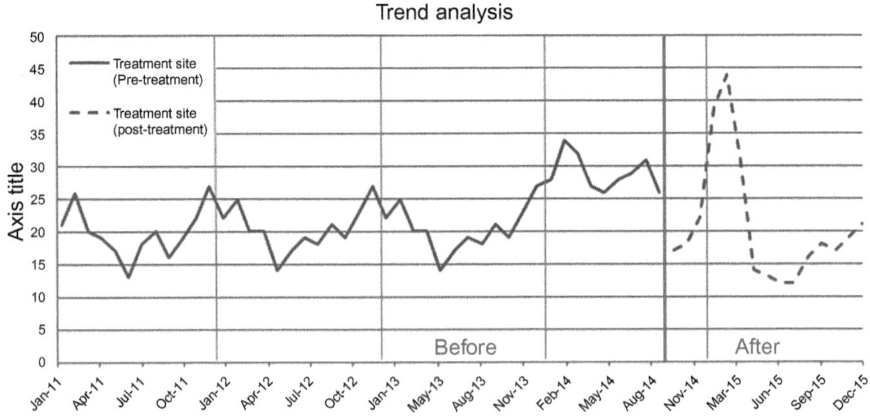

Figure 6.13 Standalone treatment evaluation at a single site.

[15] Hauer, E. *Observational Before-After Studies in Road Safety.* Pergamon. 1997.

As currently conducted by many practitioners, before-and-after studies using crash data suffer from four major flaws and almost always result in incorrect or misleading results. The flaws are (A) the failure to correct for differences in *time* period between the before-and-after period, (B) the failure to correct for differences in *exposure* to traffic (i.e., volume) between the before-and-after period, (C) the failure to account for *unanticipated events* such as historical or seasonality effects, and (D) the failure to account for *regression to the mean*, the most common problem in all before-and-after evaluations. Using the example presented earlier, Figure 6.14 provides a graphical representation of the four possible flaws circled and labeled A through D. A trend line is also provided with the before-and-after data to help illustrate these concepts.

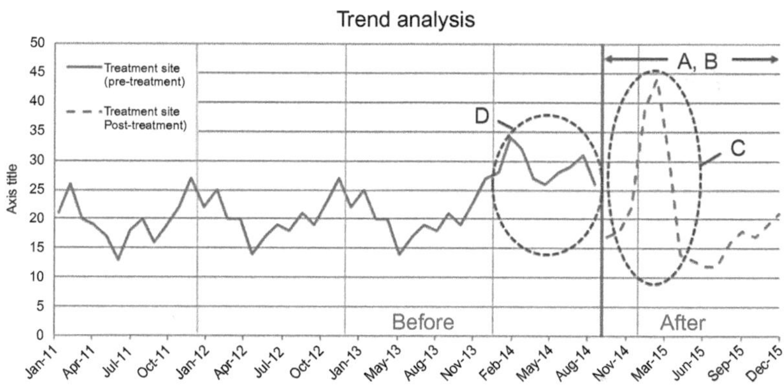

Figure 6.14 Four major flaws in safety studies (A–D).

More detail is provided on these four potential biases in safety studies in the following sections. The goal here is to recognize the assumptions an analyst is making when conducting one of these study types. Studies that remove as much bias as possible will provide much better results regarding countermeasure effectiveness that can be used by agencies when considering installing the same countermeasure elsewhere. The flaws are expounded on in letters A through D, corresponding to the letters in the graph shown in Figure 6.14.

Time

The before-and-after time periods are rarely equal for direct comparison. In many cases, the after period is shorter in length and requires adjustment. Unlike trend analysis where the analyst compares average monthly crash frequency for the before-and-after periods, the analyst should compare total crash frequency. However, the time periods prevent an appropriate comparison. In A in Figure 6.14, a ratio of time in the before-and-after periods (ratio denoting duration of time, $r_d = 45/15$ months $= 3$) could be used to provide common time periods for analysis. Here, the total crashes in the after period would be multiplied by 3 to help remove potential bias from differences in time. The analyst should note that fluctuations in crashes during various seasons are not accounted for using this method. For instance, the 15-month-after period includes collision data over two major holiday seasons where crashes are shown to naturally increase: Thanksgiving and Christmas. This season is accounted for twice in the after period, whereas the lower-crash-frequency spring and summer months are accounted for only once.

Exposure

It is commonly accepted that traffic collisions are highly correlated to the number of vehicles on the road: the more traffic is on the road, the more likely a crash is to take place, and vice versa. Exposure to traffic is accounted for in B, in a similar manner to time period in Figure 6.14, where a ratio of traffic volumes in the before-and-after periods (ratio of traffic flow, $r_{tf} = 15{,}000/16{,}000$ veh/day $= 0.938$) are used to account for traffic exposure. In the example noted earlier, the total crashes in the after period would be multiplied by 0.938 for direct comparison with total crashes in the before period.

Unanticipated Events

This bias is introduced when a major event in time is introduced during the study period being evaluated—most commonly seasonality and historical effects. This effect can be introduced at the treatment site during any time period of the evaluation. For instance, in the example provided in Figure 6.14, a severe and very unusual series of weather events culminated in a large spike in collisions in January and February, 2015, circled C. Simple trend analysis or accounting for time or exposure do not provide a means of removing this bias, and in this case, the analysis would underestimate the effect of the treatment on safety.

To combat this problem, a comparison group of similar roadway types could be used to account for these unusual spikes in collisions. Each of the sites must be tested to determine they follow a similar pattern during the before period. Once proven similar, the sites can be used to predict what would have been the true number of collisions in the after period had the countermeasure not been installed. This can be compared to the actual collisions after countermeasure deployment to determine the net effect of the countermeasure on safety.

Using the previous example, an updated graph is provided in Figure 6.15 that shows the summation of several comparison sites plotted over time next to the treatment group. Notice the comparison group (roughly) follows the same trends in the before period (with the exception of D, which is explained in the next section), helping negate the increase in collisions during this after period, which were overestimating crashes using previous methods. *Note:* The previous two biases (time and exposure) are accounted for more directly using this analysis method as time periods are the same for both treatment and comparison groups.

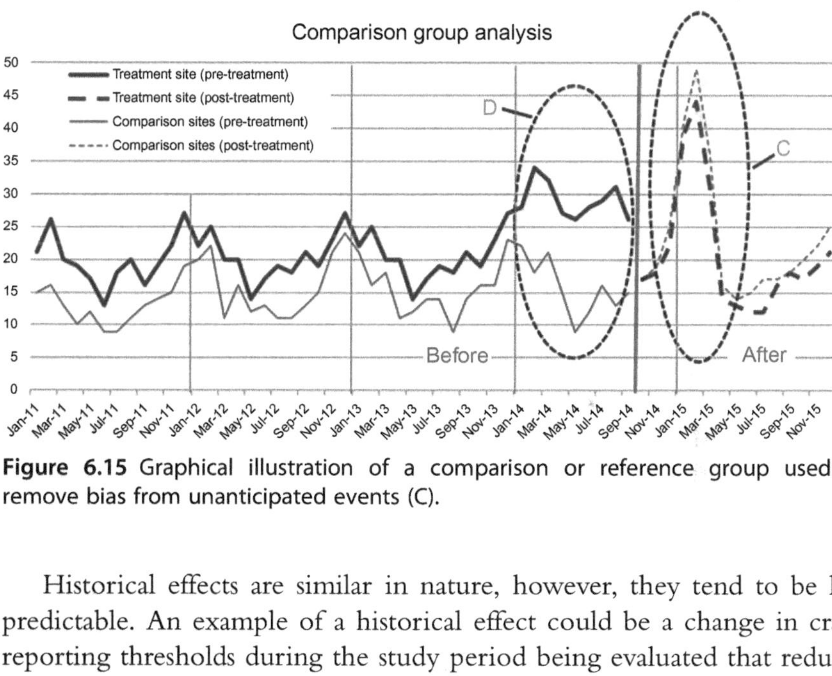

Figure 6.15 Graphical illustration of a comparison or reference group used to remove bias from unanticipated events (C).

Historical effects are similar in nature, however, they tend to be less predictable. An example of a historical effect could be a change in crash reporting thresholds during the study period being evaluated that reduces the number of collisions being reported. In this case, the effect is more sustained over time; however, the comparison or reference group would also follow the same reporting threshold.

Comparison groups are very good at accounting for many biases in safety studies, however, they do not account for regression to the mean, discussed in the next section. For this reason, they are most appropriately used when evaluating sites that were chosen based on (1) operational improvements or (2) where the site was chosen randomly from a pool of sites chosen for safety, allowing the other sites to populate the comparison group.

Regression to the Mean

Regression to the mean (RTM) occurs when locations with high collision counts during one time period experience more normal counts during the next time period, even if no causative factor changes. Shown graphically as D in Figure 6.15, there was a severe spike in collisions in 2014. Regardless of the fact that a countermeasure was installed, the site would have likely returned to a normal pattern of collisions shortly after this time period. If RTM was present at this site and unaccounted for in the analysis, the safety effect of the countermeasure would be overpredicted. For this reason, a group of several reference sites can be used to control for RTM. Similar to comparison sites, the reference sites are used to derive a safety performance function (SPF) based on Bayesian statistics, often referred to as the "empirical Bayes" (EB) method.

SPFs are cost- and time-prohibitive to develop, and for this reason they are rarely developed by agencies that have limited funding. Instead, these SPFs are usually developed at the state or national level by expert safety analysts very familiar with the EB methodology. Although the analysis methods are difficult and time-prohibitive for safety analysts to derive, there are methods for using those SPFs already developed. This method is described in the following section.

Applying Safety Performance Functions in the Highway Safety Manual

Noted earlier, there are several methods that can be used to evaluate countermeasures deployed along roadways and at intersections. To account for impacts of bias in safety evaluations, it is desirable to account for all possible bias, especially RTM. Development of SPFs is a complex procedure requiring significant time and resources; therefore, the focus of this section is application of previously developed SPFs in the *Highway Safety Manual* (*HSM*)[16] to predict the average expected crash frequency at a site, comparing it to the actual number of crashes that took place following countermeasure implementation.

An SPF is a regression equation that typically uses traffic volume and segment length as independent variables. The *HSM* provides prediction methods for the following road types:

- Roadway segments
  - Rural two-lane roads
  - Rural four-lane divided and undivided roads

[16] Transportation Research Board (TRB) (2010). *Highway Safety Manual*, 1st edition. American Association of State Highway and Transportation Officials (AASHTO), Washington, D.C.

- Urban and suburban arterials with two lanes
- Urban and suburban arterials with three lanes (with center two-way left-turn lane)
- Urban and suburban arterials with four lanes divided and undivided
- Urban and suburban arterials with five lanes (with center two-way left-turn lane)

- Intersection types
  - Rural two-lane roads with three- and four-leg stop control
  - Rural two-lane roads with four-legged signalized intersections
  - Rural four-lane roads with three- and four-leg stop control
  - Rural four-lane roads with four-legged signalized intersections
  - Urban and suburban two-lane roads with three- and four-leg stop control
  - Urban and suburban two-lane roads with signalized intersections

A detailed 18-step procedure for utilizing previously developed SPFs to estimate the average expected crash frequency at a site is provided in the *HSM*. Here, the steps of the predictive method are summarized as:

- Steps 1 and 2: Decide which facilities and roads will be used in the predictive process and for what period of time.
- Steps 3–8: Identify homogeneous sites and assemble geometric conditions, crash data, and AADT data for the sites to be used.
- Steps 9–11: Apply the appropriate SPF, any applicable crash modification factors (CMFs), and a calibration factor if available.
- Steps 12–15: Apply site or project-specific EB method if applicable.
- Steps 16–18: Repeat for all sites and years, sum, and compare results.

A site in the *HSM* is defined as an intersection or homogeneous roadway segment. The focus of this section will be on Steps 1 through 11.

Three components are used to predict the average crash frequency at a site: (1) the base model, called an SPF, (2) CMFs to adjust the estimate for additional site specific conditions (discussed in Section 4.3.3), (3) and a calibration factor to adjust the estimate for accuracy in the state or local area. These components are used in the general form in the following equation:

$$N_{predicted} = SPF_{base} \times (CMF_1 \times CMF_2 \times \ldots \times CMF_n) \times C_x$$

where

$N_{predicted}$ = predicted average crash frequency for a specific year for site type x

N_{spf} = predicted average crash frequency determined for base conditions of the SPF developed for site type x

CMF_n = crash modification factors specific to the SPF for site type x

C_x = calibration factor to adjust the SPF for local conditions for site type x

Each SPF is specific to a facility or site type (e.g., urban four-lane divided segments) and/or a specific year. Similarly, each CMF must be applied directly when the original condition is not the same as the CMF base case. As an example, consider lanes where the base condition CMF (CMF = 1.0) would be for a 12-ft lane. If an existing roadway has 9-ft lanes (CMF = 1.5) and the agency is considering expanding to 11-ft lanes (CMF = 1.05), an analyst would need to take the ratio of the updated condition to the existing condition CMFs, or 1.05/1.5 to get a CMF for the treatment of 0.70.

One major flaw with applying CMFs is that the method assumes countermeasures and roadway characteristics act independently of one another. At this time, the combined effects of multiple treatments and characteristics of a site are generally not known and further research is needed to fully understand how they interact. As a general rule, it is always best to try to separate out collision types applicable to one treatment from another for analysis; however, that is usually not possible. Instead, analysts must often multiply the combination of two or more CMFs to get the combined effect. An example is provided in the following example to show how to apply an SPF, CMFs, and a calibration factor for the predictive method.

EXAMPLE 6.13 Calculating Average Expected Crash Frequency Using the *HSM* Predictive Method

You are asked to determine the net effect of a new curb design aimed at reducing the number of collisions in curves along a rural four-lane divided roadway segment. One year after constructing this new curb section along 10 miles of roadway, the crash rate was found to be 22.3 crashes per year. The following information is given about the site:

- 10-mile segment
- 12-ft lane width
- 6-ft paved right shoulder
- 15,000 AADT
- No roadway lighting
- 80-ft traversable median with no barrier
- No automated enforcement
- Calibration factor is 0.96

(*Continued*)

EXAMPLE 6.13 Calculating Average Expected Crash Frequency Using the *HSM* Predictive Method—(Continued)

$$N_{4RD} = e^{a+b\times\ln(AADT)+\ln(L)}$$

where

N_{4RD} = predicted average crash frequency determined for base conditions of the SPF developed for rural four-lane divided highways (*HSM* Equation 11-9)

AADT = annual average daily traffic (vehicles/day) on the roadway segment

L = length of roadway segment

a, b = regression coefficients (appropriate values to be selected from Equation 11-5)

Solution

Steps 1–8: Because this example is directed at applying the predictive method to a single preselected segment with existing data, steps 1 through 8 are not necessary.

Step 9: Using the SPF for this example yields the following base number of collisions:

$$N_{spf(4RD)} = e^{-9.025+1.049\ x\ln(15{,}000)+\ln(10.0)}$$
$$N_{spf(4RD)} = 28.9\ crashes\ per\ year$$

Step 10: The *HSM* procedure for rural divided highways involves five CMFs:

1. Lane width (12 ft), $CMF_{1RD} = 1.0$
2. Right shoulder width (6 ft), $CMF_{2RD} = 1.04$
3. Median width (80 ft), $CMF_{3RD} = 0.95$
4. Lighting (none), $CMF_{4RD} = 1.0$
5. Automated enforcement (none), $CMF_{5RD} = 1.0$

The combined CMF value is calculated in the following equation.

$$CMF_{combined} = 1.0 \times 1.04 \times 0.95 \times 1.0 \times 1.0 = 0.99$$

Step 11: The calibration factor for your state is provided as 0.96. The average expected crash frequency, $N_{predicted}$, is calculated as follows.

$$N_{predicted} = SPF_{base} \times (CMF_1 \times CMF_2 \times \ldots \times CMF_n) \times C_x$$
$$N_{predicted} = 28.9 \times 0.99 \times 0.96 = 27.5\ crashes\ per\ year$$

Therefore, the expected safety improvement provided by the new curb and gutter treatment is expected to reduce the average collision rate by 4.3 crashes per year (27.5–23.2), or a 15.6% reduction in crashes per year.

6.5 PRACTICE PROBLEMS

Problem 6.1

For the intersection shown, how many vehicle-to-pedestrian conflict points are present if pedestrians are allowed to cross all streets?

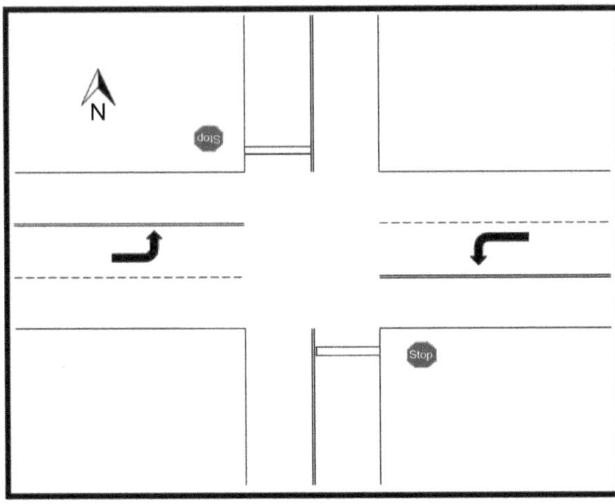

Solution

Draw the individual movements for eastbound and westbound directions (Step 1), followed by the northbound and southbound directions (Step 2). There are 18 pedestrian-vehicle conflict points.

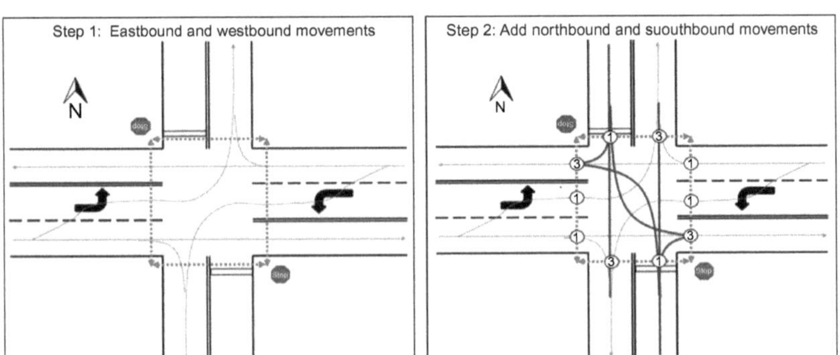

Problem 6.2

For the T-intersection shown, how many vehicle-to-vehicle conflict points are present?

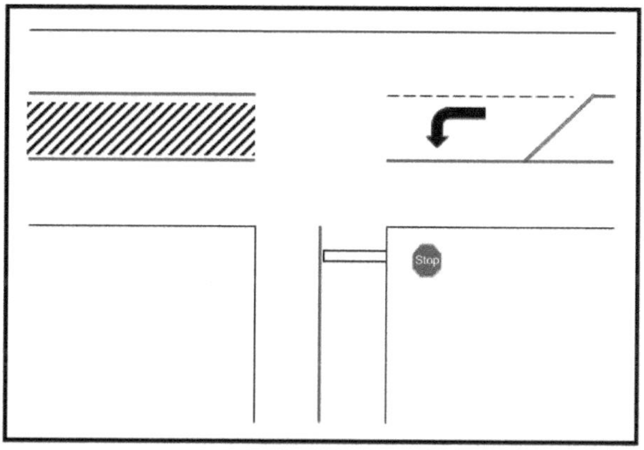

Solution

Draw the individual movements for eastbound and westbound directions (Step 1), followed by the northbound direction (Step 2). There are nine vehicle-to-vehicle conflicts.

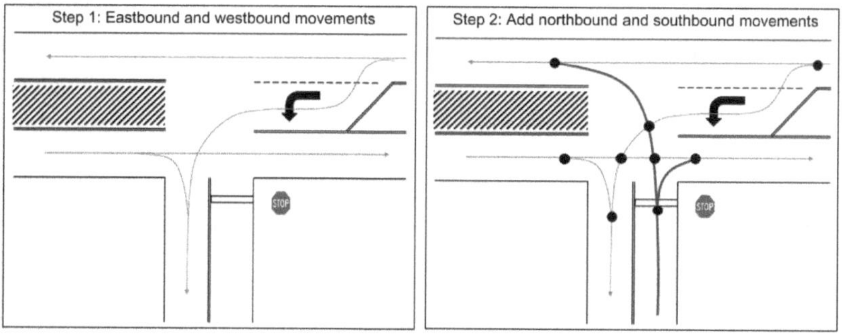

Problem 6.3

Several members of the general public have made calls of complaint about a signalized intersection located just outside of your city. Your supervisor says the signalized intersection was recently installed to accommodate a new subdivision to the east and a large shopping center to the west. The new intersection was installed on a major arterial. You ask your supervisor if she knows what the problem is and she says that the complaints do not seem to correlate with one type of safety problem. Because the intersection was recently installed, you have no collision data to conduct an analysis. You resort to a conflict study method and you take your colleague with you to study the site for a full day. You use ADTs collected by the DOT to determine the approach volumes. The frequency and rate of conflicts are provided in the following table and graph. What can you determine based on your assessment?

Solution

Using conflict magnitude alone for this analysis would not be wise as the traffic volumes are considerably different for each of the approaches. This would result in all of your recommendations made for the mainline only. Using the rate of conflicts per 1000 vehicles, the graph shows that the majority of conflicts are taking place on the southbound and westbound approaches, but these do not appear to be significantly greater than the other approaches. When looking at conflict type for each approach, there appears to be a major safety problem for the first three conflict types, which are all related and all very dangerous. Looking back at your field notes and condition diagram, you determine that there is a problem with sight distance to signalized intersection along the mainline, resulting in several people accidentally running the red light or sliding into the intersection and nearly hitting opposing vehicles. The new intersection was installed in a large radius curve. Another major issue is right-turn conflicts from the side street with U-turning vehicles and pedestrians. This is especially true for the heavier westbound movement. You recommend considering a treatment that makes the crosswalk right of way more prominent to drivers, such as signage, a no-right-turn-on-red sign, or even changing the right-turn protected/permitted signal to a right-turn flashing yellow arrow.

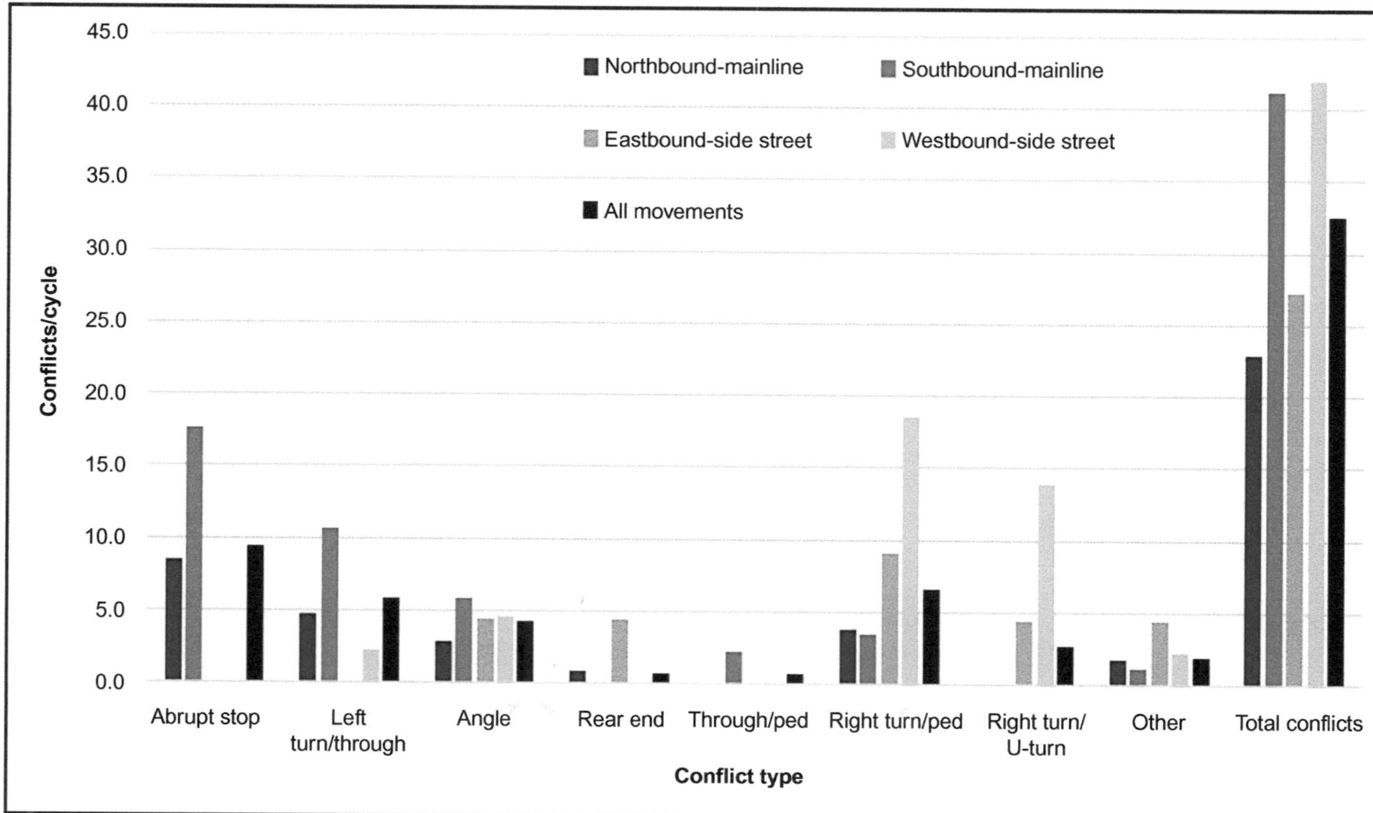

Conflicts/cycle
45.0
40.0
35.0
30.0
25.0
20.0
15.0
10.0
5.0
0.0
Northbound-mainline
Southbound-mainline
Eastbound-side street
Westbound-side street
All movements
Abrupt stop
Left turn/through
Angle
Rear end
Through/ped
Right turn/ped
Right turn/ U-turn
Other
Total conflicts
Conflict type

Unit	Conflicts	1000 veh	Conflicts	1000 veh	Conflicts	1000 veh	Conflicts	1000 veh	Conflicts	1000 veh
Abrupt Stop	9	8.6	15	17.6	0	0.0	0	0.0	24	9.4
Left Turn/ Through	5	4.8	9	10.6	0	0.0	1	2.3	15	5.9
Angle	3	2.9	5	5.9	1	4.5	2	4.7	11	4.3
Rear End	1	1.0	0	0.0	1	4.5	0	0.0	2	0.8
Through/Ped	0	0.0	2	2.4	0	0.0	0	0.0	2	0.8
Right Turn/Ped	4	3.8	3	3.5	2	9.1	8	18.6	17	6.7
Right Turn/U-Turn	0	0.0	0	0.0	1	4.5	6	14.0	7	2.7

Problem 6.4

The DOT has procured federal funding that must be used to install countermeasures on its top six rural two-lane highways. The DOT has identified 10 2-mile segments with a high frequency of severe collisions in rural areas with combined AADTs less than 10,000 vehicles per day (see the following table). To make sure that trends are consistent over time, a 5-year period of data was utilized. Which six sites make the cut?

Segment	Severe (F + A) collisions	AADT
1	18	9,500
2	15	7,500
3	14	6,250
4	11	9,000
5	16	9,250
6	30	9,750
7	17	7,950
8	22	9,050
9	24	9,500
10	15	8,000

Solution

The collision rates are calculated in the following table. Segments 2, 3, 6, 7, 8, and 9 make the cut.

Segment	Severe (F + A) collisions	AADT	$R_{segment}$
1	18	9500	0.519
2	15	7500	0.548
3	14	6250	0.614
4	11	9000	0.335
5	16	9250	0.474
6	30	9750	0.843
7	17	7950	0.586
8	22	9050	0.666
9	24	9500	0.692
10	15	8000	0.514

Problem 6.5

An agency is looking at identifying sites that may have high numbers of rear-end collisions to install a newly identified countermeasure. The following five intersections are being considered at this time from data from the previous three years.

Intersection	Rear-end collisions	AADT	
		Road 1	Road 2
1	25	10,000	20,000
2	22	6200	15,400
3	15	32,000	8500
4	33	12,150	10,250
5	26	13,900	20,000

The average rear-end crash rate for sites of similar traffic volume and surrounding land use for use in analyzing the intersections is $R_{a\ (rear\ end)} = 1.8$ crashes per million entering vehicles. Using the rate quality control method, which intersections should be considered for possible implementation of the rear-end countermeasure? What would be your next recommendation for analysis prior to countermeasure implementation?

Solution

The critical crash rate will vary depending on the traffic volume at each site. For this reason, expanding the previous table will be the easiest method.

Intersection	Rear-end collisions	AADT		R_a	M	K	$R_{critical}$	R_{intx}
		Road 1	Road 2					
1	50	10,000	20,000	1.8	32.85	1.645	2.20	1.52
2	64	8000	10,000	1.8	19.71	1.645	2.32	3.25
3	65	32,000	8500	1.8	44.35	1.645	2.14	1.47
4	58	12,150	10,250	1.8	24.53	1.645	2.27	2.36
5	52	13,900	20,000	1.8	37.12	1.645	2.18	1.40

Intersections 2 and 4 should be considered for implementation as R_{intx} is greater than $R_{critical}$. Drawing a collision diagram is recommended to determine if the countermeasure is applicable to the location the rear-end collisions are taking place.

Problem 6.6

Collision data from five crash-prone intersections in the city of Okato are provided in the following table. Citizens have complained that Intersection 2 is hazardous. The collision data for these five intersections are provided in the following table, with data from the past three years.

Intersection	Collision type					AADT	
	F	A	B	C	PDO	Road 1	Road 2
1	0	1	6	8	20	10,000	20,000
2	3	2	8	6	15	20,000	10,000
3	1	1	5	8	13	32,000	8,500
4	0	1	2	10	24	12,150	10,250
5	0	0	13	16	25	13,900	20,000

A. What are the collision rates for the following five sites?

Solution

$$R_1 = \frac{[(1+6+8+20)\times 1,000,000]}{[365\times 3\times(10,000+20,000)]} = 1.07$$

$$R_2 = \frac{[(3+2+8+6+15)\times 1,000,000]}{[365\times 3\times(20,000+10,000)]} = 1.04$$

$$R_3 = \frac{[(1+1+5+8+13)\times 1,000,000]}{[365\times 3\times(32,000+8,500)]} = 0.63$$

$$R_4 = \frac{[(1+2+10+24)\times 1,000,000]}{[365\times 3\times(12,150+10,250)]} = 1.51$$

$$R_5 = \frac{[(13+16+25)\times 1,000,000]}{[365\times 3\times(13,900+20,000)]} = 1.45$$

B. Based solely on this analysis, would you agree or disagree that this site is dangerous?

Solution

Disagree, the collision rate is approximately average compared to that of the other four sites.

C. What other analysis might you look at conducting?

Solution

This site is a high-volume site and collision rate biases toward low-volume sites. A better analysis method that considers frequency also (such as EPDO) is recommended.

Problem 6.7

Relevant collision data for the last three years in the town of Zumba are given in the following table. Citizens have complained that Intersection 2 is hazardous.

Intersection	Number of injuries by type					AADT	
	F	A	B	C	PD	Road 1	Road 2
1	0	3	0	3	18	6,100	8,950
2	2	4	5	4	12	6,550	18,200
3	0	1	4	18	32	13,650	8,000
4	0	3	3	16	41	12,900	18,850
5	0	0	0	8	19	5,600	14,400

You have decided to analyze Intersection 2 for possible countermeasures using equivalent property damage only analysis based on the Kentucky formula.

The average EPDO crash rate for sites of similar traffic volume and surrounding land use for analyzing these intersections has been calculated and is:

$R_{a\ (EPDO)} = 1.87$ EPDO crashes per million entering vehicles

A. Using the Kentucky Formula, what is the number of EPDO collisions expected at Intersection 2?

Solution

$$\text{EPDO} = 9.5(\text{F} + \text{A}) + 3.5(\text{B} + \text{C}) + \text{PDO}$$
$$\text{EPDO} = 9.5(2 + 4) + 3.5(5 + 4) + 12$$
$$\text{EPDO} = 76 + 31.5 + 12 = \underline{119.5}$$

A. 120
B. 110
C. 105
D. 130

B. The EPDO crash rate (R_{EPDO}) per MEV for Intersection 2 is most nearly what?

$$R_{EPDO(n)} = \frac{[(EPDO) \times 1{,}000{,}000]}{[365 \times T \times (AADT_1 + AADT_2)]}$$

$$R_{EPDO(2)} = \frac{[(120) \times 1{,}000{,}000]}{[365 \times 3 \times (6550 + 18{,}200)]} = 4.43$$

Solution

A. 13.29
B. 4.43
C. 8.26
D. 1.92

C. Using the rate quality control method, the appropriate EPDO critical crash rate per MEV for Intersection 2, based on 95% confidence level, is what?

Solution

A. 1.96
B. 1.63
C. 2.91
D. 2.32

$$R_{critical} = R_a + \frac{0.5}{M} + K\sqrt{\frac{R_a}{M}}$$

where

$$R_a = 1.87\frac{collisions}{MEV}(given)$$

$$M = \frac{(6550 + 18,200\ vpd) * (365\ days) * (3\ years)}{1,000,000}, or\ 27.1\ MEV$$

$$K = 1.645$$

$$R_{critical} = 1.87 + \frac{0.5}{27.1} + 1.645\sqrt{\frac{1.87}{27.1}}$$

$$R_{critical} = 2.32\ collisions\ per\ MEV$$

D. Using R_{EPDO} and $R_{critical}$, what action would you most likely take regarding Site 2?

Solution

Because R_{EPDO} exceeds $R_{critical}$, the analyst should investigate the site for possible remedial measures.

Problem 6.8

Concerned citizens of Utopolis have recently complained about a recent signalized intersection installation into the newly constructed shopping center off of Curvy Junction Road. Crash reports were unavailable following the installation of the shopping center as the site was still new. You visit the site and draw a condition diagram, noting special features that look like safety problems. Based on your site observations and pictures from the site (see the following figure), the roadway is posted at 45 mph and the intersection is posted in a sharp curve. There is a slight downgrade from the west to the east of approximately 2%. On the eastbound approach (west side), skid marks were noted across the stop bar along with some debris. On the westbound approach (east side), there is a railroad bridge with a column in the median. This column, along with trees on both approaches, *may* cause problems with intersection sight distance.

Note: The left and right pictures are east and westbound, respectively.

Because no collision data were available, the analyst conducted a quick conflict analysis during am and pm peaks on the following day. Based on this analysis, there appeared to be several late-braking events on the eastbound approach. Several drivers were seen braking suddenly, and some even entering the intersection, on this approach. Although not measured directly, vehicle speed appeared to be a problem for this downhill approach about 10% of the time. In addition, there were several near miss rear-end collisions for the second through fifth vehicles joining the queue at the red signal. Although there did appear to be some sight distance obstructions on the westbound approach, no major conflicts were out of the ordinary. Based on these findings, what recommendations would you consider implementing at this site?

Solution

Based on the conflict study findings, condition diagram, and photo log, the analyst should determine that the primary problem at this site was the eastbound approach. Specifically, restricted sight distance was a severe problem with a minor focus on excessive speeds.

Possible countermeasures for these causal factors are determined and the CMF Clearinghouse is used to document the expected benefit of installing each countermeasure. A summary is provided in the following table:

Possible countermeasure	Expected benefit (%)	Quality	Source
Remove sight obstructions (increase triangle sight distance)	37	***	Elvik (2004)
	48	***	Rodegerdts (2004)
Install/improve warning signs	35	**	Polanis (1999)
Install dynamic signal warning flashers	18–25	****	Srinivasan (2011)
Install rumble strips	6	**	Elvik (2004)

Source: The FHWA CMF Clearinghouse available at http://www.cmfclearinghouse.org/

Based on these findings, several recommendations could be made. Considering the low net benefit and quality of the study, rumble strips do not appear to be an option. However, based on the three remaining countermeasures, the analyst would likely recommend removing trees or tree branches to improve sight distance in the sight triangle. The analyst could observe conflicts again following removal of issues and follow up with additional, more costly treatments as necessary.

Problem 6.9

A state DOT requested that 14 jurisdictions provide their 2 most dangerous rural signalized intersections for consideration of pooled funds for safety improvements. Crash data were collected over a 5-year period for each of the 28 intersections, and the 28 intersections of the reference population had an RSI of $41,600 per collision. The DOT believes it has enough pooled funds to make safety improvements at no more than 3 sites. During a meeting with your peers, you review the safety data using rate and frequency. You all agree that 6 of these sites have many more collisions than the other 22 and recommend looking further at these to see which 3 should possibly be considered for funding. Crash data for the 6 sites are provided in the following table.

Signalized intersection	Crash type						
	Rear end	Side swipe	Angle	Head on	Fixed object	Other	Total
Cost	\$7500	\$10,000	\$215,000	\$950,000	\$16,000	\$17,000	
1	7	2	3	1	22	37	72
2	11	1	5	0	16	21	54
3	7	0	4	2	18	22	53
4	6	3	8	0	42	21	80
5	12	1	7	0	17	22	59
6	4	0	2	0	26	35	67
Total	47	7	29	3	141	158	385

A. Using the RSI method, determine the cost per collision for Site 1.

Solution

First the total cost of each collision type over the 5-year period must be calculated for each of crash types. The calculations for each crash type at Intersection 1 are given in the following list, followed by a calculation of the RSI.

Rear end: 7 × \$7500 = \$52,500
Side swipe = 2 × \$10,000 = \$20,000
Angle: 3 × \$215,000 = \$645,000
Head on = 1 × \$950,000 = \$950,000
Fixed object: 22 × \$16,000 = \$352,000
Other: 37 × \$17,000 = \$629,000

$$RSI_n = \frac{\sum_{j=1}^{n} RSI_j}{N_{0,n}}$$

$$RSI_1 = \frac{(\$52,500 + \$20,000 + \$645,000 + \$0 + \$950,000 + \$352,000 + \$629,000)}{(7 + 2 + 3 + 1 + 22 + 37)}$$

$$RSI_1 = \frac{(\$2,648,500)}{(72)} = \$36,785$$

B. The RSI for the remaining 5 intersections is calculated in the following. Based on the RSI for Intersections 1 (calculated above) through 6, what recommendation might you make to your supervisor?

Solution

The top three intersection RSIs are Intersections 1, 3, and 5; however, only one site has a higher average crash cost than the population

Signalized intersection	Crash type and costs						Total	RSI
	Rear end	Side swipe	Angle	Head on	Fixed object	Other		
Cost	$7500	$10,000	$215,000	$950,000	$16,000	$17,000		
1	—	—	—	—	—	—	—	—
2	$82,500	$10,000	$1,075,000	$0	$256,000	$357,000	$1,780,500	$32,972
3	$52,500	$0	$860,000	$1,900,000	$288,000	$374,000	$3,474,500	$65,557
4	$45,000	$30,000	$1,720,000	$0	$672,000	$357,000	$2,824,000	$35,300
5	$90,000	$10,000	$1,505,000	$0	$272,000	$374,000	$2,251,000	$38,153
6	$30,000	$0	$430,000	$0	$416,000	$595,000	$1,471,000	$21,955

RSI of $41,600. Based on this method, the analyst should only recommend Site 3 for funding. The analyst would be wise to recommend looking at the original pool of sites using another method, such as collision rate, and then revisiting the RSI method. Although collision frequency was lower at the other 22 sites, based on the population RSI, it is highly likely that most costly crashes are occurring at the other sites.

Problem 6.10

The State DOT is looking to improve the operations and safety of several interchange merge areas along its highways and freeways. Many countermeasures are being considered, however, the primary countermeasure is ramp metering. The state is looking at 1-mile segments starting at the merge point of all interchanges along highway and interstate routes with posted speed limits equal to or greater than 55 mph. Based on the total collisions at all sites across the state, 10 were selected based on highest frequency of collisions. Two newly constructed interchanges are included: Site 2 includes 6 months of data while Site 6 includes a full year. Which of the 10 sites should the state consider for further evaluation?

Site	Annual average 2012–2014					Total collisions			Special notes
	Side swipe	Run off the road	Fixed object	Other collisions	Freeway or highway ADT	2011	2012	2013	
1	16	8	9	14	100,000	19	14	14	
2	5	2	11	3	95,000	—	—	21	Site rebuilt in spring 2013
3	10	16	6	39	59,000	26	21	24	
4	10	18	12	32	48,000	21	27	24	
5	22	18	27	19	160,000	22	19	45	
6	12	12	25	3	55,000	—	—	52	Site constructed late 2012
7	13	6	12	26	31,000	20	21	16	
8	15	9	15	16	10,5000	16	20	19	
9	14	7	14	9	63,000	14	16	14	
10	19	16	4	14	86,000	19	13	21	

Solution

The scenario just presented meets each of the five conditions for SWP. The following table is referenced for several of the conditions.

Site	Side swipe + run off the road + fixed object	Frequency per year	Rate per million vehicles
1	33	15.7	0.43
2	18	21.0	1.21
3	32	23.7	1.10
4	40	24.0	1.37
5	67	28.7	0.49
6	49	52.0	2.59
7	31	19.0	1.68
8	39	18.3	0.48
9	35	14.7	0.64
10	39	17.7	0.56
—	Mean (similar sites)	18.0	0.8
Standard deviation (similar sites)	4.8	0.5	

The solutions for each condition are shown in the following.

Condition A: The department is trying to reduce collisions related to merging behavior on limited access roadways with posted speed limits of 55 mph or greater. These safety problems are most directly associated with side-swipe, run-off-the-road, and fixed-object crashes. Looking at the frequency of these collisions in the previous table (excluding newly constructed sites), Site 5 looks the most promising with 67 collisions per year from 2011–2013.

Condition B: Compare newly constructed Sites 2 and 6 with comparable sites (1, 3–5, and 7–10) using total collisions. The mean frequency per year and standard deviation of the remaining comparable sites are calculated in the previous table (18.0 ± 4.8 crashes). Note that Site 2 total crash frequency was doubled (from 21 to 42) in the table. When looking at total collisions compared to the reference population, both Sites 2 and 6 have very high collision frequencies. In fact, Sites 2 and 6 have differences in collisions that are 5 and 7.1 standard deviations, which is very significant!

Site 2:

$$\frac{42 - 18.0}{4.8} = 5.0$$

Site 6:

$$\frac{52 - 18.0}{4.8} = 7.1$$

Condition C: Looking at the collisions from 2011–2013, crash frequency at Site 5 increased drastically in 2013 by a factor of more than 2. It would be worthwhile to look at crash trends over a longer period at this site to determine if this increase was isolated.

Condition D: Looking at collision rates for sites over the 3-year period (excluding Sites 2 and 9) in the table, Sites 4 and 7 are higher than any of the other similar sites. Note that rates here are calculated for "segments" and not "spots," and that total collisions were using the calculation of "rate per million vehicles."

Condition E: Similar to Condition B, compare newly constructed Sites 2 and 6 with the remaining comparable intersections. Sites 2 and 6 have differences in collision rates that are 0.8 and 3.58 standard deviations. Site 6 is the only site more than 1 standard deviation, making it a promising site when looking at collision rate.

Site 2:

$$\frac{1.2 - 0.8}{0.5} = 0.8$$

Site 6:

$$\frac{2.6 - 0.8}{0.5} = 3.58$$

Summary

Site 5 has very high collision frequency and rate and should be looked at closely to determine problems. Not only does this site have a high frequency of collisions, they appear to increase drastically in the last year. Both newly constructed interchanges (Sites 2 and 6) appear to have high frequencies of collisions; however, Site 6 also has a very high rate of collisions compared to the reference population. Site 6 should definitely be evaluated more closely, while Site 2 should be reevaluated when more data are available. This is because this site was only evaluated with data over the winter months, which are more often the higher collision months from a seasonal standpoint. Last, collision rates for Sites 4 and 7 were high and should be looked at more closely. A condition and collision diagram would be helpful in determining the exact safety countermeasure to implement, if any, at each of the sites noted.

Problem 6.11

Last year, your jurisdiction installed flashing right-turn arrows (FRTA) at a four-legged signalized intersection test site where several pedestrian-vehicle collisions had taken place over several years. This treatment was chosen because collision data seemed to indicate the primary problem was with right-turning drivers. After further investigation at the site, it appeared that motorists were not appropriately yielding to pedestrians when given a permitted signal indication (green ball). You have been asked to determine the net effect of the FRTA. Two years after installing the FRTA, the crash rate was found to be one crash per year. The following information is given about the site:

- $AADT_{minor} = 20{,}000$
- $AADT_{major} = 25{,}000$
- 2500 pedestrians crossing per day
- 4 bus stops within 1000 ft of intersection
- 6 establishments that sell alcohol within 1000 ft of intersection
- Maximum crossing distance of 5 lanes
- Calibration factor is 1.62

$$N_{ped} = e^{\left[a+[b \times \ln(AADT_{total})]+\left[c \times \ln\left(\frac{AADT_{min}}{AADT_{maj}}\right)\right]+[d \times \ln(PedVol)]+(e \times n_{lanes(x)})\right]}$$

where

N_{ped} = predicted average crash frequency determined for base conditions of the SPF developed for urban and suburban signalized intersections (*HSM* Equation 12-29)

$AADT_{total}$ = sum of average daily traffic volumes (vehicles/day) for the major and minor roads

$PedVol$ = sum of the daily pedestrian volumes (pedestrians/day) crossing all intersection legs

$n_{lanes(x)}$ = maximum number of traffic lanes crossed by a pedestrian in any crossing maneuver at the intersection considering the presence of refuge islands

a,b,c,d,e = regression coefficients (appropriate values to be selected from the Highway Safety Manual).

Solution

Steps 1–8: Because this example is directed at applying the predictive method to a single preselected segment with existing data, steps 1 through 8 are not necessary.

Step 9: Using the SPF for this example yields the following base number of collisions:

$$N_{ped} = e^{\left[-9.53+[0.4 \times \ln 45{,}000]+\left[0.26 \times ln\left(\frac{20{,}000}{25{,}000}\right)\right]+[0.45 \times \ln(2500)]+(0.04 \times 5)]\right]}$$

$$N_{ped} = 0.206 \textit{ crashes per year}$$

Step 10: The *HSM* procedure for urban and suburban arterials involves three CMFs, noted in the following:

- 4 bus stops within 1000 ft of intersection, $CMF_{1p} = 4.15$
- 1 school within 1000 ft of intersection, $CMF_{1p} = 1.35$
- 6 alcohol sales establishments within 1000 ft of intersection, $CMF_{3p} = 1.12$

The combined CMF value is calculated as follows.

$$CMF_{combined} = 4.15 \times 1.35 \times 1.12 = 6.28$$

Step 11: The calibration factor for your state is 1.62. The average expected crash frequency, $N_{predicted}$, is calculated as follows.

$$N_{predicted} = SPF_{base} \times (CMF_1 \times CMF_2 \times \ldots \times CMF_n) \times C_x$$

$$N_{predicted} = 0.206 \times 6.28 \times 1.62 = 2.09 \textit{ crashes per year}$$

Therefore, the expected safety improvement provided by the new FRTA is expected to reduce the average collision rate by 1.09 crashes per year (2.09 − 1.0), or a 48% reduction in crashes per year.

PART 7

Geotechnical

Thomas H. Brown, Jr., PhD, PE
Institute for Transportation Research and Education, North Carolina State University

Contents

7.1 INTRODUCTION

Geotechnical engineering involves many topics; however, this major topic in civil engineering is limited here to the analysis of soil that is the foundation of major highway elements, such as roadways and bridges, and elements that pass under roadways, such as culverts. In terms of it being an engineering material, soil has several characteristic attributes:

1. It is a material that has been around for a long time. Not only has it been the foundation of many magnificent structures, it was the material used to make the structures themselves. Much of what we know today was learned by trial and error, and our knowledge of this ancient material is continually added to by modern science and technology.
2. It occurs naturally. It sounds almost trite to say that, but soil is not a material created by man, like steel or aluminum or even concrete. We must deal with what nature has provided, which highlights the art of geotechnical engineering. Its properties are so variable, even in the same area. And soil can act in unpredictable ways when used as a construction material.

3. Soil is used everywhere. No highway project can avoid making understanding the soil conditions a first priority. Roadways and bridges must obviously be supported properly; however, even the surface contours along roadways must be designed for proper drainage of storm water. If the soil structure associated with these elements is not right, or is not made right, serious consequences can and will result.

The civil engineer Arthur Casagrande (1902–1981), who developed the soil mechanics program at Harvard University in the 1930s, said of the importance of soil sampling and analysis:

> *Soil has been used as a construction material since the dawn of civilization and yet its employment as such is more hazardous than the use of any other construction material. More money is expended upon soil construction than upon any other construction, and its failures cause a greater loss of life and property than any other construction failure. This is true in spite of the accumulated knowledge and experiences of centuries.*

7.2 SOIL SAMPLING AND TESTING ANALYSIS

7.2.1 Purpose of Sampling

A highway, and certainly the supports for bridges, cannot be designed and constructed properly without accurate and trusted soil information. The cost to do the soil analysis right the first time is considerably lower than the cost to fix settlement problems after construction is complete and the facility is in full operation. Therefore, the process of taking soil samples, the first step in the process, must be approached with considerable care. Cutting corners at this stage is counterproductive and unwise. The cost of taking the proper number of samples in the appropriate locations is minor compared to the consequences if the knowledge of the soil conditions is incomplete or incorrect. That said, the number and location of soil samples required for a particular project in a particular location is an art, not a science. Considerable experience is required and cannot be overemphasized.

7.2.2 First Things First

Before sampling can actually occur, it is customary to investigate all the various sources of information available to the geotechnical engineer. These include, in no particular order, the following:

- Scope of work for the project
- Topographic maps from the U.S. Geodetic Survey (USGS)

- Soil survey maps from the Natural Resources Conservation Service (NRCS)
- Groundwater reports
- Aerial photographs

Obtaining as much information as possible from these sources will make determining the location and number of samples to be taken as efficient and effective as possible.

7.2.3 Once in the Field

Even before taking soil samples, much can be gained by surveying the project area. Things to notice are, in no particular order:

- Lay of the land, meaning the topography, steep, or level
- Types of vegetation, trees to soil cover, based on the season of the year
- Evidence of storm-water runoff, and the presence of adverse erosion
- Qualities and types of soil, and especially the presence and type of rock
- Taste (only kidding), though this might be done if the truth be known

Of course, if the current project is near former projects, much can be learned from the documents and experience gained. However, it is not always safe to assume that conditions are the same. Many major failures of structures have occurred when this false assumption is made. Testing the current site cannot be avoided.

One of the major decisions that must be made is the number, type, and location of the soil samples to be taken. Two major goals of the sampling are to determine (1) the layering of the various soil and rock types, and (2) properties of the soil, such as strength, compressibility, and water retention. Both require specific equipment. Information on the first goal, layering of the soil strata, must provide:

- Thickness of each type of soil
- Depth to the water table
- Depth to any rock

For the second goal, cataloging of the various soil samples is imperative so that the information needed to determine the settlement and rate of settlement can be determined with the proper level of confidence.

7.2.4 Soil-Sampling Equipment

Drilling methods, as well as the equipment used, continue to evolve. Traditional methods include auguring, bentonite mud drilling, and rock core drilling. Equipment has evolved to obtain the right type of sample. These include augers, split-spoon samplers, and Shelby tubes. Much depends on whether an undisturbed sample is required, or whether a disturbed sample is sufficient. Procedures for using soil-sampling equipment are provided by the American Association of State Highway and Transportation Officials (AASHTO) and the American Society for Testing and Materials (ASTM). These assure that the samples are obtained with appropriate accuracy and consistency from one project to another, across the country. In this age of the internet, images of all the various types of drilling equipment and soil-sampling devices abound.

7.2.5 Soil-Sampling Records

For each sample taken, a boring log is generated. An example is shown in Figure 7.1. The information from multiple borings can be assembled into a soil profile, as shown in Figure 7.2. The details of how these diagrams are generated, or interpreted, are beyond the scope of this work. It is sufficient to say that here again is where the art of geotechnical engineering is manifested so clearly. As a physician looks at an electrocardiogram (ECG), so a geotechnical engineer looks at a boring log or a soil profile. Much is riding on his or her judgment and experience.

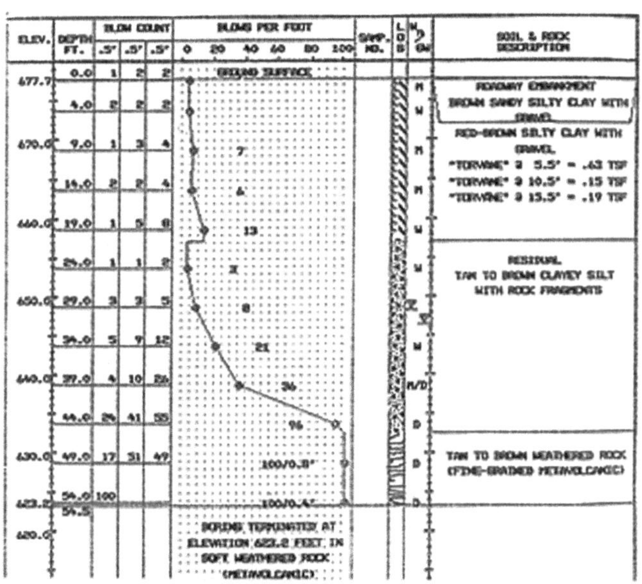

Figure 7.1 Boring log.

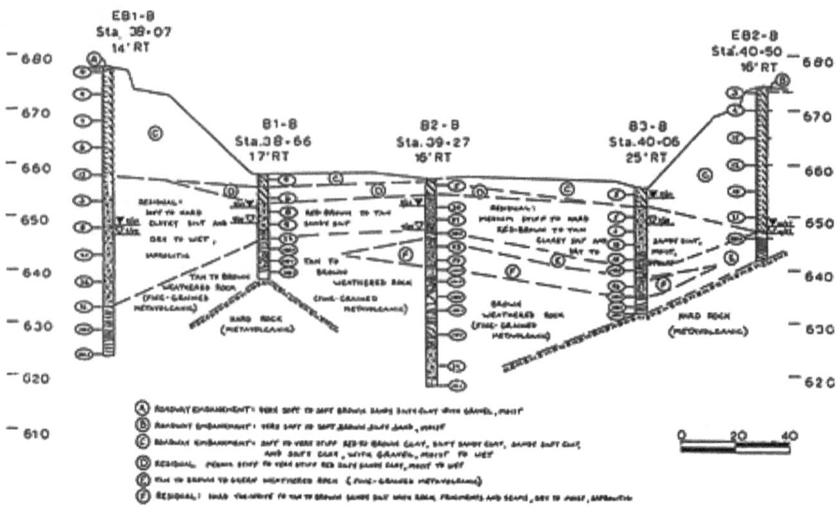

Figure 7.2 Soil profile.

7.3 SOIL CLASSIFICATION

7.3.1 Introduction

There are two systems for classifying soil, one by AASHTO and one by the Natural Resources Conservation Service. The one used for highway design is, not unexpectedly, the AASHTO system. Both use similar soil criteria, specifically soil gradation and Atterberg limits.

7.3.2 Soil Gradation

Soil gradation is the separation of a soil sample into its various sizes. Sizes in soil can obviously be from boulders to powder, called the *fines*. At first, it could be assumed that the more uniform a soil, the better; however, it is just the opposite. The more varied the sizes in the soil, the greater the ability of the smaller particles to fill the voids created by the larger ones. In this way, the soil has a greater density, meaning less air, and is supported by considerably more contact points. It also reduces the amount of water that can be absorbed by the soil in the voids.

Figure 7.3 shows a grain size chart to illustrate the point that a well-graded soil is the best. The chart has semilog scales: along the bottom, 1 million grain sizes in millimeters can be plotted. Along the vertical axis is the "percent finer by weight" scale. At the top are the subjective divisions of the possible sizes, from boulders to fines. Below these

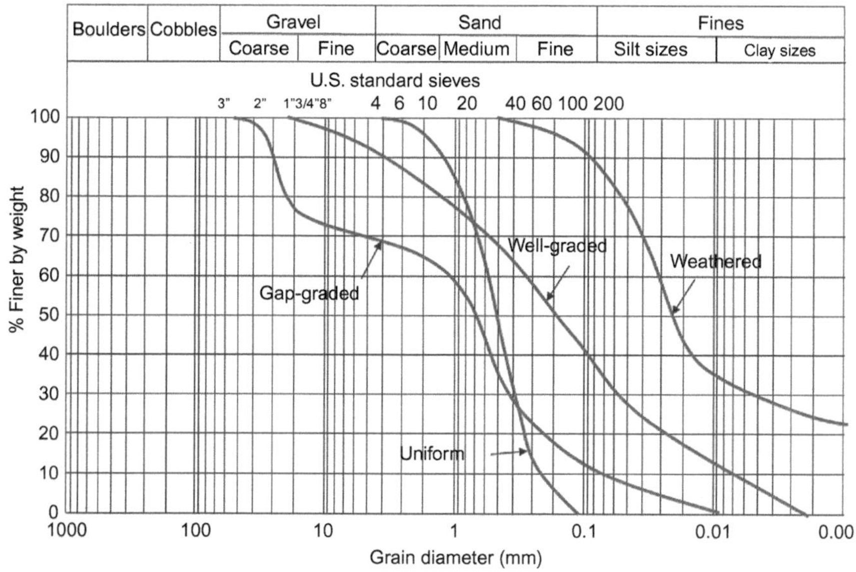

Figure 7.3 Soil gradation chart.

identifications are the U.S. standard sieve sizes, from 3 inches to a 200 sieve. As will be seen shortly, the "percent passing the 200 sieve" is one of the quantities used to classify soils according to the AASHTO system.

Note in Figure 7.3 the four different gradation curves. The most undesirable would be the "weathered" curve. It contains mostly fines, and sizes that fall into clay sizes. The next most undesirable is the "uniform" curve. It has soil that is almost all the same size. This soil allows for the greatest percentage of voids, which can allow unwanted water, and allows for the greatest slipping between soil layers. Not bad, but not particularly desirable is "gap-graded" soil. It has a wide variety of particle sizes; however, it has gaps in several key sizes. The best soil is "well-graded." It has a wide range of sizes and no gaps. This soil will compact nicely and, once settled, will see minimal future settlement over time.

There are two primary ways in which the particle sizes of soil samples are determined. One is by the use of vibratory sieves. A sieve is typically a circular pan with a grid defined by either the size of the opening, such as 1 in, or the number of openings per inch, such as a #4 sieve, which has 4 openings per inch. A table of sieve sizes from #4 down to a #200 is

Table 7.1 Standard sieve sizes

Standard sieve size	**Opening**	
	Inches	**Millimeters**
#4	0.1870	4.760
#8	0.0937	2.380
#10	0.0661	1.680
#20	0.0331	0.840
#40	0.0167	0.425
#60	0.0098	0.250
#100	0.0059	0.149
#200	0.0029	0.074

provided in Table 7.1, both in U.S. customary and International System of Units (SI/metric) measurements. Note that the size of the wire creating the grid must be taken into account in determining the exact size of the opening. For example, a #4 sieve has 4 openings per inch, but the opening is smaller than a quarter of an inch.

These sieves can separate very large rocks down to particles that pass a #200 sieve. However, the second method, a sedimentation analysis, is typically used on particles that pass the #10 sieve. The opening for a #10 sieve is about 1/16 in. The sedimentation analysis involves placing the soil sample in a column of water, shaking thoroughly, then allowing it to settle to the bottom over a predetermined time. A hydrometer is used to measure the change in density of the water and soil mixture, which can then be used to provide soil gradation information. Exactly how this is currently performed in the lab is, as they say, beyond the scope of this work. However, suffice it to say that the engineer Arthur Casagrande (quoted earlier) was the originator of this method, although the time to perform the analysis has been greatly reduced through modern improvements in the technique.

As mentioned previously, in this internet world, there are a considerable number of images showing both vibratory sieves and the sedimentation analysis, which can be found by searching "soil gradation by sedimentation test."

Table 7.1 lists the standard sieve sizes, noting the opening size in both inches and millimeters. Remember, the actual opening is slightly smaller than the theoretical opening due to the diameter of the wire that creates the grid. And while there are sieves down to #200, most labs use the sedimentation analysis for particles that pass the #10 sieve.

EXAMPLE 7.1 Plot Gradation Curve for a Soil Sample

Plot the gradation curve for the following soil sample with the given particle size distribution. (As will be seen shortly, the particle size distribution greater than a #4 sieve is not needed to classify a soil sample using the AASHTO system.)

Sieve	Retained weight (g)
#4	46
#10	113
#40	207
#200	572
Pan	538
Total	1476

Solution

From the given data, the following chart can be generated.

Sieve	Retained weight (g)	% of total	Cumulative %	% finer
#4	46	3	3	97
#10	113	8	11	89
#40	207	14	25	75
#200	572	39	64	36
Pan	538	36	100	—
Total	1476	100	—	—

The first two columns from the left are the given information. The third column is the standard distribution by percentage of the total sample weight for each particle size, which adds up to 100%. Note that the particles passing the #200 sieve are called the *pan*, and no further information is available on the particle size distribution. All of the particles in the pan could be dust, or just under the opening size of the #200 sieve.

The next column is called the cumulative %, where the first row is the same as the first row in the third column. The next row is the sum of the first and second rows of the third column, then the third row of the third column is added to make the third row of the cumulative %. Then the fourth row of the third column is added the make the fourth row of the cumulative %. Finally, the fifth row of the third column is added so that the fifth row of the cumulative % is 100%.

The last column of the chart is the cumulative % subtracted from 100%. This gives the % finer, meaning, for example, 97% of the particles passed the #4 sieve. These are the values that will be plotted, summarized in the following chart.

(Continued)

EXAMPLE 7.1 Plot Gradation Curve for a Soil Sample—(Continued)

Sieve	Diameter in mm	% finer
#4	4.76	97
#10	1.68	89
#40	0.425	75
#200	0.074	36

Plotting these values on a semilog grid gives the following diagram.

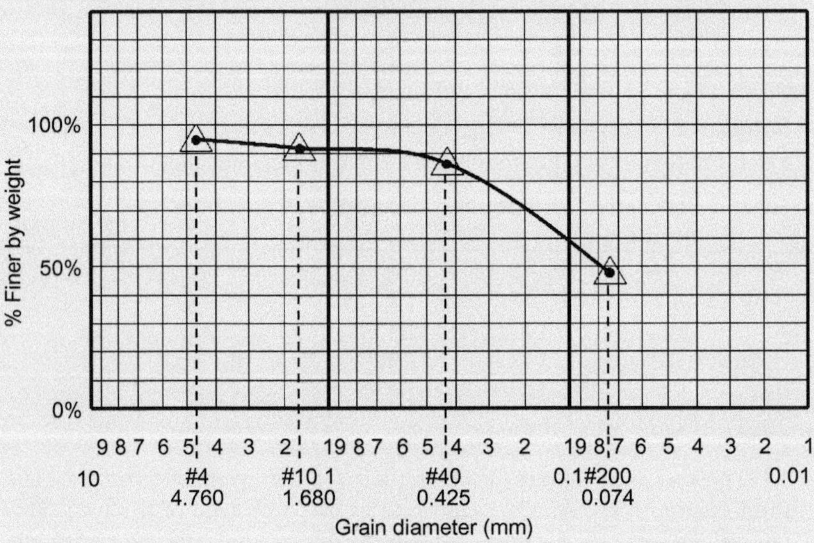

Observe that the curve starts at the #4 sieve and stops at the #200 sieve. No information is known about the particle sizes either greater than the #4 or smaller than the #200.

7.3.3 Atterberg Limits

Fine-grained soils, such as clays and silts, display significant changes in behavior and strength depending on the water content. To quantify these changes in behavior, Swedish soil scientist Albert Atterberg developed a series of "limits" relative to the water content of these types of soils. The actual device to describe and measure these limits was developed by Arthur Casagrande, mentioned earlier. Note that only a portion of a soil sample is tested for its Atterberg limits. While only the fines are affected, that is, those passing the #200 sieve, it is customary to use particles that pass the #40 sieve in the tests because this was the finest sieve in existence during Atterberg's investigations.

The Atterberg limits system establishes four states of the consistency of these fine-grained soils based on water content: (1) solid, (2) semisolid, (3) plastic, and (4) liquid. The water content (w) is defined as the weight of the water in the soil sample (W_w) divided by the weight of the soil sample without any water (W_s), which is multiplied by 100 to give a percentage. In equation form this gives:

$$w = \frac{W_w}{W_s} \times 100 \tag{7.1}$$

More about the water content (w) will be presented in the next section.

There are three important Atterberg limits: shrinkage limit (SL), plastic limit (PL), and liquid limit (LL). The shrinkage limit is the water content at which the volume of the soil starts to increase. It is also at the point where the soil is 100% saturated. The plastic limit is when a thread of soil rolled out on a nonporous surface begins to crumble when it reaches 1/8 in in diameter. The liquid limit is when a specific width groove in a sample contained in a standard pan closes up under the repeated dropping of that pan onto a hard rubber surface. A specific number of these drops establishes the liquid limit. Photos of the device and the tests conducted can be found on the internet under a search for "Atterberg limits." The specifics of these tests are defined by ASTM D4943 for the shrinkage limit, and by ASTM D4318 for the plastic and liquid limits.

In the classification of soils for use in highway design, the liquid limit and the plasticity index (PI) are the most important. The plasticity index gives the best measure of the plasticity of the soil. It is the difference between the liquid limit and the plastic limit, given by the equation:

$$\mathrm{PI} = \mathrm{LL} - \mathrm{PL} \tag{7.2}$$

Even though the % symbol is not used, the limits are percentages. So a PI of 35, means 35%. The following is a correlation between the plasticity index and a soil sample's level of plasticity:

- (0–3): nonplastic
- (3–15): slightly plastic
- (15–30): medium plastic

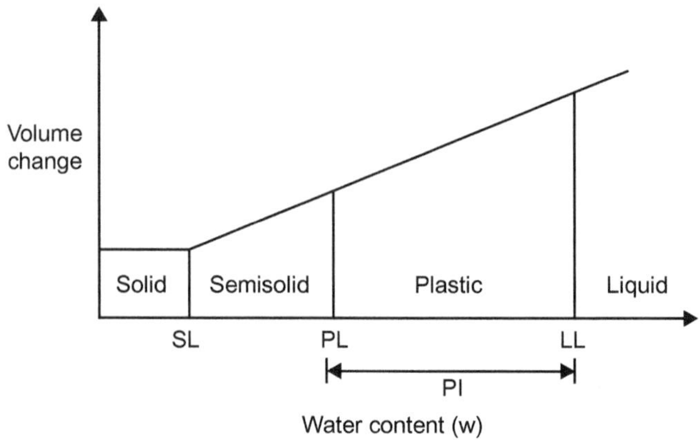

Figure 7.4 Atterberg limits.

- (> 30): highly plastic

Figure 7.4 shows the relationship between the water content and the volume of the soil sample. Note that the volume of the solid state doesn't change until all the voids are filled with water. Also note the graphical representation of the plasticity index.

If a soil is classified as nonplastic (NP), then the Atterberg limit tests cannot be performed. A nonplastic soil does not absorb water and, therefore, crumbles immediately during any of the tests mentioned. Soils that are granular, such as sand, are typically nonplastic.

7.3.4 AASHTO Classification

The AASHTO Soil Classification System was developed by the U.S. Bureau of Public Roads around 1928. As might be expected, the system is based on the performance of soils supporting highway pavements and their associated structures, such as bridges and culverts. The various designations are grouped based on their load-carrying capability. Soil gradation information and Atterberg limits are used to differentiate these classifications.

There are seven major groups, labeled A-1 through A-7. Several of these groups are further divided to provide basically 12 distinct classifications. The AASHTO system is shown in Table 7.2.

Note that the best soil is A-1 and the poorest is A-7. Also note that an A-3 soil, which is typically a nonplastic soil-like sand, is better than an A-2 soil.

Table 7.2 AASHTO classification system

General classification	Granular materials (35% or less passing the #200 sieve)							Silt-clay materials (> 35% passing #200 sieve)			
Group classification	A-1		A-3	A-2				A-4	A-5	A-6	A-7
	A-1-a	A-1-b		A-2-4	A-2-5	A-2-6	A-2-7				***
Sieve analysis % passing											
#10	50 max	-----	-----	-----	-----	-----	-----	-----	-----	-----	-----
#40	30 max	50 max	51 min	-----	-----	-----	-----	-----	-----	-----	-----
#200	15 max	25 max	10 max	35 max	35 max	35 max	35 max	36 min	36 min	36 min	36 min
Characteristics of fraction passing #40 Liquid limit plasticity index	-----		-----	40 max	41 min	40 max	41 min	40 max	41 min	40 max	41 min
	6 max		NP	10 max	10 max	11 min	11 min	10 max	10 max	11 min	11 min
Usual types of other materials	Stone, sand and gravel		Sand	Silty or clayey gravel and sand				Silty soils		Clayey soils	
General rating as subgrade	Excellent to Good							Fair to Poor			

***If PI $\leq$ LL − 30, then soil is A-7-5. If PI > LL − 30, then soil is A-7-6.

EXAMPLE 7.2 Classify Soil Sample Using AASHTO System

Using the following information, classify the soil sample according to the AASHTO system. The percent passing the #200 sieve is 37%, the liquid limit is 29%, and the plastic limit is 19%.

Solution

First, from the liquid limit and the plastic limit, calculate the plasticity index using Eq. (7.2) to obtain PI = 10%.

With 37% passing the #200 sieve, this eliminates groups A-1, A-3, and A-2 because all three limit this value to 35%.

With the liquid limit = 29%, this eliminates groups A-5 and A-7 because both require this value to be at least 41%.

With the plasticity index = 10%, this eliminates group A-6 because this value must be a minimum of 11%.

Therefore, this soil would be classified as a group A-4 soil, which would not be acceptable to use.

7.4 PHASE RELATIONS

7.4.1 Phase Diagrams

Matter occurs as a solid, liquid, or gas, or a combination. For soil, the solids are earth, containing a multitude of possible types and sizes; the liquid is water; and the gas is air. In analyzing soil samples it has been found that completing a phase diagram will facilitate obtaining important quantities needed to predict the settlement of the soil in question, and its rate of settlement. A skeleton of a phase diagram is shown in Figure 7.5.

There are 15 quantities that make up a phase diagram. In the four blanks on the right will be the weights, and in the seven blanks on the

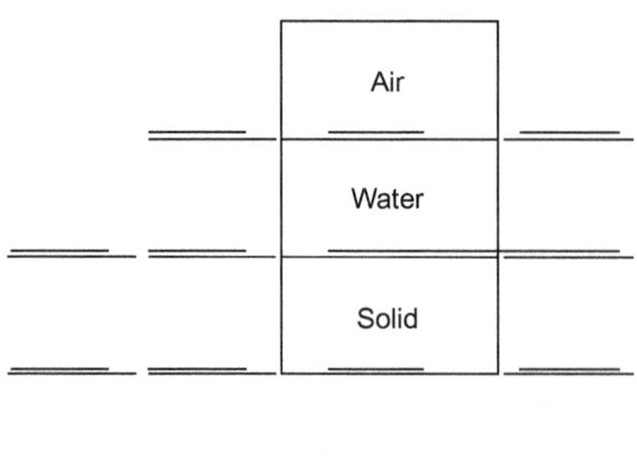

Figure 7.5 Phase diagram framework.

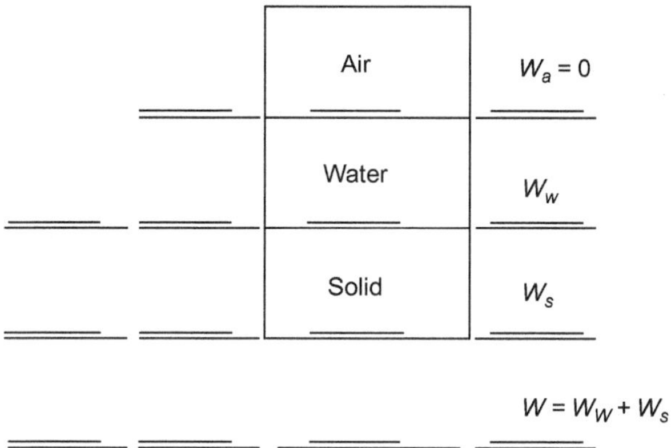

Figure 7.6 Phase diagram: weights.

left will be the various volumes. In the rectangle will be the three specific weights, two of which are always known, air and water. The last blank below the rectangle will contain the overall specific weight of the sample, but is not the sum of the three individual specific weights, which is why they are isolated in the rectangle.

Considering each of these groups separately, the four weights are the weights of the air, water, and solids, then the total. The subscripts used are "a," "w," and "s" respectively, where there is no subscript on the total. This notation will follow for the group of volumes as well. Figure 7.6 shows these four weights in the phase diagram. Note that the weight of the air is always considered zero in comparison to the weights of the solids and water, however the volume of the air is not necessarily zero, as will be seen shortly.

The next group is the volumes associated with the weights. As stated, the volume of the air is not necessarily zero; however, it is if the sample is completely saturated with water. And because the specific weight of water is known, if either the weight of the water or the volume of water is known, then the other is easily calculated. These volumes are shown in Figure 7.7.

Two of the volumes can be combined into a volume of the voids in the soil sample. In fact, as will be seen shortly, the volume of the voids will be part of the definition of three important parameters of a soil sample. So, noting this volume, along with the volume of the solids, the total volume is shown in Figure 7.8.

Note that two of the volumes are duplicated in the phase diagram, the volume of the solids and the total volume. And because three other

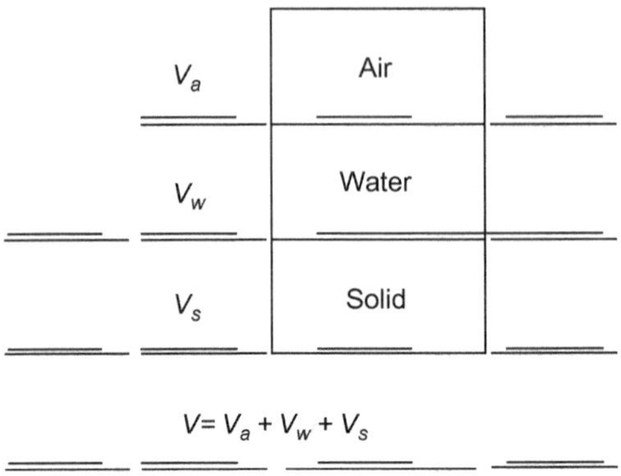

Figure 7.7 Phase diagram: volumes.

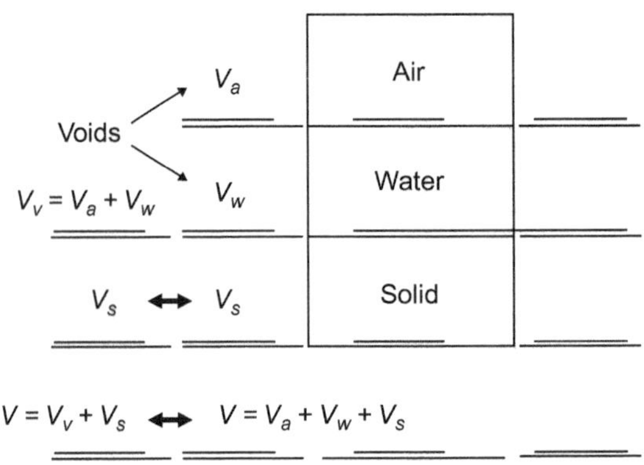

Figure 7.8 Phase diagram: volume of the voids.

quantities are always known—the weight and specific weight of air, and the specific weight of water—of the 15 quantities making up a phase diagram, only 10 are actually unknown.

The last four quantities making up a phase diagram are the specific weights. It is a property that represents the weight per volume of a substance. They are given the Greek symbol, gamma (γ). They are noted in the rectangle, and, as said earlier, are not added up to give the overall specific weight of the soil sample. Once the total weight and total volume have been determined, then it can be calculated. Note that the specific weight of air is zero and the specific weight of water is 62.4 lb/ft^3. The specific weight of the solids is typically an unknown. These are summarized in Figure 7.9.

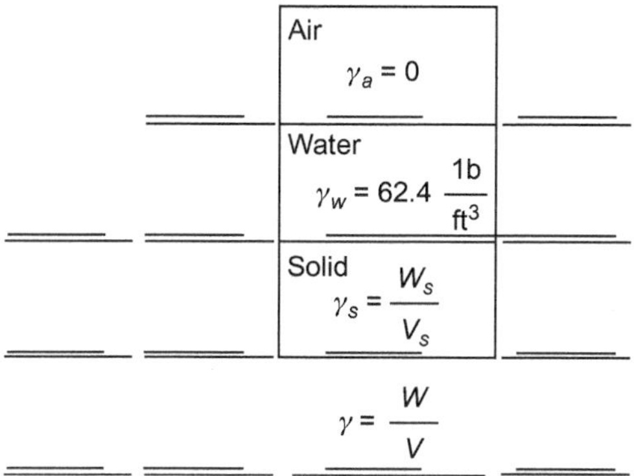

Figure 7.9 Phase diagram: specific weights.

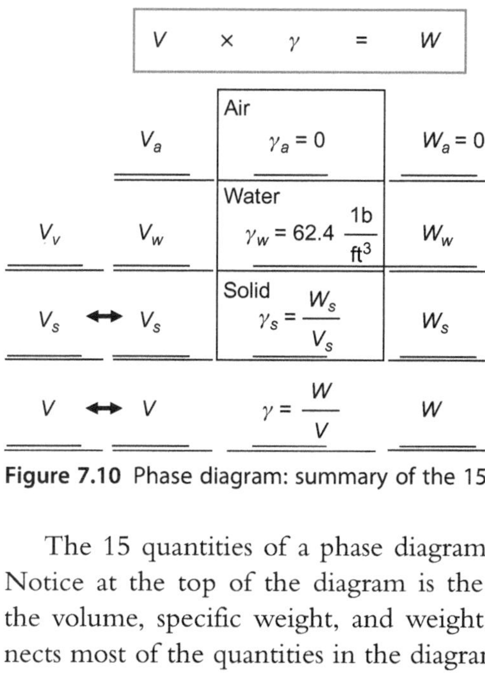

Figure 7.10 Phase diagram: summary of the 15 quantities.

The 15 quantities of a phase diagram are summarized in Figure 7.10. Notice at the top of the diagram is the governing relationship between the volume, specific weight, and weight. It is this relationship that connects most of the quantities in the diagram.

An important property of a substance is the specific gravity. It is the ratio of the specific weight of that substance to the specific weight of water. This property is calculated by Eq. (7.3), denoted here as (G).

$$G_{\text{substance}} = \frac{\gamma_{\text{substance}}}{\gamma_w} \tag{7.3}$$

So for air, the specific gravity is zero, and for water it is 1. These are summarized in Figure 7.11, along with the equations for the specific

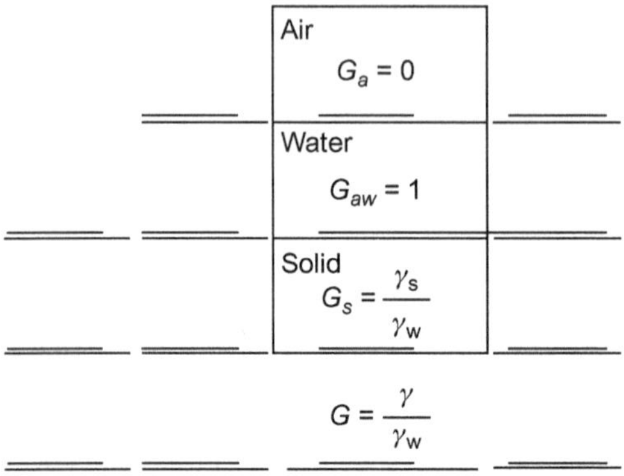

Figure 7.11 Phase diagram: specific gravities.

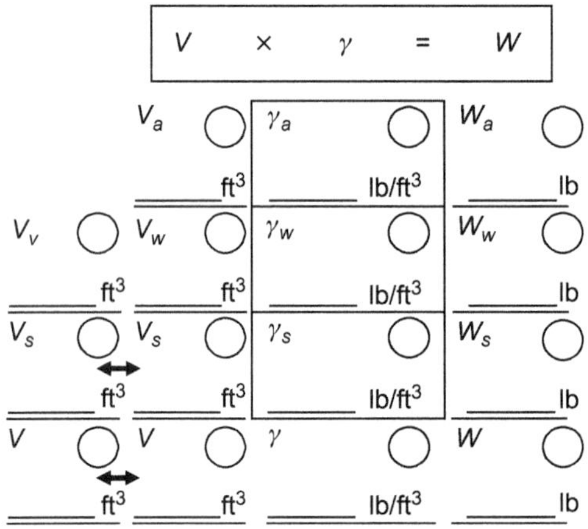

Figure 7.12 Phase diagram worksheet.

gravity of the solids and the specific gravity of the overall sample, though it is not as important.

If standard U.S. customary units are used, pound (lb) for the weights, cubic feet (ft^3) for the volumes, and pound per cubic foot (lb/ft^3) for the specific weights, a phase diagram worksheet evolves, as shown in Figure 7.12.

The 15 circles next to each of the 15 quantities allow for the order of how they were determined to be documented. This is very helpful in explaining to others how a phase diagram is completed from the given and known information. It will be used shortly in an example, and then later in a problem at the end of the section.

7.4.2 Special Properties

There are several special properties of soil important in the determination of its future settlement and rate of settlement. One has already been introduced: the water content (w) defined by Eq. (7.1). It can be used to establish two additional equations as follows.

For the first equation, start with the relationship between the weights in a soil sample:

$$W = W_w + W_s$$

Divide through by the weight of the solids to give:

$$\frac{W}{W_s} = \frac{W_w}{W_s} + \frac{W_s}{W_s}$$

The first term on the right side is the water content (w) and the second is 1.

$$\frac{W}{W_s} = w + 1$$

Rearranging gives:

$$W_s = \frac{W}{1 + w} \tag{7.4}$$

For the second equation, start with the definition for the water content (w) from Eq. (7.1):

$$w = \frac{W_w}{W_s}$$

Solve for the weight of the water,

$$W_w = w \times W_s$$

Substitute the weight of the solids from Eq. (7.4),

$$W_w = w \times \frac{W}{1 + w}$$

Then combine terms in the numerator to give:

$$W_w = \frac{w(W)}{1 + w} \tag{7.5}$$

There are three additional properties that need to be introduced. The first is the void ratio, designated e, and given by the following definition:

$$e = \frac{V_v}{V_s} \tag{7.6}$$

The second is the porosity, designated n, and given by the following definition:

$$n = \frac{V_v}{V} \tag{7.7}$$

The third relates the amount of water in the voids, in percent, called the saturation ratio, designated S_r, and given by the following definition:

$$S_r = \frac{V_w}{V_v} \times 100 = \% \tag{7.8}$$

Without showing the details, the void ratio (e) can be related to the water content (w), the specific gravity of the solids (G_s), and the saturation ratio (S_r), as:

$$e = \frac{w(G_s)}{S_r} \tag{7.9}$$

And again, without showing the details, the porosity (n) can be related to the void ratio (e) as:

$$n = \frac{e}{(1+e)} \tag{7.10}$$

A property of a soil sample that is very important in the next section, compaction, is defined in an unusual way. It is called the "dry unit weight," denoted γ_d, and is basically a specific weight defined as follows:

$$\gamma_d = \frac{W_s}{V} \tag{7.11}$$

Substitute the weight of the solids from Eq. (7.4),

$$\gamma_d = \frac{W/1+w}{V}$$

Rearrange to put the total volume (V) in the numerator,

$$\gamma_d = \frac{W/V}{1+w}$$

Then use the definition of the specific weight of the entire sample, also called the "unit weight," to give:

$$\gamma_d = \frac{\gamma}{1 + w} \tag{7.12}$$

Again, this will be a very important property in the next section.

EXAMPLE 7.3 Completing a Phase Diagram

The total weight of a moist sample is 330 lb and its volume is 3 ft^3. The water content was found to be 27% and the specific gravity of the solids is 2.72. Find the unit weight, void ratio (e), the porosity (n), and the degree of saturation (S_r).

Solution

Before attempting to determine any of the requested quantities, first complete a phase diagram for the sample. The phase diagram worksheet shown in Figure 7.12 will be used. Three quantities never change: the weight of the air and the specific weights of the air and water. Two quantities are given in the statement of the problem, the total weight (W) and the total volume (V). The total volume is duplicated, and so the following six steps are already completed:

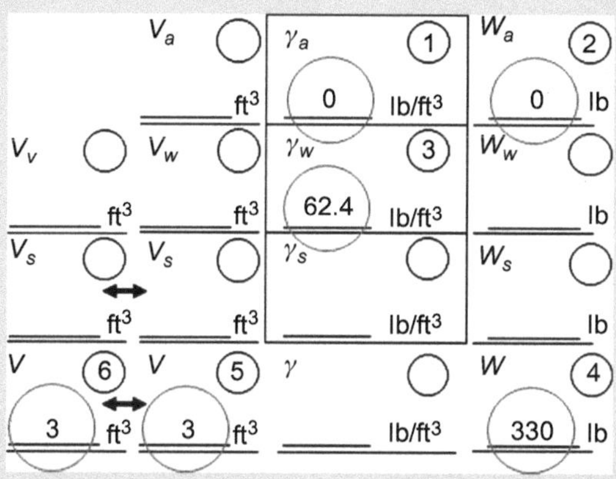

Next, the unit weight can then be calculated from the total weight and the total volume:

$$\gamma = \frac{W}{V} = \frac{330 \text{ lb}}{3 \text{ ft}^3} = 110 \text{ lb/ft}^3$$

(Continued)

EXAMPLE 7.3 Completing a Phase Diagram—(Continued)

From the given water content and the total weight, the weight of the solids can be found:

$$W_s = \frac{W}{1+w} = \frac{330\text{ lb}}{1+0.27} = 259.8\text{ lb}$$

Using the relationship between the weights, the weight of the water can be found:

$$W_w = W - W_s = 330\text{ lb} - 259.8\text{ lb} = 70.2\text{ lb}$$

The phase diagram now looks like the following:

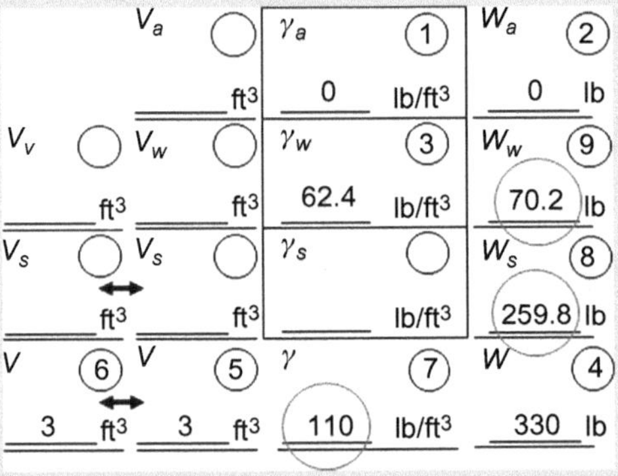

Now that the weight of the water is known, the volume of the water can be determined as:

$$V_w = \frac{W_w}{\gamma_w} = \frac{70.2\text{ lb}}{62.4\text{ lb/ft}^3} = 1.125\text{ ft}^3$$

Using the given specific gravity of the solids and the specific weight of water gives the specific weight of the solids:

$$\gamma_s = G_s\gamma_w = (2.72)(62.4\text{ lb/ft}^3) = 169.7\text{ lb/ft}^3$$

With this specific weight of solids and the weight of the solids, the volume of the solids becomes:

$$V_s = \frac{W_s}{\gamma_s} = \frac{259.8\text{ lb}}{169.7\text{ lb/ft}^3} = 1.531\text{ ft}^3$$

The volume of the solids is duplicated. The volume of the air can now be determined as:

$$V_a = V_v - V_w = 1.469\text{ ft}^3 - 1.125\text{ ft}^3 = 0.344\text{ ft}^3$$

(Continued)

EXAMPLE 7.3 Completing a Phase Diagram—(Continued)

The volume of the voids could have been found before the volume of the air, however, at this point the phase diagram falls nicely into place, so the order of steps is no longer important. Finding the volume of the voids one way makes the other way a check. So, adding the volumes of the air and the water gives:

$$V_v = V_a + V_w = 0.344\ \text{ft}^3 + 1.125\ \text{ft}^3 = 1.469\ \text{ft}^3$$

As the check, subtract the volume of the solids from the total volume to give:

$$V_v = V - V_s = 3\ \text{ft}^3 - 1.531\ \text{ft}^3 = 1.469\ \text{ft}^3$$

The completed phase diagram becomes:

	V_a (15) 0.344 ft³	γ_a (1) 0 lb/ft³	W_a (2) 0 lb
V_v (14) 1.469 ft³	V_w (10) 1.125 ft³	γ_w (3) 62.4 lb/ft³	W_w (9) 70.2 lb
V_s (13) 1.531 ft³	V_s (12) 1.531 ft³	γ_s (10) 169.7 lb/ft³	W_s (8) 259.8 lb
V (6) 3 ft³	V (5) 3 ft³	γ (7) 110 lb/ft³	W (4) 330 lb

At this point, the quantities needed become very straightforward to calculate. In no special order, the "dry" unit weight becomes:

$$\gamma_d = \frac{W_s}{V} = \frac{259.8\ \text{lb}}{3\ \text{ft}^3} = 86.6\ \text{lb/ft}^3$$

The void ratio becomes:

$$e = \frac{V_v}{V_s} = \frac{1.469\ \text{ft}^3}{1.531\ \text{ft}^3} = 0.96$$

The porosity becomes:

$$n = \frac{V_v}{V} = \frac{1.469\ \text{ft}^3}{3\ \text{ft}^3} = 0.49$$

And finally, the saturation ratio can be found to be:

$$S_r = \frac{V_w}{V_v} \times 100 = \frac{1.125\ \text{ft}^3}{1.469\ \text{ft}^3} \times 100 = 0.766 \times 100 = 76.6\%$$

What is interesting is that once the phase diagram is complete, there is a more comprehensive appreciation for the makeup of the soil sample than if the required quantities were found from a series of complex equations.

We now turn our attention to the significance of the "dry" unit weight and the water content in terms of a soil's behavior under the process called *compaction*.

7.5 COMPACTION

7.5.1 Purpose

The purpose of compaction is to force out the air trapped in the voids of the soil. In this way the specific weight will increase, but more importantly, the strength of the soil will increase. Other benefits of compaction are as follows:

- The settlement of the soil over time under a permanent load will decrease.
- The ability of the soil to retain water, called permeability, will be reduced.

Various types of equipment are used for compaction, from small individually operated tampers to large vibrating drum rollers. An internet search on compaction equipment will yield a complete range of equipment currently being used.

7.5.2 Dry Unit Weight versus Water Content

Compaction is primarily a function of the water content in the soil. In the previous section, the water content was given by Eq. (7.1),

$$w = \frac{W_w}{W_s} \tag{7.1}$$

and the "dry unit weight" was given by Eq. (7.11).

$$\gamma_d = \frac{W_s}{V} \tag{7.11}$$

Then the dry unit weight was reformulated using the definition of water content to provide Eq. (7.12).

$$\gamma_d = \frac{\gamma}{1 + w} \tag{7.12}$$

If samples of the soil under investigation have varying amounts of water added, so that the Atterberg limits are somewhere between the

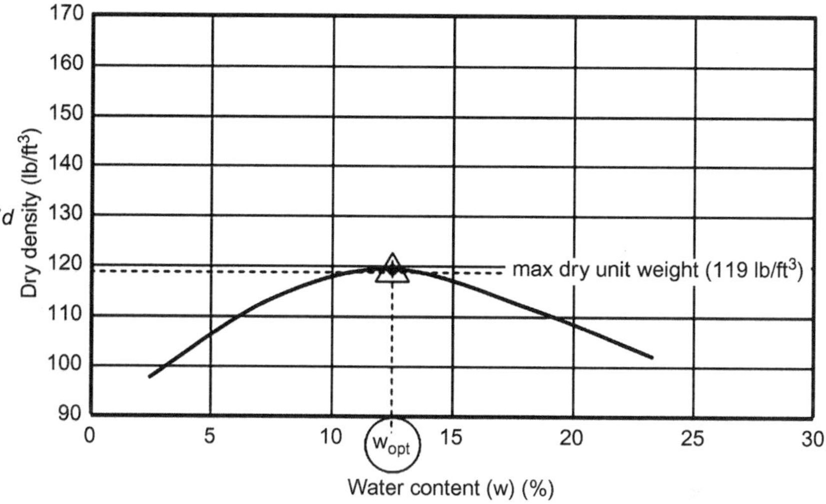

Figure 7.13 Dry unit weight versus water content.

plastic limit (PL) and the liquid limit (LL), and then compacted with the same compactive effort, a curve of the water content to dry unit weight can be developed for those samples. Figure 7.13 shows such a plot.

What is important is that when soil is compacted to its maximum dry density, then it has a minimum volume of voids, and thus is at its maximum strength. For the plot of Figure 7.13, the maximum dry unit weight occurs at 119 lb/ft^3, for an optimum water content of 12.5%. However, in the field it is not possible to meet these maximum values accurately, so a 95% criteria is typically established. This 95% compaction line cuts the curve in Figure 7.13 in two places, giving rise to a maximum and minimum combination of water content and dry unit weight. These are shown in Figure 7.14.

As can be seen from Figure 7.14, 95% of the maximum dry unit weight of 119 lb/ft^3 becomes 113 lb/ft^3, occurring for both a minimum water content of 7% and a maximum water content of 18%. As might be expected, this is a very realizable range in the field.

Again, if 100% of the voids are filled with water, meaning a saturation ratio (S_r) equal to 100%, then this gives rise to a zero voids unit weight (γ_z), given by the following equation:

$$\gamma_z = \frac{\gamma_w}{w + \left(\frac{1}{G_s}\right)} \tag{7.13}$$

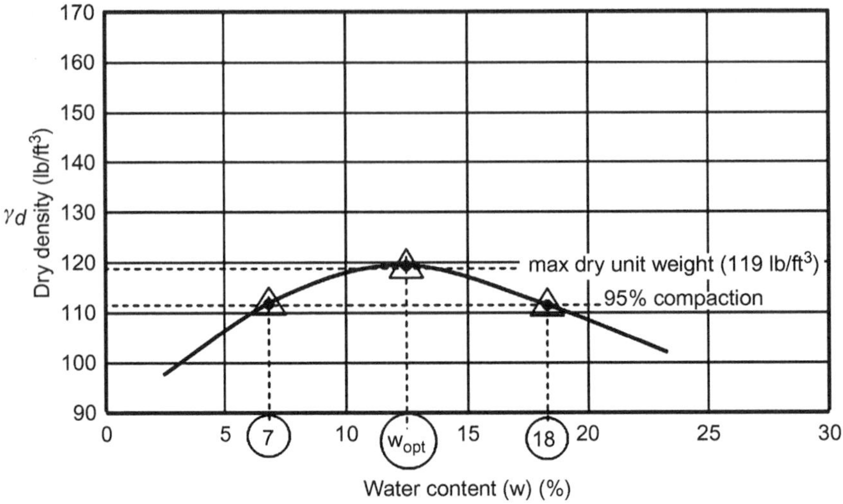

Figure 7.14 95% values: maximum and minimum range.

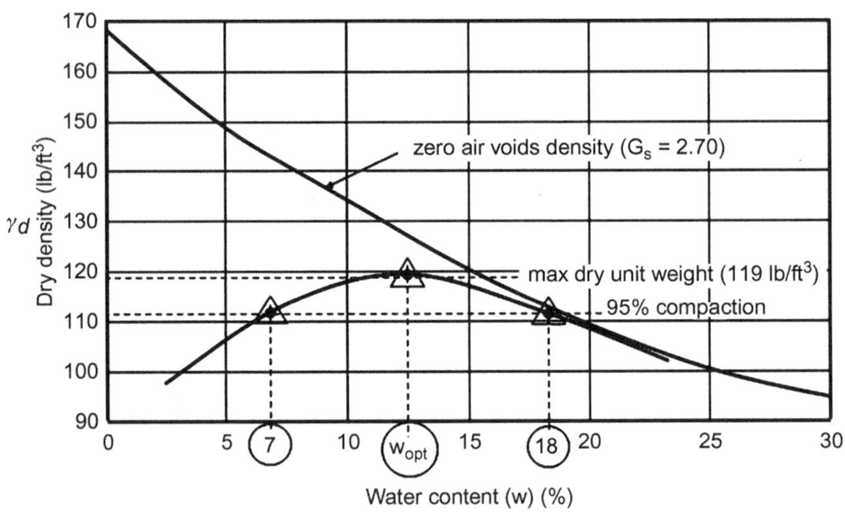

Figure 7.15 Zero air voids for $g_s = 2.70$.

Using a specific gravity (G_s) of 2.70, this zero voids curve is shown in Figure 7.15 superimposed on the previous plots.

While the actual compaction process varies depending on the situation, basically it involves three phases:

1. Assuming the existing soil is not satisfactory by itself, a suitable fill soil is located and samples are tested in the lab to determine a

range of water contents to provide 95% of the maximum dry unit weight.
2. The fill material is combined with the existing soil and compacted with the addition of appropriate amounts of water to achieve the specified values.
3. The compacted soil is then tested in the field to determine if the specifications have been met before the project can proceed.

7.5.3 Testing in the Lab

The soil samples removed from both the existing soil and the proposed fill soil are first tested in the lab using one of two Proctor tests, so named after R. R. Proctor. One is called the standard Proctor compaction test and the other the modified Proctor compaction test. Each are intended to produce the same compactive effort that will be applied in the field.

Both tests use a 4-in-diameter cylinder that can hold 1/30 ft^3 of a soil sample. Typically the soil is added in five layers and then 25 blows are applied. The main difference in the two tests are that the standard Proctor test uses a 5.5-lb hammer dropping a distance of 12 in, whereas the modified Proctor uses a 10-lb hammer dropping 18 in. The difference is to simulate either light rollers or tamping for the standard Proctor, and heavy rollers for the modified Proctor. Consequently, the dry unit weight is larger and the water content is lower for the modified Proctor.

7.5.4 Testing in the Field

In the field, testing the compacted soil is usually conducted according to procedures established by each state Department of Transporation (DOT). As with the Proctor tests in the lab, a specific volume of soil is obtained and weighed. Then the water is pressed out, and its new volume and weight are recorded. The difference in weights and volumes for several samples will yield sufficient pairs of dry unit weight and water content to determine if the compacted soil meets the specifications established from the lab tests based on the project.

7.5.5 Sample Calculations and Plotting

The following example illustrates the testing calculations associated with compaction.

EXAMPLE 7.4 Determine Maximum Dry Unit Weight and Optimum Water Content

Using the following table of information, determine the maximum dry unit weight and its associated optimum water content. (Note that all that is needed to develop a very accurate plot of dry unit weight vs. water content is four data points.)

Test number	Wet weight (g)	Wet weight (lb/ft^3)	Wet weight sample (g)	Wet dry sample (g)	Moisture (%)	Dry weight (lb/ft^3)
1	1970.0		100.0	91.8		
2	2039.0		100.0	90.3		
3	2078.0		100.0	88.8		
4	2045.0		100.0	87.3		

Solution

First, using the weight of wet soil given in the second column, determine the wet unit weight for each test. The volume of the sample will be that used in the Proctor tests, 1/30 ft^3. A sample calculation is provided for the first set of values.

Test 1: wet weight = 1970 grams

Convert grams to pounds in the following calculation:

$$\text{wet weight} = 1970.0\text{ g} \times \frac{1\text{ kg}}{1000\text{ g}} \times \frac{2.205\text{ lb}}{1\text{ kg}} = 4.344\text{ lb}$$

Calculate the wet unit weight using a volume of 1/30 ft^3 and place in column 3:

$$\text{wet unit weight} = \frac{4.344\text{ lb}}{1/30\text{ ft}^3} = 130.3\text{ lb/ft}^3$$

Using the wet and dry weights in columns 4 and 5, calculate the water content in % and place in column 6, labeled "moisture":

$$w = \frac{W_w}{W_s} = \frac{100\text{ g} - 91.8\text{ g}}{91.8\text{ g}} \times 100 = 0.089 = 8.9\ \%$$

Finally, using the wet unit weight and the water content, calculate the dry unit weight using Eq. (7.12) and place in column 7:

$$\gamma_d = \frac{\gamma}{1 + w} = \frac{130.3\text{ lb/ft}^3}{1 + 0.089} = 119.7\text{ lb/ft}^3$$

(*Continued*)

EXAMPLE 7.4 Determine Maximum Dry Unit Weight and Optimum Water Content—(Continued)

Do similar calculations for the remaining three tests to obtain the results provided in the following completed table.

Test number	Wet weight (g)	Wet weight (lb/ft^3)	Wet weight sample (g)	Wet dry sample (g)	Moisture (%)	Dry weight (lb/ft^3)
1	1970.0	130.0	100.0	91.8	8.9	119.7
2	2039.0	134.9	100.0	90.3	10.7	121.9
3	2078.0	137.5	100.0	88.8	12.6	122.1
4	2045.0	135.3	100.0	87.3	14.6	118.1

Now plot the water content in column 6 and the dry unit weight in column 7. The following graph should result.

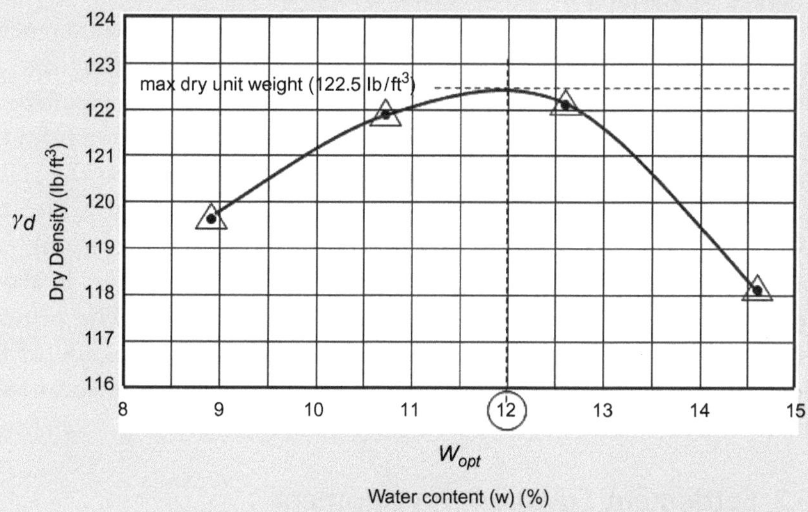

From the graph, it appears the maximum dry unit weight is 122.5 lb/ft^3 at an optimum water content of 12%.

Our remaining topic is to pull together what has already been determined about the soil in question, along with additional information and soil properties, applied to the specific project, to estimate the total settlement that can be expected and the rate of that settlement. This analysis is called consolidation.

7.6 CONSOLIDATION

7.6.1 Causes of Settlement

The primary factors influencing the amount of settlement in the context of supporting structures are the soil's properties, including the number and types of layers, and the actual load itself. Layers of cohesive soils such as clays and silts are especially susceptive to settlement, which is why the calculations that will be presented focus on the clay layers. Noncohesive soils, such as sand, settle mainly through a horizontal movement of particles, rather than a vertical movement that occurs for cohesive soils.

Keep in mind that settlement cannot be eliminated, just reduced. Also, the rate of settlement, particularly the time it takes to reach a 95% value, will be important to determine. In addition, the evenness of the settlement over the area of the project is important, primarily to reduce the effects of warping in the structural elements.

The process of settlement begins initially with any trapped air being expelled. When all the voids are only filled with water, meaning the soil is saturated, further settlement can only occur if the water is expelled. As the load on the clay layer increases, eventually the load is shifted from the water to the soil. This transfer of the load from the water to the soil is called "consolidation." As might be expected, the path through which the expelled water must travel is important. It is dependent mainly on whether the layer below the clay layer is pervious, meaning it allows water to pass, or nonpervious, typically referring to rock. The primary mechanism for settlement is therefore a reduction in the volume of the voids, causing the load to be carried by the soil. The more soil-to-soil contact and the less soil carried by water, the less settlement.

7.6.2 Settlement Equation Development

One equation for calculating the total amount of settlement that can be expected for a particular soil and a particular load can be expressed as:

$$S_c = H\left(\frac{\Delta e}{1 + e_0}\right) \tag{7.14}$$

where

S_c = total settlement
H = thickness of the clay layer
Δe = change in the void ratio
e_o = initial void ratio

The problem with using this equation is determining the "Δe," the change in the void ratio. It is important to use the two fundamental loadings that cause settlement. First, the loading caused by the various layers above the clay layer, and second, the additional loading caused by the structure itself.

Recall that the void ratio was defined in Section 7.4, specifically by Eq. (7.6):

$$e = \frac{V_v}{V_s} \tag{7.6}$$

Noting that the volume of the voids is the total volume minus the volume of the solids, we get:

$$e = \frac{V_v}{V_s} = \frac{V - V_s}{V_s} \tag{7.15}$$

This equation is the basis of a "consolidometer," used to determine the change in the void ratio due to the two loadings. Tests are made on undisturbed soil samples taken from the field, the two loadings are applied, and a graph of the void ratio is developed. Figure 7.16 shows such a plot, which requires a semilog grid.

Note that the slope from Point 1 to Point 2 is negative. And that once the undisturbed soil has been released from a load, it does not return to its original volume. Also, the slope of the decompression and recompression are basically the same.

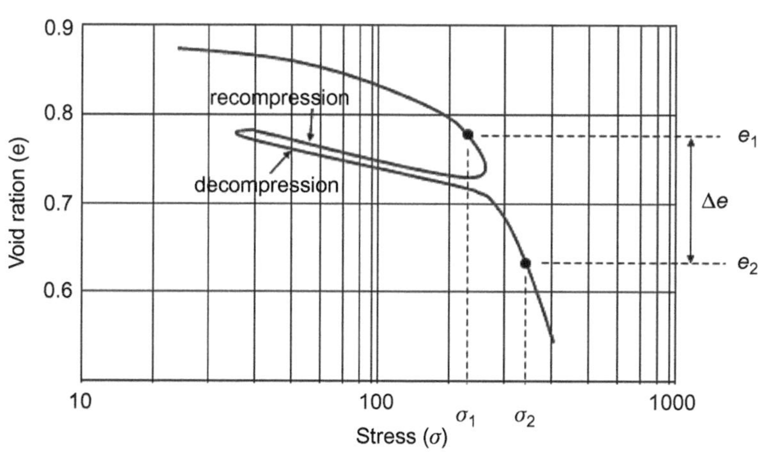

Figure 7.16 Void ratio versus loading diagram.

The slope of the curve from Points 1 to 2 is called the "compression index," denoted C_c, and given by the equation:

$$C_c = \frac{e_1 - e_2}{\log(\sigma_1) - \log(\sigma_2)}$$

However, the numerator is the change in the void ratio (Δe), and because the difference in the denominator is negative, the equation for the compression index can be reformulated using logarithm algebra to be:

$$C_c = \frac{\Delta e}{\log\left(\frac{\sigma_2}{\sigma_1}\right)}$$

Solving for the change in the void ratio gives:

$$\Delta e = C_c \log\left(\frac{\sigma_2}{\sigma_1}\right) \tag{7.16}$$

The stress (σ_1) is the stress due to the weight of soil above the midpoint of the clay layer, called the "initial vertical stress" and denoted (σ_0). The stress (σ_2) is the sum of the initial vertical stress and the stress felt at the midpoint of the clay layer due to a structure at the surface, called the "change in vertical stress" and denoted ($\Delta\sigma$). So therefore,

$$\sigma_1 = \sigma_0$$
$$\sigma_2 = \sigma_0 + \Delta\sigma$$

Substitute these two stresses into the expression for change in void ratio, Eq. (7.16), to give:

$$\Delta e = C_c \log\left(\frac{\sigma_2}{\sigma_1}\right) = C_c \log\left(\frac{\sigma_0 + \Delta\sigma}{\sigma_0}\right) \tag{7.17}$$

Finally, substitute for the change in the void ratio from Eq. (7.17) into Eq. (7.14) to provide the equation that will be used to calculate the expected total settlement:

$$S_c = \frac{H}{1 + e_0} C_c \log\left(\frac{\sigma_0 + \Delta\sigma}{\sigma_0}\right) \tag{7.18}$$

So, the next task is to determine the process for finding these two stresses.

7.6.3 Initial Vertical Stress

The initial vertical stress is caused by the weight of the layers of soil above a particular point in question, which is typically the midpoint of the clay

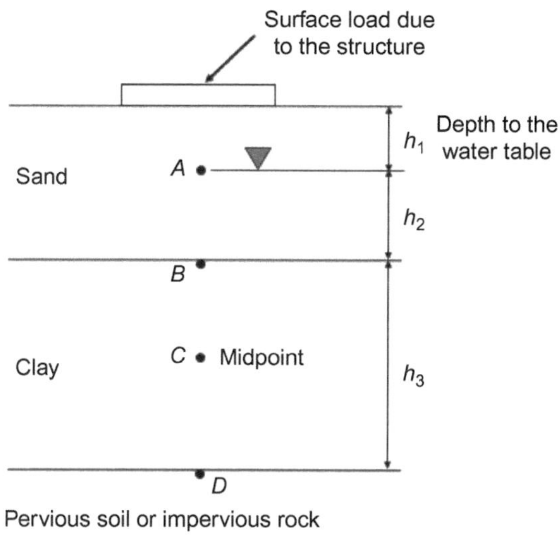

Figure 7.17 Simplified soil profile.

layer. This stress is actually a pressure, similar to the pressure felt under the surface of a liquid. Essentially, the contribution to the total pressure, or stress, is the sum of the unit weights of each layer of soil times the thickness of the layer. The stress (σ) or pressure (p) developed by a single layer is expressed in Eq. (7.19) as:

$$\sigma_{\text{layer}} = p_{\text{layer}} = \gamma_{\text{layer}} \times h_{\text{layer}} \tag{7.19}$$

where (γ) is the unit weight of the soil in the layer and (h) is the thickness of the layer.

Figure 7.17 shows a very simplified soil profile, where there is a layer of sand above a layer of clay. Obviously, in a real situation there would be a great many layers, however, considering only these two simple layers will avoid undue complications in describing the process.

Note several key items on Figure 7.17. First, there is the assumption that there will eventually be a load due to the project's structure. Its contribution to the total stress at the midpoint of the clay layer will be discussed in the next section. Second, the depth to the water table is indicated. This is important as the net unit weight will be different above and below this level. Third, it will be important later as to what is below the clay layer, either pervious soil or impervious rock. This will specify how far trapped water will have to travel to be removed from the clay layer as the stress increases and the final settlement is achieved.

Figure 7.18 shows the buildup of stress that develops in a soil profile. For this simple two-layer profile, three stresses must be calculated as shown.

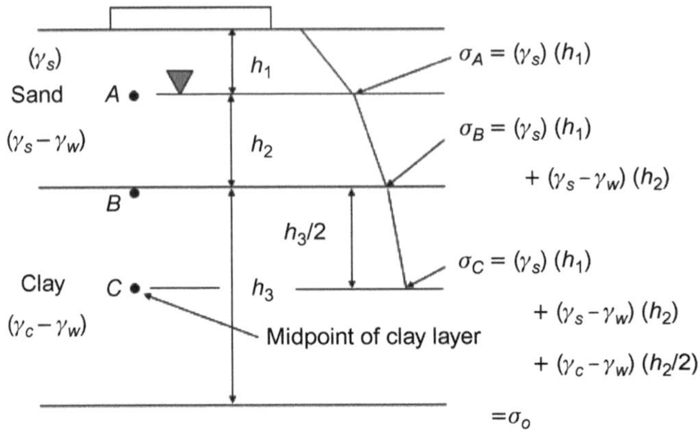

Figure 7.18 Stress buildup in a soil profile.

First, the stress down to the water table, Point A, is calculated due to the weight of the dry sand. Second, the stress from the water table to the interface between the sand and the clay layer, Point B, is calculated. Note that the specific weight of water must be subtracted from the unit weight of the sand to account for the buoyancy effect of the water. Third, the stress down to the midpoint of the clay layer, Point C, is calculated. Again, the specific weight of water is subtracted from the unit weight of the clay. Also, as shown, the stress at Point C is equal to the initial vertical stress (σ_0).

EXAMPLE 7.5 Calculating the Initial Vertical Stress

For the two-layer soil profile shown, determine the initial vertical stress (σ_0) at the midpoint of the clay layer, Point C.

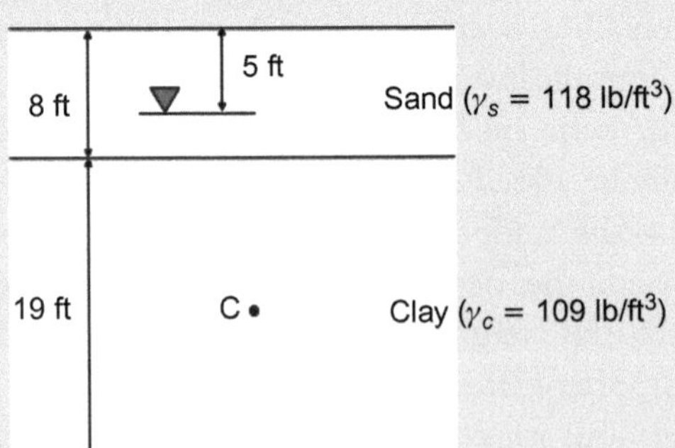

(Continued)

EXAMPLE 7.5 Calculating the Initial Vertical Stress—(Continued)

Solution

For the sand above the water table:

$$\sigma_{\text{sand}}^{\text{above}} = (118\ \text{lb/ft}^3) \times (5\ \text{ft}) = 590\ \text{lb/ft}^2$$

For the sand below the water table:

$$\sigma_{\text{sand}}^{\text{below}} = (118\ \text{lb/ft}^3 - 62.4\ \text{lb/ft}^3) \times (3\ \text{ft}) = 167\ \text{lb/ft}^2$$

For the clay above its midpoint:

$$\sigma_{\text{clay}} = (109\ \text{lb/ft}^3 - 62.4\ \text{lb/ft}^3) \times (9.5\ \text{ft}) = 443\ \text{lb/ft}^2$$

The total initial vertical stress becomes:

$$\sigma_0 = (590\ \text{lb/ft}^2) + (167\ \text{lb/ft}^2) + (443\ \text{lb/ft}^2) = 1200\ \text{lb/ft}^2$$

The next step in the process is to calculate the change in vertical stress due to the structure.

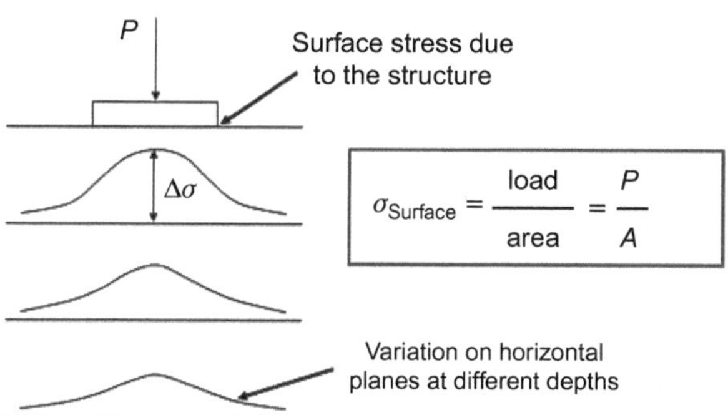

Figure 7.19 Influence of surface stress.

7.6.4 Change in Vertical Stress

The change in vertical stress ($\Delta\sigma$) is caused by the load at the surface of the soil profile. As shown in Figure 7.19, the total load (P) causes a surface stress that is felt at all points below the surface. As indicated, this influence is strongest at the surface and tapers off as the depth under the soil profile increases. Also, note that the influence at any particular layer is greatest along the centerline of the structure and tapers off asymptotically as the distance from the centerline increases.

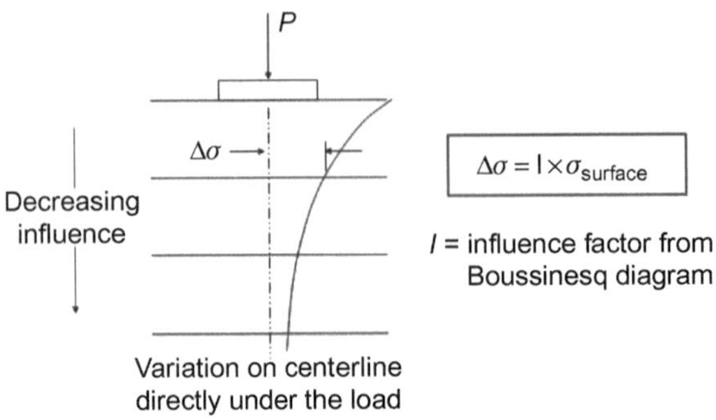

Figure 7.20 Change in vertical stress along centerline of load.

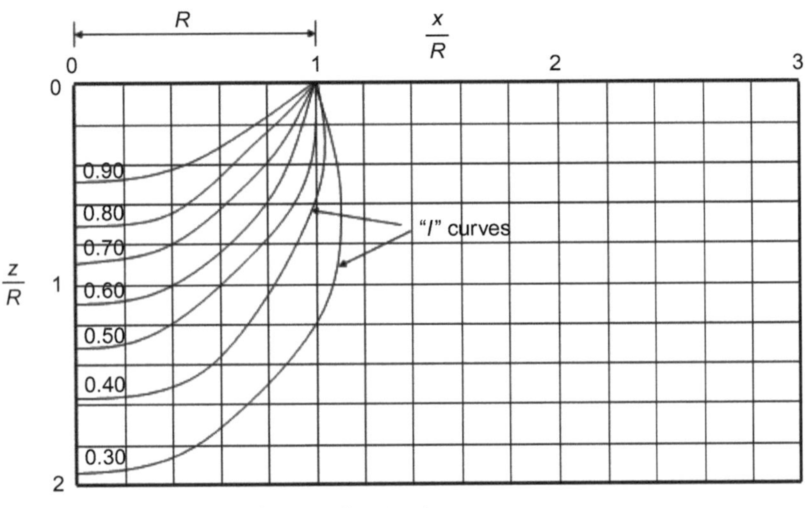

Figure 7.21 Boussinesq diagram for circular areas.

Because the maximum influence of the surface stress will be felt along the centerline, the change in vertical stress can be represented along the centerline by the curve shown in Figure 7.20.

As indicated, the change in vertical stress ($\Delta\sigma$) is determined by multiplying the surface stress by an influence factor (I) obtained from a Boussinesq diagram. Figure 7.21 shows a Boussinesq diagram for circular areas. There are Boussinesq diagrams for rectangular areas, however, they are much more complicated. Therefore, to keep the description of the process as simple as possible, only circular areas will be considered. (*Note*: For a simple rectangle, and certainly for a square area, using an

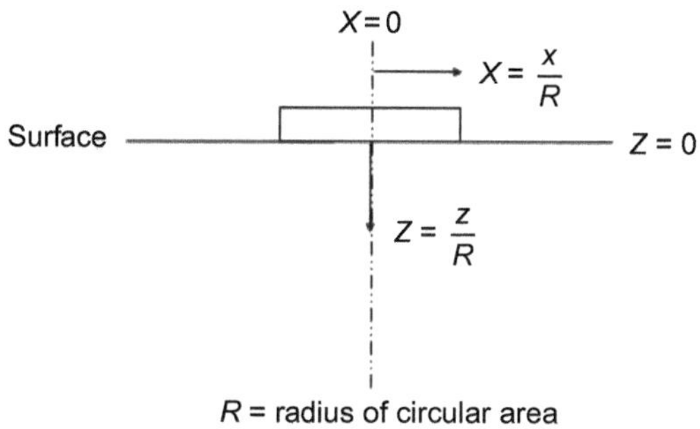

Figure 7.22 Nondimensional vertical and horizontal distances.

enclosed circular area works well and avoids the complications of the more detailed diagrams.)

As can be seen, the distance downward (z) and the distance outward (x) have been made nondimensional by dividing each by the radius of the circular load (R). This can be more clearly seen on Figure 7.22.

Because we are interested in the change in vertical stress ($\Delta\sigma$) at the midpoint of the clay layer, then this is the distance (z). And because the maximum influence will occur along the centerline, the distance (x) will be zero. The best way to show this process is to work an example.

EXAMPLE 7.6 Determine the Change in Vertical Stress

Using the soil profile of Example 7.5, determine the change in vertical stress at the midpoint of the clay layer if a 65-ft-diameter tank weighing 750 tons is planned for this site.

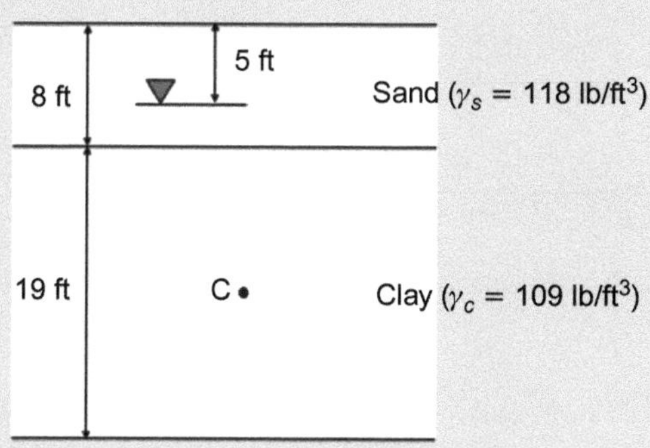

(*Continued*)

EXAMPLE 7.6 Determine the Change in Vertical Stress—(Continued)

Solution

From the soil profile, the distance from the surface to the midpoint of the clay layer is 17.5 ft (8 + 9.5 ft). So the nondimensional vertical distance becomes:

$$Z = \frac{z}{R} = \frac{17.5\ \text{ft}}{32.5\ \text{ft}} = 0.54$$

Because the maximum value will occur along the centerline, the horizontal distance (X) is zero. Therefore, plotting these two points on the Boussinesq diagram for circular loads gives an influence factor (I) equal to 0.88, as shown in the following diagram.

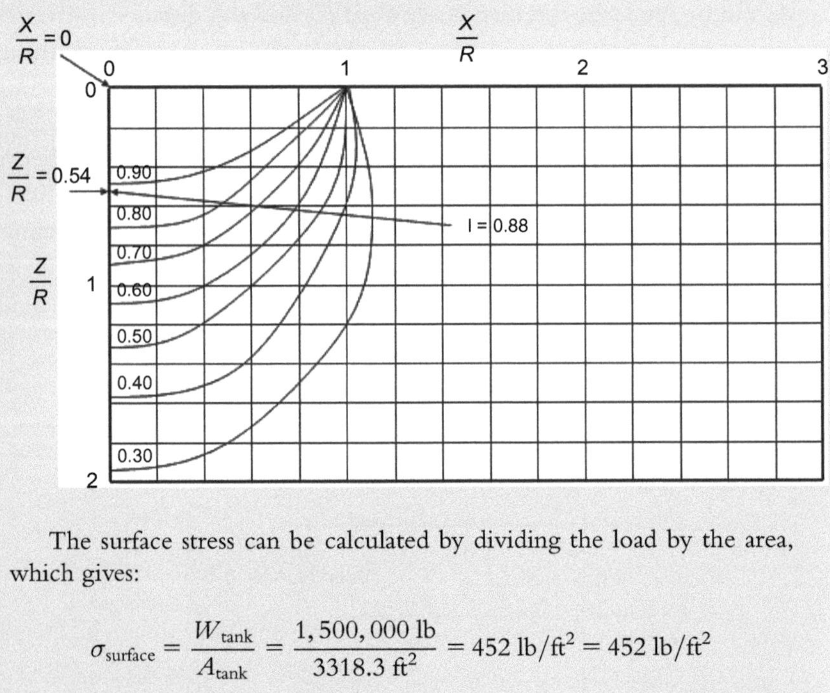

The surface stress can be calculated by dividing the load by the area, which gives:

$$\sigma_{\text{surface}} = \frac{W_{\text{tank}}}{A_{\text{tank}}} = \frac{1,500,000\ \text{lb}}{3318.3\ \text{ft}^2} = 452\ \text{lb/ft}^2 = 452\ \text{lb/ft}^2$$

Multiplying this surface stress by the influence factor gives the change in vertical stress as:

$$\Delta\sigma = I \times \sigma_{\text{surface}} = (0.88)(452\ \text{lb/ft}^2) = 398\ \text{lb/ft}^2$$

At this point the initial vertical stress (σ_0) and the change in vertical stress ($\Delta\sigma$) are known. So along with the thickness of the clay layer (H), the initial void ratio (e_o), and the compression index (C_c), the total settlement can be calculated.

7.6.5 Calculation of the Total Settlement

Recall from Section 7.6.2, Eq. (7.18), which presented the equation for the total settlement (S_c).

$$S_c = \frac{H}{1+e_0} C_c \log\left(\frac{\sigma_0 + \Delta\sigma}{\sigma_0}\right) \tag{7.18}$$

EXAMPLE 7.7 Determine the Total Settlement

Using the soil profile of Examples 7.5 and 7.6, the initial vertical stress (σ_0) found to be 1200 lb/ft^2 in Example 7.5, and the change in vertical stress ($\Delta\sigma$) found to be 398 lb/ft^2 in Example 7.6, determine the total settlement (S_c). Use an initial void ratio of 0.8 and a compression index (C_c) of 0.7.

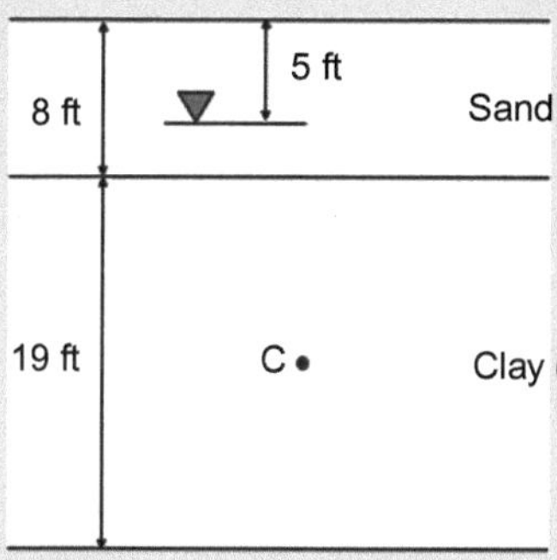

Solution

For the stresses obtained from Examples 7.5 and 7.6, and the soil properties given, the expected total settlement is determined to be:

$$\begin{aligned} S_c &= \frac{H}{1+e_0} C_c \log\left(\frac{\sigma_0 + \Delta\sigma}{\sigma_0}\right) \\ &= \frac{19\ \text{ft}}{1+0.8}(0.7)\log\left(\frac{1200\ \text{lb/ft}^2 + 398\ \text{lb/ft}^2}{1200\ \text{lb/ft}^2}\right) \\ &= 0.92\ \text{ft} \approx 11\ \text{in} \end{aligned}$$

7.6.6 Calculation of the Rate of Settlement

Just as important as determining the total settlement is the rate of settlement, both how long it takes to reach 95% of the total settlement and how much settlement will take place over a specific amount of time.

It is traditional to label the ratio of the partial settlement that takes place, (S_t), to the total settlement (S_c) as the symbol (U), where:

$$U = \frac{S_t}{S_c} \tag{7.20}$$

The factor (U) is determined from the graph in Figure 7.23, where the time factor (T) is determined from the equation:

$$T = \frac{C_v t}{(h_{dr})^2} \tag{7.21}$$

Note that the value of (U) is a percentage and decreases on the vertical axis as the time factor (T) increases along the horizontal axis. Also, in Eq. (7.21), the (C_v) is the coefficient of consolidation determined in the lab from several of the properties of the soil. Its units are (length^2/time), typically (ft^2/day). The time (t) is therefore given in days. The term (h_{dr}) is called the "drainage distance," and is the distance the water must travel to be expelled from the clay layer during the settlement process. This distance (typically determined in feet) depends on whether the layer under

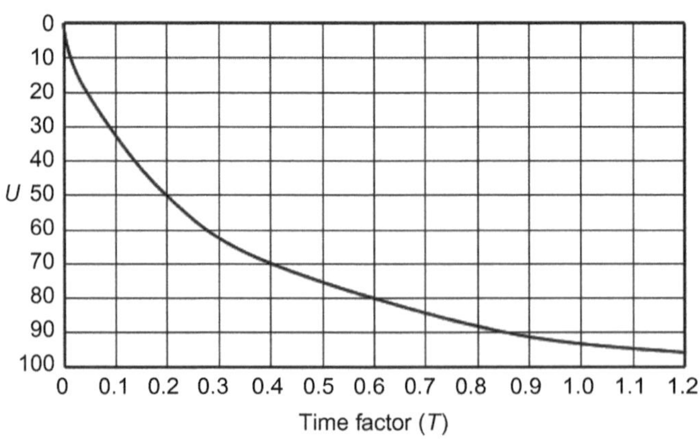

Figure 7.23 Settlement ratio (u) to the time factor (T).

the clay layer is pervious soil or impervious rock. For pervious soil, the drainage distance is half the thickness of the clay layer (H):

$$h_{dr} = \frac{H}{2}$$

For impervious rock, the drainage distance is the full thickness of the clay layer (H):

$$h_{dr} = H$$

Figure 7.24 shows these two distances and why they are different.

It can be seen on Figure 7.23 that as the value for (U) gets closer to 1.0, the longer the time and the more difficult it is to read the chart. Table 7.3 is useful for determining values of (U) for specific values of the time factor (T), and vice versa.

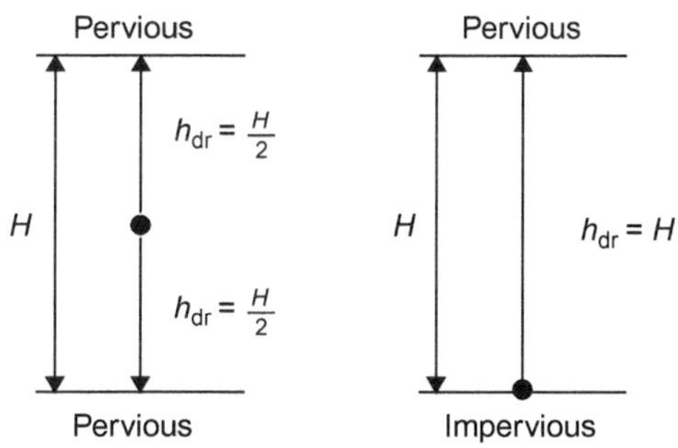

Figure 7.24 Drainage distances for pervious soil versus impervious rock.

Table 7.3 Table of values for the ratio U and the time factor T

U	T
0.1	0.008
0.2	0.031
0.3	0.071
0.4	0.126
0.5	0.197
0.6	0.287
0.7	0.403
0.8	0.567
0.9	0.848
0.95	1.163
1.0	∞

Note that the chart and table reflect the fact that for 100% settlement it would take an infinite amount of time.

EXAMPLE 7.8 Determine the Rate of Settlement

Using the soil profile of Examples 7.5, 7.6, and 7.7, and the total settlement (S_c) found in Example 7.7 of 11 in, determine how long it would take for 95% of that settlement to take place. Also, determine how much settlement will take place in 6 months or 180 days. Consider that the layer below the clay layer is pervious soil. Assume the coefficient of consolidation (C_v) is 0.35 ft²/day.

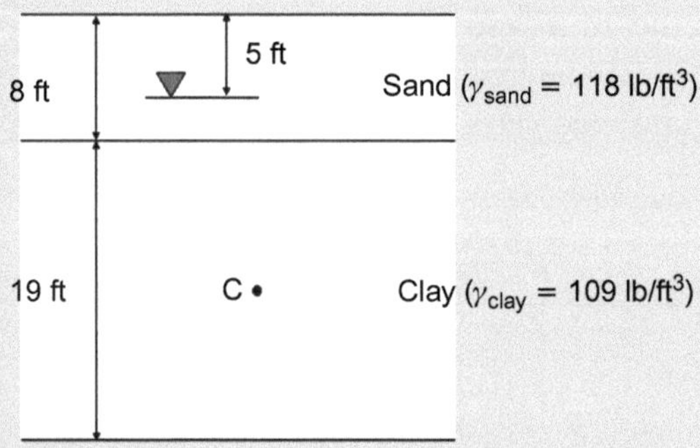

Solution

For a value of (U) equal to 95%, or 0.95, Table 7.3 indicates that the time factor (T) is 1.163. For the layer below the clay layer being pervious soil, the drainage distance (h_{dr}) is half the thickness of the clay layer, or 9.5 ft. Solving for the time (t) in Eq. (7.21), and substituting these values and the coefficient of consolidation, gives:

$$t = \frac{T(h_{dr})^2}{C_v} = \frac{(1.163)(9.5\ \text{ft})^2}{0.35\ \text{ft}^2/\text{day}} = 300\ \text{days}$$

For a time (t) of 180 days, this gives a time factor (T) from Eq. (7.21) of:

$$T = \frac{C_v t}{(h_{dr})^2} = \frac{(0.35\ \text{ft}^2/\text{day})(180\ \text{days})}{(9.5\ \text{ft})^2} = 0.70$$

(Continued)

EXAMPLE 7.8 Determine the Rate of Settlement—(Continued)

Using the graph in Figure 7.23 for this value of the time factor (T) gives a value of (U) equal to 85%, as shown in the following graph:

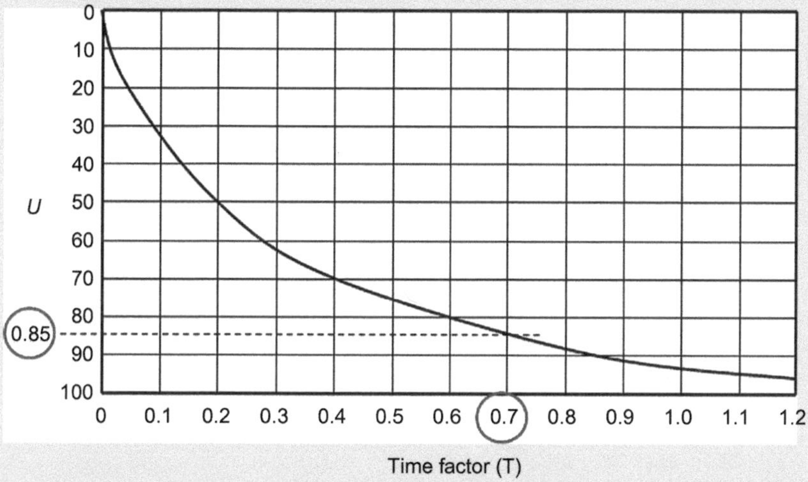

Solving for the partial settlement (S_t) for 180 days in Eq. (7.20), with the total settlement of 11 in, gives:

$$S_t = U\ S_c = (0.85)(11\ \text{in}) = 9.35\ \text{in}$$

7.7 SUMMARY

It should now be clear that the process of determining the total settlement and rate of settlement is highly dependent on the analysis of carefully chosen and properly obtained soil samples for the project in question. From the boring logs, gradation analysis, phase diagram properties, and compaction and consolidation test information, at every step the geotechnical engineer must rely on his or her knowledge, experience, and confidence in the process. If the soil fails to properly support the structure being designed over its useful life, then the efforts of others working on the project will be undermined and will usually result in unnecessary resources being expended to correct the deficiency. A successful design will always provide a source of pride for all involved.

7.8 PRACTICE PROBLEMS

Problem 7.1: Plot Gradation Curve for a Soil Sample

Plot the gradation curve for the following particle size distribution.

Sieve	Retained weight (g)
#4	61
#10	29
#40	48
#200	375
Pan	476
Total	989

Solution

From the given data, the following chart can be generated.

Sieve	Retained weight (g)	% of total	Cumulative %	% finer
#4	61	6	6	94
#10	29	3	9	91
#40	48	5	14	86
#200	375	38	52	48
Pan	476	48	100	—
Total	989	100	—	—

As presented earlier, the first two columns from the left are the given information. The third column is the standard distribution by percentage of the total sample weight for each particle size, which adds up to 100%.

The next column is the cumulative % where the fifth row should add up to 100%, providing a check.

The last column of the chart is the cumulative % subtracted from 100%. These are the values that will be plotted, summarized in the following chart.

Sieve	Diameter (mm)	% finer
#4	4.76	94
#10	1.68	91
#40	0.425	86
#200	0.074	48

Plotting these values on semilog grid gives the following diagram.

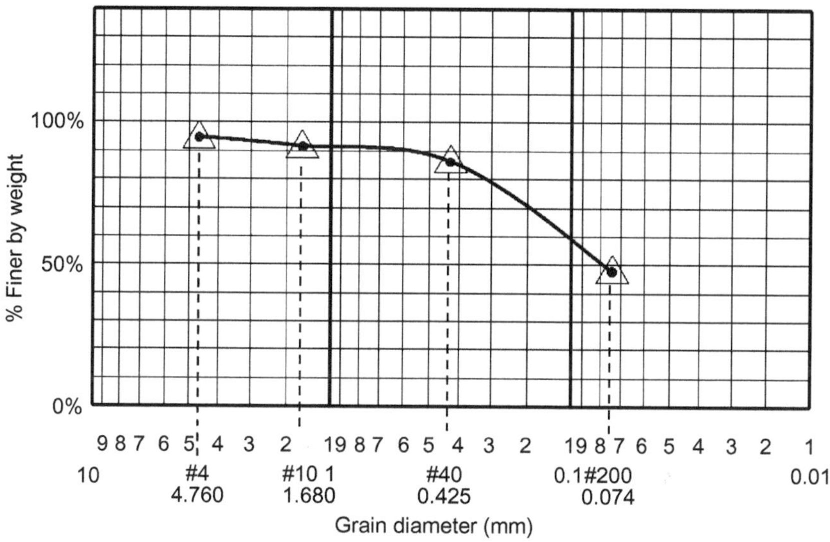

Again, the curve starts at the #4 sieve and stops at the #200 sieve. No information is known about the particle sizes either greater than the #4 or smaller than the #200.

Problem 7.2: Classify a Soil Sample Using the AASHTO System

Using the following information, classify the soil sample according to the AASHTO system. The percent passing the #200 sieve is 33%, the liquid limit is 37%, and the plastic limit is 23%.

Solution

First, from the liquid limit and the plastic limit, calculate the plasticity index using Eq. (7.2) to obtain PI = 14%.

With 33% passing the #200 sieve, this eliminates groups A-1, A-3, A-4, A-5, A-6, and A-7, leaving only group A-2. For groups A-1 and A-3, 33% is too large, and for groups A-4, A-5, A-6 and A-7, 33% is too small.

With the liquid limit = 37%, this eliminates groups A-2-5 and A-2-7 because both require this value to be at least 41%.

With the plasticity index = 14%, this eliminates group A-2-4 because this value must be 10% or less.

Therefore, this soil would be classified as a group A-2-6 soil, which would be acceptable to use.

Problem 7.3: Completing a Phase Diagram

A cubic foot of saturated soil weighs 130 lb. After the soil is oven dried, it weighs 109 lb. Compute the void ratio (e), the porosity (n), the water content (w), and specific gravity of solids (G_s).

Solution

As stated earlier, before attempting to determine any of the requested quantities, first complete a phase diagram for the sample. The phase diagram worksheet shown in Figure 7.12 will again be used. Three quantities never change, namely, the weight of the air, and the specific weights of the air and water. Three quantities are given in the statement of the problem, the total volume (V) which is duplicated, the total weight (W), and the dried weight, which is the weight of the solids (W_s). Don't miss the term "saturated," which means all the voids are filled with water, meaning the saturation ratio (S_r) is 100%.

Because the total volume and total weight are given, the unit weight can be calculated as:

$$\gamma = \frac{W}{V} = \frac{130\ \text{lb}}{1\ \text{ft}^3} = 130\ \text{lb/ft}^3$$

Then with the total weight and weight of the solids, the weight of the water can be found:

$$W_w = W - W_s = 130\ \text{lb} - 109\ \text{lb} = 21\ \text{lb}$$

The phase diagram now has the following nine quantities:

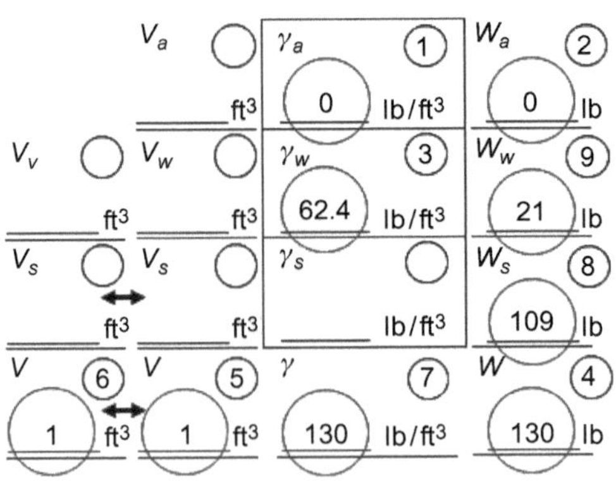

Next, because the weight of the water and its specific weight is known, the volume of the water can be determined as:

$$V_w = \frac{W_w}{\gamma_w} = \frac{21\ \text{lb}}{62.4\ \text{lb/ft}^3} = 0.34\ \text{ft}^3$$

Because the sample is saturated, the voids are filled with water, so the volume of the voids is the same as the volume of the water, making the volume of the air equal to zero.

$$S_r = 100\% \rightarrow V_v = V_w = 0.34\ \text{ft}^3 \rightarrow V_a = 0\ \text{ft}^3$$

The phase diagram now looks like the following:

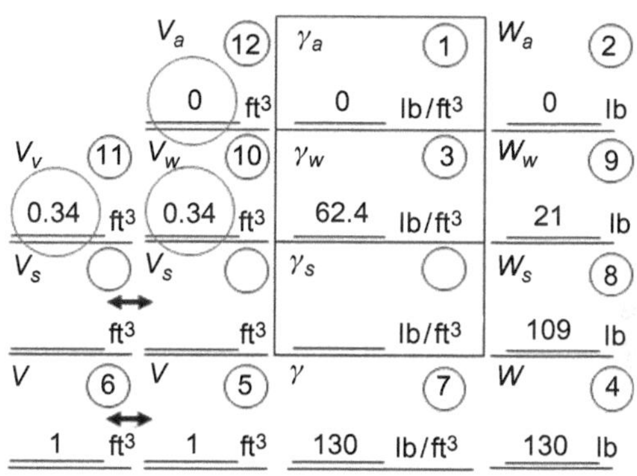

The volume of the solids can now be found, and it is duplicated:

$$V_s = V - V_v = 1\ \text{ft}^3 - 0.35\ \text{ft}^3 = 0.66\ \text{ft}^3$$

Finally, the specific weight of the solids can be found using the weight of solids and the volume of solids, to give:

$$\gamma_s = \frac{W_s}{V_s} = \frac{109\ \text{lb}}{0.66\ \text{ft}^3} = 165.2\ \text{lb/ft}^3$$

The phase diagram is now complete:

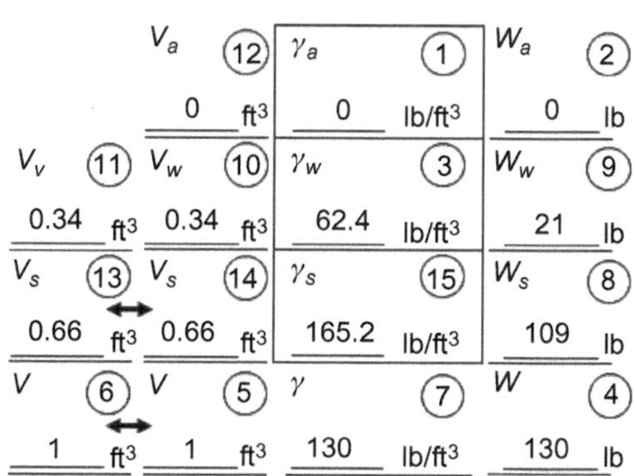

In no special order, the void ratio becomes:

$$e = \frac{V_v}{V_s} = \frac{0.34\ \text{ft}^3}{0.66\ \text{ft}^3} = 0.52$$

The porosity becomes:

$$n = \frac{V_v}{V} = \frac{0.34\ \text{ft}^3}{1\ \text{ft}^3} = 0.34$$

The water content becomes:

$$w = \frac{W_w}{W_s} = \frac{21\ \text{lb}}{109\ \text{lb}} = 0.19$$

And finally, the specific gravity of the solids can be found to be:

$$G_s = \frac{\gamma_s}{\gamma_w} = \frac{165.2\ \text{lb/ft}^3}{62.4\ \text{lb/ft}^3} = 2.65$$

Again, it cannot be overstated that completing a phase diagram first will yield not only the required parameters, but a comprehensive makeup of the sample can be seen.

Problem 7.4: Determine Maximum Dry Unit Weight and Optimum Water Content

Using the following table of information, determine the maximum dry unit weight and its associated optimum water content.

Test number	Wet weight (g)	Wet weight (lb/ft^3)	Wet weight sample (g)	Wet dry sample (g)	Moisture (%)	Dry weight (lb/ft^3)
1	1905.0		100.0	92.2		
2	1978.0		100.0	90.6		
3	2005.0		100.0	88.8		
4	1987.0		100.0	87.2		

Solution

First, using the weight of wet soil given in the second column, determine the wet unit weight for each test. The volume of the sample will be that used in the Proctor tests, 1/30 ft^3. A sample calculation is provided for the first set of values.

Test 1: wet weight = 1905 grams

Convert grams to pounds in the following calculation:

$$\text{wet weight} = 1,905.0\ \text{g} \times \frac{1\ \text{kg}}{1000\ \text{g}} \times \frac{2.205\ \text{lb}}{1\ \text{kg}} = 4.201\ \text{lb}$$

Calculate the wet unit weight using a volume of 1/30 ft^3 and place in column 3:

$$\text{wet specific weight} = \frac{4.201\ \text{lb}}{1/30\ \text{ft}^3} = 126.0\ \text{lb/ft}^3$$

Using the wet and dry weights in columns 4 and 5, calculate the water content in % and place in column 6, labeled "moisture":

$$w = \frac{W_w}{W_s} = \frac{100\ \text{g} - 92.2\ \text{g}}{92.2\ \ \text{g}} \times 100 = 0.085 = 8.5\ \%$$

Finally, using the wet unit weight and the water content, calculate the dry unit weight using Eq. (7.12) and place in column 7:

$$\gamma_d = \frac{\gamma}{1\ +\ w} = \frac{126.0\ \text{lb/ft}^3}{1 + 0.085} = 116.1\ \text{lb/ft}^3$$

Do similar calculations for the remaining three tests to obtain the results provided in the following completed table.

Test number	Wet weight (g)	Wet weight (lb/ft^3)	Wet weight sample (g)	Wet dry sample (g)	Moisture (%)	Dry weight (lb/ft^3)
1	1905.0	126.0	100.0	92.2	8.5	116.1
2	1978.0	130.8	100.0	90.6	10.4	118.5
3	2005.0	132.6	100.0	88.8	12.6	117.8
4	1987.0	131.4	100.0	87.2	14.7	114.6

Now plot the water content in column 6 and the dry unit weight in column 7. The following graph should result:

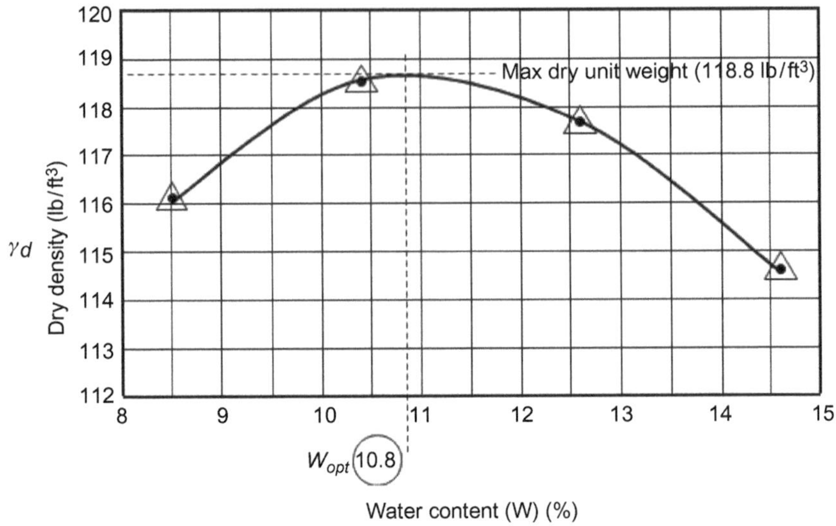

From the graph, it appears the maximum dry unit weight is 118.8 lb/ft^3 at an optimum water content of 10.8%.

Problem 7.5: Calculating the Initial Vertical Stress

For the three-layer soil profile shown, determine the initial vertical stress (σ_0) at the midpoint of the clay layer, Point C.

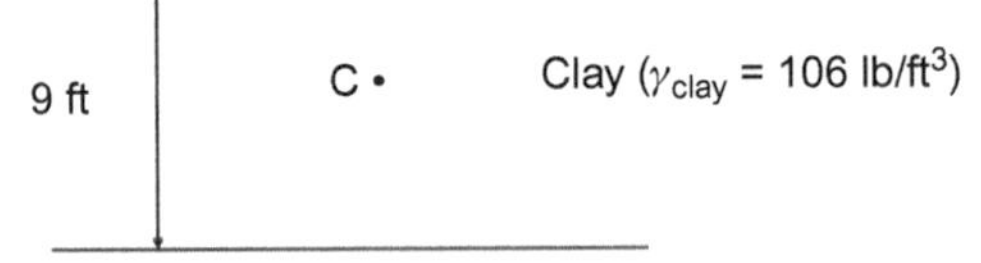

Solution

For the sand above the silt:

$$\sigma_{\text{sand}}^{\text{above}} = (121\ \text{lb/ft}^3) \times (3\ \text{ft}) = 363\ \text{lb/ft}^2$$

For the silt above the water table:

$$\sigma_{\text{silt}}^{\text{above}} = (108\ \text{lb/ft}^3) \times (2\ \text{ft}) = 216\ \text{lb/ft}^2$$

For the silt below the water table:

$$\sigma_{\text{silt}}^{\text{below}} = (108\ \text{lb/ft}^3 - 62.4\ \text{lb/ft}^3) \times (4\ \text{ft}) = 182\ \text{lb/ft}^2$$

For the clay above its midpoint:

$$\sigma_{\text{clay}} = (106\ \text{lb/ft}^3 - 62.4\ \text{lb/ft}^3) \times (4.5\ \text{ft}) = 196\ \text{lb/ft}^2$$

The total initial vertical stress becomes:

$$\sigma_0 = (363\ \text{lb/ft}^2) + (216\ \text{lb/ft}^2) + (182\ \text{lb/ft}^2) + (196\ \text{lb/ft}^2) = 957\ \text{lb/ft}^2$$

Problem 7.6: Determine the Change in Vertical Stress

Using the soil profile of Problem 7.5, determine the change in vertical stress at the midpoint of the clay layer if an 80-ft-diameter tank weighing 600 tons is planned for this site.

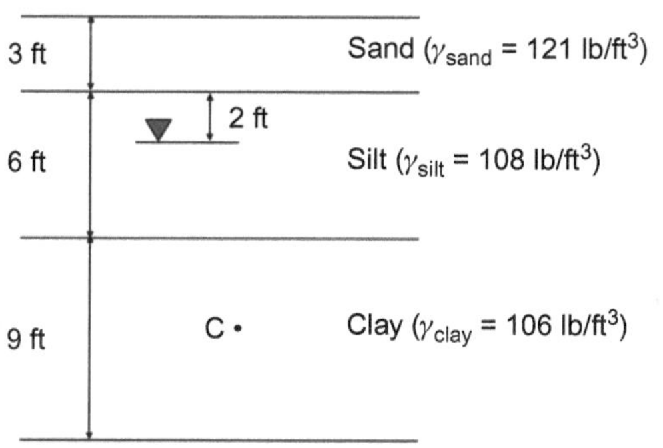

Solution

From the soil profile, the distance from the surface to the midpoint of the clay layer is 13.5 ft (3 ft + 6 ft + 4.5 ft). So the nondimensional vertical distance becomes:

$$Z = \frac{z}{R} = \frac{13.5\ \text{ft}}{40\ \text{ft}} = 0.34$$

Because the maximum value will occur along the centerline, the horizontal distance (*X*) is zero. Therefore, plotting these two points on the Boussinesq diagram for circular loads gives an influence factor (I) equal to 0.93, as shown in the following diagram.

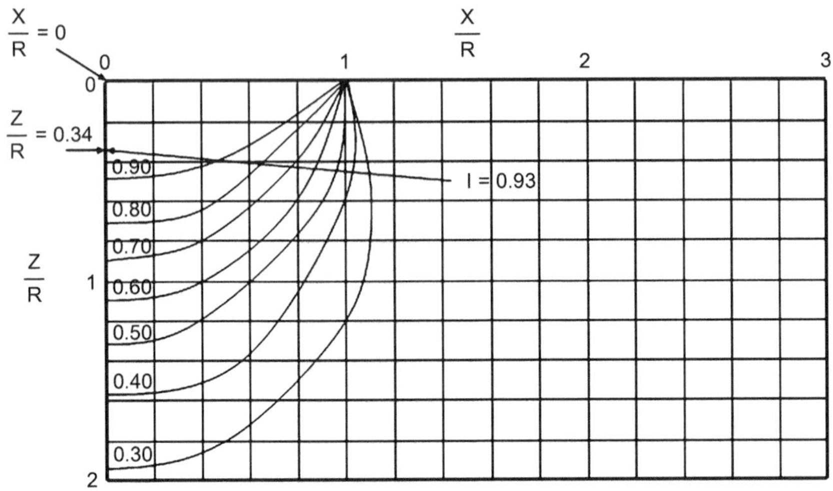

The surface stress can be calculated by dividing the load by the area, which gives:

$$\sigma_{\text{surface}} = \frac{W_{\text{tank}}}{A_{\text{tank}}} = \frac{1,200,000 \text{ lb}}{5,026.6 \text{ ft}^2} = 239 \text{ lb/ft}^2 = 239 \text{ lb/ft}^2$$

Multiplying this surface stress by the influence factor gives the change in vertical stress as:

$$\Delta\sigma = I \times \sigma_{\text{surface}} = (0.93)(239 \text{ lb/ft}^2) = 222 \text{ lb/ft}^2$$

Problem 7.7: Determine the Total Settlement

Using the soil profile of Problems 5 and 6, the initial vertical stress (σ_0) found to be 957 lb/ft^2 in Problem 7.5, and the change in vertical stress ($\Delta\sigma$) found to be 222 lb/ft^2 in Problem 7.6, determine the total settlement (Sc). Use an initial void ratio of 0.85 and a compression index (C_c) of 0.75.

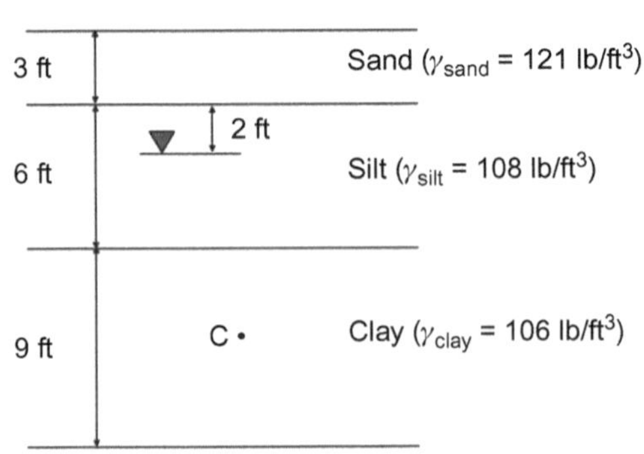

Solution

For the stresses obtained from Problems 7.5 and 7.6, and the soil properties given, the expected total settlement is determined to be:

$$\begin{aligned} S_c &= \frac{H}{1+e_0} C_c \log\left(\frac{\sigma_0 + \Delta\sigma}{\sigma_0}\right) \\ &= \frac{9\ \text{ft}}{1+0.85}(0.75)\log\left(\frac{957\ \text{lb/ft}^2 + 222\ \text{lb/ft}^2}{957\ \text{lb/ft}^2}\right) \\ &= 0.33\ \text{ft} \approx 4\ \text{in} \end{aligned}$$

Problem 7.8: Determine the Rate of Settlement

Using the soil profile of Problems 7.5, 7.6, and 7.7, and the total settlement (S_c) found in Problem 7.7 of 4 in, determine how long it would take for 95% of that settlement to take place. Also, determine how much settlement will take place in 2 months or 60 days. Consider that the layer below the clay layer is impervious rock. Assume the coefficient of consolidation (C_v) is 0.55 ft²/day.

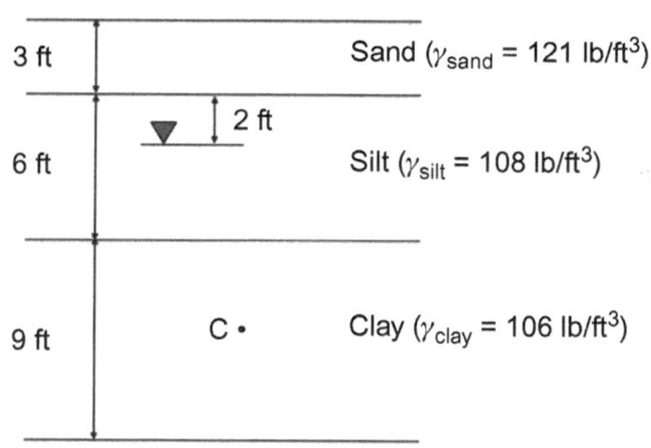

Solution

For a value of (U) equal to 95%, or 0.95, Table 7.3 indicates that the time factor (T) is 1.163. For the layer below the clay layer being impervious rock, the drainage distance (h_{dr}) is the full thickness of the clay layer, or 9 ft. Solving for the time (t) in Eq. (7.21), and substituting these values and the coefficient of consolidation, gives:

$$t = \frac{T(h_{dr})^2}{C_v} = \frac{(1.163)(9\ \text{ft})^2}{0.55\ \text{ft}^2/\text{day}} = 171\ \text{days}$$

For a time (t) of 60 days, this gives a time factor (T) from Eq. (7.21) of:

$$T = \frac{C_v t}{(h_{dr})^2} = \frac{(0.55\ \text{ft}^2/\text{day})(60\ \text{day})}{(9\ \text{ft})^2} = 0.41$$

Using the graph in Figure 7.23 for this value of the time factor (T) gives a value of (U) equal to 70%, as shown in the following graph:

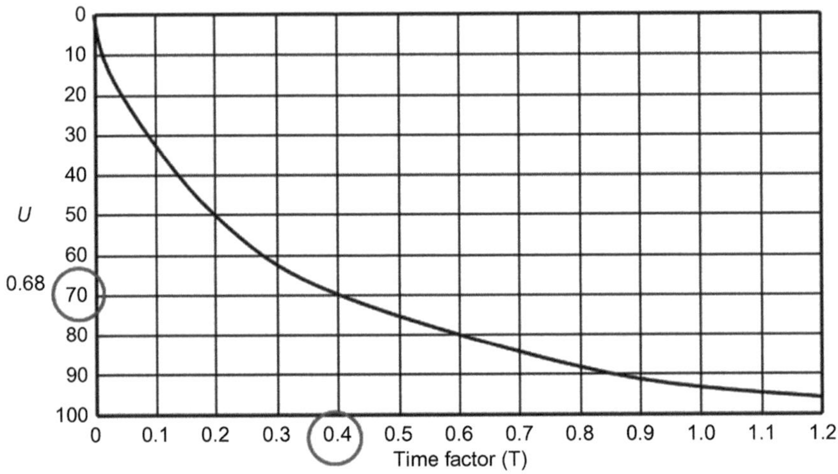

Solving for the partial settlement (S_t) for 60 days in Eq. (7.20), with the total settlement of 4 in, gives:

$$S_t = U\ S_c = (0.70)(4\ \text{in}) = 2.8\ \text{in}$$

PART 8

Structures

Thomas H. Brown, Jr., PhD, PE
Institute for Transportation Research and Education, North Carolina State University

Contents

8.1 INTRODUCTION

The scope of this part relative to structural engineering is not intended to be exhaustive, only to give the basics of what is involved in the design of the three main elements of a multispan bridge overpass: (1) the roadway deck and steel beam girders, (2) the hammerhead pier supporting the two spans, and (3) the driven pile footing for the hammerhead pier.

To provide detailed specifications and sound engineering guidance, several primary references are typically incorporated into the design of any bridge system. They are, in no order of importance, the *Standard Specifications for Highway Bridges* published by the American Association of State Highway and Transportation Officials (AASHTO), the *Manual of Steel Construction* published by the American Institute for Steel Construction (AISC), various specifications published by ASTM International, formerly the American Society for Testing and Materials (ASTM), and, most assuredly, a *Structure Design Manual* developed by each state's Department of Transportation (DOT).

The presentation of the design process will follow a top-down approach, beginning with the design of the steel girders. Depending on the length of the spans, the steel girders could be standard wide-flange (W) sizes or, as is becoming more the case with the requirement of longer and longer spans, built-up girders are constructed from plate steel. The two bridge spans sit on top of what is called a *hammerhead pier*, so-named for its distinctive shape. Because concrete cannot support tensile loads, reinforced steel rebar will be required; it is the number and size of the rebar that is the central goal of this part of the design process. Finally, the hammerhead pier must be supported, and here a driven pile footing system will be presented. Spread footings are common, however, the design process of these is almost twice as involved as that of a driven pile system, and requires sufficient soil support capabilities that are not available in all locations.

Before describing the design process, it is helpful to understand the many components of the bridge system under consideration.

8.2 COMPONENTS OF A BRIDGE SYSTEM

In general, the components of a bridge system are divided into two parts: (1) the superstructure, and (2) the substructure. The superstructure consists of the bridge spans from the steel beams up, which includes the steel girders, the diaphragm bracing, the roadway deck system, and the safety barriers. The substructure consists of the hammerhead pier and the pile footing. A simple diagram of these two major parts is shown in Figure 8.1.

The roadway deck system has several components: (1) the deck forms, (2) the reinforcing steel, and (3) the decking material itself, which is typically concrete. Figure 8.2 shows the deck forms, Figure 8.3 shows the reinforcing steel, and Figure 8.4 shows the concrete being smoothed and leveled, called *screed*, by a specialized machine.

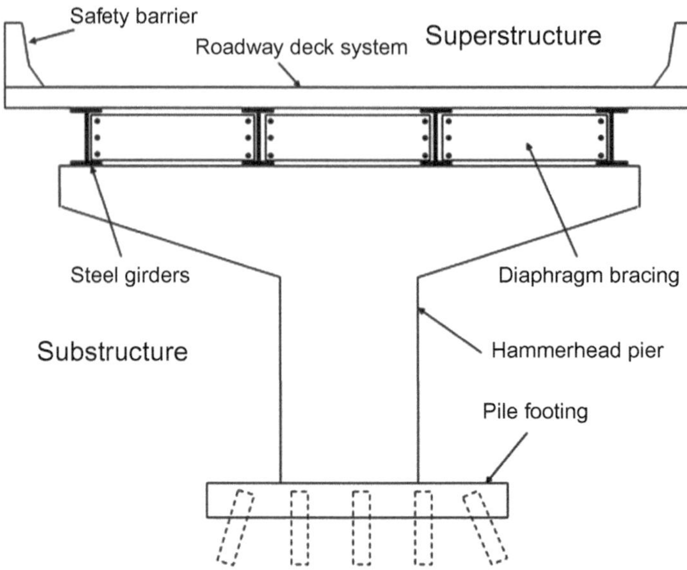

Figure 8.1 The superstructure and substructure of a bridge system.

Figure 8.2 Deck forms being installed.

Figure 8.3 Reinforcing steel imbedded in the concrete slab.

Figure 8.4 Concrete decking being screed.

In Figure 8.2 the worker is kneeling on the deck forms, which have a "V" cross section. The weight of the concrete that will eventually fill these troughs will be discussed in the next section. Also, notice the shear studs on top of the steel girders. These are installed along the top of each girder to tie the roadway deck system to the girders. There is a narrow trough along the beam formed by this system of deck forms and the shear studs, and this will eventually be filled with concrete. Its weight is also discussed in the next section.

The reinforcing steel is necessary because concrete cannot take tensile loads, by far the most important principle of concrete structure design. This principle will be addressed in the design of the hammerhead pier and the pile footing.

It can only be imagined how many workers it would take to do the job of the machine shown in Figure 8.4, and most likely it would not be as smooth and finished at the right crown needed for proper storm-water runoff.

In Figure 8.5, the diaphragm bracing is shown, which provides lateral support for the steel girders. These diaphragms are very important in making the bridge span respond to the various loads it

Figure 8.5 Diaphragm bracing for the steel girders.

Figure 8.6 Typical end bent.

Figure 8.7 Pot bearing assembly.

will see as a single unit. Notice again the shear studs installed along the top of the steel girders. These are the key to the roadway deck system and steel girders acting together to support the design loads.

While the hammerhead pier will support one end of each of the two spans, the other end of each span will be supported on what are called *end bents*. Figure 8.6 shows a typical end bent. Its design is beyond the scope of this presentation; it involves more steps in the design process than a hammerhead pier.

Important components of the bridge system are the connections between the bridge girders and either the hammerhead pier or the end bents. At the hammerhead pier, each end of the bridge spans is connected by large pin assemblies, allowing rotation but not lateral movement. However, at the end bent end of the span, the girders must be allowed to move laterally, while supporting the tremendous loads involved. Figure 8.7 shows what is called a "pot bearing." They act like a roller connection, but have limited travel.

Figure 8.8 Typical slope protection.

And lastly, an extremely important design element of a bridge system is slope protection below the end bents. Sometimes there are natural materials onsite that can be used, or the material is brought in; however, more often than not, a concrete incline is needed to resist erosion and avoid undercutting the soil beneath the end bents. Figure 8.8 shows the typical concrete slope protection. Without it, the soil beneath the bridge would not last long from the weather and runoff from the bridge. Before long, the end bent supports would be undermined and eventually lose their ability to carry the extreme loads.

The design process will be presented in the following sections in a top-down approach. The first major element to consider is the system of steel girders.

8.3 STEEL BEAM DESIGN

8.3.1 Introduction

In the design of a bridge system, it is traditionally assumed that the loads will be carried by the steel girders alone. The resistance to both shear and bending will not be carried by the roadway decking, though it is typically reinforced concrete. This assumption considerably simplifies the design process.

For each span of a multispan bridge system under consideration, the spans can be assumed to be simply-supported beams with a constant distributed load along the length of the span. This loading is called the *dead load*, abbreviated DL. To the dead load is added the loading due to vehicular traffic, called the *live load*, abbreviated LL.

In addition to the dead load and live load, secondary loads can be present. They can include wind, ice, temperature changes, and earthquakes, to name just a few. Their contribution to the loading will not be considered here; statistically, it is very improbable that all the various loadings would take place at the same time, so the factors of safety that are considered essentially account for the most likely simultaneous loadings.

There are two approaches to bridge design: (1) service load design and (2) load factor design; abbreviated LFD. Only the load factor design approach will be presented. The LFD approach uses *load factors* multiplied by either the dead load or live load, or both. AASHTO has specified these factors by groups, where here Group X will be used. More will be presented shortly as to the application of the Group X factors for both the bending load in the steel girders and the shear load on the hammerhead pier.

8.3.2 Dead Load

As mentioned earlier, the dead load is caused by the uniform distributed load made up of all the components of the bridge span. Considering the bridge span as a simply supported beam with a uniform distributed load, Figure 8.9 shows a diagram of this loading and the supports. For a simply supported beam, a pin joint is present at one end of the beam and a roller assembly at the other end.

For the beam shown in Figure 8.9, at the left end, point A, is a pin joint, which rests on the hammerhead pier, and at the right end, point B, is the roller assembly, which rests on one of the end bents. In the actual

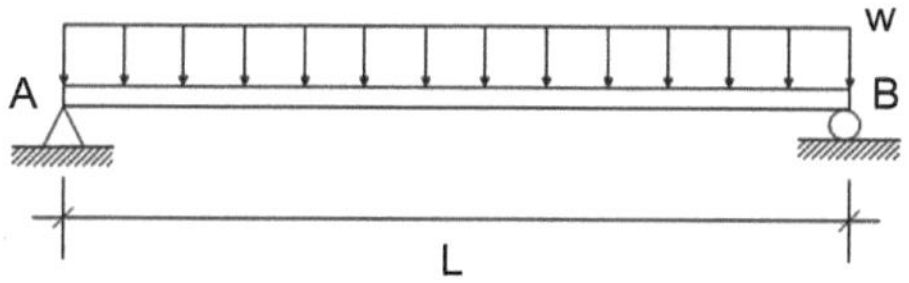

Figure 8.9 Simply supported bridge span with uniform distributed load.

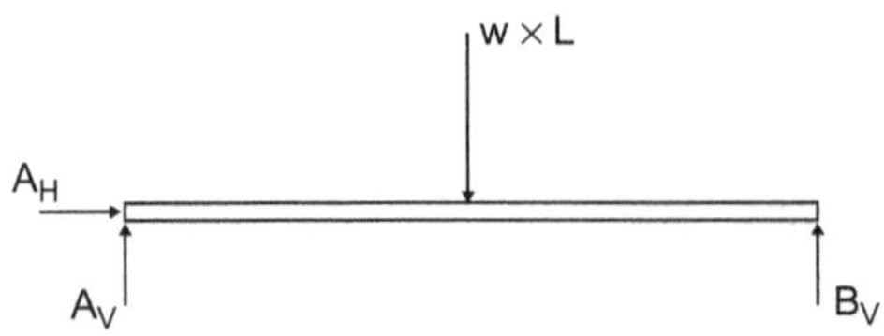

Figure 8.10 Free-body diagram of bridge span.

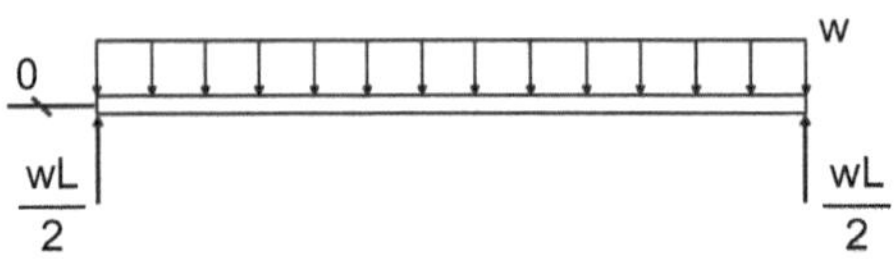

Figure 8.11 Balanced free-body diagram.

system, the roller is the pot bearing assembly, shown in Figure 8.7. The quantity labeled w is the weight per length of bridge span, usually given in kip/ft, where a kip is 1000 lb. L is the length of the span. Typically, for spans greater than 120 ft, steel girders must be used over prestressed concrete girders. Also, it is most likely that the two spans of a multispan overpass will be of unequal lengths. This will be taken into account at appropriate places in the presentation.

The process of determining the reactions at the pin and roller start with a free-body diagram, or FBD, as shown in Figure 8.10. The uniform distributed load has been replaced by a single concentrated load ($w \times L$) at the center of the span. This is a valid replacement in determining the reactions at the pin and roller, but not for any other aspect of the design of the beam. The pin joint has been replaced by two unknown reactions, one horizontal (A_H) and one vertical (A_V), and the roller assembly has been replaced by a single unknown vertical reaction, B_V.

Without showing the details, applying the equations of equilibrium gives the "balanced" free-body diagram shown in Figure 8.11. Note that the horizontal reaction A_H is zero; however, lateral loads will be present from various causes and will be included in the load factor design approach. The total distributed load of $w \times L$ is carried equally by the two vertical reactions, A_V and B_V.

Using the standard conventions on shear force (V), where positive is up on a left face, gives the shear force diagram shown in Figure 8.12. The diagram starts off at the left end of the beam ($x = 0$) with the positive shear force equal to A_V and slopes down linearly at a slope of w, to

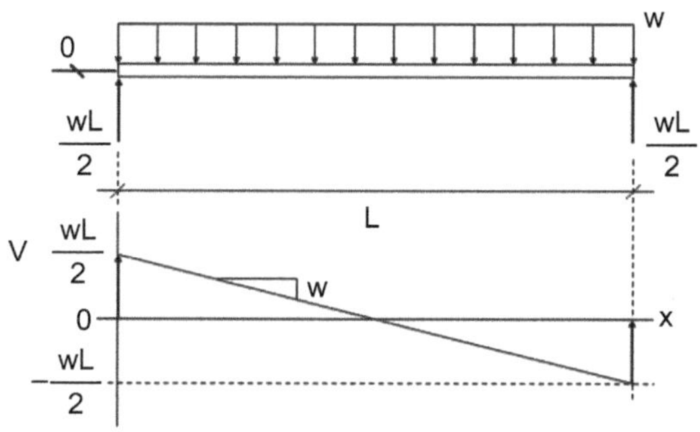

Figure 8.12 Shear force diagram.

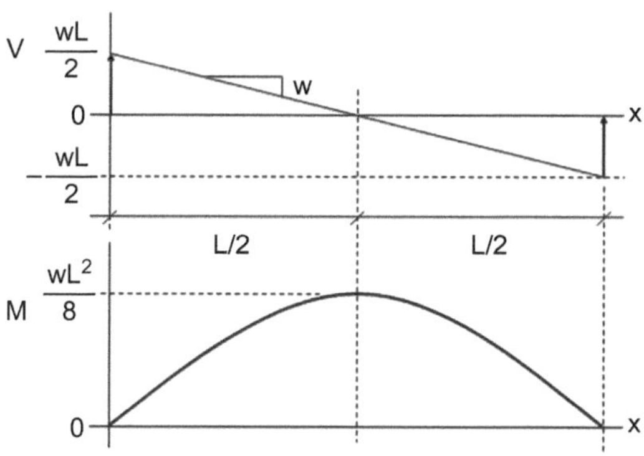

Figure 8.13 Bending moment diagram.

the negative shear force equal to B_V at the right end of the beam ($x = L$). The shear force diagram crosses the x-axis at a value of 0 at the midpoint of the beam.

Similarly, using the standard convention on bending moment (M), where concave upward is positive, gives the bending moment diagram shown in Figure 8.13. The shear force diagram is shown above it, as the slope of the bending moment at any point along the beam is the value of the shear force at that point. And the value of the bending moment at any point is the area under the shear force diagram up to that point.

Notice that the bending moment diagram starts off at 0 at $x = 0$, because the pin joint cannot resist a moment, and moves in a parabolic shape to a maximum at the midpoint of the beam, then back to 0 at the right end of the beam ($x = L$), also because the roller assembly cannot resist a moment. As stated, the slope of the bending moment diagram at any point is the value of the shear force (V) at that point.

Therefore, where the shear force is 0, the slope of the moment diagram is 0, and so the bending moment is a maximum.

The maximum bending moment is determined by calculating the area under the shear force (V) diagram from $x = 0$ to the midpoint of the beam, which is a triangle. This maximum value is shown in the calculation provided in Eq. (8.1) as:

$$M_{\max} = \underbrace{\frac{1}{2} \times \text{base} \times \text{height}}_{\text{area of a triangle}} = \frac{1}{2} \times \underbrace{\frac{L}{2}}_{\text{base}} \times \underbrace{\frac{w\,L}{2}}_{\text{height}} = \frac{w\,L^2}{8} \quad (8.1)$$

This maximum value of the bending moment will be the dead load (DL) that will be used in the load factor design approach, so:

$$DL = \frac{wL^2}{8} \quad (8.2)$$

All that is necessary now is to determine the value of w for the bridge span. That will be presented shortly.

However, when the process shifts to the hammerhead pier, the dead load (DL) will be the vertical reaction at either end of the span. For multispan bridges, this means two spans will be supported at the hammerhead pier. Therefore, the dead load on the hammerhead pier will be the sum of the loads from each span. Again, it will be assumed that the two spans of a multispan overpass are not of equal length. More will be presented on this loading in the section on hammerhead piers.

Before proceeding to show how the distributed load w is determined, as the dead load has just been discussed, it seems appropriate to discuss the live load on the bridge span.

8.3.3 Live Load

There are many criteria for determining the live load on a bridge span. Here, in the design process that follows, the HS20-44 truck will be used. There are other live load models that are used; however the design process is the same. A simple diagram of this idealized truck is shown in Figure 8.14. As will be seen shortly, not only is the total weight of the truck important, and its weight distribution at each axle, but also the loading on just one side of the truck, especially when two trucks might be side by side supported by a single girder.

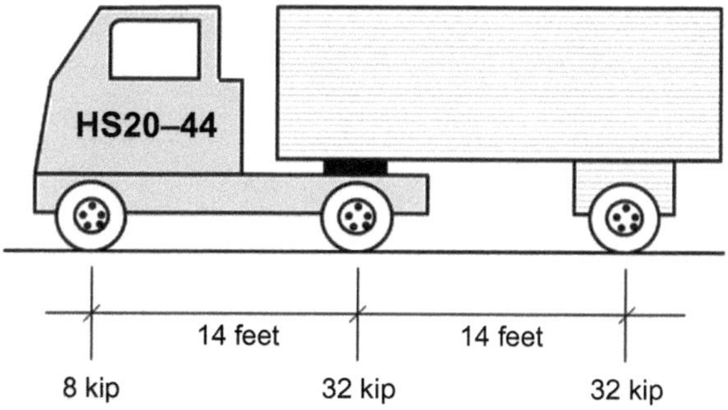

Figure 8.14 HS20-44 truck loading.

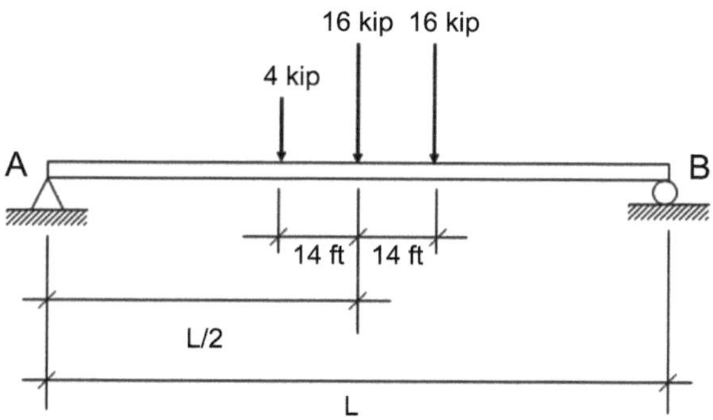

Figure 8.15 HS20-44 tire loadings.

As shown in Figure 8.14, the specifications for an HS20-44 establish that on the front axle there are 8 kip, 32 kip on the middle axle, and another 32 kip on the rear axle, for a total of 72 kip. However, AASHTO specifies that all girders of a bridge shall be designed for the worst case, which is when one side of the truck is directly over a girder. So, only half the loads on the axles are considered. For a maximum bending moment due to this live load, the HS20-44 truck will be placed on the bridge girder as shown in Figure 8.15 with the middle axle at the midpoint of the bridge span.

The free-body diagram (FBD) of this loading for determining the maximum bending moment is shown in Figure 8.16.

Applying the equations of equilibrium, the horizontal force, A_H, appears to be zero, however the HS20-44 truck is moving and so some support along the span is required. However, this additional support is accommodated later in one of the load factor design quantities that takes into account the momentum and vibration caused by the truck.

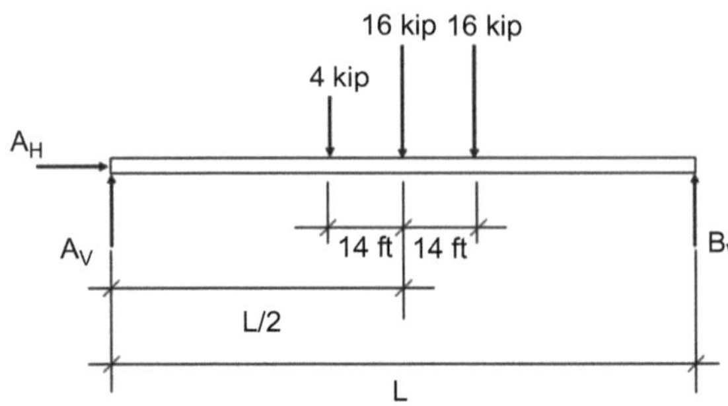

Figure 8.16 Free-body diagram of HS20-44 loading for maximum bending.

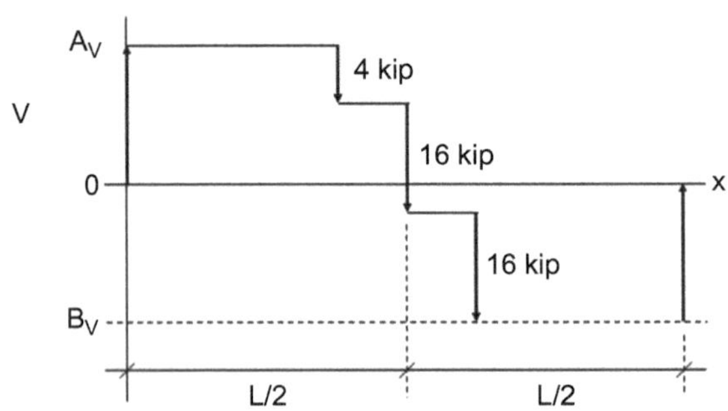

Figure 8.17 Shear force diagram for maximum bending moment.

As for the reactions at the ends of the bridge span, these are dependent on the length of the bridge span, L. Taking moments at point A, assuming clockwise is positive, the equilibrium equation becomes:

$$(4\text{ kip})\left(L/2 - 14\text{ ft}\right) + (16\text{ kip})\left(L/2\right) + (16\text{ kip})\left(L/2 + 14\text{ ft}\right) - B_V(L) = 0$$

Expanding and collecting terms, then rearranging to solve for the reaction at B, B_V, gives:

$$B_V = \frac{(18\text{ kip})\,L + 168\text{ ft}\cdot\text{kip}}{L} \tag{8.3}$$

where L, the length of the span, is in feet.

Once the reaction (B_V) has been determined, then the reaction at A, A_V, becomes:

$$A_V = 36\text{ kip} - B_V \tag{8.4}$$

Once the reactions A_V and B_V are determined, a shear force diagram can be drawn, shown in Figure 8.17.

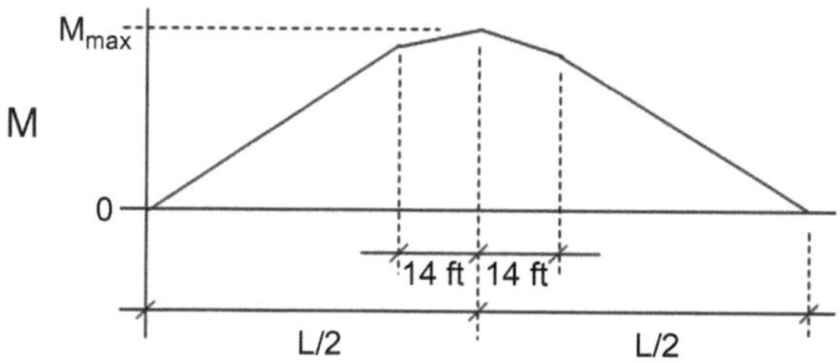

Figure 8.18 Bending moment diagram for maximum bending moment.

Using this shear force diagram, the bending moment diagram for maximum bending moment can be drawn, as shown in Figure 8.18. Notice that the maximum bending moment occurs at the midpoint of the bridge span.

From the shear force diagram, the maximum bending moment is either the positive area to the left of the midpoint of the beam or the negative area to the right of the midpoint of the beam. The positive area is associated with the reaction A_V and becomes:

$$M_{\text{max}} = (A_V)\left(L/2 - 14\text{ ft}\right) + (A_V - 4\text{ kip})(14\text{ ft})$$

Expanding terms and simplifying gives:

$$M_{\text{max}} = A_V\left(L/2\right) - 56\text{ ft}\cdot\text{kip} \tag{8.5}$$

The negative area is associated with the reaction B_V and becomes:

$$M_{\text{max}} = (B_V)\left(L/2 - 14\text{ ft}\right) + (B_V - 16\text{ kip})(14\text{ ft})$$

Expanding terms and simplifying gives:

$$M_{\text{max}} = B_V\left(L/2\right) - 224\text{ ft}\cdot\text{kip} \tag{8.6}$$

The maximum bending moment, either from Eq. (8.5) or (8.6), gives the live load (LL) that will be used shortly in the load factor design equation.

Now that the dead load and live load have been discussed, it is time to discuss how the distributed load w is determined.

8.3.4 Distributed Load

The various elements of a modern bridge span were presented in Section 8.2. These various elements constitute the distributed load w in Eq. (8.2) that will be used to calculate the dead load for bending, and this will be used to design the steel girders. It is also used in calculating the dead load for shear, which will be used to design the hammerhead pier.

As stated earlier, AASHTO requires that all girders of a bridge system must be the same. Therefore, only the distributed loading, *w*, on a single girder need be determined. Therefore, in all the calculations that follow, the number of girders will be taken into account.

The elements of a bridge system discussed here are limited to the following:

1. Concrete composite slab (weight per girder)
2. Steel girder (same for all girders in the design)
3. Future replacement surface (weight per girder)
4. Safety barriers (weight per girder)

The most obvious element missing in the list is the diaphragm bracing, which could be significant. However, including it would add an unnecessary complication as there are many possible designs. For a real design, determine the total weight of the diaphragm bracing between two interior girders and divide that weight by the bridge span (L). Add this weight per foot to the distributed load, *w*, determined from the other elements.

For the calculations that follow, the cross section shown in Figure 8.19 will be used.

It is assumed that the slab overhang on each side is equal to half the spacing between each girder. Therefore, the portion of the slab carried by an interior girder is the bridge width, BW, divided by the number of girders, G. This length is given the symbol BW_G, given by the following equation:

$$BW_G = \frac{\text{bridge width}}{\text{number of girders}} = \frac{BW}{G} \tag{8.7}$$

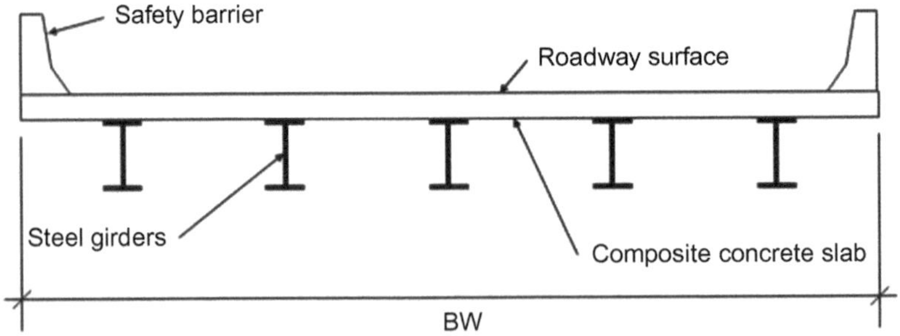

Figure 8.19 Cross section of basic bridge system elements.

The calculations associated with each of the four elements previously referenced follow.

Composite Concrete Slab

The composite concrete slab is composed of several subsystems:

1. Steel deck forms
2. Concrete in the valleys of the deck forms
3. Concrete due to the deflection of the deck forms
4. Concrete above the flange of the steel girders (covers the shear studs)
5. Concrete roadway

The standard weight of the deck forms is 3 lb/ft^2. The valleys of the deck forms are assumed to have concrete 1 in deep. Converting inches to feet and multiplying by the specific weight of concrete, 150 lb/ft^3, gives 12.5 lb/ft^2. The additional concrete present due to the deflection of the deck forms is usually taken as 5.21 lb/ft^2. Adding these three weights per square foot together gives 20.71 lb/ft^2. This value will be multiplied by the portion carried by a single girder, BW_G, given by Eq. (8.7). Therefore, the contribution to the distributed load, *w*, from these first three items becomes:

$$w_{\substack{\text{forms, valleys,} \\ \text{\& deflection}}} = (20.71\ \text{lb/ft}^2)(BW_G) \tag{8.8}$$

The concrete above the top flange of the steel girders covering the shear studs is determined by multiplying the flange width, t_f, by the height of the shear studs, both in feet, and multiplied by the specific weight of concrete. The average height of typical shear studs is about 5 in, or about 0.42 ft. Therefore, the contribution to the distributed load, *w*, from this item is (remember the flange width must be in feet):

$$w_{\substack{\text{cover shear} \\ \text{studs}}} = (t_f)(0.42\ \text{ft})(150\ \text{lb/ft}^3) \tag{8.9}$$

Lastly, the weight per foot of the concrete roadway is determined by multiplying the roadway thickness (RT) in feet times the portion of the roadway carried by a single girder, BW_G, times the specific weight of concrete. This gives a contribution to the distributed load, *w*, as:

$$w_{\text{roadway}} = (RT)(BW_G)(150\ \text{lb/ft}^3) \tag{8.10}$$

Steel Girder

Steel girders are typically wide-flange designs, designated as W-beams. Typically, the largest rolled beams available have a height of 36 in. If a larger beam is required due to the longer spans needed in today's highway systems, appropriate size girders must be assembled from steel plate.

For discussion, consider a rolled beam designated as a W36 × 300. The "W" stands for "wide-flange," in contrast to a standard I-beam. The "36" is the nominal height in inches (the actual beam may be slightly taller). The "300" is for 300 lb/ft. Therefore, if this was the chosen beam, then the contribution to the distributed load *w* would be:

$$w_{\text{steel girder}} = 300\ \text{lb/ft}\ (\text{W36} \times 300) \tag{8.11}$$

Typically, at the start of the design process, the structural engineer, from experience, selects a beam to be used. This initial selection then goes through several tests of its applicability at various stages in the design process, the most important of which is its section modulus compared to the section modulus required. If the section modulus of the selected beam is greater than the required section modulus, then the design is satisfactory. However, if not, then a larger beam will be required, usually having a larger weight per foot. The calculation for the distributed load *w* must then be redone and the process repeated until a satisfactory beam is selected.

Future Replacement Surface

At some point in the life of the bridge, the original concrete surface will require repair, consisting typically of a bituminous/asphalt surface added on top of the roadway. If this surface eventually requires repair, it is removed and a new surface is added. Therefore, only one layer needs to be considered. It is assumed that this replacement surface weighs 20 lb/ft^2.

Realizing that this replacement surface will not extend past the safety barriers on each side, these widths must be subtracted from the bridge width (BW). It will be assumed that the safety barriers are the popular Jersey barriers, which have a width of 18.5 in (1.54 ft). Therefore, the replacement surface contribution to the distributed load *w* becomes:

$$w_{\substack{\text{replacement}\\\text{surface}}} = \frac{[BW - 2(1.54\ \text{ft})](20\ \text{lb/ft}^2)}{G} \tag{8.12}$$

Safety Barriers

As mentioned, it will be assumed that the safety barriers are the popular Jersey barriers. Their standard weight per foot is 371 lb/ft. Because there are barriers on each side of the roadway, their contribution to the distributed load w is determined from the equation:

$$w_{\substack{\text{Jersey}\\\text{Barriers}}} = \frac{(2\text{ barriers})(371\text{ lb/ft})}{G} \tag{8.13}$$

Summary

The total distributed load w is, therefore, the sum of major elements considered previously, and given by Eqs. (8.8) through (8.13):

$$w_{\substack{\text{forms, valleys,}\\\text{\& deflection}}} = (20.71\text{ lb/ft}^2)(BW_G) \tag{8.8}$$

$$w_{\substack{\text{cover shear}\\\text{studs}}} = (t_f)(0.42\text{ ft})(150\text{ lb/ft}^3) \tag{8.9}$$

$$w_{\text{roadway}} = (RT)(BW_G)(150\text{ lb/ft}^3) \tag{8.10}$$

$$w_{\text{steel girder}} = 300\text{ lb/ft }(\text{W36} \times 300) \tag{8.11}$$

$$w_{\substack{\text{replacement}\\\text{surface}}} = \frac{[BW - 2(1.54\text{ ft})](20\text{ lb/ft}^2)}{G} \tag{8.12}$$

$$w_{\substack{\text{Jersey}\\\text{Barriers}}} = \frac{(2\text{ barriers})(371\text{ lb/ft})}{G} \tag{8.13}$$

Therefore, the distributed load w becomes:

$$\begin{aligned} w = w_{\substack{\text{forms, valleys,}\\\text{\& deflection}}} + w_{\substack{\text{cover shear}\\\text{studs}}} + w_{\text{roadway}} \\ + w_{\text{steel girder}} + w_{\substack{\text{replacement}\\\text{surface}}} + w_{\substack{\text{Jersey}\\\text{Barriers}}} \end{aligned} \tag{8.14}$$

EXAMPLE 8.1 Distributed Load Calculation

Determine the distributed load w for the bridge superstructure shown, where the bridge width BW is 45 ft. Assume the initial beam to be considered is a W36 × 300, which has a flange width of 16 5/8 in, or 1.39 ft. The roadway thickness is 16 in, or 1.33 ft.

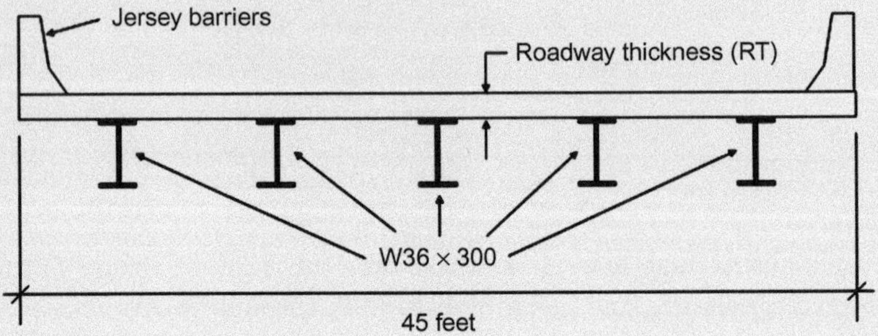

Solution

From Eq. (8.14), the distributed load w is the sum of the six values shown. However, the quantity that appears in two of the calculations is the portion of the bridge width that must be supported by an interior girder, given by Eq. (8.7) as:

$$BW_G = \frac{\text{bridge width}}{\text{number of girders}} = \frac{BW}{G} = \frac{45\ \text{ft}}{5} = 9\ \text{ft}$$

The contribution associated with the deck forms, concrete in the valleys of the forms, and the additional deflection due to the weight of the concrete in the roadway slab is determined from Eq. (8.8) as:

$$w_{\substack{\text{forms, valleys,}\\ \text{\& deflection}}} = (20.71\ \text{lb/ft}^2)(BW_G) = (20.71\ \text{lb/ft}^2)(9\ \text{ft}) = 186.39\ \text{lb/ft}$$

The contribution from the concrete covering the shear studs is found from Eq. (8.9) as:

$$w_{\substack{\text{cover shear}\\ \text{studs}}} = (t_f)(0.42\ \text{ft})(150\ \text{lb/ft}^3) = (1.39\ \text{ft})(0.42\ \text{ft})(150\ \text{lb/ft}^3) = 87.57\ \text{lb/ft}$$

The contribution from the roadway is determined from Eq. (8.10) as:

$$w_{\text{roadway}} = (RT)(BW_G)(150\ \text{lb/ft}^3) = (1.33\ \text{ft})(9\ \text{ft})(150\ \text{lb/ft}^3) = 1795.5\ \text{lb/ft}$$

(Continued)

EXAMPLE 8.1 Distributed Load Calculation—(Continued)

The contribution from the steel girder is simply that already provided in Eq. (8.11) for the W36 × 300 beam that has been chosen:

$$w_{\text{steel girder}} = 300\ \text{lb/ft}$$

The contribution from the replacement surface is found from Eq. (8.12) as:

$$\begin{aligned} w_{\substack{\text{replacement}\\\text{surface}}} &= \frac{[BW - 2(1.54\,\text{ft})](20\,\text{lb/ft}^2)}{G} = \frac{[45\,\text{ft} - 2(1.54\,\text{ft})](20\,\text{lb/ft}^2)}{5} \\ &= 167.68\,\text{lb/ft} \end{aligned}$$

Finally, the contribution from the Jersey barriers is determined from Eq. (8.13) as:

$$\begin{aligned} w_{\substack{\text{Jersey}\\\text{Barriers}}} &= \frac{(2\ \text{barriers})(371\ \text{lb/ft})}{G} = \frac{(2\ \text{barriers})(371\ \text{lb/ft})}{5} \\ &= 148.4\ \text{lb/ft} \end{aligned}$$

Using Eq. (8.14), the total distributed load w becomes:

$$\begin{aligned} w &= w_{\substack{\text{forms,}\\\text{valleys,}\\\text{\& deflection}}} + w_{\substack{\text{cover}\\\text{shear}\\\text{studs}}} + w_{\text{roadway}} + w_{\text{steel girder}} + w_{\substack{\text{replacement}\\\text{surface}}} + w_{\substack{\text{Jersey}\\\text{Barriers}}} \\ &= (186.39 + 87.57 + 1795.5 + 300 + 167.68 + 148.4)\,\text{lb/ft} \\ &= 2685.54\,\text{lb/ft} \\ &\approx 2.7\,\text{kip/ft} \end{aligned}$$

This distributed load w can now be used to calculate the reactions on the hammerhead pier and end bents, as well as the maximum bending moment in the interior girder due to the dead load.

8.3.5 Load Factor Design Equation

As stated earlier, the load factor design approach is used to determine the required resistance to both the dead load (DL) and the live load (LL). While many factors can be taken into account, only a few will be considered here. Based on AASHTO guidelines, the following load factor design equation is used:

$$M_{\mathrm{LFD}} = 1.3[DL + 1.67(DF_{\mathrm{bending}})(1 + I)(LL)] \tag{8.15}$$

The "1.3" and the "1.67" are parts of the load factor design approach. The dead load (DL) is determined from Eq. (8.2) using the distributed load w just discussed as:

$$DL = \frac{wL^2}{8} \tag{8.2}$$

The live load (LL) is determined from either Eq. (8.5) or (8.6), depending on which reaction is chosen.

$$M_{\max} = A_V(L/2) - 56\ \mathrm{ft \cdot kip} \tag{8.5}$$

$$M_{\max} = B_V(L/2) - 224\ \mathrm{ft \cdot kip} \tag{8.6}$$

The reactions B_V or A_V are found from Eqs. (8.3) and (8.4), respectively.

$$B_V = \frac{(18\ \mathrm{kip})\ L + 168\ \mathrm{ft \cdot kip}}{L} \tag{8.3}$$

$$A_V = 36\ \mathrm{kip} - B_V \tag{8.4}$$

All that is left are the quantities DF_{bending}, the distribution factor for bending, and the impact factor, I.

Distribution Factor for Bending

The distribution factor takes into account the possibility that two trucks could be side by side near the same interior girder. Figure 8.20 shows just such a situation. The two trucks could be going in the same direction or in opposite directions.

The distance S is the girder spacing. The distribution factor for bending, DF_{bending}, becomes:

$$DF_{\mathrm{bending}} = \frac{S}{5.5} \tag{8.16}$$

Note that in the calculations for the distributed load, the amount of the bridge being supported by a single interior girder was defined by

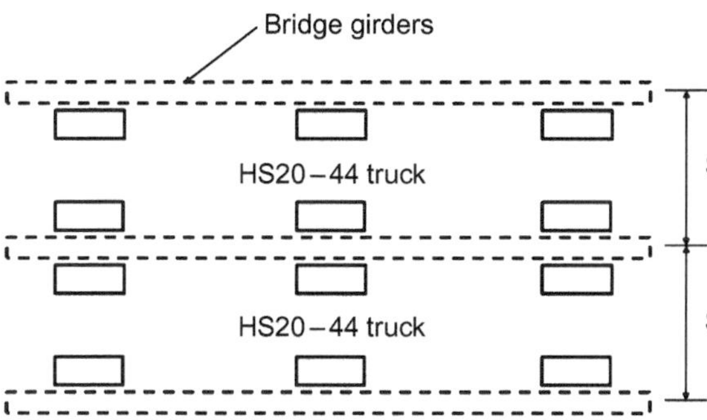

Figure 8.20 Trucks side-by-side–distribution factor.

Eq. (8.7) and labeled BW_G. For some bridge designs, this distance is not the same as the girder spacing.

Impact Factor

To account for the vibration and momentum that vehicles produce on the superstructure, an impact factor I is used to increase the effect of the live load, LL. It is given by the following equation, where L is the bridge span in feet.

$$I = \frac{50}{L + 125} \tag{8.17}$$

Note that the maximum value for the impact factor is 0.3.

8.3.6 Load Factor Design Calculations

Now that all the various components of the load factor design equation have been addressed, the bending moment in the interior girder can be determined. From that value and the allowable yield stress of the steel girder, a required section modulus can be calculated. The value is compared to the value available from the selected girder, and if the required section modulus is less than the available section modulus, the design is considered satisfactory.

To put these statements into useable equations, recall the formula for the bending stress σ in a beam under pure bending:

$$\sigma = \frac{My}{I}$$

where M is the bending moment, y is the distance from the neutral axis to the point of interest, and I is the area moment of inertia of the cross section of the beam about the neutral axis. The maximum bending stress σ_{max} occurs at the maximum value of y, so that this equation becomes:

$$\sigma_{max} = \frac{My_{max}}{I}$$

If the cross section of the beam is symmetrical, the area moment of inertia and the maximum value of y can be combined into a quantity called the section modulus. Normally, the section modulus is given the symbol S, however S is already being used to specify the girder spacing. Therefore, the symbol Z will be used, so that:

$$Z = \frac{I}{y_{max}}$$

The equation for the maximum bending stress σ_{max} therefore becomes:

$$\sigma_{max} = \frac{My_{max}}{I} = \frac{M}{I/y_{max}} = \frac{M}{Z}$$

Solving for the section modulus Z gives:

$$Z = \frac{M}{\sigma_{max}}$$

If the load factor design moment M_{LFD} is substituted for the bending moment M, and the allowable yield stress of the steel girders f_y is substituted for the maximum bending stress σ_{max}, then the section modulus required, labeled Z_{Need}, becomes:

$$Z_{Need} = \frac{M_{LFD}}{f_y} \tag{8.18}$$

Therefore, if the available section modulus Z_{Have} is greater than the required section modulus, determined from Eq. (8.18), then the design is satisfactory. Simply stated, a satisfactory design is where:

$$Z_{Have} > Z_{Need} \tag{8.19}$$

The best way to illustrate the many steps so far discussed in the load factor design process is to provide an example.

EXAMPLE 8.2 Load Factor Design Calculations

Determine if the bridge design of Example 8.1 is satisfactory for a span of 72 ft. The girder spacing of 9 ft is shown in the following diagram.

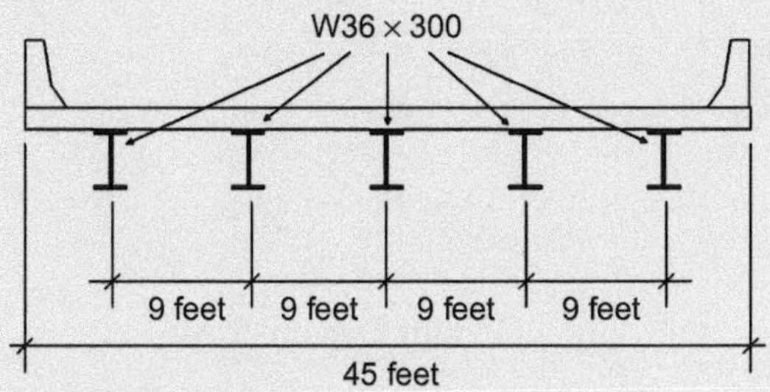

In Example 8.1, the distributed load was found to be 2.7 kip/ft. The steel girder was specified to be a W36 × 300, which has a section modulus of 1260 in^3. Assume the allowable yield stress of the steel girder is 50,000 psi, or 50 ksi.

Solution

The design will be satisfactory if Eq. (8.19) is valid.

$$Z_{\text{Have}} > Z_{\text{Need}} \tag{8.19}$$

Z_{Have} is given as 1260 in^3. Z_{Need} is determined from Eq. (8.18), where the allowable yield stress of the steel girder is given as 50 ksi.

$$Z_{\text{Need}} = \frac{M_{\text{LFD}}}{f_Y} \tag{8.18}$$

The load factor design moment M_{LFD} is found from Eq. (8.15).

$$M_{\text{LFD}} = 1.3[DL + 1.67(DF_{\text{bending}})(1 + I)(LL)] \tag{8.15}$$

The dead load DL is found from Eq. (8.2), where the distributed load w was found from Example 8.1 to be 2.7 kip/ft.

$$DL = \frac{wL^2}{8} \tag{8.2}$$

Substituting for the distributed load w of 2.7 kip/ft, and the span L of 72 ft, gives a dead load DL equal to:

$$DL = \frac{wL^2}{8} = \frac{(2.7\ \text{kip/ft})(72\ \text{ft})^2}{8} = 1749.6\ \text{ft}\cdot\text{kip}$$

Using a girder spacing S of 9 ft, the distribution factor for bending, DF_{bending}, can be found from Eq. (8.16) as

$$DF_{\text{bending}} = \frac{S}{5.5} = \frac{9}{5.5} = 1.64$$

(*Continued*)

EXAMPLE 8.2 Load Factor Design Calculations—(Continued)

The impact factor I can be found using the span L of 72 ft from Eq. (8.17) as:

$$I = \frac{50}{L + 125} = \frac{50}{72 + 125} = 0.254$$

The live load LL can be found from Eq. (8.6), where the reaction B_V is found from Eq. (8.3).

$$M_{max} = B_V(L/2) - 224\ \text{ft}\cdot\text{kip} \tag{8.6}$$

$$B_V = \frac{(18\ \text{kip})\ L + 168\ \text{ft}\cdot\text{kip}}{L} \tag{8.3}$$

Substituting for the span L of 72 ft in Eq. (8.3) gives:

$$B_V = \frac{(18\ \text{kip})L + 168\ \text{ft}\cdot\text{kip}}{L} = \frac{(18\ \text{kip})(72\ \text{ft}) + 168\ \text{ft}\cdot\text{kip}}{72\ \text{ft}} = 20.33\ \text{kip}$$

Substituting this value of B_V into Eq. (8.6) gives the live load LL as:

$$LL = M_{max} = B_V(L/2) - 224\ \text{ft}\cdot\text{kip} = (20.33\ \text{kip})(72\ \text{ft}/2) - 224\ \text{ft}\cdot\text{kip} = 508\ \text{ft}\cdot\text{kip}$$

Substituting the various quantities found into Eq. (8.15) gives the load factor design moment M_{LFD} as:

$$\begin{aligned} M_{LFD} &= 1.3[DL + 1.67(DF_{bending})(1 + I)(LL)] \\ &= 1.3[(1749.6\ \text{ft}\cdot\text{kip}) + 1.67(1.64)(1 + 0.254)(508\ \text{ft}\cdot\text{kip})] \\ &= 1.3[(1749.6\ \text{ft}\cdot\text{kip}) + (1744.7\ \text{ft}\cdot\text{kip})] \\ &= 1.3[3494.3\ \text{ft}\cdot\text{kip}] \\ &= 4542.6\ \text{ft}\cdot\text{kip} \end{aligned}$$

Substitute this value for the load factor design moment M_{LFD} into Eq. (8.18) to determine the required section modulus, Z_{Need}. Note the conversion of feet to inches, to obtain the correct units on the section modulus.

$$Z_{Need} = \frac{M_{LFD}}{f_y} = \frac{(4542.6\ \text{ft}\cdot\text{kip})(12\ \text{in/ft})}{50\ \text{ksi}} = 1090\ \text{in}^3$$

Because the required section modulus of 1090 in^3 is less than the available section modulus of 1260 in^3, the design is satisfactory. If the design had not been satisfactory, several options are available to the structural engineer. One option is to increase the number of girders. The other option is to increase the size of each of the girders; that would involve having the girders custom manufactured. As longer and longer bridges are needed, the use of custom girders has actually become commonplace.

8.3.7 Designing against Flange or Web Failure

Once a satisfactory steel girder size has been determined to meet the required section modulus, the flanges and web of the girder must meet certain criteria to avoid failure. For the discussion that follows, refer to the diagram in Figure 8.21, which shows a typical girder cross section with its associated dimensions. Note that the depth d is not the same as the designation, such as W36 is a nominal 36-in-high beam, but its actual distance, d, is 36.74 in.

For the flanges, the ratio of the overhang distance, b_o, to the flange thickness, t_f, must be less than the right side of following equation:

$$\frac{b_o}{t_f} \le \frac{2055}{\sqrt{f_y}} \tag{8.20}$$

Meeting the criteria of Eq. (8.20) will guard against the flanges failing in shear.

For the web, the ratio of the distance T between the flanges and the web thickness, t_w, must be less than the right side of the following equation:

$$\frac{T}{t_w} \le \frac{19,230}{\sqrt{f_y}} \tag{8.21}$$

Meeting the criteria of Eq. (8.21) will guard against the web buckling between the flanges.

For standard wide flange beams, their dimensions automatically satisfy these two criteria. However, for custom manufactured beams, careful attention to these two criteria must be addressed by the bridge designer.

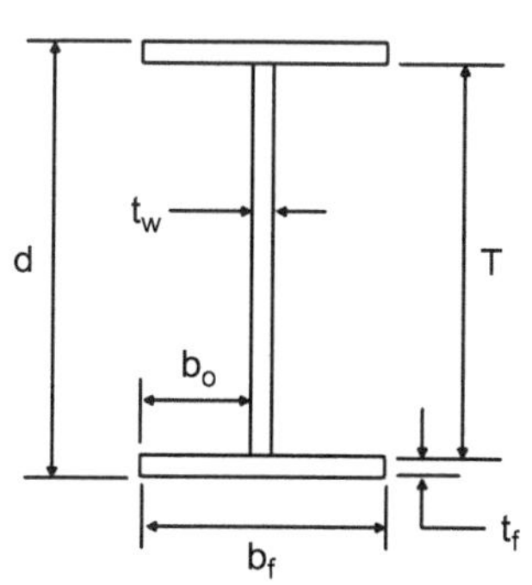

Figure 8.21 Steel girder cross section.

8.4 HAMMERHEAD DESIGN

8.4.1 Introduction

The top-down design process continues with what the bridge girders rest on. For single-span bridges, the ends of the bridge girders rest on end bents. However, for multispan bridges, and in particular for two spans, one end of each span rests on an end bent and the other end of both spans rest on an *interior bent.* These interior bents are usually of two types: multiple column designs and hammerhead designs. The multiple column designs involve statically indeterminate analysis; therefore the simpler hammerhead design will be presented here.

Figure 8.22 shows a simple diagram of a two-span bridge arrangement, where the two spans are of unequal length.

Like the loading on the bridge girders, the loading on the hammerhead pier is from both the dead weight of the bridge spans and from the live loading from vehicular traffic. This loading results in concentrated loads at each girder of both spans on the hammerhead pier.

8.4.2 Load Factor Design Equation

As with the design of the bridge girders, the load factor design approach will be used to design the hammerhead pier. Again, based on AASHTO guidelines, the following load factor design equation will be used, where R_{LFD} is the factored reaction at each girder:

$$R_{LFD} = 1.3[DL + 1.67(DF_{shear})(1 + I)(LL)] \qquad (8.22)$$

Again, the "1.3" and the "1.67" are part of the load factor design approach. The distribution factor DF_{shear}, like the one for bending, is determined from the girder spacing, *S*, in feet; however it is found using the table of values provided in Table 8.1.

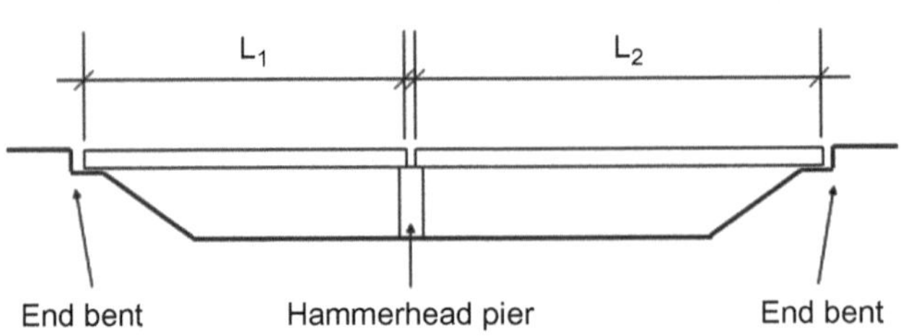

Figure 8.22 Diagram of a two-span bridge arrangement.

Table 8.1 Distribution factors for shear loading

S (ft)	10	11	12	13	14	15	16
DF_{shear}	2.00	2.27	2.50	2.69	2.86	3.07	3.25

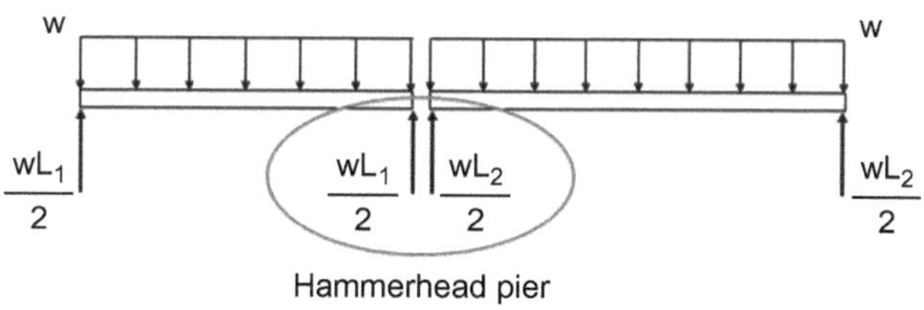

Figure 8.23 Reactions on hammerhead pier.

The reason the subscript on the distribution factor is labeled "shear" is that the shear stress in the bridge girder dominates at its end points, whereas the bending stress dominates at the midpoint of the girder.

The impact factor I is determined for the bridge girder from Eq. (8.17). However, if the bridge spans are different, the shortest span should be used to obtain the largest value; the designer must remember the maximum value for the impact factor is limited to 0.3.

$$I = \frac{50}{L + 125} \tag{8.17}$$

Dead Load

The hammerhead pier supports half the distributed load from one span plus half the distributed load from the other span, as shown in Figure 8.23.

It will be assumed that both spans have the same distributed loading per foot, w, so that the total loading on the hammerhead pier, which is the dead load DL, is given by the following equation:

$$DL = \frac{wL_1}{2} + \frac{wL_2}{2} \tag{8.23}$$

where L_1 and L_2 are the lengths of the two spans.

Live Load

As was the case for the design of the bridge girders, the standard HS20-44 truck will be used to determine the live load, LL, acting on the

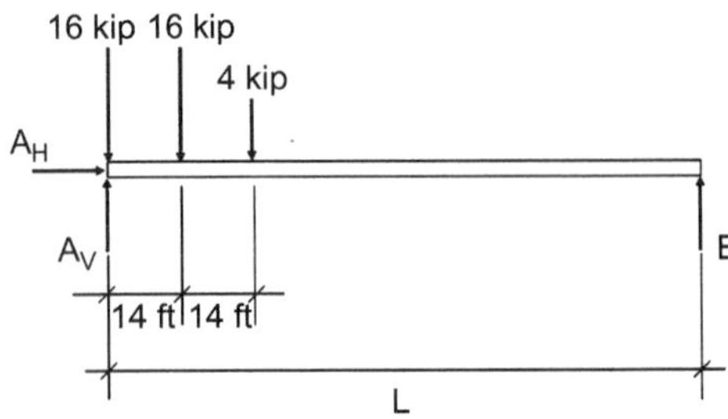

Figure 8.24 FBD of HS20-44 loading on hammerhead pier.

hammerhead pier. However, for this loading, the rear axle of the truck is placed over the hammerhead pier support, as shown in the free-body diagram (FBD) of Figure 8.24.

Notice that the HS20-44 truck is located on only one of the two bridge spans. The greater reaction will be on the longer of the two spans.

Applying the equations of equilibrium, the horizontal force, A_H, is again zero; however, the HS20-44 truck is moving and so some support along the span is required. However, this additional support is accommodated in the impact factor I.

As for the reactions at the ends of the bridge span, they are dependent on the length L of the bridge span. Because the reaction at point A is the reaction that must be supported by the hammerhead pier, applying the moment equation of equilibrium at point B, assuming clockwise is positive, gives the following equation:

$$A_V(L) - (4\text{ kip})(L - 28\text{ ft}) - (16\text{ kip})(L - 14\text{ ft}) - (16\text{ kip})(L) = 0$$

Expanding and collecting terms, then rearranging to solve for the reaction at A, A_V, gives:

$$A_V = \frac{(36\text{ kip})L - 336\text{ ft}\cdot\text{kip}}{L} \tag{8.24}$$

where L, the length of the span, is in feet.

Once the reaction A_V has been determined, the reaction at B, B_V, though not needed, becomes:

$$B_V = 36\text{ kip} - A_V \tag{8.25}$$

The following example shows how all this comes together to provide the loading on the hammerhead pier.

8.4.3 Load Factor Design Calculations

At this point an example will pull together all the elements of this process in calculating the load factor design reaction.

EXAMPLE 8.3 Load Factor Design Calculations for Hammerhead Pier

Determine the load factor design reaction, R_{LFD}, for the bridge design shown, where one span of a two-span bridge arrangement is 60 ft and the other is 70 ft. Assume the distributed loading per foot w is 4 kip/ft for both spans.

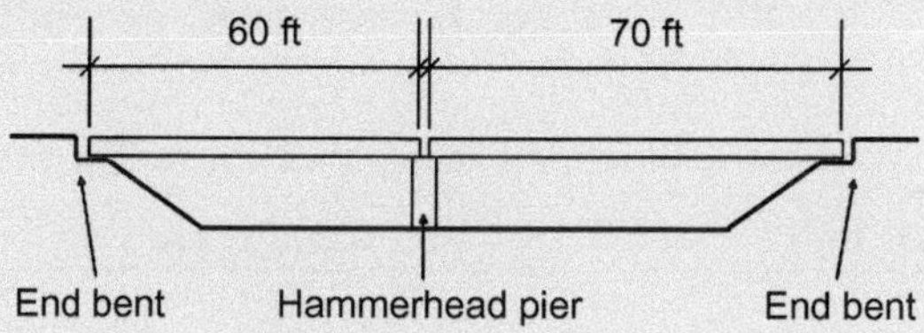

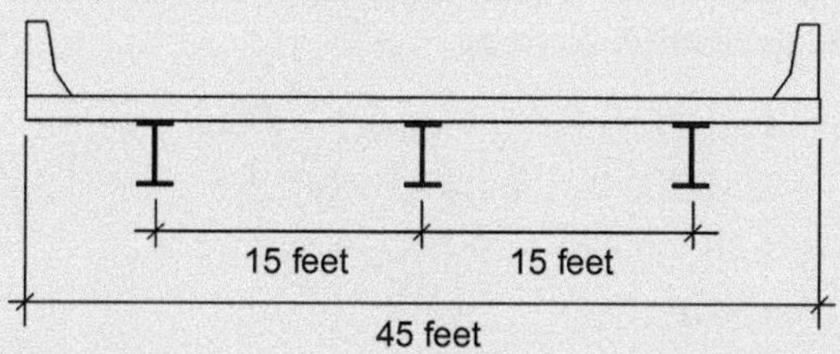

Note that the bridge has only three girders, while an actual design would have at least five or more. However, as will be seen, restricting the example to only three girders simplifies the analysis that follows for the hammerhead pier.

Solution

The load factor design reaction R_{LFD} is found from Eq. (8.22).

$$R_{LFD} = 1.3[DL + 1.67(DF_{shear})(1 + I)(LL)] \quad (8.22)$$

The dead load DL is found from Eq. (8.23) as:

$$DL = \frac{wL_1}{2} + \frac{wL_2}{2} \quad (8.23)$$

where L_1 and L_2 are the lengths of the two spans.

Substituting for the distributed load w of 4 kip/ft, and the lengths of the two spans L_1 and L_2 of 60 ft and 70 ft, respectively, gives a dead load DL equal to:

$$DL = \frac{wL_1}{2} + \frac{wL_2}{2} = \frac{(4 \text{ kip/ft})(60 \text{ ft})}{2} + \frac{(4 \text{ kip/ft})(70 \text{ ft})}{2} = 260 \text{ kip}$$

(Continued)

EXAMPLE 8.3 Load Factor Design Calculations for Hammerhead Pier—(Continued)

Using a girder spacing S of 15 ft, the distribution factor for shear, DF_{shear}, from Table 8.1 is 3.07.

The impact factor I can be found using the shorter of the two spans of 60 ft from Eq. (8.17) as:

$$I = \frac{50}{L + 125} = \frac{50}{60 + 125} = 0.270$$

The live load, LL, which is the reaction A_V, can be found from Eq. (8.24) using the longer of the two spans as:

$$A_V = \frac{(36\ \text{kip})\ L - 336\ \text{ft} \cdot \text{kip}}{L} \tag{8.24}$$

Substituting for the span of 70 ft in Eq. (8.24) gives the live load LL as:

$$LL = A_V = \frac{(36\ \text{kip})\ L - 336\ \text{ft} \cdot \text{kip}}{L} = \frac{(36\ \text{kip})(70\ \text{ft}) - 336\ \text{ft} \cdot \text{kip}}{70\ \text{ft}} = 31.2\ \text{kip}$$

Substituting the various quantities found into Eq. (8.22) gives the load factor design reaction R_{LFD} as:

$$\begin{aligned} R_{LFD} &= 1.3\ [DL + 1.67(DF_{shear})(1 + I)(LL)] \\ &= 1.3\ [(260\ \text{kip}) + 1.67(3.07)(1 + 0.270)(31.2\ \text{kip})] \\ &= 1.3[(260\ \text{kip}) + (203\ \text{kip})] \\ &= 1.3[463\ \text{kip}] \\ &= 602\ \text{kip} \end{aligned}$$

8.4.4 Components of the Hammerhead Pier

Recall that a bridge system is composed of two main parts: (1) the superstructure and (2) the substructure. As was shown in Figure 8.1, the superstructure is composed of the girders, typically steel, upward to the roadway decking and the safety barriers. The substructure is composed of either a hammerhead pier or a multiple column support, and a pile or spread footing. The end bents would also be included in the components of the substructure, as shown in Figure 8.22.

The hammerhead pier is composed of two major components: (1) the cap, and (2) the column, as shown in Figure 8.25. The design of the column will be presented in the next section, followed by the presentation of the design of a pile footing in the last section.

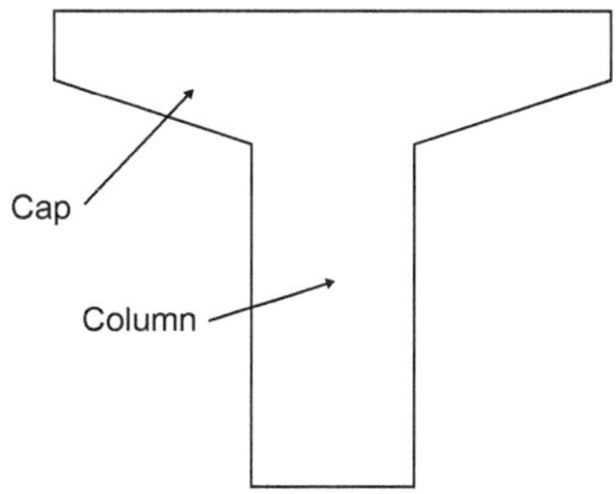

Figure 8.25 Components of a hammerhead pier.

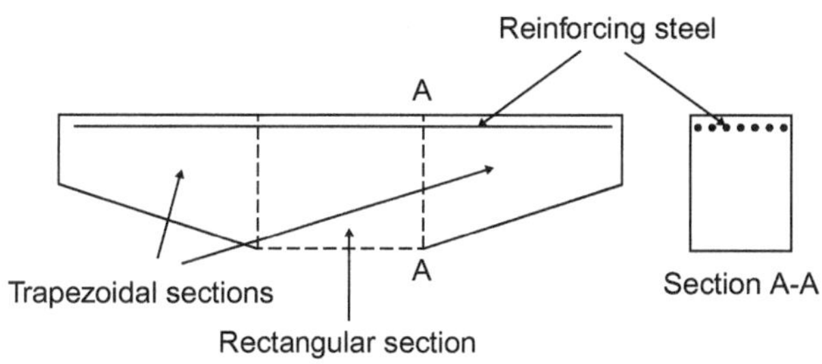

Figure 8.26 Components of the cap for a hammerhead pier.

The cap is composed of two trapezoidal sections connected by a rectangular section, forming what might be called the "head" of the hammerhead pier. The column is a rectangular section, which is connected to the pile footing below. However, the design of the cap involves the loading imparted from the bridge girders and the weight of the trapezoidal sections themselves. Because concrete cannot support tensile loads, reinforcing steel is needed along the top of the cap, as shown in Figure 8.26.

The design of the cap, therefore, involves calculating the bending moment caused by the bridge girders resting on the trapezoidal sections, along with the dead weight of the trapezoidal sections themselves. Thus the design of the hammerhead piers reduces to the resistance of this bending moment produced by the compressive force of the concrete at the bottom of A-A on Figure 8.26, and the tensile force produced by the reinforcing steel at the top. The determination of these quantities is presented in the next section.

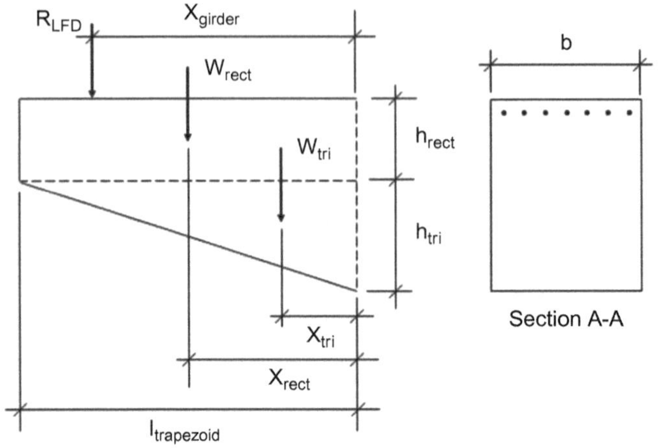

Figure 8.27 Loads on the trapezoidal sections of a hammerhead pier.

8.4.5 Loads on the Hammerhead Pier

The focus of the design of a hammerhead pier is on the loads resting on the trapezoidal sections of the cap. Earlier, we limited the calculation of the load factor design reaction to a bridge with only three girders. It was stated that this would make the presentation of the design of the hammerhead pier less complicated than if there were more than three girders. The three basic loadings are shown in Figure 8.27.

The load factor design reaction R_{LFD} is found from Eq. (8.22), and would be the same value for all girders being supported by the trapezoidal section.

$$R_{\mathrm{LFD}} = 1.3[DL + 1.67(DF_{\mathrm{shear}})(1 + I)(LL)] \qquad (8.22)$$

The weights of the rectangular and triangular sections are given by the following equations, where the width of the pier is b, and the specific weight of concrete is taken as 150 lb/ft^3.

$$W_{\mathrm{rect}} = (h_{\mathrm{rect}} \times l_{\mathrm{trapezoid}} \times b)(150\ \mathrm{lb/ft^3}) \qquad (8.26)$$

$$W_{\mathrm{tri}} = 1/2(h_{\mathrm{tri}} \times l_{\mathrm{trapezoid}} \times b)(150\ \mathrm{lb/ft^3}) \qquad (8.27)$$

Thus the bending moment about A-A (Figure 8.26) caused by these three loads is given by the following equation:

$$M_{\mathrm{Loads}} = (R_{\mathrm{LFD}} \times x_{\mathrm{girder}}) + (W_{\mathrm{rect}} \times x_{\mathrm{rect}}) + (W_{\mathrm{tri}} \times x_{\mathrm{tri}}) \qquad (8.28)$$

where

$$x_{\text{rect}} = 1/2\,(l_{\text{trapezoid}}) \tag{8.29}$$

$$x_{\text{tri}} = 1/3\,(l_{\text{trapezoid}}) \tag{8.30}$$

8.4.6 Capacity Reduction Factor

To account for inaccuracies in the layout and construction of the hammerhead pier and its reinforcing steel, a *capacity reduction factor*, ϕ, is used to increase the magnitude of the bending moment due to the loads. This factor is typically taken as 0.9 for concrete in bending, so the moment M_{Loads} calculated in Eq. (8.28) must be increased by dividing by this factor. This bending moment will be labeled M_{n} where the "n" means "nominal," and is calculated as follows:

$$M_n = \frac{M_{\text{Loads}}}{0.9} \tag{8.31}$$

8.4.7 Cracking Moment

To account for the possibility of rupture of the concrete, the design moment must be at least 1.2 times the *cracking moment*, M_{cr}, and given by Eq. (8.32) as:

$$M_{\text{cr}} = \frac{f_r\, I}{y_t} \tag{8.32}$$

where f_{r} is the modulus of rupture for concrete, I is the area moment of inertia of A-A, and y_{t} is the distance from the neutral axis to the extreme fiber in tension. These three quantities are determined by the following equations:

$$f_r = 7.5\sqrt{f_c} \tag{8.33}$$

$$I = \frac{bh^3}{12} \tag{8.34}$$

$$y_t = \frac{h}{2} \tag{8.35}$$

where b is the width of the pier and h is the height of A-A.

Therefore, it is possible that 1.2 times M_{cr} is larger than the nominal moment M_n determined from Eq. (8.31). So the design bending moment could be given by:

$$M_n = 1.2\ M_{cr} \tag{8.36}$$

As stated earlier, this moment, whether the nominal moment M_n found from Eq. (8.31) or from Eq. (8.36), is opposed by the compressive force produced by the concrete at the bottom of A-A, and the tensile force produced by the reinforcing steel at the top of A-A. The compressive force is the resultant of a complex stress distribution; however it is simplified by the concept of a Whitney stress block, discussed next.

8.4.8 Whitney Stress Block

Opposing the nominal moment M_n, which is the greater of the two determined from Eqs. (8.31) and (8.36), are the two forces, T and C, shown in Figure 8.28, that form a couple as they must be equal to satisfy equilibrium horizontally. The force T is the tensile force produced by the reinforcing steel, and the force C is the compressive force produced by the concrete. This force C is determined by calculating the rectangular volume of what is called the Whitney stress block, which simplifies the actual stress distribution over the lower part of A-A. One side of this 3D block is a reduced value of the compressive strength f_c of concrete, one side is a length a, which is yet to be determined, and the third side is the width b of the pier.

The distance between the forces T and C is the distance d minus half the length a of the Whitney stress block. The distance d is the height h of the trapezoidal section minus the required cover depending on the

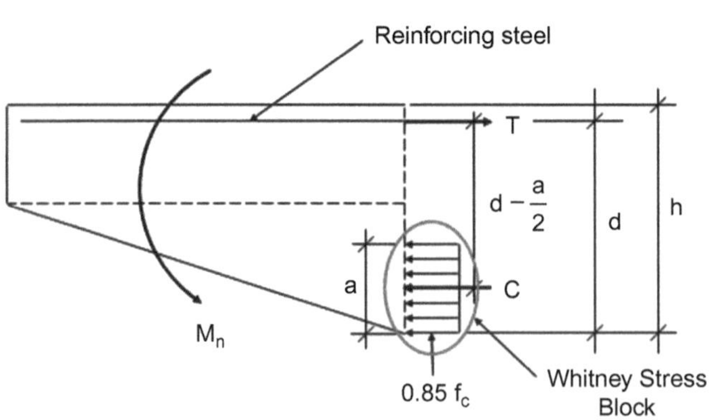

Figure 8.28 Whitney stress block.

environment in which the hammerhead pier will be constructed. For normal environments, the cover is usually 3 in, and for corrosive environments 4 in is usually specified.

Equating the tensile force T to the compressive force C will provide an expression to determine the required cross-sectional area of the steel, A_s. The tensile force T is determined from the following equation:

$$T = f_y A_s \tag{8.37}$$

The reinforcing steel is usually grade 60, meaning its yield strength f_y is taken as 60 ksi.

The compressive force C is determined from the following equation:

$$C = (0.85 f_c)(a)(b) \tag{8.38}$$

Equating T and C gives:

$$T = C$$
$$f_y A_s = (0.85 f_c)(a)(b)$$

Solve for the required area of the reinforcing steel A_s to give:

$$A_s = \frac{(0.85 f_c)(a)(b)}{f_y} \tag{8.39}$$

All that is needed is the length a of the Whitney stress block. This is determined by equating the couple produced by the forces T and C to the nominal moment M_n determined from the larger of the values determined from Eqs. (8.31) or (8.36).

Evaluating the couple of the forces T and C about a point on the line of action of the tensile force T gives:

$$M_n = (C)\left(d - \frac{a}{2}\right) = \left[(0.85 f_c)(a)(b)\right]\left(d - \frac{a}{2}\right)$$

Expanding the right side of this equation, collecting terms in a^2 and a, and rearranging to put in standard quadratic form gives:

$$a^2 - 2da + \frac{2M_n}{(0.85 f_c)(b)} = 0 \tag{8.40}$$

Solving this quadratic equation will give two roots for the length a, however, one of them will be extremely large, so it is rejected. Using the other root in Eq. (8.39) will give the required area of reinforcing steel. From this, there are many options for the size of each reinforcing bar and the corresponding number needed.

Table 8.2 Table of information on standard reinforcing steel

Bar number	Diameter (in)	Area (in^2)
4	0.500	0.20
6	0.750	0.44
8	1.000	0.79
9	1.128	1.00
11	1.411	1.56
14	1.693	2.25

Table 8.2 provides a table of sizes and areas for standard reinforcing steel. The numbered designation represents increments of an eighth of an inch in diameter; however the actual dimensions are slightly different.

Notice that a #9 bar should be 1.125 in in diameter but is in fact 1.128 in. Also, the #9 bar has a cross-sectional area of exactly 1.00 in^2.

8.4.9 Calculations for the Required Reinforcing Steel

At this point an example will pull together all the elements of the process that will ultimately provide the required reinforcing steel for the cap of the hammerhead pier.

EXAMPLE 8.4 Determining the Required Reinforcing Steel for a Hammerhead Pier

Using the load factor design reaction R_{LFD}, found in Example 8.3 to be 602 kip, determine the required area of reinforcing steel for the hammerhead pier shown. The pier will be located in a corrosive environment. Assume the strength of the concrete f_c is 3 ksi and the reinforcing steel is grade 60, which has a yield strength f_y of 60 ksi.

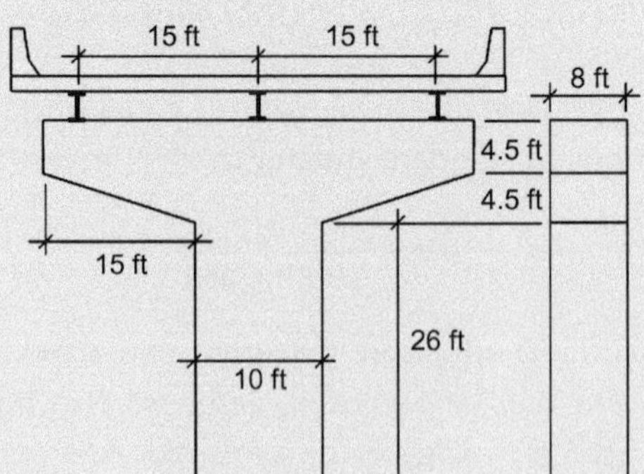

(Continued)

EXAMPLE 8.4 Determining the Required Reinforcing Steel for a Hammerhead Pier—(Continued)

Notice that the heights of the rectangular and triangular sections of the trapezoidal sections are the same value, to simplify the calculations. Also, consider the length of the trapezoidal sections to be the same, 15 ft. The fact that this dimension matches the girder spacing is only coincidental.

Solution

The first step is to determine the moment due to the loading caused by the bridge span and the weight of the trapezoidal section. The load factor design reaction R_{LFD} is 602 kip located 10 ft from the face of A-A of the trapezoidal section. The weights of the rectangular and triangular sections are determined from Eqs. (8.26) and (8.27) as:

$$W_{rect} = (h_{rect} \times l_{trapezoid} \times b)(150\ \text{lb/ft}^3)$$
$$= (4.5\ \text{ft})(15\ \text{ft})(8\ \text{ft})(150\ \text{lb/ft}^3)$$
$$= 81,000\ \text{lb} = 81\ \text{kip}$$
$$W_{tri} = 1/2(h_{tri} \times l_{trapezoid} \times b)(150\ \text{lb/ft}^3)$$
$$= 1/2(4.5\ \text{ft})(15\ \text{ft})(8\ \text{ft})(150\ \text{lb/ft}^3)$$
$$= 40,500\ \text{lb} = 40.5\ \text{kip}$$

The locations of these weights are found from Eqs. (8.29) and (8.30) as:

$$x_{rect} = 1/2\ (l_{trapezoid}) = 1/2(15\ \text{ft}) = 7.5\ \text{ft}$$
$$x_{tri} = 1/3\ (l_{trapezoid}) = 1/3(15\ \text{ft}) = 5\ \text{ft}$$

These three loads and their locations are shown on the following diagram.

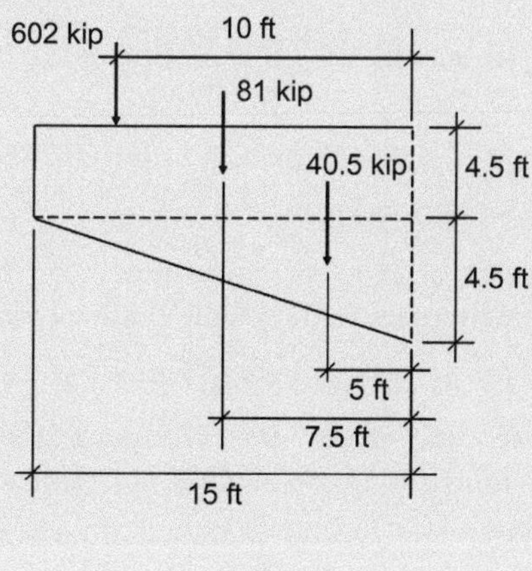

(Continued)

EXAMPLE 8.4 Determining the Required Reinforcing Steel for a Hammerhead Pier—(Continued)

Thus, the bending moment about A-A caused by these three loads is determined by Eq. (8.28) as:

$$\begin{aligned} M_{\text{Loads}} &= (R_{\text{LFD}} \times x_{\text{girder}}) + (W_{\text{rect}} \times x_{\text{rect}}) + (W_{\text{tri}} \times x_{\text{tri}}) \\ &= (602\ \text{kip})(10\ \text{ft}) + (81\ \text{kip})(7.5\ \text{ft}) + (40.5\ \text{kip})(5\ \text{ft}) \\ &= 6830\ \text{ft}\cdot\text{kip} \end{aligned}$$

This moment must be increased by the load capacity factor using Eq. (8.31) to give a nominal moment M_n equal to:

$$M_n = \frac{M_{\text{Loads}}}{0.9} = \frac{6830\ \text{ft}\cdot\text{kip}}{0.9} = 7589\ \text{ft}\cdot\text{kip}$$

However, the cracking moment, M_{cr}, must be calculated from Eq. (8.32) and then multiplied by 1.2 to see if it is greater than the nominal moment. The modulus of rupture is calculated from Eq. (8.33) as:

$$f_r = 7.5\sqrt{f_c} = 7.5\sqrt{3000\ \text{psi}} = 410.8\ \text{psi}$$

The area moment of inertia of A-A is found from Eq. (8.34), where the height h is 9 ft or 108 in, and the width b is 8 ft or 96 in:

$$I = \frac{bh^3}{12} = \frac{(96\ \text{in})(108\ \text{ft})^3}{12} = 10,077,696\ \text{in}^4$$

The distance y_t is found from Eq. (8.35) as:

$$y_t = \frac{h}{2} = \frac{108\ \text{in}}{2} = 54\ \text{in}$$

Substituting these values into Eq. (8.32) gives the cracking moment M_{cr} as:

$$\begin{aligned} M_{cr} &= \frac{f_r I}{y_t} = \frac{(410.8\ \text{psi})(10,077,696\ \text{in}^4)}{54\ \text{in}} \\ &= 76,665,139\ \text{in}\cdot\text{lb} = 6,388,762\ \text{ft}\cdot\text{lb} \\ &= 6389\ \text{ft}\cdot\text{kip} \end{aligned}$$

Multiply this value of the cracking moment by 1.2 as specified in Eq. (8.36) to give an alternate value for the nominal moment M_n as:

$$M_n = 1.2\ M_{cr} = 1.2(6389\ \text{ft}\cdot\text{kip}) = 7667\ \text{ft}\cdot\text{kip}$$

Comparing the two values for M_n gives the value obtained from the cracking moment to be slightly larger. So, the value of the

(Continued)

EXAMPLE 8.4 Determining the Required Reinforcing Steel for a Hammerhead Pier—(Continued)

nominal moment M_n to be used to obtain the length a of the Whitney stress block is:

$$M_n = 7667 \text{ ft}\cdot\text{kip}$$

The last quantity needed for Eq. (8.40) to find the length a is the distance d. It is the height of A-A minus the cover. Because the hammerhead pier will be located in a corrosive environment, the cover should be 4 in. Therefore, the distance d becomes:

$$d = h - (4 \text{ in}) = 108 \text{ in} - 4 \text{ in} = 104 \text{ in}$$

Substituting all the values now known into Eq. (8.40) gives

$$a^2 - 2da + \frac{2M_n}{(0.85 f_c)(\text{b})} = 0$$

$$a^2 - 2(104 \text{ in})(a) + \frac{2(7667 \text{ ft}\cdot\text{kip})(12 \text{ in}/1 \text{ ft})}{(0.85)(3 \text{ ksi})(96 \text{ in})} = 0$$

$$a^2 - 208a + 752 = 0$$

Using the quadratic equation, the two roots of the length a become 203.5 in and 4.5 in. The larger root is rejected, so the length a used to find the required area of reinforcing steel is 4.5 in.

Substituting this value of the length a into Eq. (8.39) gives:

$$A_s = \frac{(0.85 f_c)(a)(b)}{f_y} = \frac{(0.85)(3 \text{ ksi})(4.5 \text{ in})(96 \text{ in})}{(60 \text{ ksi})}$$

$$= 18.36 \text{ in}^2$$

Therefore, from Table 8.2 there are several choices for size and number of rebar. The simplest combination is 19 #9 bars to give exactly 19.00 in^2, or 9 #14 bars to give 20.25 in^2. The choice is governed by layout and construction constraints, and, of course, depends on the experience of the structural engineer.

8.5 COLUMN DESIGN

8.5.1 Introduction

The previous section examined the design of the cap of the hammerhead pier. This section considers the column on which that cap rests. The loads on the column are from the dead load from bridge spans, the live loads from the vehicular traffic, and the weights of the cap and column. While the column is primarily concrete, a certain percentage of the gross area of the column is reinforcing steel. Whatever percentage steel is chosen, it is then verified that the design is satisfactory. If not, a higher percentage of steel will be selected, and the process repeated. In some cases, the entire design of the hammerhead pier might need changing to safely support the loads that will be present.

8.5.2 Load Capacity

The load carrying capacity, P_{max}, of the column is determined from the following equation:

$$P_{max} = \phi\,[(0.85 f_c)(A_g - A_s) + f_y A_s] \tag{8.41}$$

where ϕ is the "capacity reduction factor" for compression. In the previous section, this factor was 0.9 for bending. Here it will be 0.75 for compression. Also, in the previous section, the strength of the concrete, f_c, was taken as 3 ksi, however, here it will be taken as 4 ksi. The reinforcing steel will still be grade 60, with a yield strength f_y of 60 ksi. The gross area of the column is labeled A_g and the area of the reinforcing steel is labeled A_s. AASHTO specifies that a minimum of 1% of the gross area be reinforcing steel, up to a maximum of 8%. Here we will start with 1% and increase it only if the design proves unsatisfactory.

Once the required area of reinforcing steel is determined, an appropriate size and number of rebar will be specified. This reinforcing steel will be placed primarily along the outside perimeter of the column, leaving an appropriate cover depending on the environment in which the hammerhead pier will be located.

8.5.3 Load Factor Design Equation

The compressive load on the column is determined from the following load factor design equation:

$$P_{LFD} = 1.3[DL_{Total} + (\#\text{ of girders})(1.67)(DF)(1 + I)(LL)] \tag{8.42}$$

where the total dead load is made up of the weight of the column plus the dead load from the bridge spans, expressed as:

$$DL_{Total} = DL_{Column} + (\#\text{ of girders})DL_{Spans} \tag{8.43}$$

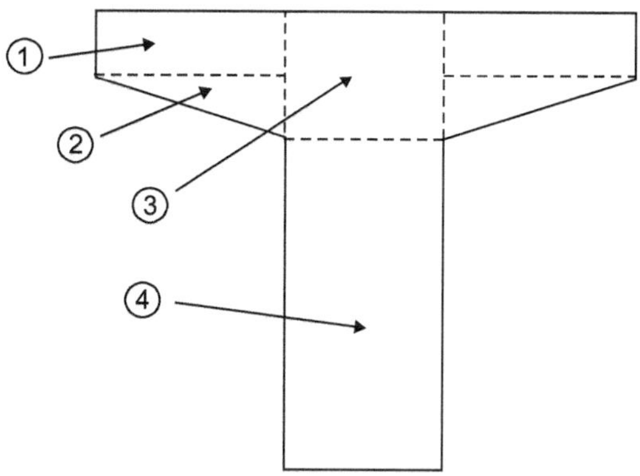

Figure 8.29 Elements of the volume of the column.

The dead load of the column is the volume of the column times the specific weight of concrete. The weight of the reinforcing steel is neglected. Note that the volume of the column includes the volume of the cap. Therefore, the volume of the column consists of the six sections shown in Figure 8.29 (where sections (1) and (2) are duplicated).

Therefore, the volume of the column is simply:

$$\text{Vol}_{\text{Column}} = 2 \times \text{Vol}_1 + 2 \times \text{Vol}_2 + \text{Vol}_3 + \text{Vol}_4 \tag{8.44}$$

Note that the volume of section (1) was part of the calculation to determine the weight of the rectangular section and the volume of section (2) was part of the calculation to determine the weight of the triangular section. Here, the total volume of the column will be calculated, then multiplied by the specific weight of concrete, to obtain the final dead load of the column.

Once the compressive load factor design requirement P_{LFD} is determined, it is compared to the maximum compressive load P_{max}. If it is less than the maximum, the design is satisfactory, and the area of reinforcing steel selected will be the design value. From this area, a selection of size and number of rebar can be specified. An example will highlight these steps.

8.5.4 Calculations for the Required Reinforcing Steel

An example will provide the calculations involved in the process of determining the required reinforcing steel for the column of the hammerhead pier.

EXAMPLE 8.5 Determining the Required Reinforcing Steel for the Column

In Example 8.3, the load factor design reaction R_{LFD} for the bridge design shown was determined, where one span of a two-span bridge arrangement is 60 ft and the other is 70 ft. The distributed loading per foot, *w*, was given as 4 kip/ft for both spans.

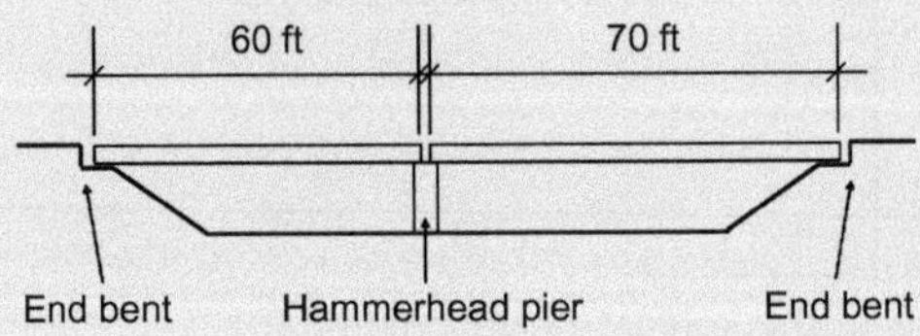

In Example 8.3, the dead load from the bridge spans was found to be 260 kip per girder, and the live load 31.2 kip per girder. Using the hammerhead design given in Example 8.4, determine the required reinforcing steel for the column. Because the concrete is in compression, assume the strength of the concrete f_c is 4 ksi. The reinforcing steel is grade 60, which has a yield strength f_y of 60 ksi.

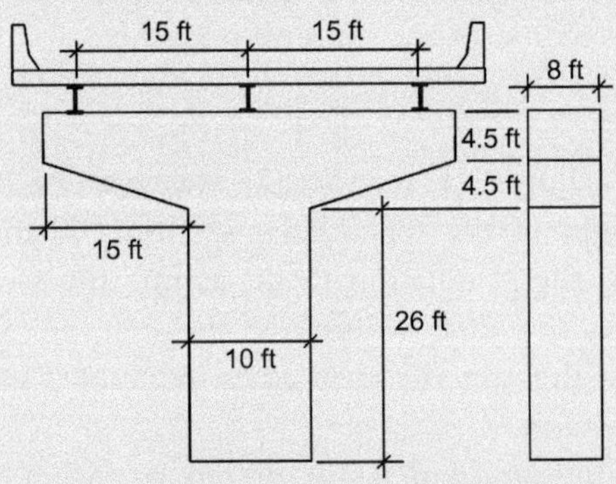

Based on the girder spacing, the distribution factor is 3.07, and, using the shorter of the two spans, 60 ft, the impact factor is 0.270.

Solution

First, determine the maximum compressive load P_{max} from Eq. (8.41), where the capacity reduction factor ϕ is taken as 0.75.

$$P_{max} = \phi[(0.85 f_c)(A_g - A_s) + f_y A_s] \tag{8.41}$$

The gross area of the column is given by:

$$A_g = (10\text{ ft})(8\text{ ft}) = 80\text{ ft}^2 = 11,520\text{ in}^2$$

Therefore, select the area of the reinforcing steel to be 1% of this value, or:

$$A_s = (1\%)A_g = (0.01)(11,520\text{ in}^2) = 115.2\text{ in}^2$$

(*Continued*)

EXAMPLE 8.5 Determining the Required Reinforcing Steel for the Column—(Continued)

Substitute the areas and the other known values to give P_{max} as:

$$\begin{aligned} P_{max} &= (0.75)[(0.85)(4\ \text{ksi})(11,520\ \text{in}^2 - 115.2\ \text{in}^2) + (60\ \text{ksi})(115.2\ \text{in}^2)] \\ &= (0.75)[38,776\ \text{kip} + 6912\ \text{kip}] = (0.75)[45,688\ \text{kip}] \\ &= 34,266\ \text{kip} \end{aligned}$$

Next, calculate the volume of concrete in the column from Eq. (8.44). Using the dimensions given, this volume becomes:

$$\begin{aligned} \text{Vol}_{\text{Column}} &= 2 \times \text{Vol}_1 + 2 \times \text{Vol}_2 + \text{Vol}_3 + \text{Vol}_4 \\ &= 2[(4.5\ \text{ft})(15\ \text{ft})(8\ \text{ft})] + 2\left[1/2(4.5\ \text{ft})(15\ \text{ft})(8\ \text{ft})\right] \\ &\quad + (9\ \text{ft})(10\ \text{ft})(8\ \text{ft}) + (10\ \text{ft})(8\ \text{ft})(26\ \text{ft}) \\ &= 4420\ \text{ft}^3 \end{aligned}$$

Therefore, using a specific weight of concrete of 150 lb/ft^3, or more conveniently for the units involved, 0.15 kip/ft^3, gives the dead load due to the column as:

$$\begin{aligned} DL_{\text{Column}} &= \text{Vol}_{\text{Column}}(0.15\ \text{kip/ft}^3) = (4420\ \text{ft}^3)(0.15\ \text{kip/ft}^3) \\ &= 663\ \text{kip} \end{aligned}$$

With the dead load due to the bridge spans determined in Example 8.3 as 260 kip, the total dead load, DL, from Eq. (8.43) becomes:

$$\begin{aligned} DL_{\text{Total}} &= DL_{\text{Column}} + (\#\ \text{of girders})DL_{\text{Spans}} \\ &= 663\ \text{kip} + (3)(260\ \text{kip}) \\ &= 1443\ \text{kip} \end{aligned}$$

The compressive load on the column can now be determined from Eq. (8.42) as:

$$\begin{aligned} P_{LFD} &= 1.3[DL_{\text{Total}} + (\#\ \text{of girders})(1.67)(DF)(1 + I)(LL)] \\ &= 1.3[1443\ \text{kip} + (3)(1.67)(3.07)(1 + 0.270)(31.2\ \text{kip})] \\ &= 1.3[1443\ \text{kip} + 609\ \text{kip}] = 1.3[2052\ \text{kip}] \\ &= 2668\ \text{kip} \end{aligned}$$

Comparing this value of the compressive load P_{LFD} to the maximum P_{max}, it is clear that the design is satisfactory.

$$P_{\text{max}} > P_{\text{LFD}}$$
$$34,266\ \text{kip} > 2668\ \text{kip}$$

So the choice of 1% of the gross area for the reinforcing steel meets the requirements. Selecting a size and number of rebar is a matter of local layout and construction constraints, and also dependent on the experience of the structural engineer.

8.6 PILE FOOTING DESIGN

8.6.1 Introduction

The final step in the bridge design process is the design of the footing for the hammerhead pier. Here, of the two possible types, spread or pile, the design of a pile footing will be presented. The goal of the analysis is to determine the number of piles needed and an appropriate pattern for them to be located within the overall footing. How the footing and the column are to be connected will also be discussed.

While the bridge spans and hammerhead pier will be subjected to both primary and secondary loads, in this analysis only vertical loads will be considered. This includes both dead loads and live loads. The footing will be assumed to be subjected to only axial loads, and these loads are uniform over the surface area of the footing. This means any differential settling will be neglected.

8.6.2 Load Capacity

The load carrying capacity P_{max} of the footing is determined from the following equation:

$$P_{\text{max}} = \phi[(0.85f_c)(A_g)] \tag{8.45}$$

where ϕ is the capacity reduction factor for compression. In the previous section, this factor was specified to be 0.75. However, the strength f_c of the concrete for the footing is taken to be 3 ksi. The term A_g is the gross area of the column, as defined and used in the previous section.

Also, the reinforcing steel that will be used to connect the column to the footing will still be grade 60, with a yield strength f_y of 60 ksi.

8.6.3 Load Factor Design Equation

The compressive load on the piles that make up the footing is determined from the load factor design equation that follows (this equation is very similar to Eq. (8.42); however the total dead loads are different):

$$P_{\text{LFD}} = 1.3\left[DL_{\text{Total}} + (\#\text{ of girders})(1.67)(DF)(1+I)(LL)\right] \tag{8.46}$$

The total dead load is made up of the weight of the column plus the dead load from the bridge spans and the dead load of the footing itself, expressed as:

$$DL_{\text{Total}} = DL_{\text{Column}} + (\#\text{ of girders})DL_{\text{Spans}} + DL_{\text{footing}} \tag{8.47}$$

The dead load of the footing is taken as 8% of what is called the "unfactored" loads, which are determined from the following equation:

$$P_{\text{unfactored}} = DL_{\text{Column}} + (\#\text{ of girders})DL_{\text{Spans}} + (\#\text{ of girders})(DF)(LL) \quad (8.48)$$

Therefore, the dead load for the footing, DL_{footing}, is determined as:

$$DL_{\text{footing}} = 8\% \times P_{\text{unfactored}} \quad (8.49)$$

Once the compressive load factor design requirement (P_{LFD}) is determined, it is compared to the maximum compressive load (P_{max}). If it is less than the maximum, the design is satisfactory and the analysis can continue with the determination of the number of piles needed.

8.6.4 Determination of the Required Number of Piles

When the load factor design load (P_{LFD}) has been established, the number of piles needed can be determined. There are various choices for piles, however, the one that will be used here is referred to as an "HP12 × 53." The "H" stands for the shape, which means it is as wide as it is tall; "12" is the nominal width and height dimension of 12 in; "P" is "pile" ; and "53" represents its weight per foot.

The allowable loading for an HP12 × 53 pile is 45 kip. Therefore, the number of piles needed is determined by dividing the load factor design load (P_{LFD}) by 45 kip, and rounding to the next whole number. In fact, the number of piles chosen should form a logical pattern. For example, if the hammerhead pier is 8 ft wide and 6 ft deep, then 8 × 6 is the logical pattern if 48 piles are sufficient to carry the load. If not, this ratio of width to depth can be increased until the required number is reached.

8.6.5 Layout of the Piles

Suppose that 48 piles are needed and will be arranged in an 8 × 6 pattern. Figure 8.30 shows such a layout. Notice that a cover of 15 in, or 1.25 ft, is provided to the edge of the footing around the perimeter of the layout, and that there is a typical spacing between the piles of 2.5 ft.

This gives a footing with the following width W, not to be confused with weight, and a depth L, not to be confused with the length of a bridge span.

$$W = (2)(1.25\text{ ft}) + (\#\text{ of spaces wide})(2.5\text{ ft}) \quad (8.50)$$

$$L = (2)(1.25\text{ ft}) + (\#\text{ of spaces deep})(2.5\text{ ft}) \quad (8.51)$$

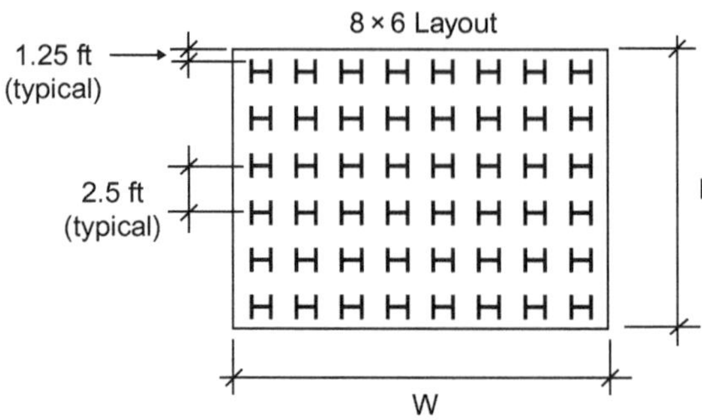

Figure 8.30 Layout of an 8 × 6 pattern of piles.

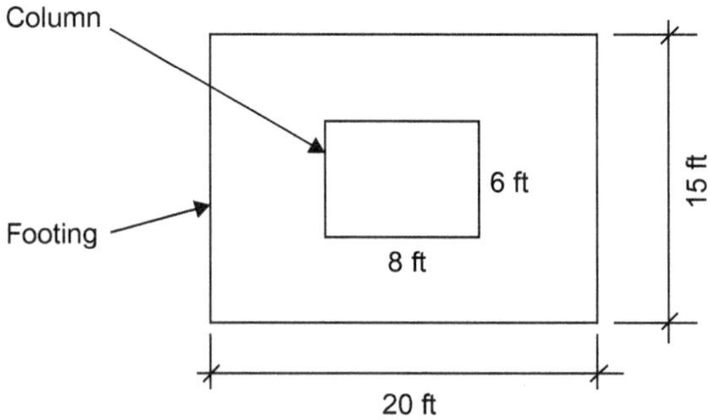

Figure 8.31 Footprint of column and footing.

Note that the number of spaces in either direction is the number of piles in that direction minus 1.

For an 8 × 6 pattern of 48 piles, the dimensions of the footing would be:

$$\begin{aligned} W &= (2)(1.25\text{ ft}) + (\#\text{ of spaces wide})(2.5\text{ ft}) \\ &= (2)(1.25\text{ ft}) + (8-1)(2.5\text{ ft}) = 20\text{ ft} \\ L &= (2)(1.25\text{ ft}) + (\#\text{ of spaces deep})(2.5\text{ ft}) \\ &= (2)(1.25\text{ ft}) + (6-1)(2.5\text{ ft}) = 15\text{ ft} \end{aligned}$$

This would give an actual footprint of an 8 × 6 ft column, and the footing size as shown in Figure 8.31.

8.6.6 Reinforcing Steel between Column and Footing

To connect the hammerhead pier to the pile footing, the minimum area of reinforcing steel, A_s, needed is specified to be 0.5% of the cross-sectional area of the column, A_g. The minimum length, L_{min}, of this reinforcing steel that should extend up into the column from the footing is the largest of the following three values:

$$\begin{aligned} &1. \quad \frac{(0.02)(f_y)(d_b)}{\sqrt{f_c}} \\ &2. \quad (0.0003)(f_y)(d_b) \\ &3. \quad 8\text{ in} \end{aligned} \tag{8.52}$$

Typically, the largest length is found from expression (1); however the others should always be checked.

8.6.7 Calculations for the Required Number of Piles

As in the previous sections, an example will pull together the various elements of the design for the footing.

EXAMPLE 8.6 Determining the Required Number of Piles for the Footing

For the bridge system presented in Examples 8.3, 8.4, and 8.5, determine the number of piles required for the footing, the layout of the piles and the size of the footing, and the amount and length of reinforcing steel to connect the hammerhead column to the footing.

From Example 8.3, the dead load from the bridge spans was found to be 260 kip per girder, and the live load 31.2 kip per girder. For the 15-ft spacing on the three girders, the distribution factor is 3.07, and the impact factor I using the shorter of the two spans was found to be 0.270.

From Example 8.5, the 10 × 8 ft column has a gross area of 11,520 in^2. The total dead weight of the hammerhead pier was found to be 663 kip.

The piles are HP12 × 53, with an allowable loading of 45 kip. Assume the strength of the concrete, f_c, is 3 ksi and the reinforcing steel is grade 60 with a yield strength f_y of 60 ksi.

Solution

First, determine the load carrying capacity P_{max} of the footing from Eq. (8.45):

$$P_{max} = \phi[(0.85 f_c)(A_g)] \tag{8.45}$$

where ϕ is taken to be 0.75, the strength f_c of the concrete is 3 ksi, and the gross area of the column, A_g, is 11,520 in^2. Substituting gives:

$$\begin{aligned} P_{max} &= \phi[(0.85 f_c)(A_g)] = (0.75)[(0.85)(3\text{ ksi})(11,520\text{ in}^2)] \\ &= 22,032\text{ kip} \end{aligned}$$

(Continued)

EXAMPLE 8.6 Determining the Required Number of Piles for the Footing—(Continued)

Next, calculate the compressive load, P_{LFD}, on the piles that make up the footing from Eq. (8.46).

$$P_{LFD} = 1.3[DL_{Total} + (\# \text{ of girders})(1.67)(DF)(1 + I)(LL)] \quad (8.46)$$

The total dead load, DL_{Total}, is determined from Eq. (8.47). The other information has already been determined in previous examples.

$$DL_{Total} = DL_{Column} + (\# \text{ of girders})DL_{Spans} + DL_{footing} \quad (8.47)$$

The dead load of the column is 663 kip, the number of girders is 3, the dead load of the spans is 260 kip per girder, and the dead load of the footing is found from Eq. (8.49) as 8% of the unfactored load, $P_{unfactored}$, which is found from Eq. (8.48).

$$P_{unfactored} = DL_{Column} + (\# \text{ of girders})DL_{Spans} + (\# \text{ of girders})(DF)(LL) \quad (8.48)$$

$$DL_{footing} = 8\% \times P_{unfactored} \quad (8.49)$$

Substituting known values into Eq. (8.48) gives:

$$\begin{aligned} P_{unfactored} &= DL_{Column} + (\# \text{ of girders})DL_{Spans} + (\# \text{ of girders})(DF)(LL) \\ &= 663 \text{ kip} + (3)(260 \text{ kip}) + (3)(3.07)(31.2 \text{ kip}) \\ &= 1730 \text{ kip} \end{aligned}$$

The dead load for the footing can now be determined from Eq. (8.49) as:

$$\begin{aligned} DL_{footing} &= 8\% \times P_{unfactored} = (0.08)(1730 \text{ kip}) \\ &= 138 \text{ kip} \end{aligned}$$

The total dead load can now be determined from Eq. (8.47) as:

$$\begin{aligned} DL_{Total} &= DL_{Column} + (\# \text{ of girders})DL_{Spans} + DL_{footing} \\ &= 663 \text{ kip} + (3)(260 \text{ kip}) + 138 \text{ kip} \\ &= 1581 \text{ kip} \end{aligned}$$

Substitute all the known values in Eq. (8.46) to give the load factor design load as:

$$\begin{aligned} P_{LFD} &= 1.3[DL_{Total} + (\# \text{ of girders})(1.67)\,(DF)(1 + I)(LL)] \\ &= 1.3[1581 \text{ kip} + (3)(1.67)(3.07)(1 + 0.270)(31.2)] \\ &= 1.3[1581 \text{ kip} + 609 \text{ kip}] = 1.3[2190 \text{ kip}] \\ &= 2847 \text{ kip} \end{aligned}$$

(Continued)

EXAMPLE 8.6 Determining the Required Number of Piles for the Footing—(Continued)

Because the compressive load factor design load P_{LFD} is less than the maximum axial load, $P_{\max}$, the design is satisfactory and the number of piles can now be determined.

$$P_{\max} > P_{LFD}$$
$$22,032 \text{ kip} > 2847 \text{ kip}$$

Using the maximum allowable loading for a HP12 × 53 of 45 kip, divide this value into the load factor design load to give:

$$\text{\# of piles} = \frac{P_{\mathrm{LFD}}}{45 \text{ kip/pile}} = \frac{2847 \text{ kip}}{45 \text{ kip/pile}} = 63.3 \text{ piles} \rightarrow 64 \text{ piles minimum}$$

The column is 10 × 8 ft, so an 8 × 8 layout would work. Therefore, the dimensions of the footing can be found from Eqs. (8.50) and (8.51).

$$W = (2)(1.25 \text{ ft}) + (\text{\# of spaces wide})(2.5 \text{ ft}) \quad (8.50)$$

$$L = (2)(1.25 \text{ ft}) + (\text{\# of spaces deep})(2.5 \text{ ft}) \quad (8.51)$$

Because the layout is symmetrical, the width W and the depth L will be the same. Substuting (8−1) for the number of spaces gives:

$$W = L = (2)(1.25 \text{ ft}) + (8-1)(2.5 \text{ ft}) = 20 \text{ ft}$$

Lastly, the area of reinforcing steel A_s needed to connect the column to the footing is 0.5% of the gross cross-sectional area of the column. Therefore:

$$A_s = 0.5\% \times A_g = (0.005)(11,520 \text{ in}^2)$$
$$= 57.6 \text{ in}^2$$

One choice would be 58−60 #9 bars, which have 1.00 in^2 each. The minimum length needed to extend up into the column is found from the largest of the three values obtained from Eq. (8.52).

$$\begin{aligned} &1.\ \frac{(0.02)(f_y)(d_b)}{\sqrt{f_c}} \\ &2.\ (0.0003)(f_y)(d_b) \\ &3.\ 8 \text{ in} \end{aligned} \quad (8.52)$$

From the first expression, using #9 bar with a diameter of 1.128 in (see Table 8.2), the minimum length would be

$$\frac{(0.02)(f_y)(d_b)}{\sqrt{f_c}} = \frac{(0.02)(60{,}000 \text{ psi})(1.128 \text{ in})}{\sqrt{3000 \text{ psi}}} = 24.7 \text{ in}$$

(Continued)

EXAMPLE 8.6 Determining the Required Number of Piles for the Footing—(Continued)

From the second expression the minimum length would be

$$(0.0003)(f_y)(d_b) = (0.0003)(60,000 \text{ psi})(1.128 \text{ in}) = 20.3 \text{ in}$$

Clearly, the value from the first expression governs. To make the layout as straightforward as possible, most likely a length to the nearest half foot would be specified. In this case, specify 30 in, or 2.5 ft.

The footing is now completely specified.

8.7 SUMMARY

The process of designing a bridge system clearly involves a great many considerations. The approach taken here has been to proceed from a top-down perspective. The four major calculations have been to determine: (1) the size and number of steel girders, (2) the size and number of rebar in the cap of the hammerhead pier, (3) the size and number of rebar in the column of the hammerhead pier, and (4) the number of piles, their layout, and the required steel to connect the column to the footing. To determine these four major design elements, the dead loads and live loads had to be determined, and several important factors associated with the loading on the bridge system accounted for. The details of how a modern slab-type deck is constructed was presented, and the determination of all the associated weights given. A multispan bridge design was used throughout the examples (and subsequent problems), as it tends to be the most common design. The hammerhead pier was chosen over a multicolumn design, as it is the more common design currently in use. A pile-type footing was chosen as it is also more common than a spread footing, which requires a solid surface soil to withstand the greater loads of modern vehicular traffic.

It is hoped that the process presented provides insight and instils confidence for the design of a bridge system that will last for the life expected of a modern highway system.

8.8 PRACTICE PROBLEMS

Problem 8.1: Distributed Load Calculation

Determine the distributed load w for the bridge superstructure shown, where the bridge width, BW, is 40 ft. Assume the initial beam to be considered is a W36 × 260, which has a flange width of 16.5 in, or 1.38 ft. The roadway thickness is 14 in, or 1.17 ft.

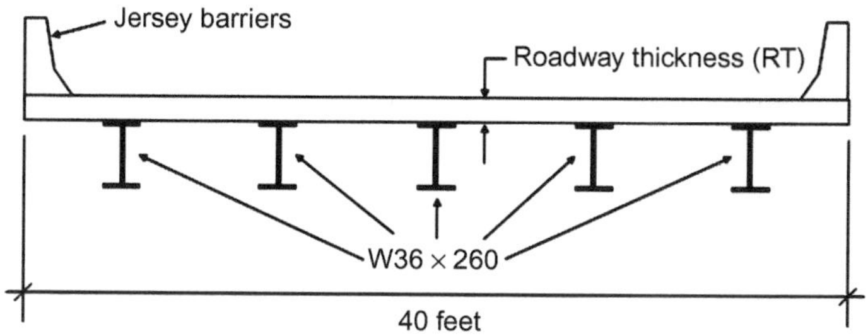

Solution

From Eq. (8.14), the distributed load w is the sum of the six terms shown. The quantity that appears in two of the calculations is the portion of the bridge width that must be supported by an interior girder, given by Eq. (8.7) as:

$$BW_G = \frac{\text{bridge width}}{\text{number of girders}} = \frac{BW}{G} = \frac{40\text{ ft}}{5} = 8\text{ ft}$$

The contribution associated with the deck forms, concrete in the valleys of the forms, and the additional deflection due to the weight of the concrete in the roadway slab is determined from Eq. (8.8) as:

$$w_{\substack{\text{forms, valleys,}\\ \text{\& deflection}}} = (20.71\text{ lb/ft}^2)(BW_G) = (20.71\text{ lb/ft}^2)(8\text{ ft})$$

$$= 165.68\text{ lb/ft}$$

The contribution from the concrete covering the shear studs is found from Eq. (8.9) as:

$$w_{\substack{\text{cover shear}\\ \text{studs}}} = (t_f)(0.42\text{ ft})(150\text{ lb/ft}^3) = (1.38\text{ ft})(0.42\text{ ft})(150\text{ lb/ft}^3)$$

$$= 86.94\text{ lb/ft}$$

The contribution from the roadway is determined from Eq. (8.10) as:

$$w_{\text{roadway}} = (RT)(BW_G)(150\text{ lb/ft}^3) = (1.17\text{ ft})(8\text{ ft})(150\text{ lb/ft}^3)$$
$$= 1404\text{ lb/ft}$$

The contribution from the steel girder is found simply from the designation W36 × 260 as:

$$w_{\text{steel girder}} = 260\ \text{lb/ft}$$

The contribution from the replacement surface is found from Eq. (8.12) as:

$$\begin{aligned} w_{\substack{\text{replacement}\\ \text{surface}}} &= \frac{[BW - 2\,(1.54\ \text{ft})](20\ \text{lb/ft}^2)}{G} \\ &= \frac{[40\ \text{ft} - 2\,(1.54\ \text{ft})](20\ \text{lb/ft}^2)}{5} \\ &= 147.68\ \text{lb/ft} \end{aligned}$$

Finally, the contribution from the Jersey barriers is determined from Eq. (8.13) as:

$$\begin{aligned} w_{\substack{\text{Jersey}\\ \text{barriers}}} &= \frac{(2\ \text{barriers})(371\ \text{lb/ft})}{G} = \frac{(2\ \text{barriers})(371\ \text{lb/ft})}{5} \\ &= 148.4\ \text{lb/ft} \end{aligned}$$

Using Eq. (8.14), the total distributed load w becomes:

$$\begin{aligned} w &= w_{\substack{\text{forms,}\\ \text{valleys,}\\ \text{\&deflection}}} + w_{\substack{\text{cover}\\ \text{shear}\\ \text{studs}}} + w_{\text{roadway}} + w_{\substack{\text{steel}\\ \text{girder}}} + w_{\substack{\text{replacement}\\ \text{surface}}} + w_{\substack{\text{Jersey}\\ \text{barriers}}} \\ &= (165.68 + 86.94 + 1404 + 260 + 147.68 + 148.4)\,\text{lb/ft} \\ &= 2212.70\,\text{lb/ft} \\ &\approx 2.2\,\text{kip/ft} \end{aligned}$$

This distributed load w can now be used to calculate the reactions on the hammerhead pier and end bents, and the maximum bending moment in the interior girder due to the dead load.

Problem 8.2: Load Factor Design Calculations

Determine if the bridge design of Problem 8.1 is satisfactory for a span of 60 ft. The girder spacing of 8 ft is shown in the following diagram.

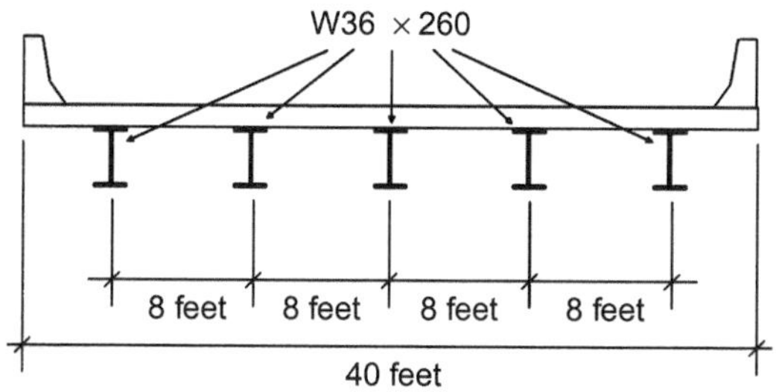

In Problem 8.1, the distributed load was found to be 2.2 kip/ft. The steel girder was specified to be a W36 × 260, which has a section modulus of 1080 in^3. Assume the allowable yield stress of the steel girder is 50,000 psi, or 50 ksi.

Solution

The design will be satisfactory if Eq. (8.19) is valid.

$$Z_{\text{Have}} > Z_{\text{Need}} \tag{8.19}$$

Z_{Have} is given as 1080 in^3. Z_{Need} is determined from Eq. (8.18), where the allowable yield stress of the steel girder is given as 50 ksi.

$$Z_{\text{Need}} = \frac{M_{\text{LFD}}}{f_y} \tag{8.18}$$

The load factor design moment M_{LFD} is found from Eq. (8.15).

$$M_{\text{LFD}} = 1.3[DL + 1.67(DF_{\text{bending}})(1 + I)(LL)] \tag{8.15}$$

The dead load DL is found from Eq. (8.2), where the distributed load w was found from Problem 8.1 to be 2.2 kip/ft.

$$DL = \frac{wL^2}{8} \tag{8.2}$$

Substituting for the distributed load w of 2.2 kip/ft, and the span L of 60 ft, gives a dead load DL equal to:

$$DL = \frac{wL^2}{8} = \frac{(2.2\ \text{k/ft})(60\ \text{ft})^2}{8} = 990\ \text{ft}\cdot\text{kip}$$

Using a girder spacing S of 8 ft, the distribution factor for bending, DF_{bending}, can be found from Eq. (8.16) as

$$DF_{\text{bending}} = \frac{S}{5.5} = \frac{8}{5.5} = 1.45$$

The impact factor I can be found using the span L of 60 ft from Eq. (8.17) as:

$$I = \frac{50}{L + 125} = \frac{50}{60 + 125} = 0.270$$

The live load LL can be found from Eq. (8.6), where the reaction B_V is found from Eq. (8.3).

$$M_{\max} = B_V\left(L/2\right) - 224\ \text{ft}\cdot\text{kip} \tag{8.6}$$

$$B_V = \frac{(18\ \text{kip})L + 168\ \text{ft}\cdot\text{kip}}{L} \tag{8.3}$$

Substituting for the span L of 60 ft in Eq. (8.3) gives:

$$B_V = \frac{(18\,\text{kip})L + 168\,\text{ft}\cdot\text{kip}}{L} = \frac{(18\,\text{kip})\,(60\,\text{ft}) + 168\,\text{ft}\cdot\text{kip}}{60\,\text{ft}} = 20.80\,\text{kip}$$

Substituting this value of B_V into Eq. (8.6) gives the live load LL as:

$$\begin{aligned} LL = M_{\max} &= B_V\left(L/2\right) - 224\ \text{ft}\cdot\text{kip} \\ &= (20.80\ \text{kip})\left(60\ \text{ft}/2\right) - 224\ \text{ft}\cdot\text{kip} = 400\ \text{ft}\cdot\text{kip} \end{aligned}$$

Substituting the various quantities found into Eq. (8.15) gives the load factor design moment M_{LFD} as:

$$\begin{aligned} M_{\text{LFD}} &= 1.3[DL + 1.67(DF_{\text{bending}})(1 + I)(LL)] \\ &= 1.3[(990\ \text{ft}\cdot\text{kip}) + 1.67(1.45)(1 + 0.270)(400\ \text{ft}\cdot\text{kip})] \\ &= 1.3[(990\ \text{ft}\cdot\text{kip}) + (1230.1\ \text{ft}\cdot\text{kip})] \\ &= 1.3[2220.1\ \text{ft}\cdot\text{kip}] \\ &= 2886.1\ \text{ft}\cdot\text{kip} \end{aligned}$$

Substitute this value for the load factor design moment M_{LFD} into Eq. (8.18) to determine the required section modulus, Z_{Need}.

Note the conversion of feet to inches, to obtain the correct units on the section modulus.

$$Z_{\text{Need}} = \frac{M_{\text{LFD}}}{f_y} = \frac{(22886.1 \text{ ft}\cdot\text{kip})(12 \text{ in/ft})}{50 \text{ ksi}} = 693 \text{ in}^3$$

Because the required section modulus of 693 in^3 is less than the available section modulus of 1080 in^3, the design is satisfactory.

Problem 8.3: Load Factor Design Calculations for Hammerhead Pier

Determine the load factor design reaction R_{LFD} for the bridge design shown, where one span of a two-span bridge arrangement is 50 ft and the other is 60 ft. Assume the distributed loading per foot w is 3 kip/ft for both spans.

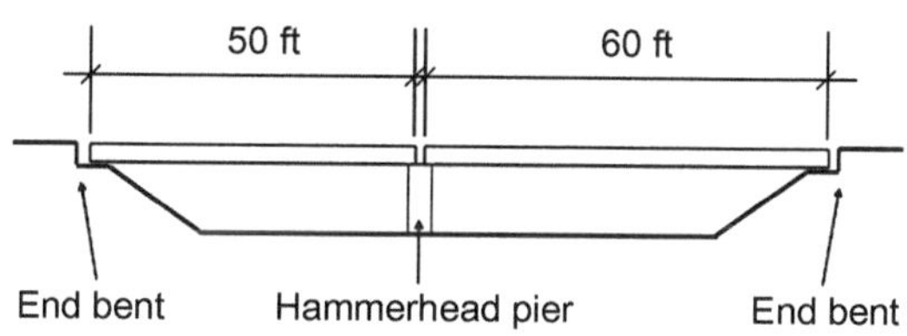

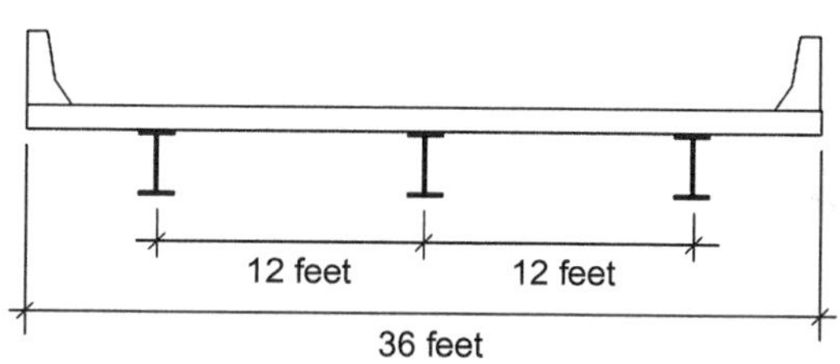

Again, the bridge has only three girders, while an actual design would have at least five or more. However, restricting the problem to only three girders simplifies the analysis that follows for the hammerhead pier.

Solution

The load factor design reaction R_{LFD} is found from Eq. (8.22).

$$R_{\text{LFD}} = 1.3[DL + 1.67(DF_{\text{shear}})(1 + I)(LL)] \tag{8.22}$$

The dead load DL is found from Eq. (8.23) as:

$$DL = \frac{wL_1}{2} + \frac{wL_2}{2} \tag{8.23}$$

where L_1 and L_2 are the lengths of the two spans.

Substituting for the distributed load w of 3 kip/ft, and the lengths of the two spans, L_1 and L_2, of 50 ft and 60 ft, respectively, gives a dead load DL equal to:

$$DL = \frac{wL_1}{2} + \frac{wL_2}{2} = \frac{(3\text{ kip/ft})(50\text{ ft})}{2} + \frac{(3\text{ kip/ft})(60\text{ ft})}{2} = 165\text{ kip}$$

Using a girder spacing S of 12 ft, the distribution factor for shear, DF_{shear}, from Table 8.1 is 2.50.

The impact factor I can be found using the shorter of the two spans of 50 ft from Eq. (8.17) as:

$$I = \frac{50}{L + 125} = \frac{50}{50 + 125} = 0.286$$

The live load LL, which is the reaction A_V, can be found from Eq. (8.24) using the longer of the two spans as:

$$A_V = \frac{(36\text{ kip})L - 336\text{ ft}\cdot\text{kip}}{L} \tag{8.24}$$

Substituting for the span of 60 ft in Eq. (8.24) gives the live load LL as:

$$LL = A_V = \frac{(36\text{ kip})L - 336\text{ ft}\cdot\text{kip}}{L} = \frac{(36\text{ kip})(60\text{ ft}) - 336\text{ ft}\cdot\text{kip}}{60\text{ ft}} = 30.4\text{ kip}$$

Substituting the various quantities found into Eq. (8.22) gives the load factor design reaction R_{LFD} as:

$$\begin{aligned} R_{\text{LFD}} &= 1.3[DL + 1.67(DF_{\text{shear}})(1 + I)(LL)] \\ &= 1.3[(165\text{ kip}) + 1.67(2.50)(1 + 0.286)(30.4\text{ kip})] \\ &= 1.3[(165\text{ kip}) + (163\text{ kip})] \\ &= 1.3[328\text{ kip}] \\ &= 426\text{ kip} \end{aligned}$$

Problem 8.4: Determining the Required Reinforcing Steel for a Hammerhead Pier

Using the load factor design reaction R_{LFD} found in Problem 8.3 to be 426 kip, determine the required area of reinforcing steel for the hammerhead pier shown. The pier will be located in a standard environment. Assume the strength of the concrete f_c is 3 ksi and the reinforcing steel is grade 60, which has a yield strength f_y of 60 ksi.

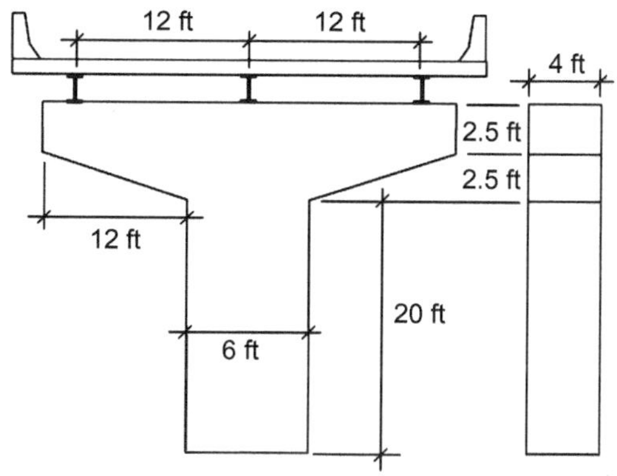

Notice that the heights of the rectangular and triangular sections of the trapezoidal sections are the same value, to simplify the calculations. Also, consider the length of the trapezoidal sections to be the same, 12 ft. The fact that this dimension matches the girder spacing is only coincidental.

Solution

The first step is to determine the moment due to the loading caused by the bridge span and the weight of the trapezoidal section. The load factor design reaction R_{LFD} is 426 kip located 9 ft from the face of A-A of the trapezoidal section. The weights of the rectangular and triangular sections are determined from Eqs. (8.26) and (8.27) as:

$$\begin{aligned} W_{\mathrm{rect}} &= (h_{\mathrm{rect}} \times l_{\mathrm{trapezoid}} \times b)(150\ \mathrm{lb/ft^3}) \\ &= (2.5\ \mathrm{ft})(12\ \mathrm{ft})(4\ \mathrm{ft})(150\ \mathrm{lb/ft^3}) \\ &= 18{,}000\ \mathrm{lb} = 18\ \mathrm{kip} \end{aligned}$$

$$\begin{aligned} W_{\mathrm{tri}} &= 1/2(h_{\mathrm{tri}} \times l_{\mathrm{trapezoid}} \times b)(150\ \mathrm{lb/ft^3}) \\ &= 1/2(2.5\ \mathrm{ft})(12\ \mathrm{ft})(4\ \mathrm{ft})(150\ \mathrm{lb/ft^3}) \\ &= 9000\ \mathrm{lb} = 9\ \mathrm{kip} \end{aligned}$$

The locations of these weights are found from Eqs. (8.29) and (8.30) as:

$$x_{\mathrm{rect}} = 1/2(l_{\mathrm{trapezoid}}) = 1/2\ (12\ \mathrm{ft}) = 6\ \mathrm{ft}$$

$$x_{\mathrm{tri}} = 1/3\ (l_{\mathrm{trapezoid}}) = 1/3\ (12\ \mathrm{ft}) = 4\ \mathrm{ft}$$

These three loads and their locations are shown on the following diagram:

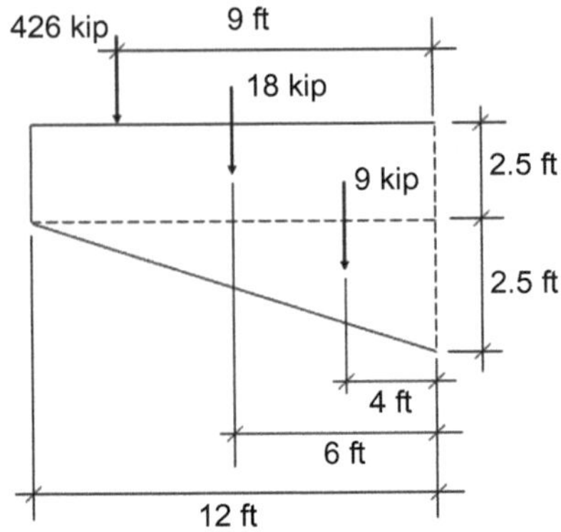

The bending moment about A-A caused by these three loads is therefore determined by Eq. (8.28) as:

$$\begin{aligned} M_{\text{Loads}} &= (R_{\text{LFD}} \times x_{\text{girder}}) + (W_{\text{rect}} \times x_{\text{rect}}) + (W_{\text{tri}} \times x_{\text{tri}}) \\ &= (426\ \text{kip})(9\ \text{ft}) + (18\ \text{kip})(6\ \text{ft}) + (9\ \text{kip})(4\ \text{ft}) \\ &= 3978\ \text{ft}\cdot\text{kip} \end{aligned}$$

This moment must be increased by the load capacity factor using Eq. (8.31) to give a nominal moment M_n equal to:

$$M_n = \frac{M_{\text{Loads}}}{0.9} = \frac{3978\ \text{ft}\cdot\text{kip}}{0.9} = 4420\ \text{ft}\cdot\text{kip}$$

However, the cracking moment, M_{cr}, must be calculated from Eq. (8.32) and then multiplied by 1.2 to see if it is greater than the above nominal moment. The modulus of rupture is calculated from Eq. (8.33) as:

$$f_r = 7.5\sqrt{f_c} = 7.5\sqrt{3000\ \text{psi}} = 410.8\ \text{psi}$$

The area moment of inertia of A-A is found from Eq. (8.34), where the height h is 5 ft or 60 in, and the width b is 4 ft or 48 in:

$$I = \frac{bh^3}{12} = \frac{(48\ \text{in})\,(60\ \text{ft})^3}{12} = 864,000\ \text{in}^4$$

The distance y_t is found from Eq. (8.35) as:

$$y_t = \frac{h}{2} = \frac{60\ \text{in}}{2} = 30\ \text{in}$$

Substituting these values into Eq. (8.32) gives the cracking moment M_{cr} as:

$$M_{cr} = \frac{f_r I}{y_t} = \frac{(410.8 \text{ psi})(864,000 \text{ in}^4)}{30 \text{ in}}$$

$$= 11,831,040 \text{ in·lb} = 985,920 \text{ ft·lb}$$

$$= 986 \text{ ft·kip}$$

Multiply this value of the cracking moment by 1.2, as specified in Eq. (8.36), to give an alternate value for the nominal moment M_n as:

$$M_n = 1.2\ M_{cr} = 1.2(986 \text{ ft·kip}) = 1183 \text{ ft·kip}$$

Comparing the two values for M_n gives the value obtained from the loadings to be significantly larger. So, the value of the nominal moment M_n to be used to obtain the length *a* of the Whitney stress block is:

$$M_n = 4420 \text{ ft·kip}$$

The last quantity needed for Eq. (8.40) to find the length *a* is the distance *d*. It is the height of A-A minus the cover. Because the hammerhead pier will be located in a standard environment, the cover should be 3 in. Therefore, the distance *d* becomes:

$$d = h - (3 \text{ in}) = 60 \text{ in} - 3 \text{ in} = 57 \text{ in}$$

Substituting all the values now known into Eq. (8.40) gives

$$a^2 - 2da + \frac{2M_n}{(0.85 f_c)(b)} = 0$$

$$a^2 - 2\,(57 \text{ in})(a) + \frac{2(4420 \text{ ft·kip})(12 \text{ in}/1 \text{ ft})}{(0.85)(3 \text{ ksi})(48 \text{ in})} = 0$$

$$a^2 - 114\ a + 867 = 0$$

Using the quadratic equation, the two roots of the length *a* become 106 in and 8 in. The larger root is rejected, so the length *a* used to find the required area of reinforcing steel is 8 in.

Substituting this value of the length *a* into Eq. (8.39) gives:

$$A_S = \frac{(0.85 f_c)(a)(b)}{f_y} = \frac{(0.85)(3 \text{ ksi})(8 \text{ in})(48 \text{ in})}{(60 \text{ ksi})}$$

$$= 16.32 \text{ in}^2$$

Therefore, from Table 8.2 there are several choices for size and number of rebar. The simplest combination is 17 #9 bars to give exactly 17.00 in^2, or 8 #14 bars to give 18.00 in^2. There are obviously many other valid choices.

Problem 8.5: Determining the Required Reinforcing Steel for the Column

In Problem 8.3, the load factor design reaction R_{LFD} for the bridge design shown was determined, where one span of a two-span bridge arrangement is 50 ft and the other is 60 ft. The distributed loading per foot w was given as 3 kip/ft for both spans.

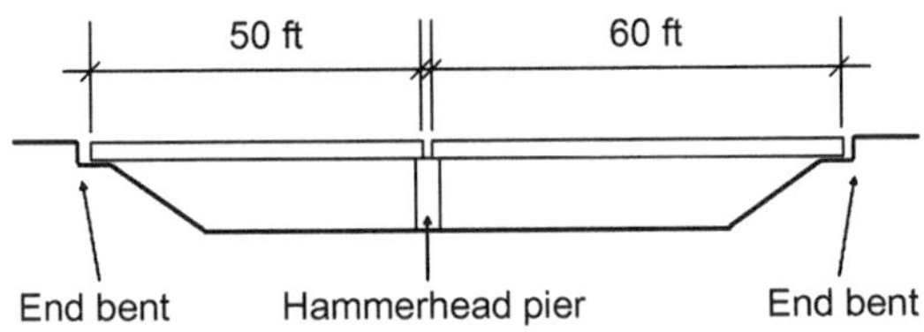

In Problem 8.3, the dead load from the bridge spans was found to be 165 kip per girder, and the live load 30.4 kip per girder. Using the hammerhead design given in Problem 8.4, provided in the following, determine the required reinforcing steel for the column. Because the concrete is in compression, assume the strength of the concrete, f_c, is 4 ksi. The reinforcing steel is grade 60, which has a yield strength f_y of 60 ksi.

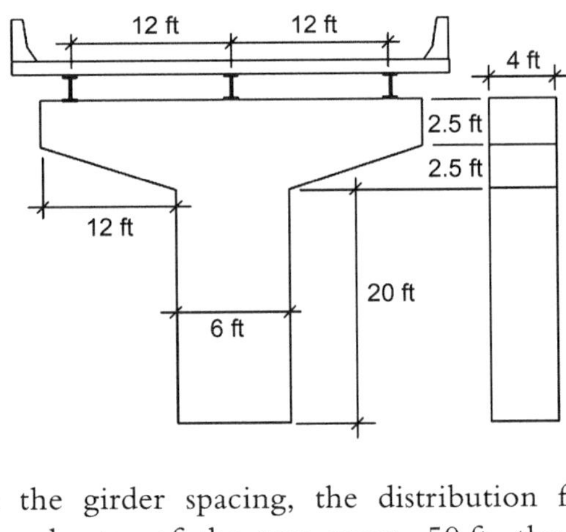

Based on the girder spacing, the distribution factor is 2.50, and using the shorter of the two spans, 50 ft, the impact factor is 0.286.

Solution

First, determine the maximum compressive load P_{max} from Eq. (8.41), where the capacity reduction factor ϕ is taken as 0.75.

$$P_{max} = \phi[(0.85 f_c)(A_g - A_s) + f_y A_s] \tag{8.41}$$

The gross area of the column is given by:

$$A_g = (6 \text{ ft})(4 \text{ ft}) = 24 \text{ ft}^2 = 3456 \text{ in}^2$$

Therefore, select the area of the reinforcing steel to be 1% of this value, or:

$$A_s = (1\%)A_g = (0.01)(3456 \text{ in}^2) = 34.56 \text{ in}^2$$

Substitute the areas and the other known values to give P_{max} as:

$$\begin{aligned} P_{max} &= (0.75)[(0.85)(4\,\text{ksi})(3456\,\text{in}^2 - 34.56\,\text{in}^2) + (60\,\text{ksi})(34.56\,\text{in}^2)] \\ &= (0.75)[11,633\,\text{kip} + 2074\,\text{kip}] = (0.75)[13,707\,\text{kip}] \\ &= 10,280\,\text{kip} \end{aligned}$$

Next, calculate the volume of concrete in the column from Eq. (8.44). Using the dimensions given, this volume becomes:

$$\begin{aligned} \text{Vol}_{\text{Column}} &= 2 \times \text{Vol}_1 + 2 \times \text{Vol}_2 + \text{Vol}_3 + \text{Vol}_4 \\ &= 2[(2.5 \text{ ft})(12 \text{ ft})(4 \text{ ft})] + 2\left[1/2(2.5 \text{ ft})(12 \text{ ft})(4 \text{ ft})\right] \\ &\quad + (5 \text{ ft})(6 \text{ ft})(4 \text{ ft}) + (6 \text{ ft})(4 \text{ ft})(20 \text{ ft}) \\ &= 960 \text{ ft}^3 \end{aligned}$$

Therefore, using a specific weight of concrete of 150 lb/ft^3, or more conveniently for the units involved, 0.15 kip/ft^3, gives the dead load due to the column as:

$$\begin{aligned} DL_{\text{Column}} &= \text{Vol}_{\text{Column}}(0.15 \text{ kip/ft}^3) = (960 \text{ ft}^3)(0.15 \text{ kip/ft}^3) \\ &= 144 \text{ kip} \end{aligned}$$

With the dead load due to the bridge spans determined in Problem 8.3 as 165 kip, the total dead load from Eq. (8.43) becomes:

$$\begin{aligned} DL_{\text{Total}} &= DL_{\text{Column}} + (\#\text{ of girders})DL_{\text{Spans}} \\ &= 144 \text{ kip} + (3)(165 \text{ kip}) \\ &= 639 \text{ kip} \end{aligned}$$

The compressive load on the column can now be determined from Eq. (8.42) as:

$$P_{LFD} = 1.3\,[DL_{Total} + (\#\text{ of girders})(1.67)\,(DF)(1+I)(LL)]$$
$$= 1.3[639\text{ kip} + (3)(1.67)(2.50)(1+0.286)(30.4\text{ kip})]$$
$$= 1.3[639\text{ kip} + 490\text{ kip}] = 1.3[1129\text{ kip}]$$
$$= 1468\text{ kip}$$

Comparing this value of the compressive load P_{LFD} to the maximum P_{max}, clearly the design is satisfactory.

$$P_{max} > P_{LFD}$$
$$10{,}280\text{ kip} > 1468\text{ kip}$$

So the choice of 1% of the gross area for the reinforcing steel meets the requirements. Selecting a size and number of rebar is a matter of local layout and construction constraints, and depends on the experience of the structural engineer.

Problem 8.6: Determining the Required Number of Piles for the Footing

For the bridge system presented in Problems 8.3, 8.4, and 8.5, determine the number of piles required for the footing, the layout of the piles and the size of the footing, and the amount and length of reinforcing steel to connect the hammerhead column to the footing.

From Problem 8.3, the dead load from the bridge spans was found to be 165 kip per girder, and the live load 30.4 kip per girder. For the 12-ft spacing on the three girders, the distribution factor is 2.50 and the impact factor I, using the shorter of the two spans, was found to be 0.286.

From Example 8.5, the 6×4 ft column has a gross area of 3456 in^2. The total dead weight of the hammerhead pier was found to be 144 kip.

The piles are HP12 × 53, with an allowable loading of 45 kip. Assume the strength of the concrete, f_c, is 3 ksi, and the reinforcing steel is grade 60 with a yield strength f_y of 60 ksi.

Solution

First, determine the load carrying capacity P_{max} of the footing from Eq. (8.45):

$$P_{max} = \phi[(0.85\,f_c)(A_g)] \tag{8.45}$$

where ϕ is taken to be 0.75, the strength f_c of the concrete is 3 ksi, and the gross area of the column, A_g, is 3456 in^2. Substituting gives:

$$P_{max} = \phi[(0.85 f_c)(A_g)] = (0.75)[(0.85)(3 \text{ ksi})(3456 \text{ in}^2)] \\ = 6610 \text{ kip}$$

Next, calculate the compressive load P_{LFD} on the piles that make up the footing from Eq. (8.46).

$$P_{LFD} = 1.3[DL_{Total} + (\# \text{ of girders})(1.67)(DF)(1 + I)(LL)] \quad (8.46)$$

The total dead load, DL_{Total}, is determined from Eq. (8.47). The other information has already been determined in previous examples.

$$DL_{Total} = DL_{Column} + (\# \text{ of girders})DL_{Spans} + DL_{footing} \quad (8.47)$$

The dead load of the column is 144 kip, the number of girders is 3, the dead load of the spans is 165 kip per girder, and the dead load of the footing is found from Eq. (8.49) as 8% of the unfactored load, $P_{unfactored}$, which is found from Eq. (8.48).

$$P_{unfactored} = DL_{Column} + (\# \text{ of girders})DL_{Spans} + (\# \text{ of girders})(DF)(LL) \quad (8.48)$$

$$DL_{footing} = 8\% \times P_{unfactored} \quad (8.49)$$

Substituting known values into Eq. (8.48) gives:

$$P_{unfactored} = DL_{Column} + (\# \text{ of girders})DL_{Spans} + (\# \text{ of girders})(DF)(LL) \\ = 144 \text{ kip} + (3)(165 \text{ kip}) + (3)(2.50)(30.4 \text{ kip}) \\ = 867 \text{ kip}$$

The dead load for the footing can now be determined from Eq. (8.49) as:

$$DL_{footing} = 8\% \times P_{unfactored} = (0.08)(867 \text{ kip}) \\ = 69 \text{ kip}$$

The total dead load can now be determined from Eq. (8.47) as:

$$DL_{Total} = DL_{Column} + (\# \text{ of girders})DL_{Spans} + DL_{footing} \\ = 144 \text{ kip} + (3)(165 \text{ kip}) + 69 \text{ kip} \\ = 708 \text{ kip}$$

Substitute all the known values in Eq. (8.46) to give the load factor design load as:

$$
\begin{aligned}
P_{\mathrm{LFD}} &= 1.3[DL_{\mathrm{Total}} + (\#\text{ of girders})(1.67)\,(DF)(1+I)(LL)] \\
&= 1.3[708\text{ kip} + (3)(1.67)\,(2.50)(1+0.286)(30.4)] \\
&= 1.3[708\text{ kip} + 490\text{ kip}] = 1.3\,[1198\text{ kip}] \\
&= 1557\text{ kip}
\end{aligned}
$$

Because the compressive load factor design load, P_{LFD}, is less than the maximum axial load $P_{\max}$, the design is satisfactory and the number of piles can now be determined.

$$
\begin{gathered}
P_{\max} > P_{\mathrm{LFD}} \\
6610\text{ kip} > 1557\text{ kip}
\end{gathered}
$$

Using the maximum allowable loading for a HP12 × 53 of 45 kip, divide this value into the load factor design load to give:

$$
\#\text{ of piles} = \frac{P_{\mathrm{LFD}}}{45\text{ kip/pile}} = \frac{1557\text{ kip}}{45\text{ kip/pile}} = 34.6\text{ piles} \rightarrow 35\text{ piles minimum}
$$

The column is 6 ft by 4 ft, so a 7 × 5 layout would work. Therefore, the dimensions of the footing can be found from Eqs. (8.50) and (8.51).

$$
W = (2)(1.25\text{ ft}) + (\#\text{ of spaces wide})(2.5\text{ ft}) \tag{8.50}
$$

$$
L = (2)(1.25\text{ ft}) + (\#\text{ of spaces deep})(2.5\text{ ft}) \tag{8.51}
$$

Using 7 − 1 spaces for the width, and 5 − 1 spaces for the depth, gives the width W and the depth L as:

$$
\begin{aligned}
W &= (2)(1.25\text{ ft}) + (7-1)(2.5\text{ ft}) = 17.5\text{ ft} \\
L &= (2)(1.25\text{ ft}) + (5-1)(2.5\text{ ft}) = 12.5\text{ ft}
\end{aligned}
$$

Lastly, the area of reinforcing steel, A_s, needed to connect the column to the footing is 0.5% of the gross cross-sectional area of the column. Therefore:

$$
\begin{aligned}
A_s &= 0.5\% \times A_g = (0.005)(3456\text{ in}^2) \\
&= 17.28\text{ in}^2
\end{aligned}
$$

One choice would be 18 #9 bars, which have 1.00 in^2 each. However, as the required area is small, maybe a smaller diameter bar would work better. From Table 8.2, select a #4 bar with a 0.20 in^2 area per bar and a diameter of 0.500 in. This means five times the number of bars would be needed, or about 90.

The minimum length needed to extend up into the column is found from the largest of the three values obtained from Eq. (8.52)

$$\begin{aligned} &1.\quad \frac{(0.02)(f_y)(d_b)}{\sqrt{f_c}} \\ &2.\quad (0.0003)(f_y)(d_b) \\ &3.\quad 8\text{ in} \end{aligned} \tag{8.52}$$

From the first expression, using #4 bar with a diameter of 0.500 in, the minimum length would be

$$\frac{(0.02)(f_y)(d_b)}{\sqrt{f_c}} = \frac{(0.02)(60,000\text{ psi})(0.500\text{ in})}{\sqrt{3000\text{ psi}}} = 10.95\text{ in}$$

From the second expression the minimum length would be

$$(0.0003)(f_y)(d_b) = (0.0003)(60{,}000\text{ psi})(0.500\text{ in}) = 9.0\text{ in}$$

Clearly, the value from the first expression governs. To make the layout as straightforward as possible, most likely a length to the nearest half foot would be specified. In this case 12 in, or 1 ft, would work.

The footing is now completely specified.

PART 9

Hydraulics

Thomas H. Brown, Jr., PhD, PE
Institute for Transportation Research and Education, North Carolina State University

Contents

9.1 INTRODUCTION

To put the process of designing a storm water runoff system for a highway facility in perspective, it has been estimated that as much as 25% of the cost of a typical highway goes into providing the proper drainage along the highway right of way. The primary concern is safety, followed by concerns relative to the effect of the runoff on adjacent properties, possibly resulting in adverse erosion, and, of course, cost. Each of these goals will be achieved by following established design steps tempered by trusted engineering experience.

The design process begins, not surprisingly, with the determination of the expected rainfall and the amount that will not either be absorbed or evaporate. Depending on the location and type of ground surface, anywhere from only 5% of the rainfall from forested areas to virtually 95% from roadways can be expected to run off. Only the rational method will be presented, though there are other methods for specific applications outside highway design.

Once the expected storm water runoff has been determined, the process of designing the collection system begins. Usually, the storm water from open highways is collected in open channels of various linings, which flow into a network of drop inlets and then into a network of underground piping to locations where it can be released without adverse effect on the environment. In multilane urban expressways and city streets, the storm water flows off the roadway into open gutters, where it then falls into a similar system of drop inlets and underground piping. For multilane expressways and city streets, there is a much higher percentage of surfaces where virtually all the rain that falls must be removed from the roadway and collected so that dangerous driving conditions are avoided. And because the flow is driven by changes in ground slope, careful attention to this aspect of the overall design cannot be overemphasized.

Before getting into a discussion of how the amount of expected storm water runoff is determined, a few design preliminaries are in order.

9.2 DESIGN PRELIMINARIES

As mentioned in the introduction, the flow of storm water runoff is completely driven by changes in ground slope. Also, one of the fundamental principles of hydraulics is that surface water flows perpendicular to contour lines. So, flow lines and contour lines form a web of mutually perpendicular lines, called an orthogonal set of equipotential lines, which are typical in many engineering fields. The best way to determine slopes and contour lines is from project drawings, which usually evolve from charts such as those from the U.S. Geological Service (USGS) topographical maps. An example of such a topographical map is shown in Figure 9.1.

The amount of slope present in a highway facility can be expressed in two related ways. One is called the *side slope*, as defined by the following equation:

$$\text{side slope} = \frac{\text{horizontal distance}}{\text{vertical distance}} = \frac{\text{run}}{\text{rise}}:1 \tag{9.1}$$

The second is called the *longitudinal slope*, or *grade*, defined by the inverse, or reciprocal, relationship:

$$\text{longitudinal slope (grade)} = \frac{\text{vertical distance}}{\text{horizontal distance}} = \frac{\text{rise}}{\text{run}} \times 100(\%) \tag{9.2}$$

Both of these terms for slope will be used in the design analysis of storm water runoff systems.

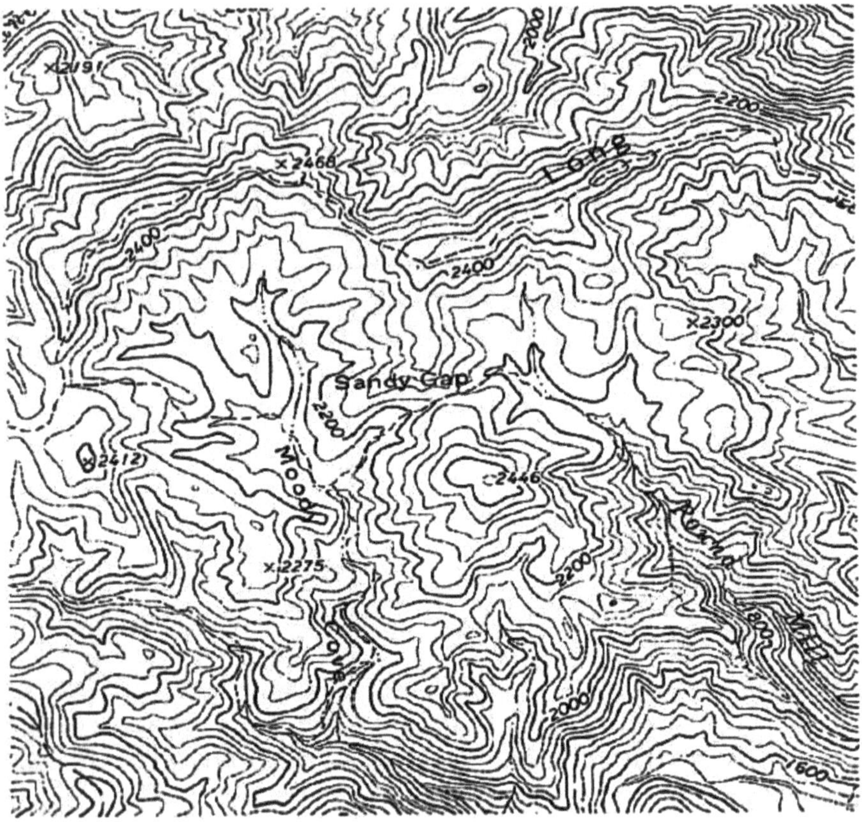

Figure 9.1 Typical topographic map.

EXAMPLE 9.1 Side Slope and Longitudinal Slope (Grade)

On a contour map, two points are 1.4 in apart, where the scale of the map is 1 in = 75 ft. The elevation of the two points is 205 ft and 210 ft mean sea level (MSL). Determine the side slope and longitudinal slope (grade) between the two points.

Solution

Using the definition of side slope from Eq. (9.1) gives:

$$\text{side slope} = \frac{\text{horizontal distance}}{\text{vertical distance}} = \frac{\text{run}}{\text{rise}}:1$$

$$= \frac{(1.4\text{ in}) \times (75\text{ ft}/1\text{ in})}{210\text{ ft} - 205\text{ ft}} = \frac{105\text{ ft}}{5\text{ ft}} = 21:1$$

(Continued)

EXAMPLE 9.1 Side Slope and Longitudinal Slope (Grade)—(Continued)

Knowing the side slope, the longitudinal slope is the reciprocal, which gives:

$$\text{longitudinal slope} = \frac{1}{\text{side slope}} \times 100 = \frac{1}{21} \times 100 = (0.0476) \times 100 \approx 5\%$$

Or, using the definition of the longitudinal slope from Eq. (9.2), gives:

$$\text{longitudinal slope (grade)} = \frac{\text{vertical distance}}{\text{horizontal distance}} = \frac{\text{rise}}{\text{run}} \times 100(\%)$$

$$= \frac{210\text{ ft} - 205\text{ ft}}{(1.4\text{ in}) \times (75\text{ ft}/1\text{ in})} = \frac{5\text{ ft}}{105\text{ ft}} \times 100(\%)$$

$$= (0.0476) \times 100 \approx 5\%$$

The longitudinal slope, or grade, is part of Manning's equation for open channel flow; the side slope is used to specify the inclination of the sides of trapezoidal or triangular channels.

9.3 RATIONAL METHOD

9.3.1 Introduction

Only three things can happen to the rain that falls onto a surface. It can (1) be absorbed into the ground, (2) evaporate, or (3) become runoff from the surface. Clearly, only the storm water runoff need be considered. However, the amount of rainfall must be known, and specifically the peak rainfall expected. This rainfall is measured typically in inches per hour (in/h) in the United States. The peak rainfall intensity increases with the *period* of the storm. For example, the rainfall is greater for a 50-year storm than for a 10-year storm. The type of highway facility determines the year storm that must be considered. Typically, any design associated with a bridge uses a 50-year storm rainfall, whereas a common storm drain is designed using a 10-year storm rainfall. Each state Department of Transportation (DOT) has its own storm period criteria for the various highway facilities.

9.3.2 Storm Water Runoff Equation

The storm water runoff equation for the rational method is given by the following expression:

$$Q = C_c I A \tag{9.3}$$

where

Q = storm water runoff (ft^3/s)
C_c = composite runoff coefficient (unitless)
I = rainfall intensity (in/h)
A = surface area of the highway facility (acres)

At first glance, the storm water runoff equation seems to lack a "number"; it is not really a pure number, but represents various unit conversions. Meaning, how can "inches per hour" be multiplied by "acres" and get "cubic feet per second"? Well, if the proper unit conversions are applied, the number that results is very close to 1, as seen in the following calculation:

$$\frac{\text{inches}}{\text{hour}} \times \text{acres} \times \frac{1\ \text{ft}}{12\ \text{in}} \times \frac{1\ \text{h}}{3600\ \text{s}} \times \frac{43{,}560\ \text{ft}^2}{1\ \text{acre}} = \frac{43{,}560}{43{,}200}\frac{\text{ft}^3}{\text{s}} \approx 1\frac{\text{ft}^3}{\text{s}}$$

Composite Runoff Coefficient

The composite runoff coefficient C_c is determined from the equation:

$$C_c = \frac{\sum C_i A_i}{\sum A_i} \tag{9.4}$$

where the summation is performed on all the various types of surfaces in the highway facility. Table 9.1 lists various runoff coefficients based on the type of surface.

Table 9.1 Runoff coefficients based on surface type

Surface	C
Asphalt	0.95
Concrete	0.95
Forested	0.20
Grass (<2% slope)	0.10
Grass (2–7% slope)	0.15
Grass (>7% slope)	0.20

Table 9.2 Runoff coefficients based on use

Use	C
Unimproved	0.30
Residential	0.50
Commercial	0.70
City	0.90
Light industry	0.80
Heavy industry	0.90

Table 9.2 lists various runoff coefficients based on the type of use.

One way to think about the composite runoff coefficient is that it represents the expected percentage of rainfall that will leave an area, which means the difference between it and 100% is the percentage that stays.

EXAMPLE 9.2 Composite Runoff Coefficient

A planned highway facility will be constructed in an unimproved area of 30 acres. When the construction is completed, 60% will be asphalt roadway, 30% will be grass mostly greater than 10% slope, and 10% will remain unimproved. Determine the composite runoff coefficient for this area after construction.

Solution

From Table 9.2, the runoff coefficient for unimproved land is 0.30, and from Table 9.1, the runoff coefficient for asphalt and grass with a 10% slope are 0.95 and 0.20, respectively. Using Eq. (9.4) for the composite runoff coefficient C_c gives:

$$C_c = \frac{\sum C_i A_i}{\sum A_i}$$

$$= \frac{(0.95)(60\% \times 30\,\text{acres}) + (0.20)(30\% \times 30\,\text{acres}) + (0.30)(10\% \times 30\,\text{acres})}{30\,\text{acres}}$$

$$= 0.66$$

This means 66% of the rainfall will leave the area, and also means 34% will stay. Also, note that because percentages of the various surface types were given, the total area of 30 acres appears in each term in the numerator so it cancels the 30 acres in the denominator. Therefore, the composite runoff coefficient could have been calculated as follows:

$$C_c = (0.95)(60\%) + (0.20)(30\%) + (0.30)(10\%) = 0.66$$

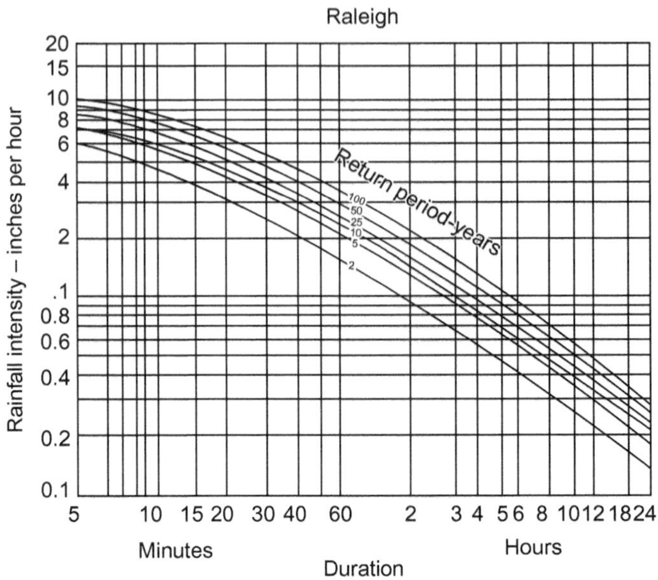

Figure 9.2 Intensity-duration-frequency chart for Raleigh, North Carolina.

Rainfall Intensity

The rainfall intensity I is determined from intensity-duration-frequency (IDF) charts developed by the National Weather Service. Figure 9.2 shows a typical IDF chart for Raleigh, North Carolina.

Along the bottom of the chart in Figure 9.2 is the duration, or "time of concentration," denoted t_c. It can be determined by various methods, however here the Kirpich equation will be used, given by the following expression:

$$t_c = \frac{\left(\frac{L^3}{H}\right)^{0.385}}{128} \tag{9.5}$$

where

t_c = time of concentration (min)

L = distance from the farthest point to the runoff exit (ft)

H = elevation change from the farthest point to the exit (ft)

and the "128" in the denominator represents unit conversions to give the time of concentration in minutes, so it is not a pure number.

For storm water runoff projects in the private sector, the minimum time of concentration is typically 5 min. However, in highway facilities projects through state DOTs, the minimum time of concentration is usually specified to be 10 min. As can be seen from the chart in Figure 9.2, a higher time of concentration will result in a lower rainfall intensity, and therefore a more conservative and less costly system. Also note that as the storm period increases, the rainfall intensity increases, as expected.

EXAMPLE 9.3 Rainfall Intensity

Suppose for the 30-acre highway facility in Example 9.2 that the farthest distance storm water would need to flow to a collection point is 1600 ft, and would have an elevation drop of 80 ft. Using the intensity-duration-frequency chart provided in Figure 9.2, determine the rainfall intensity *I*. Assume a storm period of 25 years.

Solution

Using Eq. (9.5), calculate the time of concentration t_c as:

$$t_c = \frac{\left(\frac{L^3}{H}\right)^{0.385}}{128} = \frac{\left(\frac{(1600\text{ ft})^3}{80\text{ ft}}\right)^{0.385}}{128} = 7.3\text{ min} < 10\text{ min}$$

Because the time of concentration is less than 10 min, use 10 min and the 25-year storm period on the intensity-duration-frequency chart, shown here.

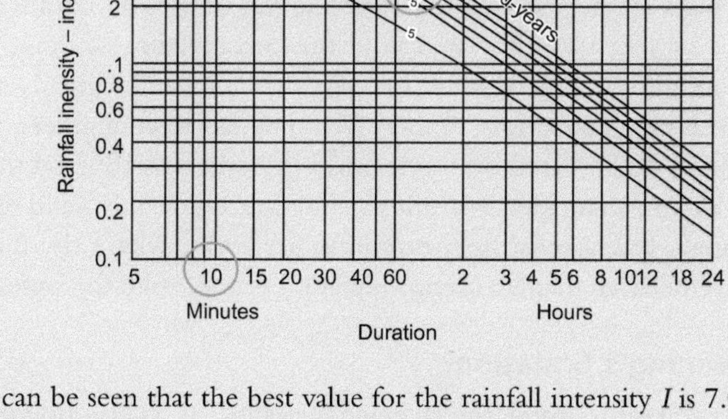

It can be seen that the best value for the rainfall intensity *I* is 7.0 in/h. This will be used in the calculation of the expected storm water runoff.

9.3.3 Storm Water Runoff Calculation

Now that all the elements of the rational method have been discussed, they can be brought together to estimate the storm water runoff for a particular highway facility.

EXAMPLE 9.4 Storm Water Runoff Calculation

Using the composite runoff coefficient found for the 30-acre highway facility in Example 9.2, namely 0.66, and the rainfall intensity found in Example 9.3, namely 7.0 in/h, determine the expected storm water runoff.

Solution

Using Eq. (9.3), calculate the storm water runoff Q as:

$$Q = C_c IA = (0.66)(7.0\ \text{in/h})(30\ \text{acres}) = 138.6\ \text{ft}^3/\text{s}$$

It seems like almost an anticlimax after all the steps to obtain the composite runoff coefficient and the rainfall intensity. It is just the nature of the rational method.

Once the expected storm water runoff is determined, the design process can move to the design of the collection system. Typically, this involves open channel flow to drop or curb inlets, then into a network of underground circular pipes to a point at which the storm water can be appropriately discharged.

9.4 OPEN CHANNEL FLOW

9.4.1 Introduction

The flow of storm water runoff is typically collected first in open channels, some of which might just be ditches. While the flow of all liquids and gases is governed by the laws associated with the conservation of mass, momentum, and energy, the flow of water in open channels can be greatly simplified by use of Manning's equation. Robert Manning, an Irish engineer, proposed his equation in 1890, based on the formula presented in 1867 by the French engineer Philippe Gauckler. Manning's equation is not only valid for flow in open channels, but also for flow in piping networks, where the flow is full. However, remember that Manning's equation is only valid for water.

9.4.2 Manning's Equation

Manning's equation gives the average velocity of water flowing in an open channel. When multiplied by the cross-sectional area of the flow, the flow rate is determined. As with most equations like Manning's equation, the various quantities in the equation must be in specific units. For the U.S. customary system, Manning's equation becomes:

$$V = \frac{1.486}{n} R^{2/3} S^{1/2} \quad (9.6)$$

where

V = average velocity of the flow (ft/s)

n = Manning's roughness coefficient (unitless)

R = hydraulic radius of the flow (ft)

S = longitudinal slope, or grade, of the channel (unitless)

and the "1.486" represents unit conversions to give the velocity in feet per second, so it is not a pure number. In some references, this number is rounded to 1.49.

To obtain the flow rate Q using the equation of continuity, multiply the velocity found from Eq. (9.6) by the cross-sectional area A of the flow to give:

$$Q = VA = \frac{1.486}{n} AR^{2/3} S^{1/2} \quad (9.7)$$

where

Q = flow rate (ft^3/s)

A = cross-sectional area of the flow (ft^2)

Again, enough cannot be said for being careful to use the units specified, particularly for the hydraulic radius R, the cross-sectional area A, and putting the grade S in decimal form. Any use of incorrect units will result in a system either entirely too large or entirely too small.

Manning's Roughness Coefficient

Manning's roughness coefficient n represents the friction between the storm water flowing in the channel and the surface of the channel. Typical values are presented in Table 9.3.

Table 9.3 Manning's roughness coefficients

Channel lining	***n***
Earth, bare and smooth	0.022
Earth, mostly weeds	0.050
Earth, well-kept grass	0.030
Concrete, troweled	0.013
Pipe, corrugated	0.024
Pipe, concrete	0.013

Notice that as the friction increases, Manning's coefficient increases, and because it is in the denominator of the equation, a greater value of n results in a lower velocity V, and subsequently a lower value for the flow rate Q. Also, while bare smooth earth has a lower Manning coefficient than well-kept grass, grass can tolerate a higher velocity than bare smooth earth without resulting in adverse erosion.

A very common channel lining is quarry stone. It is usually employed when the velocity of the flow is excessive for linings such as well-kept grass, and the cost of a concrete troweled channel is not justified. Manning's coefficient can be determined for various sizes of quarry stone by the following equation:

$$n = \frac{D_{\text{in}}^{1/6}}{44.4} \tag{9.8}$$

where

D = diameter of the quarry rock (in)

and, as expected, the "44.4" represents unit conversions to give Manning's coefficient n as dimensionless, so it is not a pure number.

Hydraulic Radius

The concept of a hydraulic radius R allows Manning's equation to be valid for any channel shape. The most common channel shape is trapezoidal, or its special case, triangular. Channels can be rectangular or partially filled circular pipes. In fact, Manning's equation is so powerful that it can also be used to design full-flowing storm water piping systems, discussed later.

The hydraulic radius R is defined as the cross-sectional area A of the flow, divided by the wetted perimeter P, as follows:

$$R = \frac{A}{P} \tag{9.9}$$

where

R = hydraulic radius (ft)

A = cross-sectional area of the flow (ft^2)

P = wetted perimeter of the flow (ft)

While the area A and the wetted perimeter P might be first determined in units of inches squared and inches, respectively, the value for the hydraulic radius R must be in feet when entered into Manning's equation.

EXAMPLE 9.5 Hydraulic Radius

Determine the general expression for the hydraulic radius of a circular pipe flowing full. Use the notation shown here:

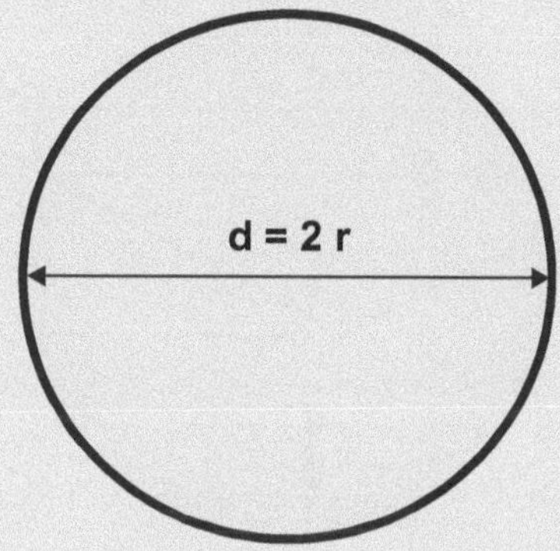

Solution

Using Eq. (9.9), substitute the standard formulas for the area of a circle, as the pipe is flowing full, and its perimeter, because the entire inside of the pipe is wetted, to give:

$$R = \frac{A}{P} = \frac{\pi r^2}{2\pi r} = \frac{r}{2}$$

This result is at first disconcerting, as it is expected that the hydraulic radius should be simply the radius of the pipe r. In fact, what is even more disconcerting is that the hydraulic radius for a pipe flowing half full is the same result, as seen by the following, where the flow area is half that of a pipe flowing full, and, similarly, the wetted perimeter is half of the flowing full value.

$$R = \frac{A}{P} = \frac{1/2\pi r^2}{\pi r} = \frac{r}{2}$$

For flows greater than half full or less than half full, the hydraulic radius is no longer half the actual radius r, but is determined from a very complex trigonometric expression. A chart presented later will get around this mathematical complexity.

Velocity and Flow Rate—Depth Known

When the depth of the flow in a channel is known, the calculations of the velocity V and flow rate Q are straightforward applications of Manning's equation. If the depth of the flow is not known, the problem is more complicated. This situation is discussed in the next sections. However, an example where the depth is known follows.

EXAMPLE 9.6 Velocity and Flow Rate—Depth Known

For the triangular channel shown, the depth is 3 ft and the width of the flow is 10 ft. The channel is lined with 8-in quarry stone. The channel has a slope of 4%. Determine the velocity V and flow rate Q.

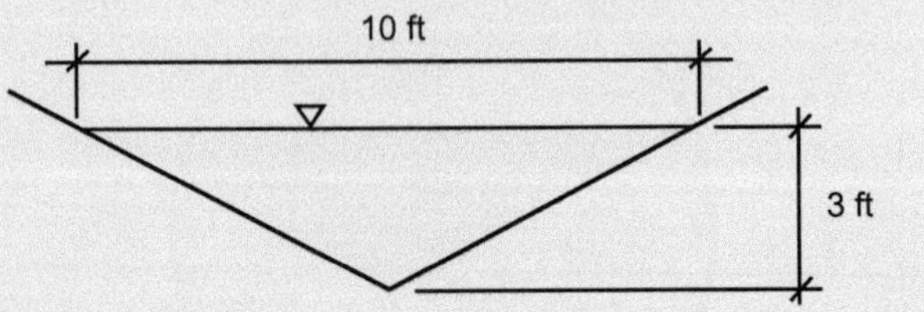

Solution

Using Eq. (9.8), calculate Manning's roughness coefficient for 8-in quarry stone as:

$$n = \frac{D_{\text{in}}^{1/6}}{44.4} = \frac{(8\ \text{in})^{1/6}}{44.4} = 0.032$$

The hydraulic radius is found from Eq. (9.9), where the wetted perimeter is the length of the inclined banks, found from the following diagram and calculation:

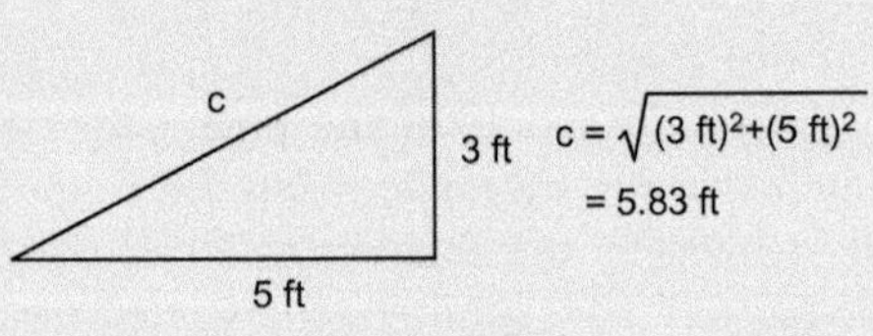

Therefore, the wetted perimeter P is twice the hypotenuse c, or 11.66 ft. The area of the flow is the area of a triangle, where the base is 10 ft and the height is 3 ft. The hydraulic radius becomes:

$$R = \frac{A}{P} = \frac{1/2(10\ \text{ft})(3\ \text{ft})}{2(5.83\ \text{ft})} = \frac{15\ \text{ft}^2}{11.66\ \text{ft}} = 1.29\ \text{ft}$$

Applying Manning's equation from Eq. (9.6), remembering to change the slope to decimal form, gives the average velocity V of the flow as:

$$V = \frac{1.486}{n} R^{2/3} S^{1/2} = \frac{1.486}{0.032}(1.29)^{2/3}(0.04)^{1/2}$$

$$= 11.0\ \text{ft/s}$$

(*Continued*)

EXAMPLE 9.6 Velocity and Flow Rate—Depth Known—(Continued)

Multiplying this velocity by the flow area A gives the flow rate Q as:

$$Q = VA = (11.0\ \text{ft/s})(15\ \text{ft}^2) = 165.0\ \text{ft}^3/\text{s}$$

This flow rate can be used to check the calculation for the flow rate Q from the rational method, as this actual measured depth of flow should reflect the composite runoff coefficient, the rainfall intensity, and the area of the highway facility.

9.4.3 Trapezoidal Channels

The most common storm water runoff channel shape is trapezoidal. Instead of calculating the hydraulic radius from scratch each time, a set of equations for the flow area A and wetted perimeter P make the calculations more manageable. Consider the general trapezoid in Figure 9.3, where typical notation has been used.

The depth of flow is labeled y, the width of the base is labeled B, and the slope of the sides is designated M, usually specified as M:1. Clearly, the greater the value of M, the flatter are the sides of the channel.

The trapezoidal area can be broken into three parts: two identical triangular pieces and one rectangular piece. They are shown in Figure 9.4.

Therefore, the total area is the sum of the areas of the three pieces, given as:

$$A = (B)(y) + (2)[1/2(My)(y)] = By + My^2 \tag{9.10}$$

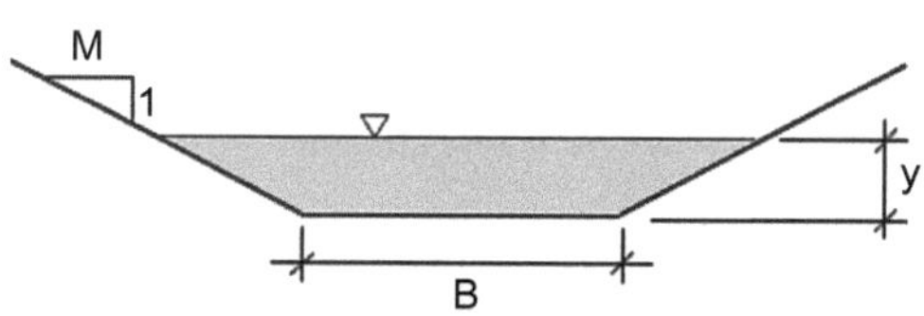

Figure 9.3 General trapezoidal channel.

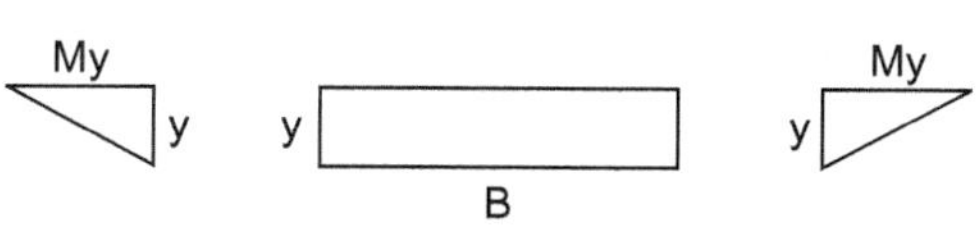

Figure 9.4 Parts of trapezoidal flow area.

Similarly, the wetted perimeter is the sum of the length to the two inclined sides plus the base *B*. The length of one of the inclined sides can be found from the Pythagorean theorem as:

$$\sqrt{(y)^2 + (My)^2} = \sqrt{y^2 + M^2y^2} = \sqrt{y^2(1 + M^2)} = y\sqrt{1 + M^2}$$

Therefore, the wetted perimeter becomes:

$$P = B + 2y\sqrt{1 + M^2} \tag{9.11}$$

Once specific values for the flow area and wetted perimeter are determined, the hydraulic radius can be found from Eq. (9.9).

EXAMPLE 9.7 Hydraulic Radius for Trapezoidal Channel

For the trapezoidal channel shown, the depth is 4 ft and the width of the base is 2 ft. The slope of the sides is 2:1. Determine the hydraulic radius.

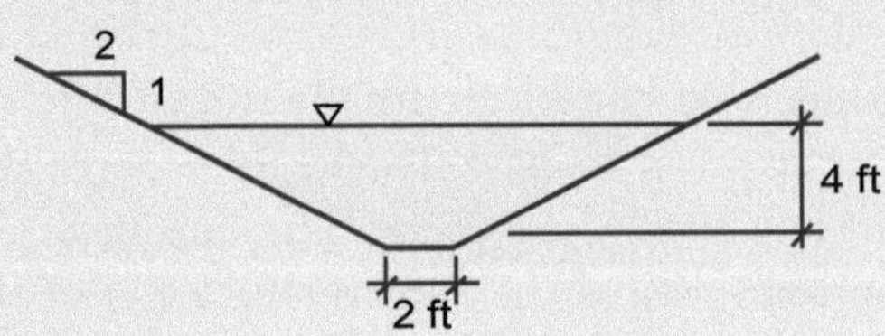

Solution

Using Eq. (9.10), calculate the flow area *A* using $M = 2$, to obtain:

$$A = By + My^2 = (2\text{ ft})(4\text{ ft}) + (2)(4\text{ ft})^2$$
$$= 40\text{ ft}^2$$

Using Eq. (9.11), calculate the wetted perimeter *P* as:

$$P = B + 2y\sqrt{1 + M^2} = (2\text{ ft}) + (2)\,(4\text{ ft})\sqrt{1 + (2)^2}$$
$$= 19.9\text{ ft}$$

Substituting for the flow area *A* and wetted perimeter *P* into Eq. (9.9), gives the hydraulic radius as:

$$R = \frac{A}{P} = \frac{40\text{ ft}^2}{19.9\text{ ft}} = 2.01\text{ ft}$$

With information about the surface type of the trapezoidal channel of Example 9.7, and its longitudinal slope, or grade, the average velocity V and flow rate Q could be determined by applying Manning's equation. However, usually the design process for a storm water collection system begins with the flow rate Q, and the depth of flow is determined. Because the depth of flow y is part of both terms in the hydraulic radius, flow area A and wetted perimeter P, the solution becomes a process of trial and error.

Four possible approaches have been established to determine the unknown information about the design of a channel, in particular for trapezoidal shapes, the most common. The four approaches are:

1. Normal depth procedure: the shape of the trapezoidal channel is given, and the expected depth is determined.
2. Depth-limited procedure: the maximum depth of flow is specified, and the width of the base of the trapezoidal channel is determined.
3. Velocity-limited procedure: the maximum velocity is specified due to the limitations of the lining of the trapezoidal channel, and the depth of flow and width of the channel are determined.
4. Best hydraulic section procedure: parameters have been established to produce the design that minimizes the excavation needed and the surface area for an appropriate lining.

Each of these four procedures will be discussed in the next sections.

9.4.4 Normal Depth Procedure

If the flow rate Q entering a trapezoidal channel from a highway facility is known, the shape is known, and the most appropriate slope has been determined, the only unknown left is the depth of flow y. However, as stated earlier, the depth of flow y is contained in both terms of the hydraulic radius R, the flow area A, and the wetted perimeter P. If the terms in Manning's equation for flow rate are rearranged to put the unknowns on the left side and the known values on the right side, then the following expression results:

$$AR^{2/3} = \frac{Qn}{1.486\sqrt{S}} \tag{9.12}$$

where for the grade, $S^{1/2} = \sqrt{S}$, and the units for the right side are $\text{ft}^{8/3}$.

The normal depth procedure proceeds by assuming a depth of flow y, calculating the flow area A, the wetted perimeter P, and then the

Table 9.4 Normal depth procedure worksheet

y (ft)	B (ft)	A (ft^2)	P (ft)	R (ft)	$AR^{2/3}$	Remarks

hydraulic radius R, then calculating the left side of Eq. (9.12) and comparing it to the value on the right side. If the left side is less than the right side, the depth of flow y chosen is too shallow, so the value must be increased and the process repeated. If the value of the left side is greater than the right side, reduce the value chosen. Continue this process until the two sides of equation are equal within the accuracy desired.

This process can be simplified by setting up a table similar to that shown in Table 9.4.

Note that for the normal depth procedure worksheet, the column labeled "B" will contain the same value for all iterations on the flow depth y. In the "Remarks" column, phrases such as "too shallow" or "too deep" or "close enough" are used. Trial and error can appear to be intimidating, however, once the value is bracketed between "too shallow" and "too deep," the process proceeds rapidly to an acceptably accurate value. The following example demonstrates the ease of convergence.

EXAMPLE 9.8 Normal Depth Procedure

For the trapezoidal channel shown, the width of the base is 3 ft and the slope of the sides is 2:1, with all surfaces bare and smooth earth. If the channel slope is 2% and the flow rate Q entering the channel from a highway facility is 75 ft^3/s, determine the expected depth of the flow y.

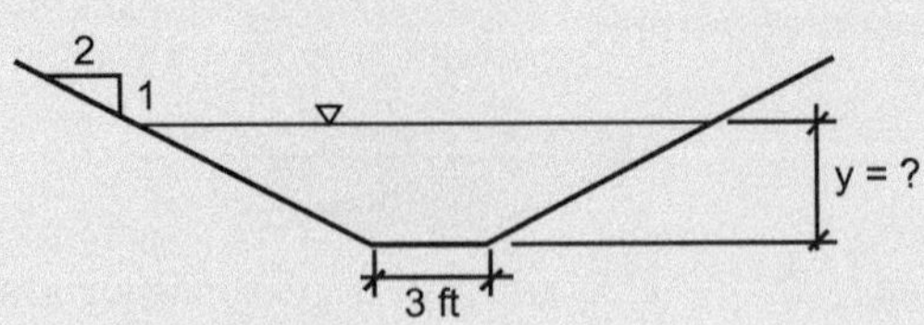

(Continued)

EXAMPLE 9.8 Normal Depth Procedure—(Continued)

Solution

Using Eq. (9.12), substitute the given information, including Manning's roughness coefficient n from Table 9.3 of 0.022, and calculate the right side as:

$$AR^{2/3} = \frac{Qn}{1.486\sqrt{S}} = \frac{(75\ \text{ft}^3/\text{s})\ (0.022)}{1.486\sqrt{0.02}}$$

$$= 7.85\ \text{ft}^{8/3}$$

Selecting an appropriate first value for the flow depth y comes from experience, however, selecting "1 ft" is always a good starting choice. Using this value, calculate the appropriate values and record them in a table like Table 9.4, to give the following:

y (ft)	B (ft)	A (ft^2)	P (ft)	R (ft)	$AR^{2/3}$	Remarks
1	3	5	7.47	0.669	3.83	Too shallow

Because the left side came out to be less than the right side, a value for the flow depth y of 1 ft is too shallow. The value obtained suggests a second calculation for y equal to 2 ft. This gives the following values, recorded in the table as follows:

y (ft)	B (ft)	A (ft^2)	P (ft)	R (ft)	$AR^{2/3}$	Remarks
1	3	5	7.47	0.669	3.83	Too shallow
2	3	14	11.94	1.172	15.56	Too deep

Because the left side came out to be greater than the right side, a value for the flow depth y of 2 ft is too deep. The value obtained, about double the right side, suggests a third calculation for y equal to 1.5 ft. This gives the following values, recorded in the table as follows:

y (ft)	B (ft)	A (ft^2)	P (ft)	R (ft)	$AR^{2/3}$	Remarks
1	3	5	7.47	0.669	3.83	Too shallow
2	3	14	11.94	1.172	15.56	Too deep
1.5	3	9	9.71	0.927	8.56	Still too deep

The process is getting close to convergence, however, the choice of depth as 1.5 ft is still too deep. The value obtained suggests a fourth calculation with y equal to 1.4 ft. This gives the following values:

y (ft)	B (ft)	A (ft^2)	P (ft)	R (ft)	$AR^{2/3}$	Remarks
1	3	5	7.47	0.669	3.83	Too shallow
2	3	14	11.94	1.172	15.56	Too deep
1.5	3	9	9.71	0.927	8.56	Still too deep
1.4	3	8.12	9.26	0.877	7.44	Close enough

(*Continued*)

EXAMPLE 9.8 Normal Depth Procedure—(Continued)

As can be seen, the value on the left side is now "close enough" to the value on the right side, 7.44 verses 7.85, respectively. There is only a 5% difference, and based on how the flow rate Q was obtained using methods such as the rational method with all the many assumptions and average values, this is indeed close enough.

Though not required, the average velocity of the flow V can be obtained by dividing the flow rate Q by the area of the flow A for the depth of 1.4 ft. This gives:

$$V = \frac{Q}{A} = \frac{75 \text{ ft}^3/\text{s}}{8.12 \text{ ft}^2} = 9.24 \text{ ft/s}$$

This value for the average velocity is at least three times the allowable value for bare and smooth earth, and double the allowable value for well-kept grass. So either quarry stone or smooth concrete would need to be used for this particular channel.

9.4.5 Depth-Limited Procedure

For some applications, the depth of flow y in a trapezoidal channel is constrained to a particular value. For this situation, the width of the base of the channel B is determined by the same trial-and-error process used in the previous section for the normal depth procedure. The only difference is that in the "remarks" column either "too narrow" or "too wide" are used. Consider the following example.

EXAMPLE 9.9 Depth-Limited Procedure

For the trapezoidal channel shown, the depth of flow y is limited to 2 ft. The slope of the sides is 1:1, with all surfaces concrete troweled. If the channel slope is 0.5%, and the flow rate Q entering the channel from a highway facility is 115 ft^3/s, determine the required width of the base B. Also, determine the average velocity of the flow V using the dimensions of the final channel.

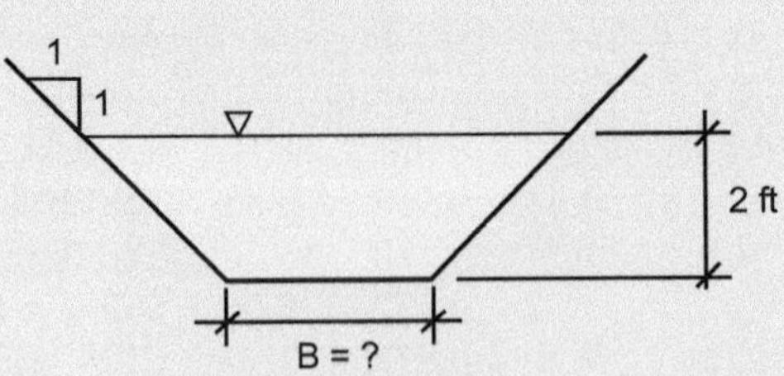

(Continued)

EXAMPLE 9.9 Depth-Limited Procedure—(Continued)

Solution

Using Eq. (9.12), substitute the given information, including Manning's roughness coefficient n from Table 9.3 of 0.013, and calculate the right side as:

$$AR^{2/3} = \frac{Qn}{1.486\sqrt{S}} = \frac{(115\ \text{ft}^3/\text{s})(0.013)}{1.486\sqrt{0.005}}$$

$$= 14.23\ \text{ft}^{8/3}$$

Just as selecting an appropriate first value for the flow depth y comes from experience, selecting a starting value for the width of the base B is also based on experience. However, selecting 1 ft is always a good starting choice. Using this value, calculate the appropriate values and record them in a table similar to Table 9.4, to give the following:

y (ft)	B (ft)	A (ft^2)	P (ft)	R (ft)	$AR^{2/3}$	Remarks
2	1	6	6.66	0.901	5.60	Too narrow

Because the left side came out to be less than the right side, a value for the width of the base B of 1 ft is too narrow. The value obtained suggests a second calculation for B equal to 2 ft. This gives the following values, recorded in the table as follows:

y (ft)	B (ft)	A (ft^2)	P (ft)	R (ft)	$AR^{2/3}$	Remarks
2	1	6	6.66	0.901	5.60	Too narrow
2	2	8	7.66	1.045	8.24	Too narrow

Because the left side came out to still be less than the right side, a value for the width of the base B of 2 ft is still too narrow. The value obtained suggests a third calculation for B equal to 4 ft. This gives the following values, recorded in the table as follows:

y (ft)	B (ft)	A (ft^2)	P (ft)	R (ft)	$AR^{2/3}$	Remarks
2	1	6	6.66	0.901	5.60	Too narrow
2	2	8	7.66	1.045	8.24	Too narrow
2	4	12	9.66	1.243	13.87	Close enough?

The value obtained for the left side is very close to the right side. To confirm, select a value of 4.1 ft to obtain the following values:

y (ft)	B (ft)	A (ft^2)	P (ft)	R (ft)	$AR^{2/3}$	Remarks
2	1	6	6.66	0.901	5.60	Too narrow
2	2	8	7.66	1.045	8.24	Too narrow
2	4	12	9.66	1.243	13.87	Close enough?
2	4.1	12.2	9.76	1.250	14.16	Close enough

(Continued)

EXAMPLE 9.9 Depth-Limited Procedure—(Continued)

As can be seen, the value on the left side is now close enough to the value on the right side, 14.16 verses 14.23, respectively. There is only a 0.5% difference, and again, based on how the flow rate Q was obtained using methods such as the rational method with all the many assumptions and average values, this is indeed close enough.

The average velocity of the flow V can be obtained by dividing the flow rate Q by the area of the flow A for the width of the base B of 4.1 ft. This gives:

$$V = \frac{Q}{A} = \frac{115\ \text{ft}^3/\text{s}}{12.2\ \text{ft}^2} = 9.43\ \text{ft/s}$$

For concrete troweled surfaces, this value for the average velocity is acceptable.

9.4.6 Velocity-Limited Procedure

As often happens, the average velocity V of the flow determined from either the normal depth or depth-limited procedures exceeds the allowable velocity of the channel lining. For this situation, the necessary channel shape can be determined by the velocity-limited procedure.

From the expected storm water runoff flow rate Q and the allowable velocity V_a for the proposed channel lining, the required cross-sectional area, designated A_x, is found from the following calculation:

$$A_x = \frac{Q}{V_a} \tag{9.13}$$

The depth of flow y is then determined from the expression that clearly has the form of the quadratic equation:

$$y = \frac{-w_2 + \sqrt{w_2^2 + 4w_1 A_x}}{2w_1} \tag{9.14}$$

where using other typical information about the channel gives:

$$w_1 = M - 2\sqrt{1 + M^2} \tag{9.15}$$

$$w_2 = \frac{A_x}{\left(\frac{V_a n}{1.486\sqrt{S}}\right)^{3/2}} \tag{9.16}$$

The terms w_1 and w_2 do not have any physical significance, they just make the calculation for the depth of flow y easier.

Once the depth of flow y is known, the width of the base B of the channel is determined from the equation:

$$B = \frac{A_X}{y} - My \tag{9.17}$$

While this procedure might seem like the place to have started, typically the depth of flow y is too shallow, and/or the width of the base of the channel is excessively wide. An example will highlight this result.

EXAMPLE 9.10 Velocity-Limited Procedure

For the trapezoidal channel in Example 9.8, the average velocity V found was three times the allowable velocity for bare and smooth earth. Using a maximum allowable velocity of 3 ft/s, determine the depth of flow y and the width of the base of the channel B from the velocity-limited procedure. The cross section is shown here, where the slope of the sides is 2:1, the channel slope is 2%, and the flow rate Q entering the channel from a highway facility is 75 ft³/s.

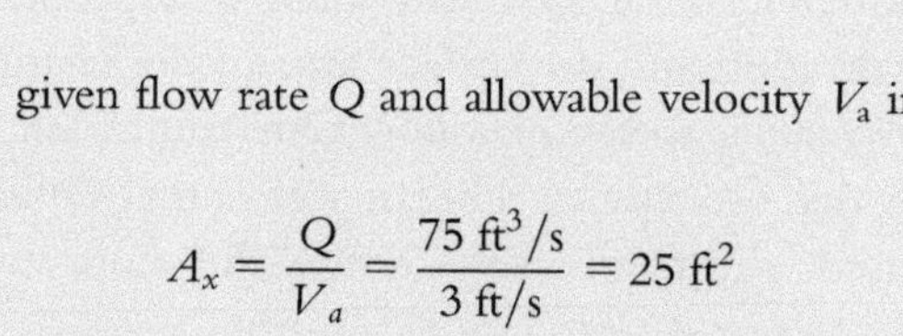

Solution

Substitute the given flow rate Q and allowable velocity V_a into Eq. (9.13) to give:

$$A_x = \frac{Q}{V_a} = \frac{75\ \text{ft}^3/\text{s}}{3\ \text{ft/s}} = 25\ \text{ft}^2$$

Using Eqs. (9.15) and (9.16), substitute this required cross-sectional area and other given information, including Manning's roughness coefficient n from Table 9.3 of 0.022, and calculate the quantities w_1 and w_2, respectively, as:

$$w_1 = M - 2\sqrt{1 + M^2} = 2 - 2\sqrt{1 + 2^2} = -2.47$$

$$w_2 = \frac{A_x}{\left(\dfrac{V_a n}{1.486\sqrt{S}}\right)^{3/2}} = \frac{25\ \text{ft}^2}{\left(\dfrac{(3\ \text{ft/s})\,(0.022)}{1.486\sqrt{0.02}}\right)^{3/2}} = 142.04$$

(Continued)

EXAMPLE 9.10 Velocity-Limited Procedure—(Continued)

Substitute these values for w_1 and w_2 into Eq. (9.14) to give:

$$y = \frac{-w_2 + \sqrt{w_2^2 + 4w_1 A_x}}{2w_1}$$

$$= \frac{-(142.04) + \sqrt{(142.04)^2 + 4(-2.47)(25)}}{2(-2.47)}$$

$$= \frac{-(142.04) + (141.17)}{-4.94} = 0.18 \text{ ft}$$

$$\approx 2 \text{ in}$$

This is clearly too shallow. To continue, use this value for the depth of flow y to determine the width of the base B from Eq. (9.17) as:

$$B = \frac{A_x}{y} - My = \frac{25 \text{ ft}^2}{0.18 \text{ ft}} - (2)(0.18 \text{ ft})$$

$$= 138.53 \text{ ft}$$

This width is clearly excessively wide. So, again, the only choice to address the potential erosion problem is to either add quarry stone or replace the bare and smooth earth with troweled concrete.

9.4.7 Best Hydraulic Section Procedure

There are established criteria that will yield a trapezoidal channel that minimizes the excavation and the surface area of the channel. The process is called the *best hydraulic section procedure*. One might ask, why didn't the process start with this procedure? The answer is that typically all the procedures will be used, and from all the results the most appropriate design is chosen.

The best hydraulic section procedure calculates the depth of flow y and the width of the base of the channel B from the following two equations:

$$y = C_m \left(\frac{Qn}{\sqrt{S}} \right)^{3/8} \tag{9.18}$$

$$B = ky \tag{9.19}$$

where the parameters C_m and k are functions of the slope of the sides M, given for typical values of M in Table 9.5.

Table 9.5 Best hydraulic section parameters

M	C_m	*k*
0.5	0.833	1.236
1	0.817	0.828
2	0.729	0.427
3	0.653	0.325
4	0.595	0.246
5	0.552	0.198

The following example highlights this procedure.

EXAMPLE 9.11 Best Hydraulic Section Procedure

For the trapezoidal channel shown in Example 9.9, the depth of flow y was limited to 2 ft, and the width of the base of the channel B was found to be about 4 ft. The slope of the sides is 1:1, with all surfaces concrete troweled. If the channel slope is 0.5%, and the flow rate Q entering the channel from a highway facility is 115 ft^3/s, determine the shape of the channel using the best hydraulic section procedure.

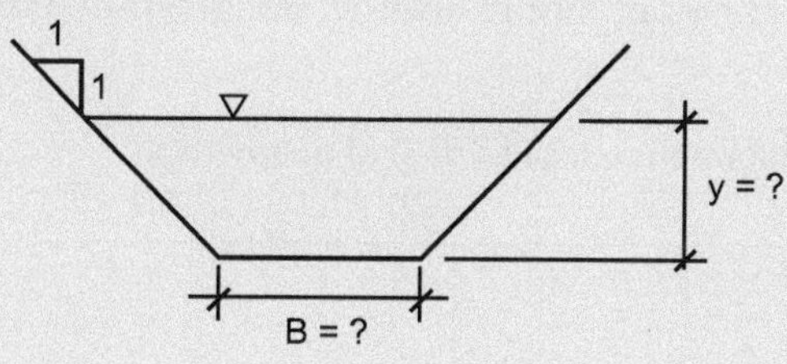

Solution

Using Eq. (9.18), substitute the given information, including Manning's roughness coefficient n from Table 9.3 of 0.013, the parameter C_m of 0.817 from Table 9.5, and calculate the depth of flow y as:

$$y = C_m \left(\frac{Qn}{\sqrt{S}} \right)^{3/8} = (0.817) \left(\frac{(115 \text{ ft}^3/\text{s})(0.013)}{\sqrt{0.005}} \right)^{3/8}$$

$$= 2.57 \text{ ft}$$

This is very close to the value specified of 2 ft. Now, calculate the width of the base of the channel B using Eq. (9.19), with the parameter k from Table 9.5 equal to 0.828.

$$B = ky = (0.828)(2.57 \text{ ft}) = 2.12 \text{ ft}$$

(Continued)

EXAMPLE 9.11 Best Hydraulic Section Procedure—(Continued)

This value is about half the value determined in Example 9.9. The size determined here would give a different velocity, obtained as follows:

$$V = \frac{Q}{A} = \frac{Q}{By + My^2} = \frac{115\ \text{ft}^3/\text{s}}{(2.12\ \text{ft})(2.57\ \text{ft}) + (1)(2.57\ \text{ft})^2} = \frac{115\ \text{ft}^3/\text{s}}{12.05\ \text{ft}^2}$$

$$= 9.54\ \text{ft/s}$$

This is only 0.1 ft/s greater, so the concrete troweled surface is still acceptable. It is also interesting that even with the significant change in shape, the flow areas are essentially the same, and so the average velocities are essentially the same.

9.4.8 Typical Design Steps

To pull these four procedures together, consider the following design problem. However, values for the allowable velocity for various sizes of quarry rock and side slope of the channel are given in Table 9.6.

Table 9.6 Allowable velocities (ft/s) of quarry rock

Size (in)	1:1	2:1	3:1	4:1
4	5.0	6.3	6.6	7.1
6	6.0	7.7	8.0	8.4
9	7.2	9.4	9.8	10.2
12	8.5	10.7	11.2	11.8

Summary Example

A trapezoidal channel is to carry 600 ft^3/s of storm water runoff from a highway facility. If the channel slope is 1% and cannot have a depth of flow greater than 4 ft, consider a design using 4-in quarry stone. The side slopes of the channel are to be 2:1.

Starting with the best hydraulic section procedure, calculate the depth of flow y and the width of the bottom B using Eqs. (9.18) and (9.19), however, first determine Manning's roughness coefficient from Eq. (9.8) as:

$$n = \frac{D_{\text{in}}^{1/6}}{44.4} = \frac{(4\ \text{in})^{1/6}}{44.4} = 0.028$$

(Continued)

Summary Example—(Continued)

The depth of flow y becomes, with the parameter C_m of 0.729 from Table 9.5:

$$y = C_m\left(\frac{Qn}{\sqrt{S}}\right)^{3/8} = (0.729)\left(\frac{(600\ \text{ft}^3/\text{s})(0.028)}{\sqrt{0.01}}\right)^{3/8}$$

$$= 4.98\ \text{ft}$$

which is greater than the maximum allowable depth. However, continue by finding the width of the base B, with the parameter k from Table 9.5 of 0.427, as:

$$B = ky = (0.427)(4.98\ \text{ft}) = 2.13\ \text{ft}$$

Using these values, the velocity is calculated to be:

$$V = \frac{Q}{A} = \frac{Q}{By + My^2} = \frac{600\ \text{ft}^3/\text{s}}{(2.13\ \text{ft})(4.98\ \text{ft}) + (2)(4.98\ \text{ft})^2} = \frac{600\ \text{ft}^3/\text{s}}{60.21\ \text{ft}^2}$$

$$= 9.96\ \text{ft/s}$$

This velocity is greater than the value specified in Table 9.6 of 6.3 ft/s. So, to continue the design options, use this allowable velocity in the velocity-limited procedure and see what channel shape would be required.

Substitute the given flow rate Q and allowable velocity V_a into Eq. (9.13) to give:

$$A_x = \frac{Q}{V_a} = \frac{600\ \text{ft}^3/\text{s}}{6.3\ \text{ft/s}} = 95.24\ \text{ft}^2$$

Using Eqs. (9.15) and (9.16), substitute this required cross-sectional area and other given information, including Manning's roughness coefficient n found previously as 0.028, and calculate the quantities w_1 and w_2, respectively, as:

$$w_1 = M - 2\sqrt{1 + M^2} = 2 - 2\sqrt{1 + 2^2} = -2.47$$

$$w_2 = \frac{A_x}{\left(\dfrac{V_a n}{1.486\sqrt{S}}\right)^{3/2}} = \frac{95.24\ \text{ft}^2}{\left(\dfrac{(6.3\ \text{ft/s})(0.028)}{1.486\sqrt{0.01}}\right)^{3/2}} = 73.64$$

Substitute these values for w_1 and w_2 into Eq. (9.14) to give:

$$y = \frac{-w_2 + \sqrt{w_2^2 + 4w_1 A_x}}{2\ w_1}$$

$$= \frac{-(73.64) + \sqrt{(73.64)^2 + 4(-2.47)(95.24)}}{2(-2.47)}$$

$$= \frac{-(73.64) + (66.95)}{-4.94} = 1.35\ \text{ft}$$

(Continued)

Summary Example—(Continued)

This is bordering on being too shallow. To continue, use this value for the depth of flow y to determine the width of the base B from Eq. (9.17) as:

$$B = \frac{A_x}{y} - My = \frac{95.24\ \text{ft}^2}{1.35\ \text{ft}} - (2)(1.35\ \text{ft})$$

$$= 67.85\ \text{ft}$$

This width is clearly excessively wide. The 4-in quarry stone is too small, so a larger size should be chosen. However, just for curiosity, the depth-limited procedure can be used to see what width of the base of the channel would result.

Using Eq. (9.12), substitute the given information, including Manning's roughness coefficient n for the 4-in quarry stone found to be 0.028, and calculate the right side as:

$$AR^{2/3} = \frac{Qn}{1.486\sqrt{S}} = \frac{(600\ \text{ft}^3/\text{s})(0.028)}{1.486\sqrt{0.01}}$$

$$= 113.06\ \text{ft}^{8/3}$$

With the depth of flow y limited to 4 ft, select 1 ft as a good starting choice for the width of the base B. Using this value, calculate the appropriate values and record them in a table similar to Table 9.4, to give the following:

y (ft)	*B* (ft)	*A* (ft²)	*P* (ft)	*R* (ft)	*AR*$^{2/3}$	Remarks
4	1	36	18.89	1.906	55.34	Way too narrow

Because the left side came out to be significantly less than the right side, a value for the width of the base B of 1 ft is way too narrow. The value obtained suggests a second calculation for B equal to 2 ft. This gives the following values, recorded in the table here:

y (ft)	*B* (ft)	*A* (ft²)	*P* (ft)	*R* (ft)	*AR*$^{2/3}$	Remarks
4	1	36	18.89	1.906	55.34	Too narrow
4	2	40	19.89	2.011	63.73	Still too narrow

Because the left side came out to still be less than the right side, a value for the width of the base B of 2 ft is still too narrow. The value obtained suggests a third calculation for B equal to 10 ft. This gives the following values, recorded in the table as follows:

y (ft)	*B* (ft)	*A* (ft²)	*P* (ft)	*R* (ft)	*AR*$^{2/3}$	Remarks
4	1	36	18.89	1.906	55.34	Too narrow
4	2	40	19.89	2.011	63.73	Still too narrow
4	10	72	27.89	2.582	135.50	Too wide

(*Continued*)

Summary Example—(Continued)

The value obtained for the left side is now greater than the right side, meaning a width of 10 ft is too wide. This value suggests a fourth value of 8 ft, which gives the following results:

y (ft)	*B* (ft)	*A* (ft²)	*P* (ft)	*R* (ft)	$AR^{2/3}$	Remarks
4	1	36	18.89	1.906	55.34	Too narrow
4	2	40	19.89	2.011	63.73	Still too narrow
4	10	72	27.89	2.582	135.50	Too wide
4	8	64	25.89	2.472	117.01	Still too wide

As can be seen, the value on the left side is still too wide, and so a fifth iteration for a value of 7 ft is used, giving:

y (ft)	*B* (ft)	*A* (ft²)	*P* (ft)	*R* (ft)	$AR^{2/3}$	Remarks
4	1	36	18.89	1.906	55.34	Too narrow
4	2	40	19.89	2.011	63.73	Still too narrow
4	10	72	27.89	2.582	135.50	Too wide
4	8	64	25.89	2.472	117.01	Still too wide
4	7	60	24.89	2.411	107.87	Still too narrow

It appears that a value in-between 7 and 8 ft will provide convergence. Use a value of 7.5 ft to confirm that it is probably close enough.

y (ft)	*B* (ft)	*A* (ft²)	*P* (ft)	*R* (ft)	$AR^{2/3}$	Remarks
4	1	36	18.89	1.906	55.34	Too narrow
4	2	40	19.89	2.011	63.73	Still too narrow
4	10	72	27.89	2.582	135.50	Too wide
4	8	64	25.89	2.472	117.01	Still too wide
4	7	60	24.89	2.411	107.87	Still too narrow
4	7.5	62	25.39	2.442	112.43	Close enough

The average velocity of the flow V can now be obtained by dividing the flow rate Q by the area of the flow A for the width of the base B of 7.5 ft. This gives:

$$V = \frac{Q}{A} = \frac{600\ \text{ft}^3/\text{s}}{62\ \text{ft}^2} = 9.68\ \text{ft/s}$$

So the 4-in quarry rock is just not acceptable. The design process would proceed to the next size, and eventually an acceptable size would be established.

9.4.9 Triangular Channels

Many channels that collect storm water runoff are essentially triangular in shape. The equations for the flow area A and the wetted perimeter P for a trapezoidal channel can be used, where the width of the base of the channel B is set equal to zero. This transforms Eqs. (9.10) and (9.11) into the following:

$$A = By + My^2 = (0)y + My^2 = My^2 \quad (9.20)$$

$$P = B + 2\,y\sqrt{1+M^2} = (0) + 2y\sqrt{1+M^2} = 2y\sqrt{1+M^2} \quad (9.21)$$

Now, the hydraulic radius R is only a function of the flow depth y, given as:

$$R = \frac{A}{P} = \frac{My^2}{2y\sqrt{1+M^2}} = \frac{My}{2\,\sqrt{1+M^2}} \quad (9.22)$$

This simplifies the design process considerably, though the algebra is more involved.

EXAMPLE 9.12 Triangular Channel

For the triangular channel shown, the flow rate from a highway facility is 60 ft^3/s. The slope of the sides is 3:1, with all surfaces concrete troweled. If the channel slope is 4%, determine the depth of flow y.

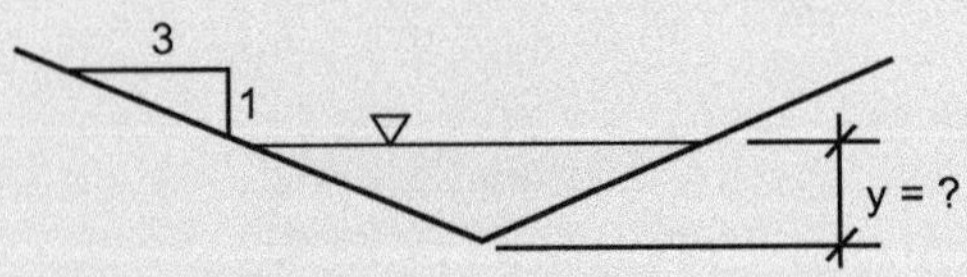

Solution

Because the only unknown in Manning's equation for the flow rate Q is the depth of flow y, substitute the given information, including Manning's roughness coefficient n from Table 9.3 of 0.013, to provide the following sequence of algebraic operations:

$$Q = \frac{1.486}{n} AR^{2/3}S^{1/2}$$

$$60\ ft^3/s = \frac{1.486}{0.013}\left[(3)y^2\right]\left[\frac{(3)y}{2\,\sqrt{1+(3)^2}}\right]^{2/3}(0.04)^{1/2}$$

$$60\ ft^3/s = (41.71)y^{8/3}$$

$$y^{8/3} = \frac{60}{41.71} = 1.438$$

$$y = (1.438)^{3/8}$$

$$y = 1.15\ ft$$

9.4.10 Summary

This concludes the design process for the various types of channels that could be used to collect the storm water runoff from a highway facility. The water in these channels could just empty into an area where it is either absorbed into the ground, or an environmentally acceptable stream. However, usually this storm water enters a network of circular pipes by dropping into several types of inlets. This can be drop inlets of various shapes, or curb-type inlets, and these can be either full flowing or partially filled. The next section considers the decisions that need to be made to handle the volume of water appropriately.

9.5 INLETS AND WEIRS

9.5.1 Introduction

Storm water runoff flowing in open channels, if not deposited directly into an area for absorption into the ground or into an environmentally acceptable stream, usually flows into a network of underground piping, which is then deposited in an area for absorption into the ground or into an environmentally acceptable stream. It is transferred from the open channel to the underground pipes by way of inlets into the piping network. These inlets are typically of two types: (1) drop or (2) curb. A typical drop inlet is shown in Figure 9.5 and a typical curb inlet in Figure 9.6.

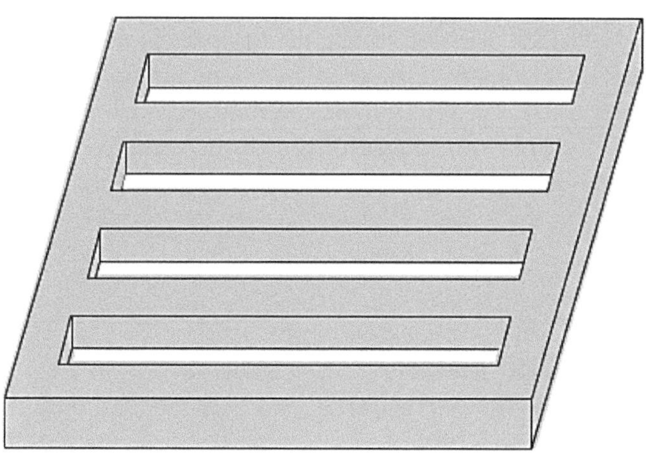

Figure 9.5 Typical drop inlet.

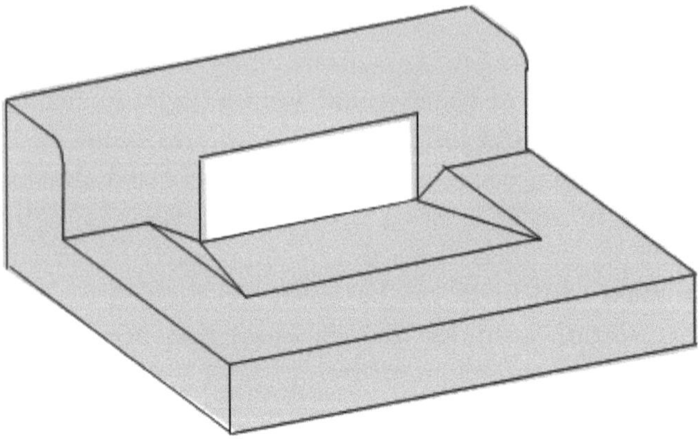

Figure 9.6 Typical curb inlet.

Drop inlets come in many designs, some with only several long openings, others have a great many small openings. The use of drop inlets that is formed by having a single large opening with a roof set several inches above the opening has been abandoned for safety reasons. When these drop inlets were employed in the median of multilane highways, they presented a very dangerous obstacle to vehicles, which were forced from the roadway. These have been replaced largely with the flush design shown in Figure 9.5.

The curb inlet shown in Figure 9.6 indicates only a vertical opening, whereas most such inlets also provide a horizontal drop inlet in front of the vertical opening. However, to simplify the discussion, only the type shown in Figure 9.6 will be considered.

Storm water runoff flows to these inlets either at a rate that floods the inlet opening, creating a "head" of water driving the flow, or the water overfalls into the opening. The flowing full condition is governed by an equation developed from the Bernoulli equation, where the pressure everywhere is atmospheric and the velocity at the surface of the storm water backup is zero. The flow rate Q into the inlet is therefore given by the equation:

$$Q = C_d A \sqrt{2gh} \tag{9.23}$$

where

Q = flow rate (ft^3/s)
C_d = coefficient of discharge (unitless)
A = cross-sectional area of the opening (ft^2)
g = acceleration of gravity (32.2 ft/s^2)
h = height of water above the centerline of the flow (ft)

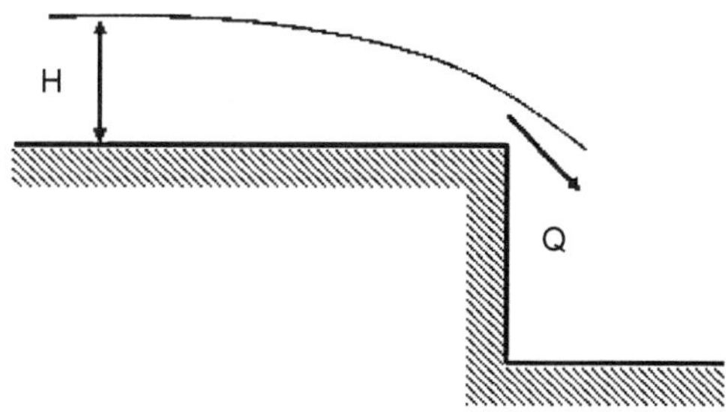

Figure 9.7 Free overfall weir.

The coefficient of discharge C_d is typically taken as 0.6, which means only 60% of the flow rate that would be expected from applying $Q = VA$ actually gets through the opening. The determination of the height h will be discussed in the next two sections. Notice that the equation is dimensionally correct, with no unit conversions necessary.

As for the case of the storm water overfalling the opening, the flow rate Q is determined from the following equation associated with the flow through weirs.

$$Q = C_W L H^{3/2} \tag{9.24}$$

where

Q = flow rate (ft^3/s)

C_w = weir coefficient (taken as 3.0 for free overfall)

L = length of the opening over which water is falling (ft)

H = driving head measured upstream of opening (ft)

The weir coefficient C_w clearly represents various unit conversions. The driving head h is shown in Figure 9.7.

The design of weirs is a very interesting topic, however, for the purposes of storm water runoff collected in open channels, the equation for free overfall is all that is needed.

9.5.2 Drop Inlets

As mentioned earlier, drop inlets can have a wide variety of designs, from several large openings to a great many smaller openings. The choice is determined by various flow conditions. The main thing to keep in mind is that the opening must accommodate the flow rate of storm water it must handle. For most of the life of the drop inlet, it will see only flow rates that result in free overfall of the water. In peak conditions, typically some level of water over the inlet is permissible. It is the balance between these two extremes that the designer must consider.

EXAMPLE 9.13 Flooded Drop Inlet

For the drop inlet shown, containing 20 8 × 2 in openings, determine the height h of storm water that will rise above the inlet if the flow rate Q from a highway facility to the location of the inlet is 15 ft^3/s. Use a coefficient of discharge C_d of 0.6.

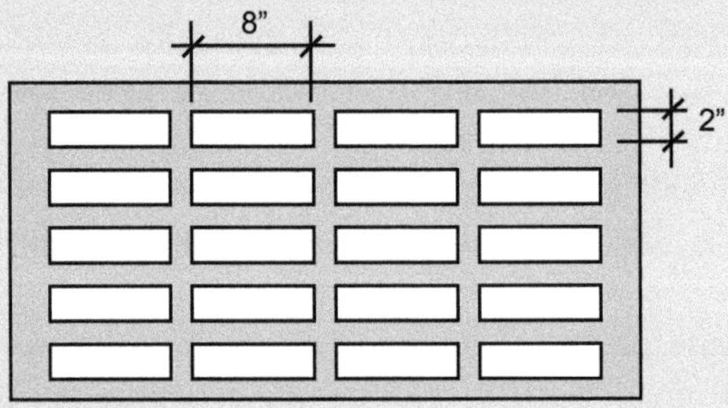

Solution

First, determine the flow area A of the drop inlet. Multiplying the number of openings by their size gives:

$$A = 20 \text{ openings} \times (8 \text{ in})(2 \text{ in}) = 320 \text{ in}^2$$
$$= 2.22 \text{ ft}^2$$

Next, solve algebraically for the depth of water h in Eq. (9.23) to give:

$$Q = C_d A \sqrt{2gh}$$

$$\frac{Q}{C_d A} = \sqrt{2gh}$$

$$\left(\frac{Q}{C_d A}\right)^2 = 2gh$$

$$h = \frac{1}{2g}\left(\frac{Q}{C_d A}\right)^2$$

Substituting the given information, gives the height h as:

$$h = \frac{1}{2g}\left(\frac{Q}{C_d A}\right)^2 = \frac{1}{2(32.2 \text{ ft/s}^2)}\left(\frac{15 \text{ ft}^3/\text{s}}{(0.6)(2.22 \text{ ft}^2)}\right)^2$$
$$= 1.97 \text{ ft} \approx 2 \text{ ft}$$

9.5.3 Curb Inlets

Curb inlets as well as drop inlets are usually flooded during peak storm conditions. This sets up a different height h for the flow rate Q through the inlet. The following example highlights this flow condition.

EXAMPLE 9.14 Flooded Curb Inlet

For the curb inlet shown, storm water is 2 in above the 6 × 18 in opening. Determine the expected flow rate through the inlet using a coefficient of discharge C_d of 0.6.

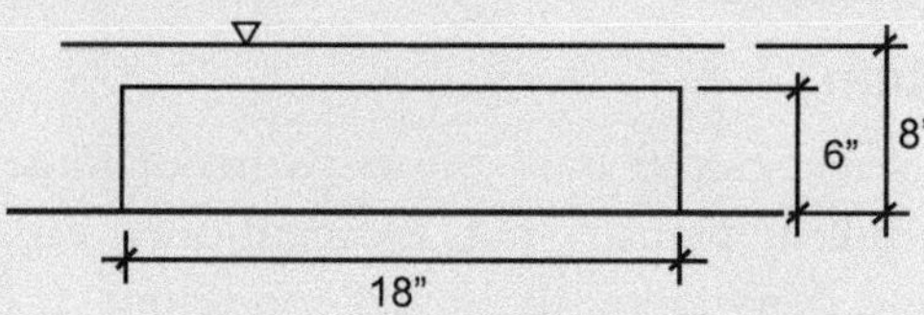

Solution

From the definition of the height h in Eq. (9.23), the distance to the centerline of the flow is half the height of the opening, which is 3 + 2 in, for a total of 5 in. This is shown in the following diagram:

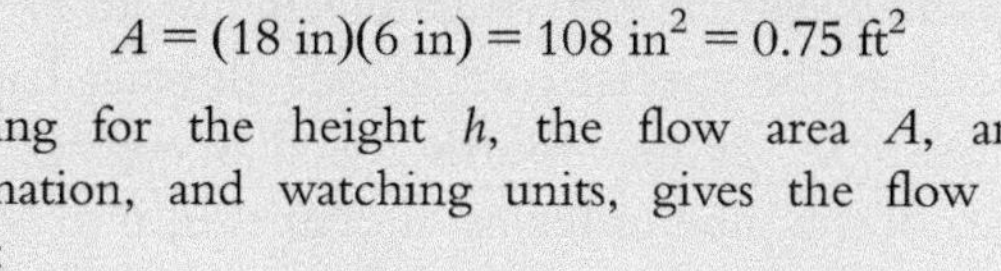

The area of the flow is the area of the opening, expressed in ft^2, as:

$$A = (18\text{ in})(6\text{ in}) = 108\text{ in}^2 = 0.75\text{ ft}^2$$

Substituting for the height h, the flow area A, and the other given information, and watching units, gives the flow rate Q from Eq. (9.23) as:

$$Q = C_d A\sqrt{2gh} = (0.6)(0.75\text{ ft}^2)\sqrt{2(32.2\text{ ft/s}^2)(5\text{ in} \times 1\text{ ft}/12\text{ in})}$$
$$= 2.33\text{ft}^3/\text{s}$$

9.5.4 Summary

At this point, the storm water runoff coming from a highway facility has been collected by open channels and funneled to a series of drop and/or curb inlets. The flow into these inlets can either be flooded or partially fill the openings. Two separate equations are used depending on which situation is occurring.

The storm water flow now enters a network of circular pipes, which, if properly designed, will typically only flow full when peak rainfall conditions exist. As will be seen, Manning's equation will be used, and the ability to determine the flow of partially filled pipes will be discussed.

9.6 FLOW IN PIPES

9.6.1 Introduction

Once the storm water runoff is in the network of underground piping, the flow is either full or partially full. As mentioned earlier, typically the flow is partial unless there are peak rainfall conditions. It is startling to see very large pipes with only a trickle of flow, knowing that they were designed for the day when the flow would be of such a magnitude that they would be flowing full.

9.6.2 Application of Manning's Equation

Manning's equation has been used so far for only open channel flow, typically in trapezoidal channels. However, it is perfectly valid for circular pipes flowing full. Of course, if the flow partially fills the pipe, Manning's equation is the only logical way to proceed.

As was determined earlier, the hydraulic radius R of a circular pipe flowing full, as well as the special case of one flowing half full, was half the radius r of the pipe, or one-fourth the diameter D. Substitute this result, and the area of the pipe in terms of diameter, into Manning's equation to give:

$$Q = \frac{1.486}{n} A R^{2/3} S^{1/2} = \frac{1.486}{n} \left(\pi/4\, D^2\right) \left(\frac{D}{4}\right)^{2/3} S^{1/2}$$
$$= (0.463) \frac{\sqrt{S}}{n} D^{8/3} \qquad (9.25)$$

Solving for the diameter D in Eq. (9.25) gives:

$$Q = (0.463)\frac{\sqrt{S}}{n}D^{8/3}$$

$$\frac{1}{0.463}\frac{Qn}{\sqrt{S}} = D^{8/3}$$

$$D = \left[\frac{1}{0.463}\frac{Qn}{\sqrt{S}}\right]^{3/8} \tag{9.26}$$

$$D = 1.33\left[\frac{Qn}{\sqrt{S}}\right]^{3/8} \quad (D \text{ in feet})$$

$$D = 16.0\left[\frac{Qn}{\sqrt{S}}\right]^{3/8} \quad (D \text{ in inches})$$

If the average velocity V is required, the flow rate Q given by Eq. (9.25) can be divided by the flow area A to give:

$$V = \frac{1.486}{n}R^{2/3}S^{1/2} = \frac{1.486}{n}\left(\frac{D}{4}\right)^{2/3}S^{1/2}$$

$$= (0.590)\frac{\sqrt{S}}{n}D^{2/3} \quad (D \text{ in feet}) \tag{9.27}$$

$$= \frac{\sqrt{S}}{8.89\,n}D^{2/3} \quad (D \text{ in inches})$$

The most common design question is what diameter pipe is needed for a particular flow rate Q. Equation (9.26) for the diameter D in inches is the most useful for this purpose. Be aware that pipe is made in only selected sizes, so always go up to the next size to avoid problems with too much friction. The typical sizes (in inches) are: 12, 15, 18, 21, 24, 30, 36, 42, 48, 54, 60, 66, and 72.

EXAMPLE 9.15 Pipe Flowing Full

If a concrete pipe is to carry 35 ft^3/s of storm water runoff, determine the required diameter. The slope of the pipe is 3%.

Solution

From Table 9.3, Manning's roughness coefficient for concrete pipe is 0.013. Using this value and the other given information in Eq. (9.26) gives the required diameter D as:

$$D = 16.0\left[\frac{Qn}{\sqrt{S}}\right]^{3/8} = 16.0\left[\frac{(35\ \text{ft}^3/\text{s})\ (0.013)}{\sqrt{0.03}}\right]^{3/8}$$

$$= 23\ \text{in}$$

Therefore, specify a 24-in diameter pipe, the next size up, to be used.

Now what happens when the pipe is not flowing full? How is the depth of flow related to the flow rate? The answer lies in the chart provided in Figure 9.8.

Along the bottom axis are the ratios of both the velocity and flow rates partially flowing full to flowing full. They are the same ratios because the flow area is the same for both cases. Along the vertical axis is the ratio of the depth of flow d to the diameter D. The letter h might have been used instead of d, however this is the notation used in most references. Note that lowercase letters are used to denote quantities for partially full pipes. An example will provide the steps to determining the necessary information.

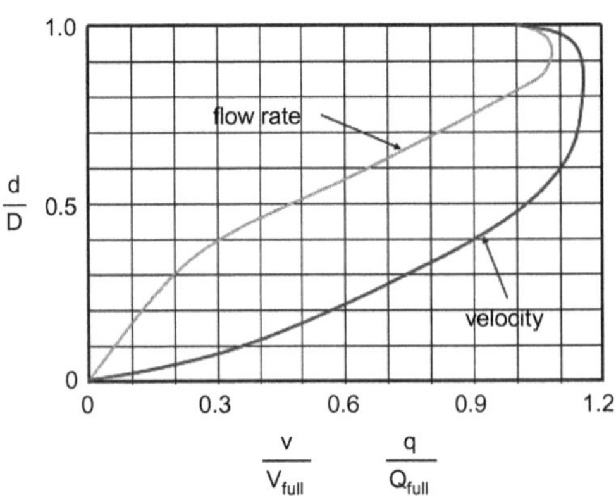

Figure 9.8 Flow parameters for partially filled pipes.

EXAMPLE 9.16 Pipe Flowing Partially Full

In Example 9.15, the required diameter was determined to be 23 in, which was rounded up to a 24-in diameter pipe. Suppose a 30-in diameter pipe was selected. What would be the depth of flow in this 30-in pipe for the given flow rate of 35 ft^3/s? The slope of the concrete pipe remains at 3%.

Solution

Using Eq. (9.25), and the same Manning roughness coefficient of 0.013; calculate the flow rate for a 30-in diameter pipe flowing full to be:

$$Q = (0.463)\frac{\sqrt{S}}{n}D^{8/3}\ (D \text{ in feet})$$

$$= (0.463)\ \frac{\sqrt{0.03}}{0.013}(2.5)^{8/3} \qquad (9.25)$$

$$= 71\ \text{ft}^3/\text{s}$$

Dividing this flow rate into the given 35 ft^3/s gives a ratio very close to 0.5. From Figure 9.8, the corresponding ratio d/D is very close to 0.5, so the pipe is basically flowing half full.

9.6.3 Head Loss Due to Friction Considerations

Up to this point, the slope of the open channels and the underground pipes has been given. However, these values must reflect the nature of the flow relative to the head loss due to friction. For any flow where the flow area A is constant, the average velocity V must be constant along the length of the channel or pipe. This means that in the Bernoulli equation, applied between any two points along the flow, the head loss, designated h_f is the difference in elevation along the pipe. Meaning, gravity must overcome the friction loss. This can be stated in the following expression:

$$z_{up} = z_{down} + h_f \qquad (9.28)$$

where z is the elevation of the same point on the channel or pipe. Typically this is the invert of the pipe, meaning the bottom, and the terms *up* and *down* refer to *upstream* and *downstream* elevations.

To determine the head loss due to friction, consider that the slope S is actually the hydraulic gradient, the ratio of h_f to L, or:

$$S = \frac{h_f}{L} \qquad (9.29)$$

where L is the length of the channel or pipe. For open channels, or pipes flowing partially full, any amount of slope will usually cause the flow of storm water to occur. However, this is not the case for pipes flowing full. If the water in the pipe "sees" too much friction, it will not flow at all. Therefore, it is critical to specify an appropriate elevation of the underground piping as it proceeds along the network.

To quantify the head loss due to friction h_f for pipes flowing full, start with Eq. (9.25), substitute for S from Eq. (9.29), and solve for the head loss as:

$$Q = (0.463)\frac{\sqrt{S}}{n}D^{8/3}$$

$$\sqrt{S} = \frac{Q\,n}{(0.463)D^{8/3}}$$

$$S = \frac{h_f}{L} = \frac{Q^2 n^2}{(0.463)^2\left(D^{8/3}\right)^2} \tag{9.30}$$

$$h_f = \frac{L\,Q^2 n^2}{(0.463)^2 D^{16/3}}$$

For Eq. (9.30), the units on the length L and flow rate Q are feet and cubic feet per second (ft^3/s or cfs), respectively, and the diameter D is in feet. However, normally you see the following equation where the diameter in inches is used:

$$h_f = 2{,}660{,}000\frac{L_{\text{ft}}\,Q_{\text{cfs}}^2\,n^2}{D_{\text{in}}^{5.33}} \tag{9.31}$$

The number 2,660,000 is a result of all the various unit conversions, and rounded from the exact value of 2,655,559, in case the reader would like to verify the conversions. It also helps to develop trust in an equation when the origin of all the terms is understood.

EXAMPLE 9.17 Head Loss Due to Friction

If the flow rate in a 36-in diameter corrugated metal pipe is 80 ft^3/s, and the flow is full, determine the head loss due to friction for every 300 ft. Also, determine the required slope of the pipe.

(Continued)

EXAMPLE 9.17 Head Loss Due to Friction—(Continued)

Solution

Using Eq. (9.31), and Manning's roughness coefficient of 0.024 found from Table 9.3, calculate the head loss due to friction h_f as:

$$h_f = 2{,}660{,}000\frac{L_{\text{ft}}Q_{\text{cfs}}^2 n^2}{D_{\text{in}}^{5.33}}$$

$$= 2{,}660{,}000\frac{(300\ \text{ft})(80\ \text{ft}^3/\text{s})^2(0.024)^2}{(36\ \text{in})^{5.33}}$$

$$= 14.9\ \text{ft}$$

Dividing this head loss by the length L gives:

$$\frac{h_f}{L} = S = \frac{14.9\ \text{ft}}{300\ \text{ft}} \approx 0.05 = 5\%$$

Realize this slope does not take into account any deterioration of the inside of the pipe over its life, so a slightly greater slope might be a prudent design decision.

9.6.4 Summary

There are certainly more considerations associated with all the various facets of the design of systems to handle storm water runoff, such as flow from one highway facility across or combined with other highway facilities, requiring the design of complex culverts and such. However, here the discussion has focused on the start of the process from the determination of the expected storm water runoff from a highway facility, the conveyance of this storm water along open channels, the collection from the open channels to underground piping through drop or curb inlets, then the flow of the water through the network of pipes that will change as more and more storm water is brought together to a single exit point for direct absorption into the ground or into an environmentally acceptable stream.

The critical steps along the way are that the peak rainfall conditions are anticipated, that the open channels do not overflow or have adverse erosions, that the drop or curb inlets do not flood beyond design expectations causing unsafe driving conditions, and that the size and slope of the underground piping is sufficient to allow the smooth flow of water to its intended point. A careful balance must be achieved between cost and functionality, realizing highway facilities must be designed for a long-term useful life.

It is hoped that the process that has been presented will allow the engineer responsible for the storm water runoff to design a system that functions as it is intended and provides years of satisfactory service as part of the overall design of the entire highway facility.

9.7 PRACTICE PROBLEMS

Problem 9.1: Side Slope and Longitudinal Slope (Grade)

On a contour map, two points are 2.6 in apart, where the scale of the map is 1 in = 50 ft. The elevation of the two points is 155 ft and 165 ft MSL. Determine the side slope and longitudinal slope (grade) between the two points.

Solution

Using the definition of side slope from Eq. (9.1) gives:

$$\text{side slope} = \frac{\text{horizontal distance}}{\text{vertical distance}} = \frac{\text{run}}{\text{rise}}{:}1$$

$$= \frac{(2.6\text{ in})\times(50\text{ ft}/1\text{ in})}{165\text{ ft} - 155\text{ ft}} = \frac{130\text{ ft}}{10\text{ ft}} = 13{:}1$$

Knowing the side slope, the longitudinal slope is the reciprocal, which gives:

$$\text{longitudinal slope} = \frac{1}{\text{side slope}}\times 100 = \frac{1}{13}\times 100 = (0.0769)\times 100 \approx 8\%$$

Or, using the definition of the longitudinal slope from Eq. (9.2), gives:

$$\text{longitudinal slope (grade)} = \frac{\text{vertical distance}}{\text{horizontal distance}} = \frac{\text{rise}}{\text{run}}\times 100(\%)$$

$$= \frac{165\text{ ft} - 155\text{ ft}}{(2.6\text{ in})\times(50\text{ ft}/1\text{ in})} = \frac{10\text{ ft}}{130\text{ ft}}\times 100(\%)$$

$$= (0.0769)\times 100 \approx 8\%$$

Problem 9.2: Composite Runoff Coefficient

A divided highway facility consists of 108 acres of asphalt roadway, 29 acres of grass with an average slope of 5% along the roadway, and in the median is 52 acres of forested-type area. Determine the composite runoff coefficient for this area.

Solution

From Table 9.1, the runoff coefficient for asphalt, grass with an average slope of 5%, and forested areas are 0.95, 0.15, and 0.20, respectively. Using Eq. (9.4) for the composite runoff coefficient C_c gives:

$$C_c = \frac{\sum C_i A_i}{\sum A_i} = \frac{(0.95)(108\text{ acres}) + (0.15)(29\text{ acres}) + (0.20)(52\text{ acres})}{(108\text{ acres}) + (29\text{ acres}) + (52\text{ acres})}$$

$$= 0.62$$

This means 62% of the rainfall will leave the area, and also means 38% will stay.

Problem 9.3: Rainfall Intensity

Suppose for the highway facility in Problem 9.2, which totaled 189 acres, that the farthest distance storm water would need to flow to a collection point is 4100 ft, and it has an elevation drop of 160 ft. Using the intensity-duration-frequency chart provided in Figure 9.2, determine the rainfall intensity *I*. Assume a storm period of 10 years.

Solution

Using Eq. (9.5), calculate the time of concentration t_c as:

$$t_c = \frac{\left(\dfrac{L^3}{H}\right)^{0.385}}{128} = \frac{\left(\dfrac{(4100\text{ ft})^3}{160\text{ ft}}\right)^{0.385}}{128} = 16.5\text{ min} > 10\text{ min}$$

Because the time of concentration is greater than 10 min, use the value calculated and the 10-year storm frequency on the intensity-duration-frequency chart shown as follows.

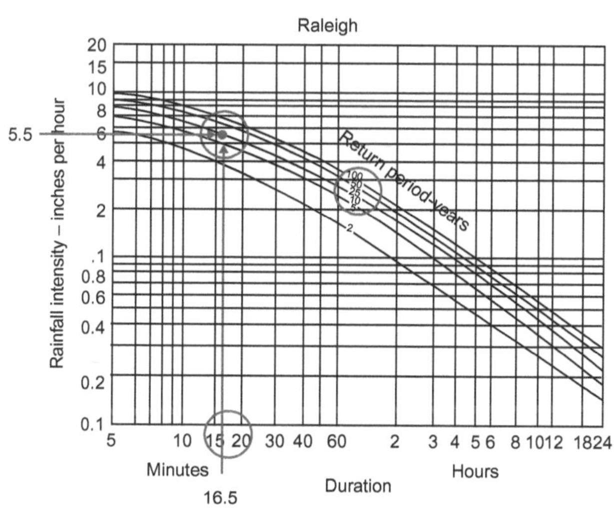

It can be seen that the best value for the rainfall intensity I is 5.5 in/h. This will be used in the calculation of the expected storm water runoff.

Problem 9.4: Storm Water Runoff Calculation

Using the composite runoff coefficient found for the 189-acre highway facility in Problem 9.2, namely 0.62, and the rainfall intensity found in Problem 9.3, namely 5.5 in/h, determine the expected storm water runoff.

Solution

Using Eq. (9.3), calculate the storm water runoff Q as:

$$Q = C_c IA = (0.62)(5.5\ \text{in/h})(189\ \text{acres}) = 644.5\ \text{ft}^3/\text{s}$$

Problem 9.5: Hydraulic Radius

Determine the hydraulic radius for the storm water flowing in the gutter shown. Treat the shape of the flow as a right triangle, where the hypotenuse is the free surface of the flow, designated by the inverted triangle. The flow extends 4 ft into the roadway, and is 3-in deep at the curb.

Solution

Using Eq. (9.9), substitute the dimensions shown for the area of a right triangle and the length of the two sides of the curb, watching units, to give:

$$R = \frac{A}{P} = \frac{1/2\,(4\ \text{ft})[(3\ \text{in})(1\ \text{ft}/12\ \text{in})]}{(4\ \text{ft}) + (3\ \text{in})(1\ \text{ft}/12\ \text{in})} = \frac{0.5\ \text{ft}^2}{4.25\ \text{ft}} = 0.118\ \text{ft}$$

Notice that the surface of the storm water is not included, the primary mistake in this calculation and other hydraulic radius calculations.

Problem 9.6: Velocity and Flow Rate—Depth Known

For the 6-ft-wide rectangular channel shown, the depth of flow is 2 ft. The channel is concrete troweled. The channel has a slope of 1%. Determine the velocity V and flow rate Q.

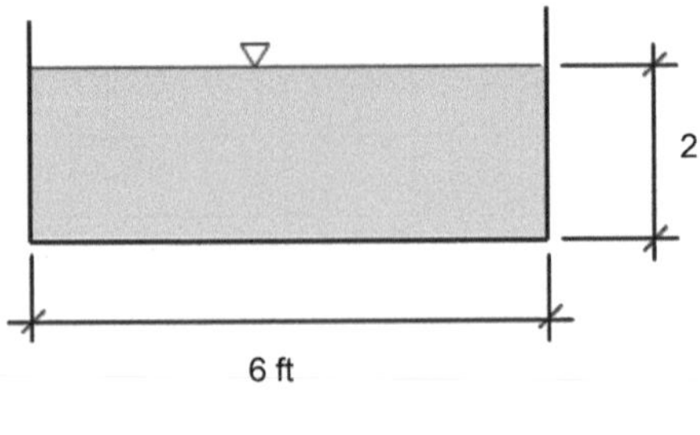

Solution

From Table 9.3, Manning's roughness coefficient for troweled concrete is given as 0.013.

The hydraulic radius is found from Eq. (9.9), where the wetted perimeter is width of the channel plus twice the depth of the flow. This gives:

$$R = \frac{A}{P} = \frac{(6\text{ ft})(2\text{ ft})}{(6\text{ ft}) + 2(2\text{ ft})} = \frac{12\text{ ft}^2}{10\text{ ft}} = 1.2\text{ ft}$$

Applying Manning's equation from Eq. (9.6), remembering to change the slope to decimal form, gives the average velocity V of the flow to be:

$$V = \frac{1.486}{n} R^{2/3} S^{1/2} = \frac{1.486}{0.013} (1.2)^{2/3} (0.01)^{1/2}$$

$$= 12.9\text{ ft/s}$$

Multiplying this velocity by the flow area A gives the flow rate Q as:

$$Q = VA = (12.9\text{ ft/s})(12\text{ ft}^2) = 154.8\text{ ft}^3/\text{s}$$

Again, this flow rate can be used to check the calculation for the flow rate Q from the rational method for the associated highway facility.

Problem 9.7: Hydraulic Radius for Trapezoidal Channel

For the trapezoidal channel shown, the depth is 3 ft and the width of the bottom is 8 ft. The slope of the sides is 3:1. Determine the hydraulic radius.

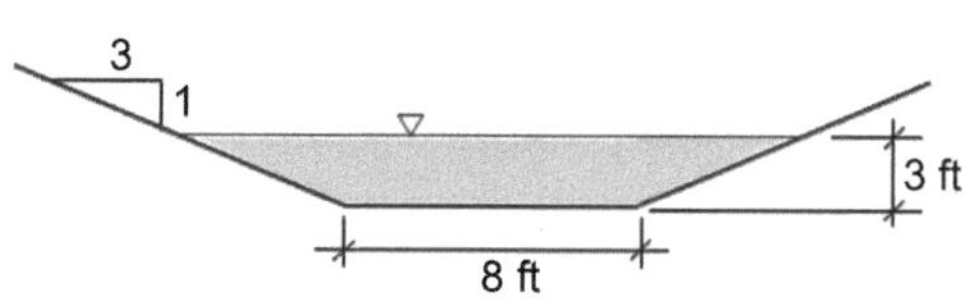

Solution

Using Eq. (9.10), calculate the flow area A using $M = 3$, to obtain:

$$A = By + My^2 = (8\text{ ft})(3\text{ ft}) + (3)(3\text{ ft})^2$$

$$= 51\text{ ft}^2$$

Using Eq. (9.11), calculate the wetted perimeter P as:

$$P = B + 2y\sqrt{1 + M^2} = (8\text{ ft}) + (2)(3\text{ ft})\sqrt{1 + (3)^2}$$

$$= 27.0\text{ ft}$$

Substituting for the flow area A and wetted perimeter P into Eq. (9.9), gives the hydraulic radius as:

$$R = \frac{A}{P} = \frac{51 \text{ ft}^2}{27.0 \text{ ft}} = 1.89 \text{ ft}$$

Problem 9.8: Normal Depth Procedure

For the trapezoidal channel shown, the width of the base is 5 ft and the slope of the sides is 3:1, with all surfaces being well-kept grass. If the channel slope is 3%, and the flow rate Q entering the channel from a highway facility is 185 ft^3/s, determine the expected depth of the flow y.

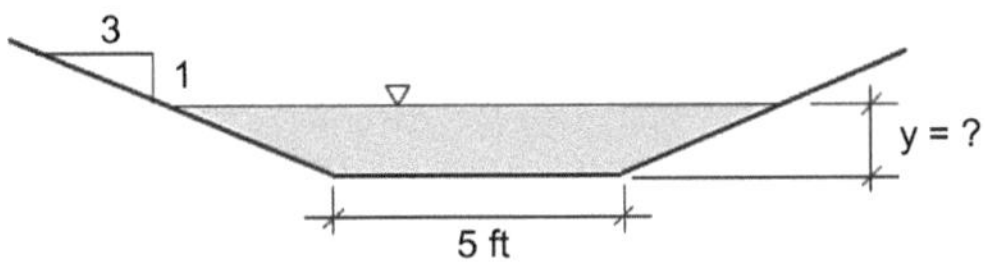

Solution

Using Eq. (9.12), substitute the given information, including Manning's roughness coefficient n from Table 9.3 of 0.030, and calculate the right side as:

$$AR^{2/3} = \frac{Qn}{1.486\sqrt{S}} = \frac{(185 \text{ ft}^3/\text{s})(0.030)}{1.486\sqrt{0.03}}$$

$$= 21.56 \text{ ft}^{8/3}$$

As was chosen in Example 9.8, select 1 ft as the starting choice for the flow depth y. Using this value, calculate the appropriate values and record them in a table similar to Table 9.4 to give the following:

y (ft)	*B* (ft)	*A* (ft^2)	*P* (ft)	*R* (ft)	*AR*$^{2/3}$	Remarks
1	5	8	11.32	0.706	6.35	Too shallow

Because the left side came out to be less than the right side, a value for the flow depth y of 1 ft is too shallow. The value obtained suggests a second calculation for y equal to 3 ft, however, sometimes it is better to "sneak up" on the value desired. So, try 2 ft as the next choice. This gives the following values, recorded in the following table:

y (ft)	*B* (ft)	*A* (ft^2)	*P* (ft)	*R* (ft)	*AR*$^{2/3}$	Remarks
1	5	8	11.32	0.706	6.35	Too shallow
2	5	22	17.65	1.247	25.48	Too deep

Because the left side came out to be greater than the right side, a value for the flow depth y of 2 ft is too deep. The value obtained suggests a third calculation for y equal to 1.9 ft. This gives the following values, recorded in the following table:

y (ft)	B (ft)	A (ft^2)	P (ft)	R (ft)	$AR^{2/3}$	Remarks
1	5	8	11.32	0.706	6.35	Too shallow
2	5	22	17.65	1.247	25.48	Too deep
1.9	5	20.33	17.02	1.195	22.89	Still a little too deep

The value obtained for a depth of 1.9 ft might be close enough. However, to make sure, do one more iteration with the depth as 1.8 ft. This gives the following values:

y (ft)	B (ft)	A (ft^2)	P (ft)	R (ft)	$AR^{2/3}$	Remarks
1	5	8	11.32	0.706	6.35	Too shallow
2	5	22	17.65	1.247	25.48	Too deep
1.9	5	20.33	17.02	1.195	22.89	Still a little too deep
1.8	5	18.72	16.38	1.143	20.46	A little too shallow

The values for 1.8 ft and 1.9 ft are about equally spaced with the right side value, so a good final answer would be a value in between, or 1.85 ft. To verify, calculate the various quantities to give:

y (ft)	B (ft)	A (ft^2)	P (ft)	R (ft)	$AR^{2/3}$	Remarks
1	5	8	11.32	0.706	6.35	Too shallow
2	5	22	17.65	1.247	25.48	Too deep
1.9	5	20.33	17.02	1.195	22.89	Still a little too deep
1.8	5	18.72	16.38	1.143	20.46	A little too shallow
1.85	5	19.52	16.70	1.169	21.66	Close enough

Clearly, the value of the flow depth y as 1.85 ft is close enough.

Though not required, the average velocity of the flow V can be obtained by dividing the flow rate Q by the area of the flow A for the depth of 1.85 ft. This gives a velocity of:

$$V = \frac{Q}{A} = \frac{185\ \text{ft}^3/\text{s}}{19.52\ \text{ft}^2} = 9.48\ \text{ft/s}$$

This value for the average velocity is at least double the allowable value for well-kept grass. So, like the trapezoid channel in Example 9.8, either quarry stone or smooth concrete would need to be used for this particular channel as well.

Problem 9.9: Depth-Limited Procedure

For the trapezoidal channel shown, the depth of flow y is limited to 1.5 ft. The slope of the sides is 4:1, with all surfaces lined with 6-in quarry stone. If the channel slope is 5% and the flow rate Q entering the channel from a highway facility is 135 ft^3/s, determine the required width of the base B. Also, determine the average velocity of the flow V using the dimensions of the final channel.

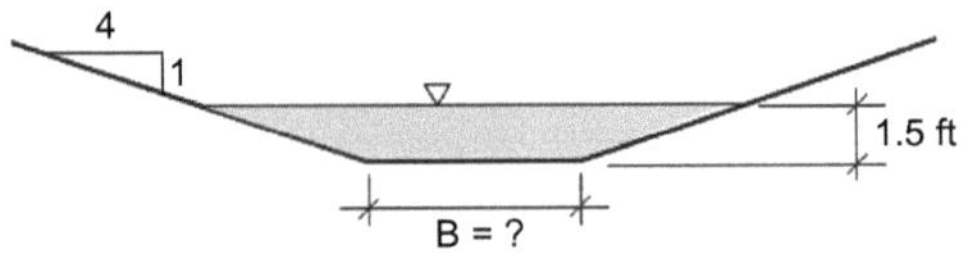

Solution

Using Eq. (9.8), calculate Manning's roughness coefficient for 6-in quarry stone as:

$$n = \frac{D_{\text{in}}^{1/6}}{44.4} = \frac{(6\text{ in})^{1/6}}{44.4} = 0.030$$

Substituting this value for Manning's roughness coefficient into Eq. (9.12), along with the other given information, calculate the right side as:

$$AR^{2/3} = \frac{Qn}{1.486\sqrt{S}} = \frac{(135\text{ ft}^3/\text{s})(0.030)}{1.486\sqrt{0.05}}$$

$$= 12.19\text{ ft}^{8/3}$$

As was chosen in Example 9.9, select 1 ft as the starting choice for the width of the base B. Using this value, calculate the appropriate values and record them in a table similar to Table 9.4 to give the following:

y (ft)	*B* (ft)	*A* (ft^2)	*P* (ft)	*R* (ft)	*AR*$^{2/3}$	Remarks
1.5	1	10.5	13.37	0.785	8.94	Too narrow

Because the left side came out to be less than the right side, a value for the width of the base B of 1 ft is too narrow. The value obtained suggests a second calculation for B equal to 2 ft. This gives the following values, recorded in the following table:

y (ft)	*B* (ft)	*A* (ft^2)	*P* (ft)	*R* (ft)	*AR*$^{2/3}$	Remarks
1.5	1	10.5	13.37	0.785	8.94	Too narrow
1.5	2	12	14.37	0.835	10.64	Still too narrow

Because the left side came out to still be less than the right side, a value for the width of the base B of 2 ft is still too narrow. The value obtained suggests a third calculation for B equal to 3 ft. This gives the following values, recorded in the following table:

y (ft)	B (ft)	A (ft^2)	P (ft)	R (ft)	$AR^{2/3}$	Remarks
1.5	1	10.5	13.37	0.785	8.94	Too narrow
1.5	2	12	14.37	0.835	10.64	Still too narrow
1.5	3	13.5	15.37	0.878	12.38	Close enough

As can be seen, the value on the left side is now close enough to the value on the right side, 12.38 versus 12.19, respectively. There is only a 1.5% difference, and again, based on how the flow rate Q was obtained, using methods such as the rational method with all the many assumptions and average values, this is indeed close enough.

The average velocity of the flow V can be obtained by dividing the flow rate Q by the area of the flow A for the width of the base B of 3 ft. This gives:

$$V = \frac{Q}{A} = \frac{135\ \text{ft}^3/\text{s}}{13.5\ \text{ft}^2} = 10.0\ \text{ft/s}$$

For 6-in quarry stone, this velocity is about 25% greater than what would be typically allowed. Therefore, the addition of larger quarry stone would be the appropriate solution.

Problem 9.10: Velocity-Limited Procedure

For the trapezoidal channel in Problem 9.8, the average velocity V found was double the allowable velocity for well-kept grass. Using a maximum allowable velocity of 5 ft/s, determine the depth of flow y and the width of the base of the channel B from the velocity-limited procedure. The cross section is shown here, where the slope of the sides is 3:1, the channel slope is 3%, and the flow rate Q entering the channel from a highway facility is 185 ft^3/s.

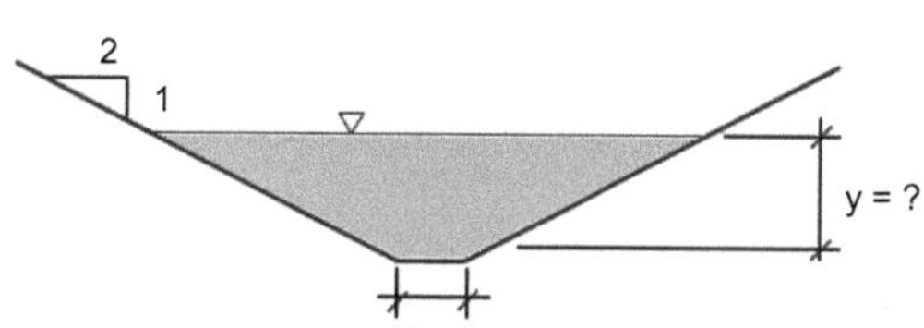

Solution

Substitute the given flow rate Q and allowable velocity V_a into Eq. (9.13) to give:

$$A_x = \frac{Q}{V_a} = \frac{185 \text{ ft}^3/\text{s}}{5 \text{ ft/s}} = 37 \text{ ft}^2$$

Using Eqs. (9.15) and (9.16), substitute this required cross-sectional area and other given information, including Manning's roughness coefficient n from Table 9.3 of 0.030, and calculate the quantities w_1 and w_2, respectively, as:

$$w_1 = M - 2\sqrt{1+M^2} = 3 - 2\sqrt{1+3^2} = -3.32$$

$$w_2 = \frac{A_x}{\left(\dfrac{V_a n}{1.486\sqrt{S}}\right)^{3/2}} = \frac{37 \text{ ft}^2}{\left(\dfrac{(5 \text{ ft/s})(0.030)}{1.486\sqrt{0.03}}\right)^{3/2}} = 83.16$$

Substitute these values for w_1 and w_2 into Eq. (9.14) to give:

$$y = \frac{-w_2 + \sqrt{w_2^2 + 4w_1A_x}}{2w_1}$$

$$= \frac{-(83.16) + \sqrt{(83.16)^2 + 4(-3.32)(37)}}{2(-3.32)}$$

$$= \frac{-(83.16) + (80.15)}{-6.64} = 0.45 \text{ ft}$$

$$\approx 5\ 1/2 \text{ in}$$

This would be considered too shallow. To continue, use this value for the depth of flow y to determine the width of the base B from Eq. (9.17) as:

$$B = \frac{A_x}{y} - My = \frac{37 \text{ ft}^2}{0.45 \text{ ft}} - (3)(0.45 \text{ ft})$$

$$= 80.87 \text{ ft}$$

This width is clearly excessively wide. So, again, the only choice to address the potential erosion problem is to either add quarry stone or replace well-kept grass with troweled concrete.

Problem 9.11: Best Hydraulic Section Procedure

For the trapezoidal channel shown in Problem 9.9, the depth of flow y was limited to 1.5 ft, and the width of the base of the channel B was found to be about 3 ft. The slope of the sides is 4:1, with all surfaces being 6-in quarry stone. If the channel slope is 5%, and the flow rate Q entering the channel from a highway facility is 135 ft^3/s, determine the shape of the channel using the best hydraulic section procedure.

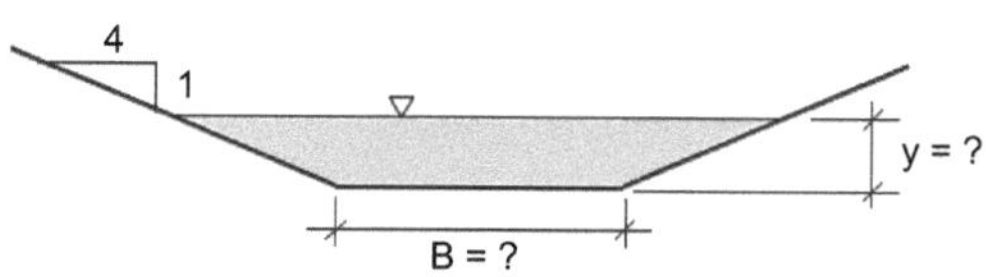

Solution

Using Eq. (9.18), substitute the given information, including Manning's roughness coefficient n from Eq. (9.8) of 0.030 and the parameter C_m of 0.595 from Table 9.5, and calculate the depth of flow y as:

$$y = C_m\left(\frac{Qn}{\sqrt{S}}\right)^{3/8} = (0.595)\left(\frac{(135\ \text{ft}^3/\text{s})(0.030)}{\sqrt{0.05}}\right)^{3/8}$$
$$= 1.76\ \text{ft}$$

This is somewhat greater than the value specified of 1.5 ft. Now, calculate the width of the base of the channel B using Eq. (9.19), with the parameter k from Table 9.5 equal to 0.246.

$$B = ky = (0.246)(1.76\ \text{ft}) = 0.43\ \text{ft}$$

This value is considerably lower than the value determined in Example 9.9. The size determined here would give a different velocity, obtained as follows:

$$V = \frac{Q}{A} = \frac{Q}{By + My^2} = \frac{135\ \text{ft}^3/\text{s}}{(0.43\ \text{ft})(1.76\ \text{ft}) + (4)(1.76\ \text{ft})^2} = \frac{135\ \text{ft}^3/\text{s}}{13.15\ \text{ft}^2}$$
$$= 10.27\ \text{ft/s}$$

This is only 0.27 ft/s greater, so the 6-in quarry stone is still 25% greater than the acceptable value. Therefore, the addition of larger quarry stone would be the appropriate solution. It is also interesting that even with the significant change in shape, the flow areas are essentially the same, and so the average velocities are essentially the same.

Problem 9.12: Triangular Channel

For the triangular channel shown, the flow rate from a highway facility is 95 ft^3/s. The slope of the sides is 2:1, with all surfaces being well-kept grass. If the channel slope is 5%, determine the depth of flow y.

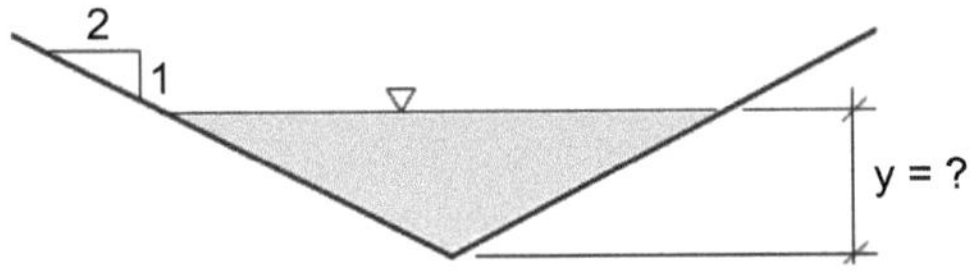

Solution

Because the only unknown in Manning's equation for the flow rate Q is the depth of flow y, substitute the given information, including Manning's roughness coefficient n from Table 9.3 of 0.030, to provide the following sequence of algebraic operations:

$$Q = \frac{1.486}{n} AR^{2/3} S^{1/2}$$

$$95 \text{ ft}^3/\text{s} = \frac{1.486}{0.030} \left[(2)y^2\right] \left[\frac{(2)y}{2\sqrt{1+(2)^2}}\right]^{2/3} (0.05)^{1/2}$$

$$95 \text{ ft}^3/\text{s} = (12.95)y^{8/3}$$

$$y^{8/3} = \frac{95}{12.95} = 7.333$$

$$y = (7.333)^{3/8}$$

$$y = 2.11 \text{ ft}$$

Problem 9.13: Flooded Drop Inlet

For the drop inlet shown, containing 30 10 × 3 in openings, determine the flow rate Q the inlet can handle if the height h of water allowed above the inlet is 15 in. Use a coefficient of discharge C_d of 0.6.

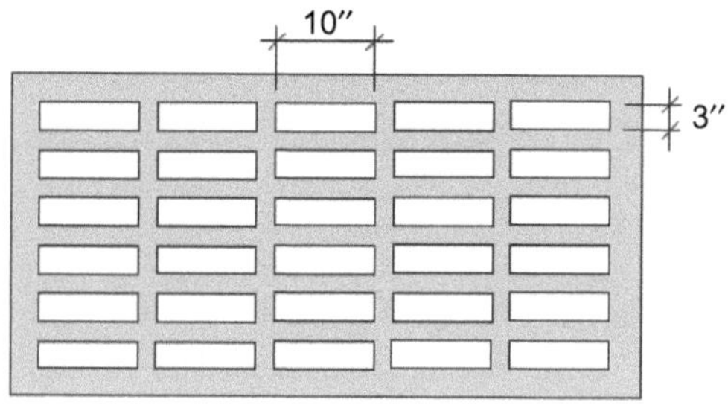

Solution

First, determine the flow area A of the drop inlet. Multiplying the number of openings by their size gives:

$$A = 30 \text{ openings} \times (10 \text{ in})(3 \text{ in}) = 900 \text{ in}^2$$
$$= 6.25 \text{ ft}^2$$

Next, substitute the given information, remembering to convert the height h of 15 in to 1.25 ft, into Eq. (9.23) to give:

$$Q = C_d A \sqrt{2gh} = (0.6)(6.25 \text{ ft}^2)\sqrt{2(32.2 \text{ ft/s}^2)(1.25 \text{ ft})}$$
$$= 33.65 \text{ ft}^3/\text{s}$$

This value can be compared to the flow rate obtained from applying the rational method for the highway facility to which the storm water is being collected by this drop inlet.

Problem 9.14: Free Fall Curb Inlet

For the curb inlet shown, storm water is overfalling the opening. If the driving head h is 6 in, determine the flow rate Q into the inlet.

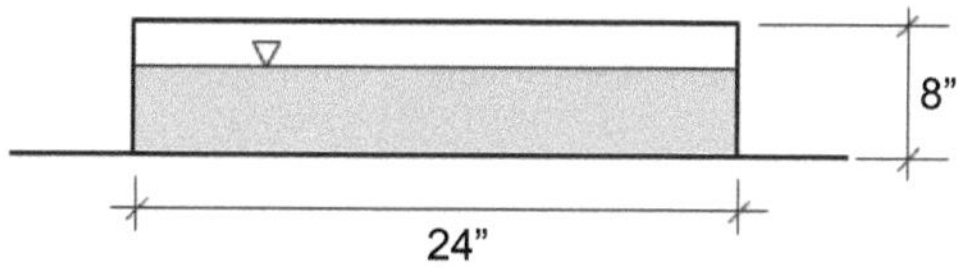

Solution

Using the given information, being sure to convert all dimensions to feet, and using a weir coefficient for free falling water of 3.0, Eq. (9.24) gives the flow rate Q as:

$$Q = C_w\, LH^{3/2} = (3.0)(24\ \text{in} \times 1\ \text{ft}/12\ \text{in})(6\ \text{in} \times 1\ \text{ft}/12\ \text{in})^{3/2}$$
$$= 2.12\ \text{ft}^3/\text{s}$$

Notice the 8-in dimension of the opening was not needed, as the flow only covered part of the opening. Because the driving head was 6 in, the depth at the opening would be less than 6 in, however, not by very much.

Problem 9.15: Pipe Flowing Full

If a corrugated metal pipe is to carry 65 ft^3/s of storm water runoff, determine the required diameter. The slope of the pipe is 2%.

Solution

From Table 9.3, Manning's roughness coefficient for corrugated metal pipe is 0.024. Using this value and the other given information in Eq. (9.26) gives the required diameter D as:

$$D = 16.0\left[\frac{Qn}{\sqrt{S}}\right]^{3/8} = 16.0\left[\frac{(65\ \text{ft}^3/\text{s})\,(0.024)}{\sqrt{0.02}}\right]^{3/8}$$
$$= 39\ \text{in}$$

Therefore, specify a 42-in diameter pipe, the next size up, to be used.

Problem 9.16: Pipe Flowing Partially Full

In Problem 9.15, the required diameter was determined to be 39 in, which was rounded up to a 42-in-diameter pipe. What would be the depth of flow in this 42-in pipe for the given flow rate of 65 ft^3/s? The slope of the corrugated metal pipe remains at 2%.

Solution

Using Eq. (9.25) and the same Manning roughness coefficient of 0.024, calculate the flow rate for a 42-in-diameter pipe flowing full to be:

$$Q = (0.463)\frac{\sqrt{S}}{n}D^{8/3}\ (D\ \text{in feet})$$

$$= (0.463)\frac{\sqrt{0.02}}{0.024}(3.5)^{8/3}$$

$$= 77\ \text{ft}^3/\text{s}$$

Dividing this flow rate into the given 65 ft^3/s gives a ratio of 0.84. From Figure 9.8, the corresponding ratio d/D is close to 0.72, so the depth of flow is 72% of the 42-in diameter, determined as:

$$\frac{d}{D} = 0.72 \rightarrow d = (0.72)D = (0.72)(42\ \text{in}) \approx 30\ \text{in}$$

Therefore, the flow is such that 12 in is open at the top of the pipe.

Problem 9.17: Head Loss Due to Friction

If the flow rate in an 18-in-diameter concrete pipe is 15 ft^3/s, and the flow is full, determine the head loss due to friction for every 100 ft. Also, determine the required slope of the pipe.

Solution

Using Eq. (9.31) and Manning's roughness coefficient of 0.013 found from Table 9.3, calculate the head loss due to friction h_f as:

$$\begin{aligned} h_f &= 2{,}660{,}000\ \frac{L_{\text{ft}} Q_{\text{cfs}}^2 n^2}{D_{\text{in}}^{5.33}} \\ &= 2{,}660{,}000 \frac{(100\ \text{ft})(15\ \text{ft}^3/\text{s})^2 (0.013)^2}{(18\ \text{in})^{5.33}} \\ &= 2.1\ \text{ft} \end{aligned}$$

Dividing this head loss by the length L gives:

$$\frac{h_f}{L} = S = \frac{2.1\ \text{ft}}{100\ \text{ft}} \approx 0.02 = 2\%$$

INDEX

Note: Page numbers followed by "*f*," "*t*," and "*b*" refer to figures, tables, and boxes, respectively.

D

G

K

L

M

S

U

V

W

Y

FSC
www.fsc.org
MIX
Papier | Fördert
gute Waldnutzung
FSC® C083411

Kolibri 360 GmbH
Ferdinand-Jühlke-Straße 7
99095 Erfurt, Deutschland
produktsicherheit@kolibri360.de